CDTE AND RELATED COMPOUNDS; PHYSICS, DEFECTS, HETERO- AND NANO-STRUCTURES, CRYSTAL GROWTH, SURFACES AND APPLICATIONS

PART I

PHYSICS, CDTE-BASED NANOSTRUCTURES, CDTE-BASED SEMIMAGNETIC SEMICONDUCTORS, DEFECTS

CDTE AND RELATED COMPOUNDS; PHYSICS, DEFECTS, HETERO- AND NANO-STRUCTURES, CRYSTAL GROWTH, SURFACES AND APPLICATIONS

PART I

PHYSICS, CDTE-BASED NANOSTRUCTURES, CDTE-BASED SEMIMAGNETIC SEMICONDUCTORS, DEFECTS

EDITED BY

ROBERT TRIBOULET

CNRS, GEMaC (Groupe d'Etude de la Matière Condensée), Meudon, France

and

PAUL SIFFERT

E-MRS General Secretary, Strasbourg, France

Published in Association with

E·MRS

EUROPEAN MATERIALS RESEARCH SOCIETY

ELSEVIER

Amsterdam • Boston • Heidelberg • London • Oxford • New York
Paris • San Diego • San Francisco • Singapore • Sydney • Tokyo

Elsevier
Radarweg 29, PO Box 211, 1000 AE Amsterdam, The Netherlands
Linacre House, Jordan Hill, Oxford OX2 8DP, UK

First edition 2010

Library of Congress Cataloging-in-Publication Data
A catalog record for this book is available from the Library of Congress

British Library Cataloguing in Publication Data
A catalogue record for this book is available from the British Library

ISBN: 978-0-08-046409-1

Printed and bound in Great Britain
10 11 12 10 9 8 7 6 5 4 3 2 1

CONTENTS

LIST OF CONTRIBUTORS

C.R. Becker
Experimentelle Physik III, Universität Würzburg, Am Hubland, D-97074 Würzburg, Germany.

M.A. Berding
SRI International, 333 Ravenswood Avenue, Menlo Park, CA 94025, USA.

V. Consonni
CEA-LETI, Minatec, 17 rue des Martyrs, 38054 Grenoble Cedex 9, France.

K. Durose
Science Laboratories, University of Durham, South Road, Durham DH1 3LE, UK.

P. Fochuk
Chernivtsi National University, 2 vul. Kotsiubinskoho, Chernivtsi 58012, Ukraine.

R.R. Galazka
Institute of Physics, Polish Academy of Sciences, Al. Lotnikow 32/46, 02-668 Warszawa, Poland.

K.A. Jones
School of Mechanical and Materials Engineering, Washington State University, P.O. Box 642920, Pullman, WA, USA.

K. Lynn
School of Mechanical and Materials Engineering, Washington State University, P.O. Box 642920, Pullman, WA, USA.

Y. Marfaing
Retired from Centre National de la Recherche Scientifique (CNRS), Groupe d'Etude de la Matière Condensée (GEMaC), 1 Place A. Briand, 92195 Meudon Cedex, France.

H. Mariette
CEA-CNRS-UJF, Laboratoire de Spectrométrie Physique, UMR 5588 CNRS/Université Joseph Fourier Grenoble, Equipe CEA-CNRS-UJF "Nanophysique et Semi-Conducteurs", 140 Avenue de la Physique, BP 87, 38402 Saint Martin d'Hères, France.

B.N. Mavrin
Institute of Spectroscopy, Russian Academy of Sciences, 142190 Troitsk, Moscow, Russia.

J.C. Moosbrugger
Department of Mechanical and Aeronautical Engineering, Center for Advanced Materials Processing, Clarkson University, Potsdam, NY 13699-5725, USA.

J.B. Mullin
EMC-HooTwo, 22 Branksome Towers, Westminster Road, Poole, Dorset BH13 6JT, UK.

J.-O. Ndap
II-VI Inc., eV Products, 373 Saxonburg Blvd., Saxonburg, PA 16056, USA.

O. Panchuk
Chernivtsi National University, 2 vul. Kotsiubinskoho, Chernivtsi 58012, Ukraine.

A. Sher
Sher Consulting, San Carlos Hills, CA 94070, USA.

E.A. Vinogradov
Institute of Spectroscopy, Russian Academy of Sciences, 142190 Troitsk, Moscow, Russia.

T. Wojtowicz
Institute of Physics, Polish Academy of Sciences, Al. Lotnikow 32/46, 02-668 Warsaw, Poland.

FOREWORD

Thirty years after the remarkable monography of K. Zanio and the numerous conferences and papers dedicated since that time to CdTe and CdZnTe, after all the significant progresses in that field and the increasing interest in these materials for their extremely attractive fundamental properties and industrial applications, the editors have thought timely to edit a book on CdTe and CdZnTe, covering all their most prominent, modern, and fundamental aspects. The subject has become so wide and enriched during these 30 years that we have decided to call in well-known specialists and experts of the field. The editors would like to thank them deeply for their valuable contributions, with special acknowledgments to Dr Henri Mariette for his pertinent recommendations and his continued help and support.

This part covers the topics Physics, CdTe-Based Nanostructures, Semimagnetic Semiconductors, and Defects. The topics Crystal Growth, Surfaces, and Applications will be covered in Part II.

R. Triboulet
P. Siffert

CHAPTER 1

Introduction

J.B. Mullin[a,b,c]

CdTe and CdZnTe are iconic examples of II–VI compounds. They epitomise the difficulties that need to be surmounted for successful device exploitation in this class of semiconductors. How can this exploitation be achieved? Put simply it can be achieved only by dominating the materials technology. Or, as Dr Sekimoto an eminent Japanese businessman and scientist so elegantly put it "Who dominates materials dominates technology". This book is about materials domination and the resulting device exploitation. It concerns the knowledge, the abundant practical experience and the valuable device technologies that have emerged in the development of CdTe and its close relative CdZnTe as a result of the impressive research efforts of countless dedicated scientists.

The II–VIs have a long evolutionary history; indeed some II–VIs have had applications from the dawn of antiquity. But our interest is their role as semiconductors. The advent of the semiconductor era (circa 1948) created a new standard in materials science – the need for semiconductor quality. This standard ideally required material that was completely single crystal and essentially defect-free and whose significant impurity content had been reduced to the ppb level. What then is the status of CdTe and its related alloys with respect to this goal?

The achievement of this goal for CdTe can be seen as a much more demanding and problematic process when compared with the group IV elements and the III–V compounds. The initial stage of semiconductor quality was reproducibly met for germanium in about a decade and within the next decade silicon had assumed a role as the dominant semiconductor. In the case of the III–Vs comparable development

[a]Electronic Materials Consultancy, 22 Branksome Towers, Poole BH13 6JT, UK
[b]Editor in Chief, Progress in Crystal Growth and Characterization of Materials
[c]Associate Editor, Journal of Crystal Growth

CDTE and Related Compounds
DOI: 10.1016/B978-0-08-046409-1.00001-0

followed a very much slower process. An exception has been InSb a semiconductor blessed with benign preparative properties. It achieved semiconductor quality status within 3–4 years following serious interest in its development around 1953. Two important controlling parameters can be identified as driving forces in the evolution of all these semiconductors, the perceived significance and importance of its device applications and the intrinsic material properties especially its preparative properties.

Why is Cadmium Telluride evolving relatively slowly as a device-quality semiconductor even though it has a readily accessible melting point and manageable vapour pressure over the melt? From a device point of view its applications have industrial significance. Thus it has gradually attracted increasing research support following its role as a solar cell material. Its close relative HgCdTe relative has received immense support in view of its infrared detector capabilities and its consequential unique military role. ZnCdTe has found a critical role as a substrate material for that difficult alloy. ZnCdTe also enjoys a crucial role as a precursor γ-ray detector material. Of course many more device applications of CdTe and its related II–VI relatives are attracting more support and study. But the dominating problem in their rate of evolution has been and to some extent still is the lack of structural material knowledge together with the intrinsic difficulties in the preparation of the compounds and alloys.

The supreme importance then of this book is that it addresses this central problem of crystal structural knowledge. This has been achieved by the editors who have brought together a most impressive selection of authors who clearly show how unwavering dedication to material science and technology can bring about the understanding and control of these most intransigent materials and the effective development of valuable device technologies.

A fundamental feature of the II–VIs is the ionic component of the bonding between adjacent atoms in contrast to the covalent bonding of the group IVs. This departure from covalent character gives rise to a range of native point defects and their associated complexes and is central to understanding and controlling the properties of these materials. The behaviour of the point defects has been studied since the pioneering work of de Nobel. The main problem however in studying native point defects is the lack of methods for their direct investigation. But whilst there is a good working knowledge of their behaviour a definitive understanding is still lacking. Nevertheless very useful progress in the control of point defects is reported.

To understand and control the II–VIs one must have a working knowledge of the properties of these native point defects. This requires the ability to control the composition with the imperative need to understand the phase diagram. Without the application of specific precursor composition control non-stoichiometric CdTe with significant concentrations of native point defects can result. An excess Te concentration typically in the region of 0.0010% has been a common problem. The knowledge assembled in the reviews on material preparation provides a splendid account of the growth procedures for dealing with this problem.

As with all semiconductors the identification and removal of impurities in and from the component elements and their compounds and alloys is an essential requirement.

Great progress is reported in this area. This leads directly to the methods for preparing the crystals of CdTe and CdZnTe. This has been handled admirably by Dr Triboulet who has spent a good proportion of his scientific career in championing the development and applications of these semiconductors. Indeed it is the *raison d'être* for the book. The established methods of melt growth, solution and vapour growth are rigorously reviewed with a fine balance between fundamental and practical considerations. This and related studies by other authors on the preparation of doped materials and their potential relationship to structural complexes have led to remarkable progress. Indeed the understanding of doped material provides a necessary key to device exploitation.

A distinctive feature of CdTe and related materials, which contrasts with the group IVs and the III–Vs, is the omnipresence of grain boundaries. This together with the ease of formation of dislocations, stacking faults and inclusions of second phases creates formidable challenges to the material scientists. The reviews give an up to date account of their identification, behaviour and control methodology without which device development would be truly problematic. Such advances have been nurtured as a result of essential research on the fundamental understanding of the optical and physical properties of CdTe and related compounds and alloys. These topics are also reviewed as are the physics of surfaces and compensation.

The advances in materials knowledge are clearly demonstrated by the development of valuable environmental, medical and opto-electronic devices. One cannot but be impressed by the range of device activities associated with CdTe CdZnTe and related materials. At the forefront of these activities the importance of CdTe as a classic thin film solar cell material is well recognised and its role is discussed in depth. The optical detector role also covers X-ray detectors, a role of increasing importance not only in the safety and security aspect of monitoring of nuclear emissions but also in its value as a detector in medical scanning systems involving computer tomography. The unique properties of doped materials such as V doped for their photo refractive properties and Mn-doped materials in connection with conventionally grown nano-structures and as semi-magnetic semiconductors. The latter application is making a significant contribution to the physics of spintronics. They offer a splendid insight into the future of these materials.

The reviews presented in this compilation provide an essential study for anyone involved in II–VI development. The extensive materials knowledge reviewed provides the key to device exploitation. Indeed the unique device applications confirm the leading roles of CdTe, CdZnTe and related materials in the history of semiconductors. The reader is cordially invited to explore and assess their fascinating role for him or herself.

CHAPTER II

Physics

Contents

CHAPTER **IIA**

Zinc Blende Alloy Materials: Band Structures and Binding Properties

A. Sher[1]

1. INTRODUCTION

This chapter concentrates on theories of the band structures and crystal structural properties of zinc blende-structured II-VI compounds and their alloys. The focus will be on CdTe-based materials, in particular $Hg_{(1-x)}Cd_xTe$.

We will begin with a brief survey of the most advanced first principles approaches. These approaches are computationally intensive, but now their predictions are remarkably accurate even for far more complex materials than zinc blende-structured semiconductor compounds. For example, they have been employed to predict wave functions, band structures including proper band gaps and effective masses, ground state crystal structures, cohesive energies, and elastic constants. These powerful methods still have not been applied to random alloys, but it is only a matter of time before someone does it.

Aspects of the bond orbital approximation (BOA) [1] will also be introduced for the insight they offer into semiconductor properties. These methods are not as accurate as the first principles methods, but they present a more transparent view into physical mechanisms.

Some results for band structures derived from the hybrid pseudopotential tight-binding (HPT) method can be found in Ref. [2]. HPT starts

SRI International and Stanford University, Stanford, (Consulting Professor). CA, USA
[1]Retired.

CDTE and Related Compounds
DOI: 10.1016/B978-0-08-046409-1.00002-2

from a universal pseudopotential for zinc blende-structured solids that includes long-ranged interactions. Then suitable local tight-binding terms are added for each semiconductor that, once selected to fit, enable the band structures for the different materials to match experimental results. After this first fitting numerous physical properties of the materials, including coherent potential approximation (CPA) treatments of alloys, are calculated in agreement with experiments with no further adjustable parameters. Following this procedure turns out to generate accurate wave functions, as indicated, for example, by subsequent predictions of transport properties [2–4], and the temperature dependence of band gaps [5]. Once the wave functions are known, matrix elements of interaction Hamiltonians can be calculated leading to transition probabilities per unit time. Then depending on the feature under investigation, sums over the whole Brillouin zone are done to reach answers. Comparing these answers with those obtained from models with parameters chosen to fit experimental results often reveals that these fitted parameters are unphysical. When this occurs it is usually because a mechanism has been overlooked in the model, which is compensated by the choice of the fitted parameters. While the HPT calculations depend on numerical methods, they are far simpler and faster that the first principles methods. Actually if one were to calculate the parameters from first principles that are obtained in the HPT method by fitting them to experimental symmetry point energies, I suspect the first principles parameters would closely agree with the fitted ones, since both theories predict results in agreement with experiments.

Finally a previously unpublished theory of mesoscopic concentration fluctuations in alloys will be presented. In alloys like $Hg_{(1-x)}Cd_xTe$, or $In_{(1-x)}Ga_xAs$ statistics guarantees that there will be mesoscopic sized regions (10-1000 Å) where the concentration x varies from the average. In lattice constant-matched alloys like $Hg_{(1-x)}Cd_xTe$, if there is a small region with a concentration x, imbedded in its surroundings with concentration $\bar{x}$, there is little strain energy penalty associated with such a fluctuation. In a lattice constant-mismatched alloy, like $In_{(1-x)}Ga_xAs$ or $Cd_{(1-x)}Zn_xTe$, there is a strain energy penalty that tends to suppress concentration fluctuations [6]. In Section 4, we will introduce the underlying theory. While the theory is incomplete, it is never-the-less obvious that these fluctuations can affect band edges at the fundamental gap, defect states, and transport properties.

2. SURVEY OF FIRST PRINCIPLES STATUS

Starting from the density functional (DF) theory of Kohn and Sham, first principles methods have advanced to a point where they reliably predict numerous properties of solids. The local density approximation (LDA) coupled to Hedin's *GW* approximation (GWA) solution methods have been around for many years, but it is only now that serious difficulties with its implementation are being overcome (*G* stands for Green's function and *W* is a screened potential). The theory now treats full atom core Coulomb potentials, and all electrons, not just the valence electrons and ionic core pseudopotentials. Generally the choice of bases states in which to expand the wave functions makes a comparatively small but significant difference in the predictions, but a huge difference in the computational time required to reach answers. For some bases choices, this presents a practical limit on the complexity of the problems that can be approached. At present the full-potential linerized muffin-tin orbitals (FP-LMTO) set is one of the fastest and most accurate choices. One difficulty most recently addressed is the number of higher state unoccupied orbitals needed to cause some predictions to converge. A more complete status of the field can be found in a paper by van Schilfgaarde et al. [7].

A typical set of results are presented in Table 1. The calculation of band gaps is one of the more difficult features to get right because it depends sensitively on the exchange interaction. Without a proper accounting of this interaction predicted band gaps from first principles are always too small. However, for example, the $GW - \Sigma_{nn'}$ values for C, Si, Ge, and GaAs are within 1.0, 12, 11, and 4.0%, respectively, of the zero point energy adjusted experimental values.

3. CRYSTAL STRUCTURES, BINDING, AND ELASTIC CONSTANTS

3.1. Crystal structures

The focus of this chapter is on crystals in the zinc blende structure. This structure consists of two interpenetrating face centered cubic sublattices with one shifted from the other by 1/4 of the distance along the (111) direction. If "a" is the length of a cube edge, then the near neighbor distance is $d = \sqrt{3}a/4$. In the group IV materials (C, Si, Ge, and gray Sn), both sublattices are occupied by the same type of atom. In III-V and II-VI (AB) compounds, one sublattice is occupied by the cations and the other by the anions. In all cases, each atom site is fourfold coordinated with near neighbors in a tetrahedral arrangement.

Pseudobinary alloys of the compounds are either cation substituted, $A_{(1-x)}B_xC$, or anion substituted $CA_{(1-x)}B_x$, where x is the fraction of B type atoms on their sublattice. When the AC (or CA) compound bond length is nearly the same as that of the BC (or CB) compound, then the system is called "lattice matched." If the bond lengths differ, they are "lattice mismatched." In lattice-mismatched alloys, the cube edge distance of the AB

Table 1 Fundamental gap (in eV)

	LDA	*GW*	*GW*	*GW*	Expt	Δ/3	*ZP*	Adj.
			Z =1	$\Sigma_{nn'}$				
C	4.09	5.48	5.74	5.77	5.49	0	0.34	5.83
Si	0.46	0.95	1.10	1.09	1.17	0.01	0.06	1.24
Ge	−0.13	0.66	0.83	0.83	0.78	0.10	0.05	0.93
GaAs	0.34	1.40	1.70	1.66	1.52	0.11	0.10	1.73
wAlN	4.20	5.83	6.24		6.28	0	0.20	6.48
wGaN	1.88	3.15	3.47	3.45	3.49	0	0.20	3.69
wInN	−0.24	0.20	0.33		0.69	0	0.16	0.85
wZnO	0.71	2.51	3.07	2.94	3.44	0	0.16	3.60
ZnS	1.86	3.21	3.57	3.51	3.78	0.03	0.10	3.91
ZnSe	1.05	2.25	2.53	2.55	2.82	0.13	0.09	3.04
ZnTe	1.03	2.23	2.55		2.39	0.30	0.08	2.77
CuBr	0.29	1.56	1.98	1.96	3.1	0.04	0.09	3.23
CdO	−0.56	0.10	0.22	0.15	0.84	0.01	0.05	0.90
CaO	3.49	6.02	6.62	6.50	~7	0		
wCdS	0.93	1.98	2.24		2.50	0.03	0.07	2.60
$SrTiO_3$	1.76	3.83	4.54	3.59	~3.3			
ScN	−0.26	0.95	1.24	0.96	~0.9	0.01		
NiO	0.45	1.1	1.6		4.3			
Cu	−2.33	−2.35	−2.23	−2.18	−2.78			
Cu	−2.33	−2.85	−2.73	−2.18	−2.78			
$Gd^{\uparrow}$	−4.6	−5.6	−6.2	−4.1	−7.9			
$Gd^{\downarrow}$	0.3	0.2	1.8	1.5	4.3			

For Gd, QPE corresponds to the position of the majority and minority *f* levels relative to E_F; for Cu, QPE corresponds to the *d* level. Low-temperature experimental data were used when available. QPEs in the *GW* column are calculated with usual GWA Eqs. (6) and (7). In the $Z=1$ column the Z factor is taken to be unity. In the $\Sigma_{nn'}$ column the off-diagonal parts of Σ are included in addition to taking $Z=1$. *k*-meshes of $8 \times 8 \times 8k$ and $6 \times 6 \times 6$ were used for cubic and hexagonal structures, respectively (symbol *w* indicates the wurtzite structure). *GW* calculations leave out spin-orbit coupling and zero-point motion effects. The former is determined from Δ/3, where Δ is the spin splitting of the Γ_{15u} level (in the zinc blende structure); it is shown in the Δ/3 column. Contributions to zero-point motion are estimated from table 2 in Ref. [45] and are shown in the *ZP* column. The "adjusted" gap adds these columns to the true gap and is the appropriate quantity to compare to *GW*.

sublattice is very nearly a concentration-weighted average of those of the AC and BC compounds, that is, following Vegard's law [2, 8]. However, the near neighbor AC and BC bond lengths remain within ~70% of their respective compound lengths. To accomplish this, the C atoms move off their ideal fcc positions to accommodate to their four neighbor local environment, for example, A_4, A_3B_1, A_2B_2, A_1B_3, or B_4. In lattice-matched alloys, all bond lengths remain nearly constant, but to the extent that there are small differences they follow Vegard's law.

3.2. Insights from the bond orbital approximation

Start by letting the crystal be in a zinc blende structure. Then the simplest form of the bond orbital method employs wave functions and local energies derived from pseudopotentials. The local cation and anion energies for their respective valence s- and p-states are denoted: ε_s^C, ε_p^C, ε_s^A, and ε_p^A. Four sp^3 hybrid wave functions pointing from one atom to each near neighbor, for example, one pointing in the (111) direction has the form:

$$|h\rangle = \big(|s\rangle + |p_x\rangle + |p_y\rangle + |p_z\rangle\big)/2. \tag{1}$$

A hybrid pointing in the (−1−1−1) direction has all the signs in front of the p-wave functions reversed. The hybrid energies of the cations and anions are given by:

$$\varepsilon_h^{\left\{\begin{smallmatrix} C \\ A \end{smallmatrix}\right\}} = \frac{\varepsilon_s^{\left\{\begin{smallmatrix} C \\ A \end{smallmatrix}\right\}} + 3\varepsilon_p^{\left\{\begin{smallmatrix} C \\ A \end{smallmatrix}\right\}}}{4}. \tag{2}$$

The interaction matrix element between two hybrids, one from an anion and the other from a near neighbor cation, Harrison calls the "covalent energy," V_2, and shows it has the form:

$$V_2 \equiv \langle h^C|H|h^A\rangle = \frac{24.5}{d^2}\ (\text{eV}), \tag{3}$$

where the bond length, d, has units of Å

Next transform to a local molecular orbital basis given by:

$$\begin{aligned} |b\rangle &= U_b^A|h^A\rangle + U_b^C|h^C\rangle, \\ |a\rangle &= U_a^A|h^A\rangle + U_a^C|h^C\rangle. \end{aligned} \tag{4}$$

Taking matrix elements of the Hamiltonian produces a 2 × 2 matrix, and solving its secular determinate yields:

$$\begin{aligned} U_b^A &= \sqrt{(1+\alpha_p)/2} = -U_a^C, \\ U_b^C &= \sqrt{(1-\alpha_p)/2} = U_a^A. \end{aligned} \tag{5}$$

where α_p is the "polarity" and is defined as:

$$\alpha_p = V_3/\sqrt{V_2^2 + V_3^2} \tag{6}$$

and the binding and antibinding energies per electron are:

$$\varepsilon_{\left\{\begin{smallmatrix} a \\ b \end{smallmatrix}\right\}} = \bar{\varepsilon}_h \left\{\begin{matrix} + \\ - \end{matrix}\right\} \sqrt{V_2^2 + V_3^2}. \tag{7}$$

Another major energy concept has now entered; it is the polar energy V_3, defined as:

$$V_3 \equiv (\varepsilon_h^C - \varepsilon_h^A)/2. \tag{8}$$

This energy tends to transfer electrons from the shallower cation to the deeper anion hybrid states. (*Note*: both hybrid energies are negative so V_3 is a positive energy.) It is ultimately responsible for the ionic contribution to the solid's binding energy.

The last contribution to the binding stems from the interaction coupling each bond to its adjacent bonds. These interactions are called the "metallization energies" and have been shown by Harrison to be:

$$\begin{aligned} V_1^C &= \left(\varepsilon_s^C - \varepsilon_p^C\right)/4 \\ V_1^A &= \left(\varepsilon_s^A - \varepsilon_p^A\right)/4. \end{aligned} \tag{9}$$

They delocalize the molecular binding and antibinding states and broaden them into the valence and conduction bands. In second-order perturbation theory, these energies contribute a metallic term, $\Delta\varepsilon_b$, to the binding energy per electron of the form:

$$\Delta\varepsilon_b = \sum_{a'} \frac{|\langle b|H|a'\rangle|^2}{\varepsilon_b - \varepsilon_{a'}}, \tag{10}$$

where $|a'\rangle$ is an antibinding state of a "molecule" adjacent to the one with binding state b.

Finally collecting terms produces the net binding energy per unit cell:

$$E_b = 2\varepsilon_b + 2\Delta\varepsilon_b + u_0 - 2\bar{\varepsilon}_{atom}, \tag{11}$$

where ε_b contains the covalent and ionic components of the binding, $\Delta\varepsilon_b$ is the metallic contribution, $u_0 = C/d^4$ is the form of the screened Coulomb repulsive energy between the ions, and $\bar{\varepsilon}_{atom}$ is the average electron energy per A and C atom. The constant C in u_0 is chosen so the E_b has its minimum at the observed equilibrium bond length d_o. The relation between the binding energy and the equilibrium bond length d_o becomes:

$$E_b - \xi = -\sqrt{V_2^2 + V_3^2} + \frac{A^2}{2d_o^4\sqrt{V_2^2 + V_3^2}} \cong \begin{cases} -\dfrac{A}{2d_o^2}, & V_2 \gg V_3 \\ -V_3, & V_2 \ll V_3 \end{cases}, \tag{12}$$

where

$$\begin{aligned} \xi &\equiv \bar{\varepsilon}_h + \nabla\varepsilon_b - 2\varepsilon_{atom} \\ A &\equiv 24.5\ (\text{eV} - (\text{Å})^2) \end{aligned}. \tag{13}$$

The second equality in Eq. (12) states that the bond orbital model predicts that the binding energy of pure covalent materials becomes progressively stronger as the inverse square of the bond length, while the binding energy of a pure ionic material is independent of the bond length. This trend is also found experimentally. Thus one expects the variation of the group IV compounds to vary fastest with bond length, the somewhat ionic III-V compound variations to be less steep, and the still more ionic II-IV compounds to be even less steep. Obviously other parameters besides bond length that impact values of ξ, and V_3, also enter.

A number of different elastic constant models are treated in detail in the "Semiconductor Alloys" book [2] and a review article [9]. Here once again because of the insight it lends, the discussion will be confined to the bond orbital model. The general relation between a displacement vector x in the unstrained material and x' in a strained material in component form is:

$$x'_\alpha = x_\alpha + \sum_\beta \eta_{\alpha\beta} x_\beta, \tag{14}$$

where $\eta_{\alpha\beta}$ are components of a 9×9 strain tensor. Then for small $\eta_{\alpha\beta}$ values the energy density U is a quadratic function of the $\eta_{\alpha\beta}$ components:

$$U = \frac{1}{2} \sum_{\alpha\beta\mu\nu} \eta_{\alpha\beta} c_{\alpha\beta\mu\nu} \eta_{\mu\nu}. \tag{15}$$

Because of rigid translation and rotation invariance there are in general only 21 independent values of $c_{\alpha\beta\mu\nu}$, the elastic stiffness tensor components. But for zinc blende compounds their symmetry reduces the number of distinct components from 21 to 3 [10]. This permits a more compact notation to be used in which there are only six components to $\eta_{\alpha\beta}$, labeled 1-6 corresponding, respectively, to xx, yy, zz, yz, xz, and xy. In this reduced notation the energy density becomes:

$$U = \frac{1}{2} \sum C_{ij} e_i e_j, \tag{16}$$

where $i = \alpha\beta$ and $j = \mu\nu$, $e_i \equiv e_{\alpha\beta} = \frac{1}{2}(\eta_{\alpha\beta} + \eta_{\beta\alpha})$, and $C_{ij} = c_{\alpha\beta\mu\nu}$. By examining three different simple distortions, relations can be found between the bulk modulus B, C_{11}-C_{12}, and C_{44} and the bond orbital parameters [2]:

$$\begin{aligned} B &= -\frac{2V_2}{\sqrt{3}d^3}\left[\alpha_c^2 - \frac{(9/8)\alpha_c^3(5\alpha_c^2 - 4)(V_{1C}^2 + V_{1A}^2)}{V_2^2 + V_3^2}\right], \\ C_{11} - C_{12} &= \frac{5.00\sqrt{3}\alpha_c^2}{d^5}, \\ \frac{9}{C_{44}} &= \frac{6}{C_{11} - C_{12}} + \frac{4}{B}. \end{aligned} \tag{17}$$

Here the covalency α_c is defined as:

$$\alpha_c \equiv \sqrt{1-\alpha_p^2} = \frac{V_2}{\sqrt{V_2^2+V_3^2}}. \tag{18}$$

The last equality for C_{44} is quite instructive. It says that in the BOA there is an analytic relation between the three elastic coefficients. Because this relationship arises from a quantum treatment that is a reasonable approximation to reality, it serves as a justification for treatments like the valence force field model (VFF) of Keating [11] that involves only two elastic constants, bond stretching and bond bending constants.

In a nearly pure covalent material where $V_2 \gg V_3$ or $\alpha_c \cong 1$, the expressions in Eq. (17) reduce to:

$$\begin{aligned} B &\cong \frac{2A}{\sqrt{3}d^5}\left[1-\frac{9}{8}\left(V_{1C}^2+V_{1A}^2\right)/V_2^2\right], \\ C_{11}-C_{12} &= \frac{5.00\sqrt{3}}{d^5}, \\ C_{44} &= \frac{3.00\sqrt{3}A\left[1-(9/8)\left(V_{1C}^2+V_{1A}^2\right)/V_2^2\right]}{2d^5\left\{1+\frac{A\left[1-(9/8)\left(V_{1C}^2+V_{1A}^2\right)/V_2^2\right]}{5d^2}\right\}}. \end{aligned} \tag{19}$$

For this nearly covalent case all the elastic constants to first order vary proportional to $1/d^5$. In the opposite extreme of a nearly ionic bonded compound, where $V_2 \ll V_3$, or $\alpha_c \cong V_2/V_3 \ll 1$, all of the elastic constants in Eq. (17) are small but the leading term is proportional to $1/d^9$.

The agreement between the experimental and BOM predictions of the elastic constant trends is good but the absolute values differ by factors of 2–4. When Coulomb interaction terms that are absent from the simple form of the BOM are added, the agreement improves. When a full calculation based on HPT band structures and wave functions are done the agreement is still better. Finally when an LMTO-based calculation is done the agreement falls within 2-3% [2].

Because the bond length difference between HgTe and CdTe is small (see Table 2), when they are alloyed into $Hg_{(1-x)}Cd_xTe$, the alloy elastic constants are very nearly a simple concentration-weighted average of the constituents. However, when alloys of CdTe or HgTe with ZnTe are formed, the stiff ZnTe bonds on the low ZnTe concentration side of alloys significantly increase the alloy elastic constants above their concentration-weighted average. This is of particular importance to dislocation densities in these alloys. Since most of the excess energy associated with dislocations

Table 2 Experimental values of the bond length and experimental and BOM results for the bonding energy and the three independent elastic constants for three compounds

	d (Å)	E_b (eV)		B		C_{11}-C_{12}		C_{44}	
Material	Exp.	Exp.	BOM	Exp.	BOM	Exp.	BOM	Exp.	BOM
ZnTe	2.643	−1.20	−1.04	5.090	1.82	3.060	1.88	3,120	1.50
CdTe	2.805	−1.10	−0.96	4.210	1.24	1,680	1.22	2.040	0.96
HgTe	2.797	−0.82	−0.48	4.759	1.23	1.817	1.33	2.259	1.11

The units for the elastic constants are 10^{11} dynes/cm^2.

is in shears, for a given stress the dislocation density tends to be lower for a short bond length material.

While the bond length difference between HgTe and CdTe is small, it still causes troublesome misfit and treading dislocations in epitaxial layers of $Hg_{(1-x)}Cd_xTe$ alloys grown on CdTe substrates. This is mitigated somewhat by shortening the CdTe bonds by adding enough Zn to a $Cd_{(1-x)}Zn_xTe$ alloy substrate material to lattice match the substrate and the epitaxial layer. In the best of circumstances threading dislocations still propagate from the substrate into the epitaxial layers, and also long-ranged cross hatch patterns persist [12]. These and other point and meso-scopic sized defects degrade the performance of devices made from these materials, especially in the long (LWIR) and very long (VLWIR) wave length spectral regions where small fluctuations have an important effect on the band gap.

A typical IR detector device structure is comprised of a conducting substrate, followed by a semi-insulating p-type base layer, a more heavily n-doped graded composition layer forming a heterojunction, and finally an insulating cap passivation layer. Performance of these devices is sensitive to the quality of the lattice matching at each interface.

4. CONCENTRATION FLUCTUATIONS

4.1. Introduction

In $Hg_{1-x}Cd_xTe$ alloys with concentration x, there is a near lattice constant match between HgTe and CdTe bonds. Thus there is little strain energy contribution to the mixing enthalpy [6]. Moreover, if there is a cluster containing N cations ($1 < N <$ many thousands) embedded in the average medium, there is no strain penalty if the cation concentration in the cluster is x, differing from the average alloy concentration $\bar{x}$. The net statistical consequence is that there must be such concentration fluctuation clusters and for each moderately large cluster of size N, they have a probability distribution [6] given approximately by:

$$P_N(x) = \frac{1}{\mathbb{N}_N} \exp\left(-\frac{(x-\bar{x})^2}{2\sigma_N^2}\right), \tag{20}$$

where the variance σ_N is:

$$\sigma_N^2 = f\frac{\bar{x}(1-\bar{x})}{N} \equiv \frac{\xi}{N}, \tag{21}$$

where f is a strain-induced reduction factor given by:

$$f \equiv \left[1 + \frac{27BV_c}{kT}x(1-x)\right]^{-1}, \tag{22}$$

with B being the bulk modulus, and V_c the primitive cell volume. The normalization factor is:

$$\begin{aligned} \mathbb{N} = \sqrt{2\pi}\sigma_N - \int_{-\infty}^{0} dx \exp(-(x-\bar{x})^2/2\sigma_N^2) \\ - \int_{1}^{\infty} dx \exp(-(x-\bar{x})^2/2\sigma_N^2) \equiv \sqrt{2\pi}\sigma_N \mathbb{N}_N^*. \end{aligned} \tag{23}$$

The quantity ξ is defined in Eq. (21), and $\mathbb{N}^*$ defined in the second equality in Eq. (22) are both of order unity. For small clusters the Bernoulli distribution must be used in place of the Gaussian. When there is a lattice constant mismatch between the constituents, there is an extra multiplicative factor f, less than unity, which enters into ξ [6]. For an alloy with a large lattice constant mismatch the extra factor can be quite small. Some representative numbers are given in Table 3. The shapes of clusters are undoubtedly erratic, but if they are assumed to be spherical then the relation between their radius r_N, and N is just:

$$N = \frac{4\pi}{3} r_N^3 \rho_c, \tag{24}$$

where ρ_c is the density of cation sites in the solid. The effect of the spherical cluster shape approximation needs to be examined in a refined

Table 3 Values of the reduction factor f for two alloys at two cutoff wave lengths

λ_c(μm)	10		20	
Alloy	x	f	x	f
$Hg_{(1-x)}Cd_xTe$	0.22	1.0	0.19	1.0
$Hg_{(1-x)}Zn_xTe$	0.18	0.39	0.15	0.42

calculation, along with ellipsoidal, cubic, and tetragonal shapes that could be more appropriate in some cases.

4.2. General statistical theory

In a refinement of the presentation in Ref. [6] for an alloy of average concentration $\bar{x}$, we can deduce how fluctuations modify average values and the mean deviation of a phenomenon whose concentration variation is represented by $G(x)$. The refinement has two major improvements. First, the coarse graining approximation is eliminated, so cluster sizes vary continually modulo a small fundamental cluster size. Second, some unphysical cluster overlaps no longer occur. Examples of phenomena sensitive to fluctuations include the band gap, the density of states at the band edges, tunneling currents in reverse-biased p-n junctions, and mobilities.

As before let N be the number of cations in a cluster, and N_{D} be the number in a large domain. Depending on the problem being addressed the domain can be the whole volume of a pixel, its area times the depletion layer thickness, or any other useful volume. Then the ratio N_{D}/N is the total number of clusters of size N in the domain. Next, define a new quantity $n(N, x)$ the number of clusters of size N and concentration x, to be [13]:

$$n(N, x) \equiv \frac{N_{\mathrm{D}}}{N} P_N(x), \tag{25}$$

where $P_N(x)$ is defined in Eq. (20) as a Gaussian. It is the probability of finding a cluster of size N with concentration x.

Now we can ask the question, how large must a domain of size M_N be to contain on average 1 cluster of size N and concentration x? Thus we require for $N_{\mathrm{D}} = M_N(x)$ that $n(N, x) \cong 1$, or:

$$M_N(x) = \frac{N}{P_N(x)}. \tag{26}$$

Notice that since $P_N(x) < 1$, we always have $M_N(x) > N$, so the class of clusters with size N and concentration x never overlap as they did in Muller and Sher [6]. However, clusters with size N' may overlap those with size N. The density $\rho_N(x)$ of a specific class of clusters is:

$$\rho_N(x) = \frac{\rho_{\mathrm{c}}}{M_N(x)}, \tag{27}$$

where ρ_c is the density of cations in the solid. Depending on the shape chosen for domains, the average separation between clusters of size N and concentration x, is $\sim(\rho_N)^{1/3}$. Small clusters are spaced more closely than large ones. A spatial variation of the concentration profile will have large volume bumps and dips with smaller volumes generally with larger amplitudes, superimposed on them.

The probability in a domain of size N_D of finding a cluster having size N and concentration x is:

$$\Xi(N,x) \equiv \frac{n(N\cdot x)}{\sum_{N'} \frac{N_D}{N'} \int dx\, P_{N'}(x)} = \frac{\frac{1}{N} P_N(x)}{\sum_{N'} \frac{1}{N'}}. \tag{28}$$

Because there will be small clusters superimposed on larger ones, there is a possibility of double counting. The normalization used in Eq. (28) eliminates the consequences of this double counting. Therefore, an average of a general phenomenon with a concentration variation $G(x)$ over all cluster sizes and concentrations is:

$$\left\langle \langle G(x) \rangle_N \right\rangle = \sum_{N=N_F}^{N_D} \int_0^1 dx \Xi(N\cdot x) G(x) = \frac{\sum_N \frac{1}{N} \int dx\, G(x) P_N(x)}{\sum_N \frac{1}{N}}, \tag{29}$$

where N_F is the size of the selected fundamental cell. One reasonable choice is a four cation, one anion cell containing 16 bonds where $N_F = 4$. Then N varies modulo 4 from a minimum of 4 to N_D. Another choice is a primitive cell where $N_F = 1$.

We still face the problems of setting functional x dependences of physical quantities for small clusters (quantum dots). There are also problems with the shapes of clusters, for example, the difference if they are nearly spheres or ellipsoids, in the bound state energies of clusters of size N that introduce local mesoscopic-sized potential wells.

4.3. Examples

Let us do a couple of simple examples to demonstrate the averaging process. In this exercise we will let the integral over x vary from minus to plus infinity rather than renormalizing $P_N(x)$.

Example (A): $G(x) = x - \bar{x}$

Then we have:

$$\langle x - \bar{x} \rangle_N = 0 \tag{30}$$

so $\langle x \rangle_N = \bar{x}$, and $\left\langle \langle x \rangle_N \right\rangle = \bar{x}$.

Example (B): $G(x) = (x - \bar{x})^2$

Now the integral becomes:

$$\left\langle (x - \bar{x})^2 \right\rangle_N = \sigma_N^2 = \frac{\bar{x}(1 - \bar{x})}{N}. \tag{31}$$

Replacing the sum over N by an integral, the second average over the pixel volume is:

$$\left\langle \left\langle (x-\bar{x})^2 \right\rangle_N \right\rangle = \int_{N_F}^{N_D} dN \frac{\bar{x}(1-\bar{x})}{N^2 \ln(N_D/N_F)} = \frac{\bar{x}(1-\bar{x})}{\ln(N_D/N_F)} \left(\frac{1}{N_F} - \frac{1}{N_D} \right). \quad (32)$$

For a pixel whose volume is $40 \times 40 \times 10\ \mu m^3$ and for $Hg_{0.79}Cd_{0.21}Te$ where $\rho_c = 7.41 \times 10^{21}\ cm^{-3}$, we find $N_D = 1.18 \times 10^{14}$, and $\ln N_D = 32.4$. If we take $N_F = 2$, the smallest value for which $\sigma_N^2/\bar{x}^2$ has the same answer for both the Gaussian and Bernoulli distributions, and $\bar{x} = 0.21$ then:

$$\frac{\left\langle \left\langle (x-\bar{x})^2 \right\rangle_N \right\rangle}{\left\langle \langle x \rangle_N \right\rangle^2} = \frac{1}{2\ln(N_D/2)} \frac{1-\bar{x}}{\bar{x}} = 5.93 \times 10^{-2}. \quad (33)$$

This says the average x variation among pixels of the above volume is 0.21 ± 0.059. This is probably too large a variation. One problem is that the averages for small N are large and they get emphasized in the conversion from a sum to an integral in the second average over the pixel. This example needs to be redone using a proper distribution for small N and retaining the sum until N is large. However, even when this is done the rms value is still too high.

As it stands, this theory only accounts for strain reductions through the f factor. Values of f are given in Table 3 for the $Hg_{(1-x)}Cd_xTe$ and $Hg_{(1-x)}Zn_xTe$ alloys for concentrations where the cut off wave lengths are 10 and 20 μm. For $Hg_{(1-x)}Cd_xTe$ the reduction factor is near unity. But it is small enough in $Hg_{(1-x)}Zn_xTe$ alloys to significantly suppress concentration fluctuations. A good part of the reason this theory predicts large rms fluctuation is that a mechanism is left out of the formalism that also reduces their amplitudes. A cluster with a concentration x differing from the average $\bar{x}$ will have a net cohesive energy and a mean deviation of its cohesive energy that is different from the average. This effect still must be built into the theory.

5. CONCLUSIONS

Starting about ~20 years ago and accelerating, materials science theory is undergoing a major change. The methods have progressed from approximations that often led to insights into the underlying cause of phenomena, but depend on parameters fitted to experiments to produce accurate predictions, to parameter free first principles computational methods. These new computational intensive methods not only help uncover the true causes of phenomena, but are proving to be reliable engineering tools that when used properly can speed developments. The combination of first principles predicted parameters, and processing and performance models are powerful device development tools.

In this chapter, three levels of theories have been referenced. The BOA relies on a simplified fundamental-based calculation with few adjusted parameters to lend insight into the caused of a broad range of phenomena, but the accuracy of the predictions is often in the range of factors of two or so. The HPT method is closer to a first principles computationally intensive method, but it does require some fitting parameters. However, once that initial fitting is done the method allows predictions of a wide variety of phenomena for compounds and alloys with no further fitted parameters to accuracies of $\sim$10%. First principles theories like the self-consistent full potential GWA method predicts the behavior of complex materials to within a few percent with no adjusted parameters.

However, even with these powerful techniques in hand it is still necessary to ask them the right questions. That brings us to the example of concentration fluctuations. The presentation above is a bare start into a study of the impact these fluctuations have on the properties of alloys. These fluctuations are expected to be especially troublesome in alloys where the band gap is small, that is, cases where the cutoff wave length of devices is in the 10-20 μm range.

I have long advocated that for these VLWIR devices the active materials should be $Hg_{(1-x)}Zn_xTe$ rather than $Hg_{(1-x)}Cd_xTe$ because in that alloy system not only concentration fluctuations, but also dislocation tend to be suppressed. Some early $\sim$20 μm cutoff, LPE grown $Hg_{(1-x)}Zn_xTe$ arrays were tested [14]. They displayed excellent bake stability and had detectivities D^* at 64 K within a few percent of BLIP ($D^*_{BLIP} = 1.3 \times 10^{11}$ Jones). Because so much effort has gone into attempting to perfect $Hg_{(1-x)}Cd_xTe$ technology (with a bit of Zn added in some cases), little effort to has gone into continuing the development of $Hg_{(1-x)}Zn_xTe$-based devices despite their promise.

ACKNOWLEDGMENTS

I would like to thank Professor Mark van Schilfgaarde for guidance on the current status of first principle calculations and for a preprint of his paper. I also wish to thank Professor An-Ban Chen for deducing Eq. (25) and other suggestions. I am indebted to Dr. E. Patten who supplied data on $Hg_{(1-x)}Zn_xTe$ alloy based devices.

REFERENCES

[1] W.A. Harrison, Electronic Structure and the Properties of Solids, W. H. Freeman and Company, San Francisco, 1980.
[2] A.-B. Chen, A. Sher, Semiconductor Alloys, Plenum Press, New York, 1995.
[3] S. Krishnamurthy, A. Sher, J. Appl. Phys. 75 (1994) 7904–7909.

[4] S. Krishnamurthy, A. Sher, J. Electronic Mater. 24 (1995) 641–646.
[5] S. Krishnamurthy, A.-B. Chen, A. Sher, M. van Schilfgaarde, J. Electronic Mater. 24 (1995) 1121–1125.
[6] M.W. Muller, A. Sher, Appl. Phys. Lett. 74 (1999) 2343–2345.
[7] M. van Schilfgaarde, T. Kotani, S.V. Faleev, Phys. Rev. B 74 (2006) 245125–245140.
[8] J.C. Mikkelsen Jr., B. Boyce, Phys. Rev. Lett. 49 (1982) 1412–1415.
[9] A-B. Chen, A. Sher, W.T. Yost, Elastic constants and related properties of semiconductor compounds and their alloys" Chapter I in "Semiconductors and semimetals, in: K.T. Faber, K. Malloy (Eds.), The Mechanical Properties of Semiconductors, vol. 37, Academic Press, Inc., Boston, 1992.
[10] J.P. Hirth, J. Lothe, Theory of Dislocations, second ed., John Wiley & Sons, New York, 1982.
[11] P.N. Keating, Phys. Rev. 149 (1966) 674–678.
[12] M.A. Berding, W.D. Nix, D.R. Rhiger, S. Sen, A. Sher, J. Electronic Mater. 29 (2000) 676–679.
[13] A.-B. Chen, 2006 (private communication).
[14] E. Patten, 1997 (private communication).

CHAPTER **IIB**

Optical Phonon Spectra in CdTe Crystals and Ternary Alloys of CdTe Compounds

B.N. Mavrin and **E.A. Vinogradov**

1. INTRODUCTION

The structure of the cubic face-centered unit cell of CdTe crystal is characterized by the space group T_d^2 ($F\bar{4}3m$) and contains four formula units. The primitive unit is one fourth as many. The vibrational representation of optical phonons consists of 1 threefold-degenerated mode F_2 which is active in IR and Raman spectra. The dipole mode F_2 is split into the transverse (TO) and longitudinal (LO) modes in the vibrational spectra.

2. PHONON SPECTRA OF CdTe

2.1. Bulk crystal

First, the optical properties of CdTe crystal in IR region from 20 to 400 μm were studied from the reflection spectra $R(\omega)$ [1–7]. The Reststrahlen band in $R(\omega)$ is at 154 cm^{-1} with the peak reflectivity of 98% at 80 K and at 150 cm^{-1} with the peak reflectivity of 78% at 273 K (Fig. 1) [8, 9]. Using the Kramers-Kronig analysis, the frequency dependences of both refraction index $n(\omega)$ and extinction coefficient $k(\omega)$ can be found from equations:

$$n(\omega) = \frac{1-R}{1+R-\sqrt{R}\cos\theta}, \quad k(\omega) = \frac{2\sqrt{R}\sin\theta}{1+R-\sqrt{R}\cos\theta}, \qquad (1)$$

Institute for Spectroscopy of Russian Academy of Sciences, Fizicheskaya Str.5, 142190, Troitsk, Moscow, Russia

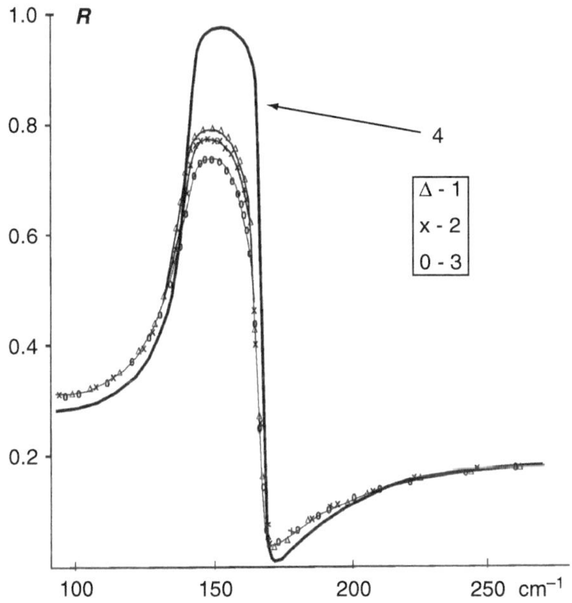

Figure 1 Infrared-reflection spectra of pure CdTe with the different polish of surface: 1—chemical treatment (273 K), 2—on pitch (273 K), 3—on cloth (273 K), 4—on pitch (80 K).

where

$$\theta(\omega) = \frac{\omega}{\pi}\int \frac{\ln R(\nu)}{\nu^2 - \omega^2}\,d\nu. \tag{2}$$

The value $n(\omega)$ and $k(\omega)$ are related with dielectric function $\varepsilon(\omega) = \varepsilon_1(\omega) + i\varepsilon_2(\omega)$ as follows:

$$\varepsilon_1(\omega) = n^2(\omega) - k^2(\omega), \quad \varepsilon_2(\omega) = 2n(\omega)k(\omega),$$
$$\mathrm{Im}\left(-\frac{1}{\varepsilon(\omega)}\right) = \frac{2n(\omega)k(\omega)}{[n^2(\omega) + k^2(\omega)]^2}. \tag{3}$$

On the other hand, using the general dispersion relation for $\varepsilon(\omega)$ with the γ damping, one can obtain:

$$\omega\varepsilon_2(\omega) = \frac{(\varepsilon_0 - \varepsilon_\infty)\omega_{TO}^2\gamma}{(\omega - \omega_{TO})^2 + \gamma^2}, \quad \omega\mathrm{Im}\left(-\frac{1}{\varepsilon(\omega)}\right) = \frac{(\varepsilon_\infty^{-1} - \varepsilon_0^{-1})\omega_{LO}^2\gamma}{(\omega_{LO} - \omega)^2 + \gamma^2}. \tag{4}$$

The maximum of the curve $\omega\varepsilon_2(\omega)$ corresponds to the TO-mode frequency ω_{TO} and the $\omega\mathrm{Im}(-1/\varepsilon(\omega))$ maximum to the LO-mode frequency ω_{LO}. The bandwidths of these curves determine the γ damping of phonons.

Three methods were used to find the oscillator strengths ($S_{TO} = \varepsilon_0 - \varepsilon_\infty$) of TO modes:

$$\varepsilon_0 = \left(\frac{\sqrt{R_0}+1}{\sqrt{R_0}-1}\right)^2 \quad \text{and} \quad \varepsilon_\infty = \left(\frac{\sqrt{R_\infty}+1}{\sqrt{R_\infty}-1}\right)^2,$$
$$\varepsilon_0 - \varepsilon_\infty = \frac{\gamma_{TO}}{\omega_{TO}}\varepsilon_2(\omega_{TO}), \quad \varepsilon_0 - \varepsilon_\infty = \frac{2}{\pi}\int_0^\infty \frac{\varepsilon_2(\omega)}{\omega}d\omega, \tag{5}$$

where R_0 and R_∞ are the reflectivities at $\omega \ll \omega_{TO}$ and $\omega \gg \omega_{LO}$, respectively.

Using Eqs. (1)–(4) and $R(\omega)$ (Fig. 1), one can obtain the dielectric function (Fig. 2) as well as TO (140 cm^{-1}) and LO (167 cm^{-1}) frequencies and dampings γ_{TO} (5.6 cm^{-1}) and γ_{LO} (6.3 cm^{-1}) for the CdTe crystal [9, 10]. The imperfections of the crystal have an effect on the optical properties. For example, the perfection of CdTe crystal lattice changes by the different treatment of the reflecting surface of crystal (Fig. 1, Table 1). In particular, from Table 1 it is seen that the frequencies of phonons increase with a decrease of phonon dampings.

The analysis of the reflectivity spectra of n-CdTe allowed one to study the plasmon-phonon interaction as well as to measure the concentrations and mobilities of free carriers [11].

The TO and LO modes are active in the Raman spectra of CdTe [12–15]. If the Raman spectra are excited in the transparency region

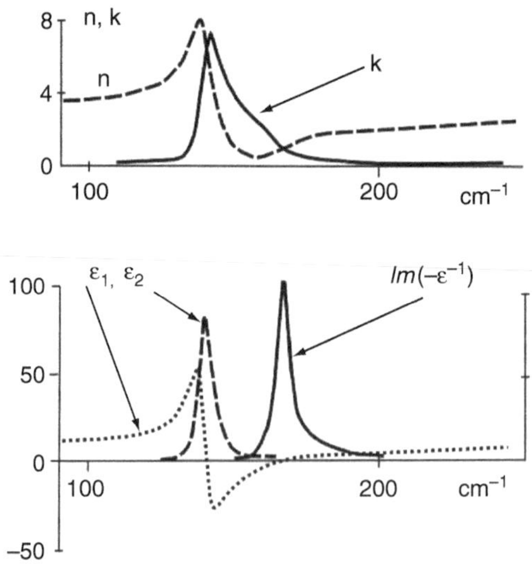

Figure 2 Optical functions of CdTe: $n(\omega)$ and $k(\omega)$, $\varepsilon_1(\omega)$, $\varepsilon_2(\omega)$-left axis, and Im($-\varepsilon^{-1}(\omega)$)-right axis, reconstructed from reflection spectra 1 of Fig. 1.

Table 1 Optical constants of CdTe single crystals after various polishes

Polish	T (K)	ω_{TO} (cm^{-1})	γ_{TO} (cm^{-1})	$4\pi\rho$	ω_{LO} (cm^{-1})	γ_{LO} (cm^{-1})	ε_0	ε_∞	$\frac{e_S^*}{e}$
On cloth	293	139	6.5	3	166	7.5	10.3	7.3	0.72
On pitch	293	140	5.6	3.15	167	6.3	10.4	7.3	0.73
Chemical	293	140.5	5	3.25	167.5	5.5	10.5	7.3	0.74
On pitch	80	143.5	2	3.30	171	1.7	10.5	7.2	0.77

(<1.5 eV), TO mode is dominant [12, 13]. But in the case of resonant Raman scattering, especially at resonance with the $E_0 + \Delta_0$ gap (~2.54 eV), LO mode and its overtones become intense [14, 15]. Besides, the bands near 120 and 140 cm^{-1} corresponding to phonons of Te on the CdTe surface can be seen in the resonant Raman spectra.

The phonons of the CdTe crystal were studied in the whole Brillouin zone by inelastic neutron scattering [16] and ab initio calculations [17]. As it follows from calculations of the density of phonon states in CdTe crystal [18], there is a gap from 123 to 128 cm^{-1} between acoustical and optical density of states and the high-frequency edge of the optical density is near 173 cm^{-1}.

2.2. Films

The direct measurements of both TO and LO phonons by infrared technique may be performed from films. Earlier it was shown [19], that one can observe an absorption by TO and LO modes at the oblique incidence of light on a film. As it is expected [19], only TO modes may be seen in the *s*-polarized light as well as both TO and LO modes in *p*-polarized light. The experimental spectra of the infrared absorption in the CdTe films on the metallic mirror [10] correspond to these predictions (Fig. 3). Moreover, the optical constants of CdTe measured from the reflection for bulk crystals (Table 1) and by the absorption in the films are similar [10]. For more detail explanation of IR optical properties of thin semiconductor (A^2B^6) films see for example [20].

The Raman spectra from CdTe films were also studied (for example, in [21]) and they had the same features as those from the bulk crystal (Table 2).

2.3. Nanostructures

The phonon spectra of nanostructures (quantum dots, quantum wells and heterostructures) can be different from those of bulk crystals and depend on the nanocrystal shape and size. The interface phonons, the folded

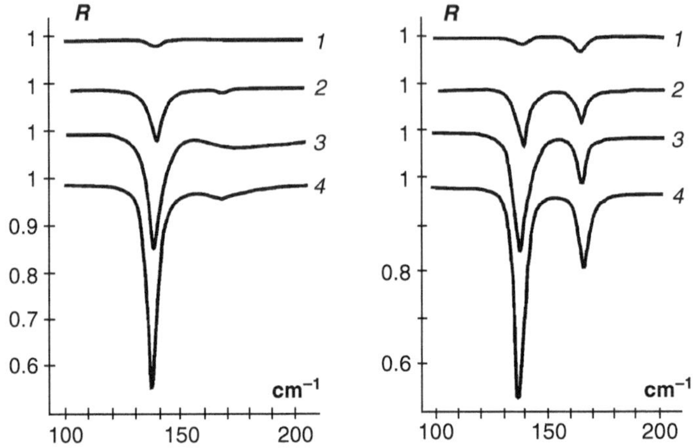

Figure 3 Reflection-absorption spectra of CdTe films on Al substrate in *s*- (left) and *p*-polarized (right) light for different thickness of the films (in μm): 1—0.5, 2—0.9, 3—1.6, 4—2.4.

Table 2 Optical constants of CdTe films (p-polarization)

Thickness, μ	ω_{TO} cm^{-1}	γ_{TO} cm^{-1}	A(TO) adsorption	ω_{LO} cm^{-1}	γ_{LO} cm^{-1}	A(LO) adsorption
0.5	140	5.5	0.01	167	6	0.03
0.8	139.5	6	0.06	167	6	0.05
0.9	139	6.5	0.1	167	6	0.06
1.6	138	7.5	0.25	167	6	0.1
2.4	137	5	0.46	167.5	5	0.16

acoustical phonons and the confined optical phonons may appear in the nanostructure spectra [22].

For example, the Raman spectra of the CdTe quantum well show a main peak near 173 cm^{-1} and smaller oscillationlike features towards lower phonon energies, which are assigned to even-order confined phonons ($m = 2, 4, \ldots, 12$) in quantum well [23]. The appearance of the oscillations is due to the quantization of the phonon momentum along the epitaxial growth normal to the layer:

$$q_m = m\pi/(d + a_0/2), \tag{6}$$

where d is the layer thickness and a_0 is the lattice constant of the bulk crystal. Using the observed dispersion ω_m (q_m) of the oscillations, it is possible not only to find the phonon dispersion at $q > 0$, but also to derive and estimate the compositional profile of the quantum well [23].

The polariton modes of a CdTe quantum-well-embedded planar microcavity are studied by resonant Raman scattering as a function of cavity-exciton detuning [24]. A maximum in the Raman efficiency is observed at the mode anticrossing, with minima in the pure photon and exciton limits.

The optical vibrations in the quantum dots are also considered to be confined to the dots with spherical shape and are assigned an equivalent wave vector:

$$q_m = \frac{\mu_m}{r}, \tag{7}$$

where μ_m is the mth node of the spherical Bessel function and r the radius of quantum dot. The strain due to the surface-free-energy appears in a crystal of finite size [25]. Hence, the optical-mode frequencies of CdTe quantum dot are determined by two competing size-dependent effects, namely, phonon confinement, causing a redshift, and compressive strain, producing a blueshift. The second effect is essential at small r.

The transmission spectra of the CdTe quantum dots with diameters of 4.2 nm deposited on Si substrate with 1 μm thickness layer were studied [26]. The center of the transmission band is at 152 cm^{-1} that was higher than that of TO mode in bulk crystal (140 cm^{-1}). In the transmission spectra of the CdTe quantum dots ($r = 4$ nm) coated by the CdSe shell with the thickness of 1-2 nm two minima were observed at 151.2 (CdTe-like) and 188 (CdSe-like) cm^{-1} [27].

The Raman spectra of the CdTe quantum dots ($r \sim 10$ nm) in borosilicate glass have shown the redshift of LO mode by 2.6 cm^{-1} [27]. The cavity-induced enhancement of the Raman efficiency from a monolayer of the CdTe quantum dots coated a dielectric microsphere was observed [28].

Far infrared reflectivity measurements of the CdTe-HgTe superlattices allowed one to find the layer composition, thickness, and the effective mass [14, 29]. Raman spectra from two high-quality CdTe/HgTe superlattices have shown CdTe, HgTe, and $Hg_{1-x}Cd_xTe$ phonon modes [30]. The locations of $Hg_{1-x}Cd_xTe$ phonon lines confirmed the degree of Hg alloying in the nominal CdTe layers, showing the utility of Raman scattering for the superlattice characterization. The IR and Raman techniques were used to study a very complex system consisting of superlattices of the CdTe quantum dots [31].

3. LOCALIZED MODES OF IMPURITIES IN CdTe

Substitutional impurities are known to affect the vibrational spectrum of a crystal. In addition to modifying the frequencies of the host lattice, new features are appeared in the spectrum [32]. When a substituent (M') of an infinitesimal amount is lighter than the host atoms (M), one expect the localized modes of the impurity with frequencies above the optic band of the host lattice. Impurities of heavier masses may have localized mode frequencies in the gap between the acoustical and optical bands (gap modes). Localized modes falling inside the optical band are known as resonance modes. The frequencies of impurity modes may be found from equation [33]:

$$1+\frac{\varepsilon\omega'^2}{N}\sum_q\sum_j\frac{|\sigma^j(q,M)|^2}{\omega_j^2(q)-\omega'^2}=0, \tag{8}$$

where $\varepsilon=(M-M')/M$, ω' is the eigenfrequency of the perturbed phonon state, N is the total number of wavevectors, $\sigma^j(q,M)$ corresponds to the atom of mass M in the j mode with the q wavevector, and $\omega_j(q)$ is the eigenfrequency of the perfect lattice. Using Eq. (8), explicit calculations of impurity modes have been carried out for a large class of zinc blende crystals [33]. The eigenvalues $\omega_j(q)$ and eigenvectors $\sigma^j(q,M)$ were determined by the modified rigid ion model. The calculated value of local mode frequency for the Se impurity in CdTe has corresponded to 171.6 cm^{-1} [33] that was in agreement with experiment [34, 35]. Several illustrations of an observation of local (Fig. 4) and gap (Fig. 5) modes in system $Cd_{1-x}Zn_xTe$ are shown. Apparently, the results of a solution of Eq. (8) depend on the lattice-dynamical model to obtain eigenvalues and eigenvectors of perfect crystal. Alternative methods of determining impurity modes using the density of phonon states are discussed in [32, 36]. The available experimental data of impurity modes in CdTe are presented in Tables 3 and 4.

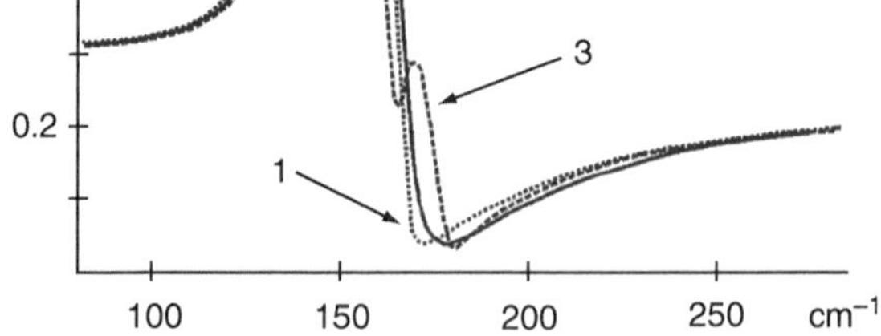

Figure 4 Reflection spectra of crystals $Cd_{1-x}Zn_xTe$ at smal x: 1—CdTe, 2—$Cd_{0.96}Zn_{0.04}Te$, 3—$Cd_{0.90}Zn_{0.10}Te$.

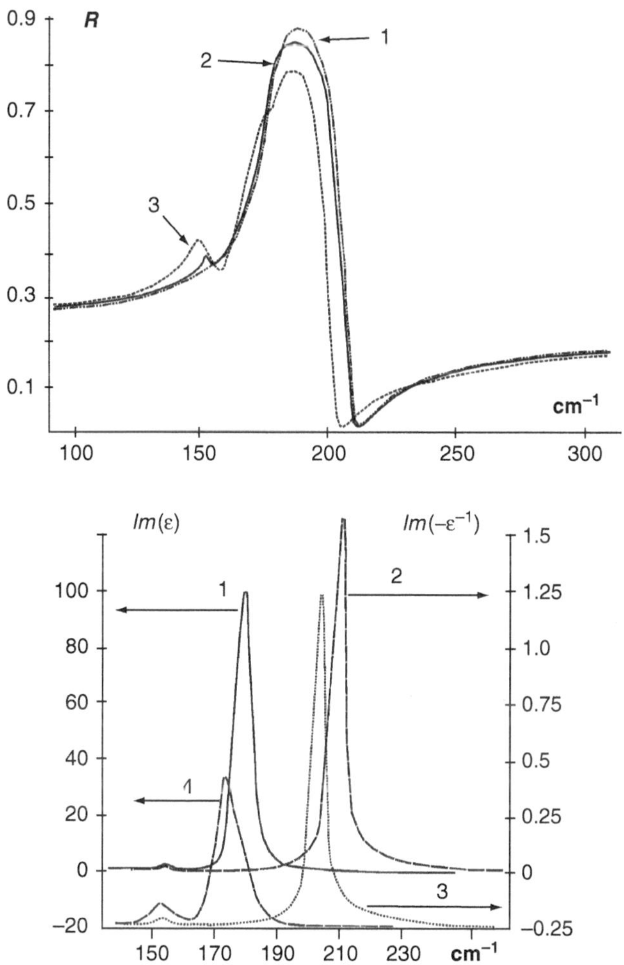

Figure 5 Reflectivity spectra of gap mode in ZnTe:Cd (A) crystals with compositions: 1—ZnTe, 2—$Zn_{0.95}Cd_{0.05}Te$, 3—$Zn_{0.80}Cd_{0.20}Te$ and their dielectric functions (B).

4. TERNARY ALLOYS OF THE CdTe COMPOUNDS

Single crystals of alloys of the CdTe compounds can be grown at almost any desired composition that allows one to control their optical and physical properties in applications. There are two extreme classes of ternary alloys $AB_{1-x}C_x$, according to the compositional dependence of their Raman and IR spectra [44, 45]. In the first class, known as one-mode crystals, the Raman and IR spectra exhibit one TO (LO) mode whose frequency varies continuously from ω_{TO}^{AB} (ω_{LO}^{AB}) at $x = 0$ to ω_{TO}^{AC} (ω_{LO}^{AC})

Table 3 Experimental value of impurity mode frequencies (in cm^{-1}) in CdTe

Compound	Frequency	Experiment	References
CdTe:Zn	167	IR	[34]
CdTe:Zn	173	IR	[37]
CdTe:Mn	195	Raman	[38]
CdTe:Fe	196	IR	[37]
CdTe:Hg	~130	Raman	[30, 39, 40]
CdTe:Mg	250	IR, Raman	[41]
CdTe:Se	173	IR	[34, 35]
CdTe:S	258.7	Raman	[42, 43]

Table 4 Optical constants of CdTe mixed crystals [8, 9, 34]

Compound	T (K)	Mode	ω_{TO} (cm^{-1})	ω_{LO} (cm^{-1})	γ_{TO} (cm^{-1})	γ_{LO} (cm^{-1})	ε_0	ε_∞	$4\pi\rho$	$\frac{e_S^*}{e}$
$CdTe_{0.94}Se_{0.06}$	293	Main	140.5	166	10.5	6	7.2	10.2	3	0.73
$CdTe_{0.94}Se_{0.06}$	293	Local	172.4	174.6	11	12	7.1	7.2	0.1	0.6
$CdTe_{0.94}Se_{0.06}$	80	Main	144	169.5	2.4	2.5	7.1	10.3	3.2	0.79
$CdTe_{0.94}Se_{0.06}$	80	Local	175	177	6.8	6.0	7.0	7.1	0.1	0.5
$CdTe_{0.9}Se_{0.1}$	293	Main	141.3	163.5	7.5	6.0	7.2	10.1	2.9	0.74
$CdTe_{0.9}Se_{0.1}$	293	Local	172.2	179.0	10.5	11.0	6.9	7.2	0.3	0.76
$Cd_{0.95}Zn_{0.05}Te$	293	Main	140.3	168.5	6.1	9.0	7.1	10.3	3.2	0.75
$Cd_{0.95}Zn_{0.05}Te$	293	Local	-	-	-	-	-	-	-	-
$Cd_{0.95}Zn_{0.05}Te$	80	Main	142.5	171.8	2.5	5.0	6.9	10.2	3.3	0.8
$Cd_{0.95}Zn_{0.05}Te$	80	Local	-	-	-	-	-	-	-	-
$Cd_{0.9}Zn_{0.1}Te$	293	Main	141.6	175.0	6.4	9.0	7.3	10.2	2.9	0.73
$Cd_{0.9}Zn_{0.1}Te$	293	Local	167.0	163.0	8.5	9.0	7.1	7.3	0.2	0.60
$Cd_{0.04}Zn_{0.96}Te$	293	Main	177.0	207.0	5.0	4.0	6.7	9.4	2.7	0.62
$Cd_{0.04}Zn_{0.96}Te$	293	Gap	154.0	154.3	3.0	4.0	9.4	9.6	0.2	0.70
$Cd_{0.2}Zn_{0.8}Te$	293	Main	173.0	201.0	9.2	4.3	6.8	9.3	2.6	0.63
$Cd_{0.2}Zn_{0.8}Te$	293	Gap	152.5	154.0	8.0	6.0	9.3	9.8	0.5	0.70
$CdSe_{0.9}Te_{0.1}$	293	Main	170.5	211.5	9.0	9.0	6.1	9.4	3.3	0.90
$CdSe_{0.9}Te_{0.1}$	293	Gap	152.5	154.0	6.6	6.0	9.4	9.6	0.2	0.50
$CdSe_{0.75}Te_{0.25}$	293	Main	169.0	205.0	12.0	8.5	6.1	8.6	2.5	0.86
$CdSe_{0.75}Te_{0.25}$	293	Gap	151.5	155.5	10.0	9.0	8.6	9.6	1.0	0.72

at $x = 1$. In the second class, two-mode crystals, the Raman and IR spectra exhibit the composition dependent TO and LO phonons of two constituents (AB-like and AC-like). Several cases have been reported [45, 46] which show some type of intermediate behavior and such crystals are difficult to classify as one- or two-mode crystals.

One the most successful models which has been used to describe the optical behavior in mixed crystals is the random element isodisplacement (REI) model [44, 45, 47]. The REI model is based on the assumption that in the long-wavelength limit the anion and cation of like species vibrate with the same phase and amplitude and that interatomic forces are determined by a statistical average of the interaction with its neighbors. A modified REI model [47] is completely defined by the macroscopic parameters of the pure end members. As distinct from REI models [44, 45], in incorporates second-neighbor force constant *f*(B-C) and a linear dependence of all force constants on the lattice parameter without resorting to microscopic fitting parameters.

At present there are no clear formal criteria for one- or two-mode behavior. The conditions given earlier [44] are not valid for ternary alloys of the CdTe compounds [8, 9]. For example, the two-mode behavior occurs in the $CdTe_{1-x}Se_x$ crystal [34] while the conditions [44] predict an one-mode behavior, since the Se atom mass is greater than the reduced mass of elements of CdTe ($m_{Se} > \mu_{CdTe}$). The $CdTe_{1-x}Se_x$ reflection spectra show two Reststrahlen bands [34]. The intensity of the low-frequency band decreases continuously from a value of pure CdTe ($x = 0$) to zero at $x \to 1$ whereas the intensity of the high-frequency band increases with the Se concentration. The dependence of optical-mode frequencies on $CdTe_{1-x}Se_x$ composition is shown in Fig. 6 and the oscillator strengths of TO and LO modes in Fig. 7. By analogy with the TO oscillator strength (Eq. (5)) the oscillator strength of LO modes was found as

$$S_{LO} = \frac{1}{\varepsilon_\infty} - \frac{1}{\varepsilon_0} = \frac{\gamma_{LO}}{\omega_{LO}} \mathrm{Im}\left(-\frac{1}{\varepsilon(\omega_{LO})}\right). \tag{9}$$

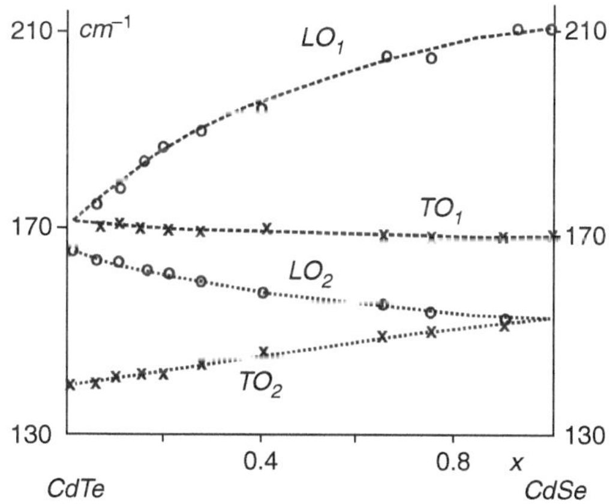

Figure 6 Variation of LO and TO frequencies in $Cd_{1-x}Se_xTe$ as a function of composition *x*. The dotted lines are the fits from a MREI model.

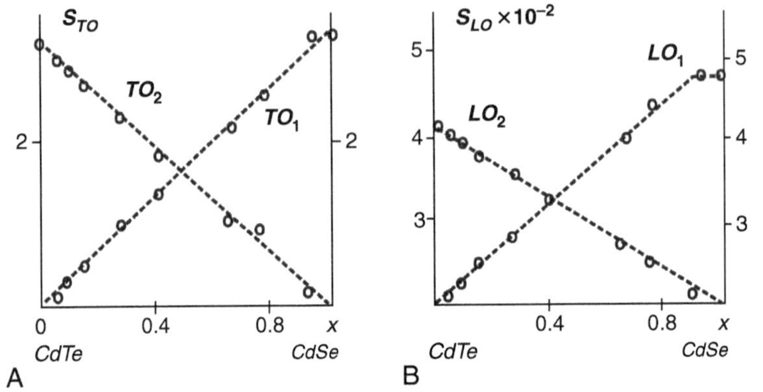

Figure 7 Oscillator strengths of TO (A) and LO (B) modes in $Cd_{1-x}Se_xTe$ versus x. The dotted lines are guided by eye.

The data in Figs. 6 and 7 are typical for two-mode behavior of optical vibrations in mixed crystal $CdTe_{1-x}Se_x$.

The mixed crystal $Cd_{1-x}Zn_xTe$ shows more complex behavior of optical modes, although $m_{Zn} > \mu_{CdTe}$. Two Reststrahlen bands are seen in the reflection spectra (Fig. 8) that is inherent to two-mode behavior [48]. The frequency dependences of optical modes are given in Fig. 9 and the oscillator strengths of TO and LO modes in Fig. 10. The oscillator

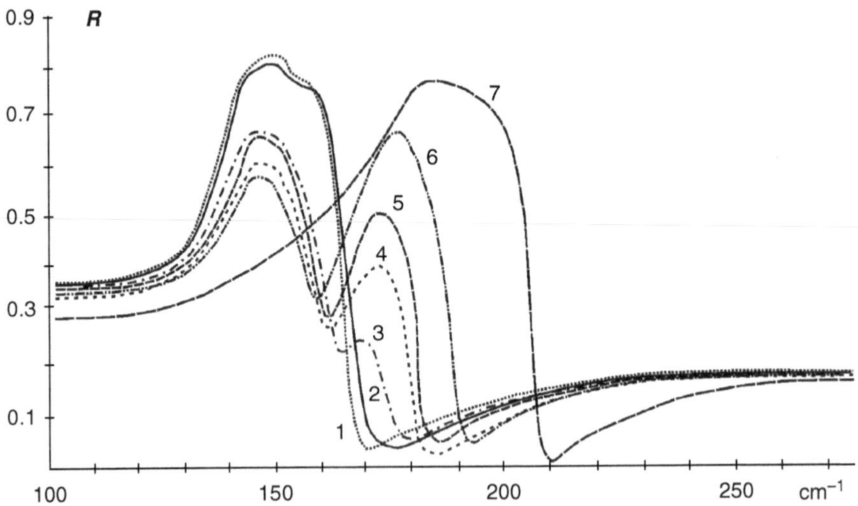

Figure 8 Representative infrared-reflection spectra for $Cd_{1-x}Zn_xTe$ crystals with composition x of 0.0 (1), 0.062 (2), 0.23 (3), 0.35 (4), 0.44 (5), 0.51 (6), and 1.0 (7).

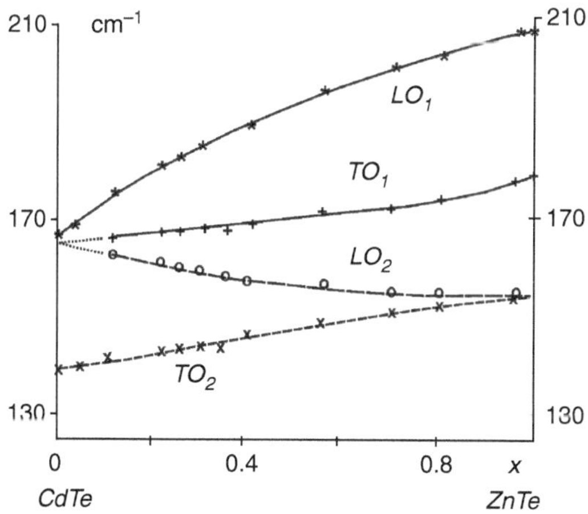

Figure 9 TO and LO frequencies in $Cd_{1-x}Zn_xTe$ versus x. The dotted lines are the fits from a MREI model.

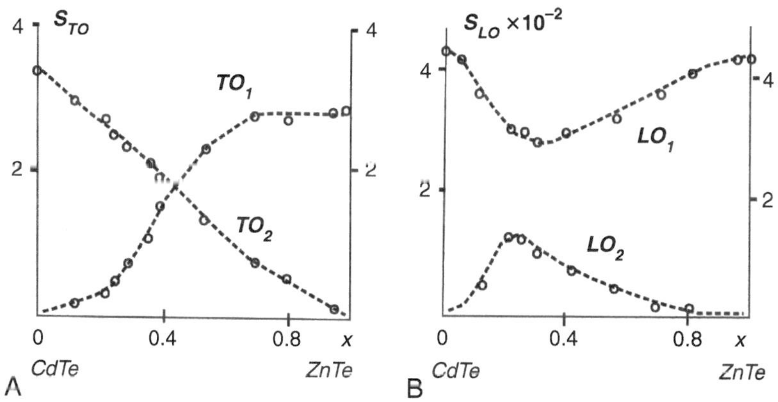

Figure 10 Oscillator strengths of TO (A) and LO (B) modes in $Cd_{1-x}Zn_xTe$ versus x. The dotted lines are guided by eye.

strengths of TO modes behave as in two-mode case (Fig. 10A). But the compositional dependence of the LO oscillator strengths (Fig. 10B) is like to one-mode behavior. Really, the high-frequency LO mode (the curve LO_1 in Fig. 10B) has the great oscillator strength in whole compositional range, whereas the low-frequency LO mode (the curve LO_2 in Fig. 10B) has small oscillator strength at $x \to 0$ and $x \to 1$ and it corresponds to an impurity LO mode. The local mode of Zn atoms in CdTe is somewhat below the LO mode of pure CdTe and really it is resonance mode. The Zn resonance mode is split into an inverted TO-LO doublet ($\omega_{TO} > \omega_{LO}$). The TO component becomes as ZnTe-like TO mode at increasing the Zn content, while the LO component turns to CdTe-like LO mode and it becomes as the Cd gap mode in ZnTe lattice at $x \to 1$. As result, the $Cd_{1-x}Zn_xTe$ spectra show a two-mode-like behavior in the compositional range of $0.5 \leq x < 1$ (ZnTe-like LO1 and TO1 modes as well as CdTe-like LO2 and TO2 impurity modes), but the one-mode compositional dependencies of both optical-mode frequencies and the oscillator strengths at $0 < x < 0.5$ (LO1 and TO2 are the main modes, TO1 and LO2 are the inverted impurity modes). Hence, our study of $Cd_{1-x}Zn_xTe$ system has shown that the measurement of optical-mode oscillator strengths is not only useful, but also definitive for the study of the compositional behavior of phonons in mixed crystals [9, 48]. The gap modes of Cd vibrations in ZnTe were investigated by IR reflection spectra of thin films $Cd_{0.05}Zn_{0.95}Te$ and strong resonance of the gap modes with cavity modes of $Cd_{0.05}Zn_{0.95}Te$ films was observed [20].

The Raman spectra of mixed crystals $Cd_{1-x}Zn_xTe$ were obtained for Zn content of $x \approx$1-0.6 at 0.6328 μm excitation for which the crystals were transparent (Fig. 11) [48]. The Raman results of TO and LO dispersion was

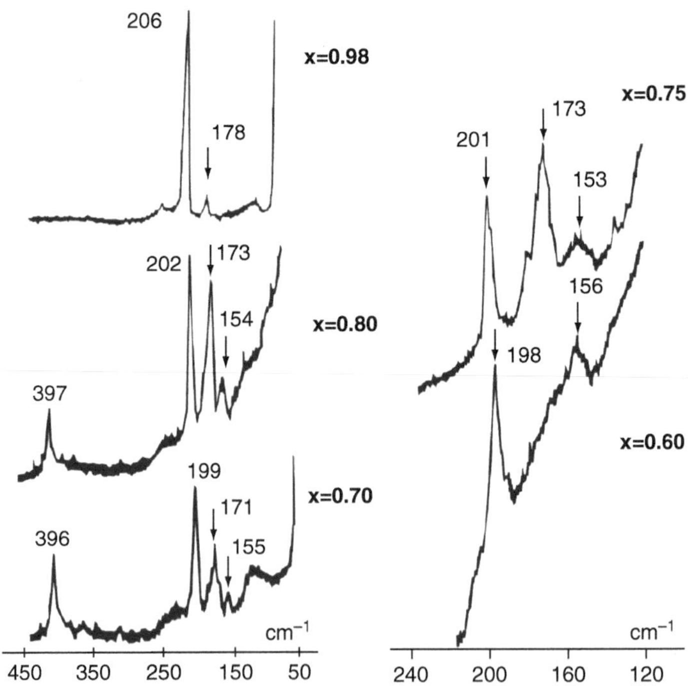

Figure 11 Raman spectra of bulk $Cd_{1-x}Zn_xTe$ for different x: $x = 0.98, 0.80, 0.70, 0.60$.

in agreement with IR data. Moreover, the Raman intensities of LO_1 and LO_2 modes for compositions up to $x = 0.55$ have been correlated with their oscillator strengths measured from IR spectra ($S_{LO_1} > S_{LO_2}$ in Fig. 10B). As to TO modes, the TO_1 oscillator strength is significantly more that of TO_2 modes (Fig. 10A) and only TO_1 modes were seen in Raman spectra (Fig. 11). The mode dispersion in $Cd_{1-x}Zn_xTe$ (Fig. 9) is mostly in agreement with other Raman and IR data [21, 49–53].

Hence, we see that the simple mass criterion [44] is not correct for mixed crystals on basis of CdTe. At a later time [54], it was argued that it is necessary to consider the optical phonon density of states (OPDOS) of the end-member crystals for the analysis of type of the mode behavior. According to [54], the criterion for one- or two-mode behavior is the nonexistence or existence of a common gap between the OPDOS of the end-member crystals. Probably, this criterion for two-mode case is realized in $CdTe_{1-x}Se_x$ system, but it does not hold in $Cd_{1-x}Zn_xTe$. Although the Reststrahlen bands of CdTe and ZnTe do not overlap, but their OPDOS are partially covered. Really, in accord with calculations [18], OPDOS of CdTe is nonzero in the frequency region of 127–172 cm^{-1}, while OPDOS of ZnTe in the region of 154–210 cm^{-1}. Practically, OPDOS of ZnTe is overlapped only with the LO band of CdTe OPDOS which is in the region of ~148–172 cm^{-1}, because the OPDOS of CdTe is split into nearly separated TO (127–148 cm^{-1}) and LO (148–172 cm^{-1}) bands [18]. Probably, the gap between CdTe-like and ZnTe-like LO DOS in $Cd_{1-x}Zn_xTe$ appears already at small x that triggers a transition of the behavior of LO modes from one- to two-mode case at increasing x.

The mixed crystals $Cd_{1-x}Mn_xTe$ possess a homogeneous crystal phase having a zink-blende structure for $0 \leq x \leq 0.7$. Their Raman spectra exhibit two pairs of sharp lines characteristic of the zone center CdTe-like and MnTe-like LO-TO modes [38, 47, 50]. In the antiferromagnetic phase of the alloys ($x > 0.6$) and in the spin-glass phase ($0.17 \leq x \leq 0.6$), a new low-energy Raman line appears that were attributed to an inelastic scattering involving the emission or absorption of a magnon [38, 50].

The compositional dependence of the CdTe-like LO modes were studied in $Cd_{1-x}Hg_xTe$ with $0 \leq x \leq 0.5$ using Raman scattering [55]. A frequency shift of LO mode linear with x was found for the whole compositional range [55]. Using far-infrared transmission and reflection spectroscopy [39, 40, 56–58], the fine structures of the CdTe-like band and the complex structure of HgTe-like band, as well as the plasmon-LO phonon coupling effect have been observed.

The Raman line shapes of the LO phonon have been analyzed for the pseudobinary alloy system $CdTe_{1-x}S_x$ over the full alloy range [42, 43]. The polycrystalline thin films were grown by pulsed laser deposition including films with x values throughout the miscibility gap ($0.06 < x < 0.97$). Peak shift, broadening, and asymmetry arising from spatial correlation effects have been yielded details of the microstructural clustering. It is found that this ternary system exhibits a two-mode behavior with CdS- and CdTe-like LO modes [42, 43, 59].

The Raman and IR spectra of $Cd_{1-x}Mg_xTe$ bulk crystals with the zinkblende structure ($x \leq 0.6$) were studied [41, 60]. The Raman spectra exhibit a classic two-mode behavior with MgTe-like and CdTe-like LO-TO pairs of zone-center optical phonons, in decreasing order of frequency. For very small x IR spectra reveal Mg^{2+} three local modes corresponding to their isotopic abundances.

REFERENCES

[1] J. Akiyoshi Mitsuishi, Phys. Soc. Jpn. 16 (1961) 533–537.
[2] A. Mitsuishi, H. Yoshinaga, K. Yata, A. Manabe, Jpn. J. Appl. Phys. 1 (Suppl. 1-1) (1964) 581–587.
[3] A. Manabe, A. Mitsuishi, H. Yoshinaga, Jpn. J. Appl. Phys. 6 (1967) 593–600.
[4] G.L. Bottger, A.L. Geddes, J. Chem. Phys. 47 (1967) 4858–4859.
[5] O.M. Stafsudd, F. Haak, K. Radisavljevic, J. Opt. Soc. Am. 57 (1967) 1475–1476.
[6] C.M. Randall, R.D. Rawcliffe, Appl. Opt. 7 (1968) 213.
[7] C.J. Johnson, G.H. Sherman, R. Weil, Appl. Opt. 8 (1969) 1667–1671.
[8] E.A. Vinogradov, The study of optical phonons in $A^{II}B^{VI}$ compounds by far-infrared spectroscopy, Diss. Ph.D. (in Russian), Moscow Institute for Physics and Technology, 1973.
[9] L.K. Vodopyanov, E.A. Vinogradov, Crystal Lattice Defects 5 (1974) 125–136.
[10] L.K. Vodopyanov, E.A. Vinogradov, V.V. Kolotkov, Yu.A. Mityagin, Sov. Phys. Solid State 16 (1974) 912–915.
[11] S. Perkowitz, R.H. Thorland, Phys. Rev. B 9 (1974) 545–550.
[12] A. Mooradian, G.B. Wright, Proceedings of the Ninth International Conference on the Physics of Semiconductors, Moscow, Nauka, Leningrad, 1020, 1968.
[13] M. Selders, E.Yi. Chen, R.K. Chang, Solid State Comm. 12 (1973) 1057–1060.
[14] D.J. Olego, J.P. Faurie, P.M. Raccah, Phys. Rev. Lett. 55 (1985) 328–331.
[15] S.S. Islam, S. Rath, K.P. Jain, S.C. Abbi, C. Julien, M. Balkanski, Phys. Rev B 46 (1992) 4982–4985.
[16] J.M. Rowe, R.M. Nicklow, D.L. Price, K. Zanio, Phys. Rev. B 10 (1974) 671–675.
[17] A.D. Corso, S. Baroni, R. Resta, S. de Gironcoli, Phys. Rev. B 47 (1993) 3588–3592.
[18] B.D. Rajput, D.A. Browne, Phys. Rev. B 53 (1996) 9052–9058.
[19] D.W. Berreman, Phys. Rev. 130 (1963) 2193–2198.
[20] E.A. Vinogradov, Physics-Uspekhi 45 (2002) 1213–1250.
[21] D.J. Olego, P.M. Raccah, J.P. Faurie, Phys. Rev. B 33 (1986) 3819–3822.
[22] B. Jusserand, M. Cardona, in: M. Cardona, G. Güntherodt (Eds.), Light Scattering Solids V, Springer, Berlin, 1989.
[23] A. Bruchhausen, A. Fainstein, B. Jusserand, R. André, Phys. Rev. B 73 (2006) 085305–085311.
[24] A. Fainstein, B. Jusserand, R. André, Phys. Rev. B 57 (1998) R9439–R9442.
[25] V. Spagnolo, G. Scamarcio, M. Lugará, G.C. Righini, Superlattices Microstruct. 16 (1994) 51–54.
[26] R.B. Vasiliev, V.S. Vinogradov, S.G. Dorofeev, S.P. Kozyrev, I.V. Kucherenko, N.N. Novikova, Phys. Solid State 49 (2007) 547–551.

[27] P.T.C. Freire, M.A. Araújo Silva, V.C.S. Reynoso, A.R. Vaz, V. Lemos, Phys. Rev. B 55 (1997) 6743–6746.

[28] Yu.P. Rakovich, J.F. Donegan, N. Gaponik, A.L. Rogach, Appl. Phys. Lett. 83 (2003) 2539–2541.

[29] L.S. Kim, S. Perkowitz, O.K. Wu, J.N. Schulman, Semicond. Sci. Techn. 5 (1990) S107–S110.

[30] Z.C. Feng, S. Perkowitz, O.K. Wu, Phys. Rev. B 41 (1990) 6057–6060.

[31] V.S. Bagaev, L.K. Vodop'yanov, V.S. Vinogradov, V.V. Zaitsev, S.P. Kozyrev, N.N. Mel'nik, E.E. Onishchenko, G. Karczewski, Phys. Solid State 46 (2004) 173–175.

[32] A.A. Maradudin, Solid State Phys. 18 (1966) 273–420.

[33] S.P. Gaur, J.F. Vetelino, S.S. Mitra, J. Phys, Chem. Solids 32 (1963) 2737–2747.

[34] L.K. Vodopyanov, E.A. Vinogradov, V.S. Vinogradov, A.E. Tsurkan, V.I. Verlan, Zhurnal Prikladnoi Spektroskopii 21 (1974) 96–104 (in Russian).

[35] S. Perkowitz, L.S. Kim, P. Becla, Phys. Rev. B 43 (1991) 6598–6601.

[36] V.S. Vinogradov, Sov. Phys. Solid State 11 (1970) 2062–2071.

[37] W. Lu, H.J. Ye, Z.Y. Yu, S.Y. Zhang, Y. Fu, W.L. Xu, S.C. Shen, W. Giriat, Phys. Stat. Solidi (b) 147 (1988) 767–778.

[38] S. Venugopalan, A. Petrou, R.R. Galazka, A.K. Ramdas, S. Rodriguez, Phys. Rev. B 25 (1982) 2681–2696.

[39] S.P. Kozyrev, L.K. Vodopyanov, R. Triboulet, Solid State Comm. 45 (1983) 383–385.

[40] S.P. Kozyrev, L.K. Vodopyanov, R. Triboulet, Phys. Rev. B 58 (1998) 1374–1384.

[41] C. Eunson Oh, I. Parks, M. Miotkowski, A.J. Dean Sciacca, A.K. Mayur, Ramdas, Phys. Rev. B 48 (1993) 15040–15046.

[42] A. Fisher, Z. Feng, E. Bykov, G. Contreras-Puente, A. Campaan, F. de Landa Castillo-Alvarado, J. Avendanŏ, A. Mason, Appl. Phys. Lett. 70 (1997) 3239–3241.

[43] A. Fisher, L. Anthony, A.D. Campaan, Appl. Phys. Lett. 72 (1998) 2559–2561.

[44] I.F. Chang, S.S. Mitra, Phys. Rev. 172 (1968) 924–933.

[45] L. Genzel, T.P. Martin, C.H. Perry, Phys. Stat. Solidi (b) 62 (1974) 83–92.

[46] E.A. Vinogradov, B.N. Mavrin, L.K. Vodopyanov, JETP 99 (2004) 749–755.

[47] D.L. Peterson, A. Petrou, W. Giriat, A.K. Ramdas, S. Rodriguez, Phys. Rev. B 33 (1986) 1160–1165.

[48] E.A. Vinogradov, L.K. Vodopyanov, Sov. Phys. Solid State 17 (1976) 2088–2091.

[49] S. Perkowitz, L.S. Kim, Z.C. Feng, P. Becla, Phys. Rev. B 42 (1990) 1455–1457.

[50] R.G. Eunsoon Oh, I. Alonso, A.K. Miotkowski, Ramdas, Phys. Rev. B 45 (1992) 10934–10941.

[51] D.N. Talwar, Z.C. Feng, P. Becla, Phys. Rev. B 48 (1993) 17064–17071.

[52] H. Harada, S. Narita, J. Phys, Soc. Jpn. 30 (1970) 1628–1639

[53] L.K. Vodopyanov, E.A. Vinogradov, A.M. Blinov, V.A. Rukavishnikov, Sov. Phys. Solid State 14 (1972) 168–270.

[54] E. Yahne, Phys. Stat. Solidi (b) 75 (1976) 221–232.

[55] A. Lusson, J. Wagner, Phys. Rev B 38 (1988) 10064–10066.

[56] A. Polian, R. Le Toullec, M. Balkanski, Phys. Rev. B 13 (1976) 3558–3565.

[57] P.M. Amirtharaj, N.K. Dhar, J. Baars, H. Seelewind, Semicond. Sci. Techn. 5 (1990) S68–S72.

[58] J.H. Chu, S.C. Shen, Semicond. Sci. Techn. 8 (1993) S86–S89.

[59] O. Goede, W. Heimbrodt, R. Muller, Phys. Status Solidi (b) 105 (1981) 543–550.

[60] S. Nakashima, T. Fukumoto, A. Mitsuishi, K. Itoh, J. Phys, Soc. Jpn. 35 (1973) 1437–1441.

CHAPTER IIC

Band Structure

C.R. Becker

1. BASIC PARAMETERS AT 300 K AND LOWER TEMPERATURES

1.1. Band structure

CdTe is a member of the II^B-VI group of semiconductors whose diverse optical and electrical properties make them excellent candidates for optoelectronic and spintronic applications. A large portion of their properties is directly related to the interaction of the localized semicore d electrons with the valence sp electrons [1]. This sp-d interaction makes band structure calculations of II^B-VI semiconductors more complicated than those for group IV and group III-V semiconductors. Even though theory for II^B-VI semiconductors is more challenging, modern *ab initio* theories are capable of describing the band structure of CdTe and other II^B-VI semiconductors.

In the mid-1960s important aspects of the band structure of CdTe became apparent primarily from experiments involving their optical properties, which are summarized by Zanio [2]. In particular, transmission measurements of Marple [3] clearly demonstrated that CdTe is a semiconductor with a direct bandgap at $\boldsymbol{k} = 0$. This was in stark contrast to the indirect band structure of the then best known semiconductors, that is, Si and Ge. Theoretical calculations based on the pseudopotential approach were successfully applied to Si and Ge and were able to reproduce the band structure of zinc blende semiconductors. At that time, the band structure of CdTe was successfully determined by the pseudopotential method of Cohen and Bergstresser [4]. However, this empirical approach ignores the semicore 4d states of Cd, which have been shown by *ab initio* methods to have a significant influence on the band structure [1]. Hence the good agreement of this pseudopotential method with experiment can be construed as an artifact or at best a good empirical result.

Experimentelle Physik III, Universität Würzburg, Am Hubland, D-97074 Würzburg, Germany

1.1.1. *Ab initio* approaches

Density functional theory (DFT) [5, 6] was first employed on semiconductors in the mid-1970s and grew in importance rapidly in the late 1970s and early 1980s. The local density approximation (LDA) of the DFT underestimates the energy gap between occupied and empty bands of II^B-VI group semiconductors to a greater extent than for other semiconductors. The shallow semicore d states are strongly *underbind* in the LDA. These long standing problems with LDA or for that matter pseudopotential or tight binding approaches, which do not take the d electrons into account properly or ignore them, have been recognized and analyzed in detail by Wei und Zunger [1]. Indeed the authors have demonstrated that the p-d repulsion and hybridization in II^B-VI semiconductors (1) lower the bandgaps, (2) reduce the cohesive energy, (3) increase the equilibrium lattice constant, (4) reduce the spin-orbit splitting, (5) alter the sign of the crystal-field splitting, (6) increase the valence band offset between common anion II^B-VI semiconductors, and (7) modify the charge distribution.

The calculated semicore d states are too high in the LDA approach and consequently the bandgap is much too small for II^B-VI group semiconductors. In the case of CdTe the bandgap is about 0.2 eV instead of 1.61 eV at 0 K. In the *GW* approach this problem is not as severe and the resulting bandgap of CdTe is about 1.2 eV. It should be noted that agreement of the *GW* results with experiment for II^B-VI group semiconductors is less satisfactory than is the case for other semiconductors. The *GW* approximation was introduced by Hedin [7, 8] in 1965 and was first applied to semiconductors in the early 1980s by Strinati et al. [9] and Hybertsen and Louie [10]. In this approach, many body effects are introduced via the screened Coulomb interaction $W(\omega)$ in a Green's function calculation of a one-electron quasiparticle. Its application has increased slowly and still is not employed as frequently as LDA due to its complexity and the required computation time.

Agreement of the perturbative application of the *GW* method can be improved by a number of schemes [11]. For example, by employing a band structure as a starting point, in which the p and d mix to a lesser extent than that of the normally employed LDA band structure. In the *GW* approximation the band structure corresponds to the poles of one particle Green's functions which are Fourier transformed to the energy-momentum domain. The electron-electron interaction consists of two terms, the electrostatic Hartree potential and the non-Hermitian, energy dependent, and spatially nonlocal operator which is called the self-energy. Calculation of the self-energy is very complex and consequently the same is true for Green's functions together with electron-electron interaction, which are therefore manageable only with the aid of approximations. In the *GW* approximation [7, 8] the self-energy Σ is represented by a product of the Green's function G and the dynamically screened Coulomb interaction W;

$$\Sigma = \mathrm{i}GW, \tag{1}$$

where

$$W = \varepsilon^{-1} \cdot \frac{1}{\mathbf{r}} \tag{2}$$

and ε is the longitudinal dielectric function of the system.

Because of the importance of the spin-orbit interaction in CdTe, it must be taken into account. This is normally done by including relativistic atomic pseudopotentials with spin-orbit interaction in the LDA or *GW* calculations of CdTe.

1.1.2. Energy dispersion

As previously mentioned, normally the *GW* approximation is perturbatively applied to the zero-order LDA band structure. The LDA band structure of a sophisticated LDA approach in which the semicore electrons, indeed essentially all electrons, were included [12] is shown in Fig. 1. As can be seen in Fig. 1, the LDA calculated energy gap at 0.21 eV is much too small. Also the *GW* calculated energy gap of 1.26 eV is less than the experimental value [12]. The *GW* approach with the random phase approximation, *GW-RPA*, used in these calculations is a realistic *GW* approach, albeit one of the simpler *GW* approaches. The calculated results for the energy gap and d electrons as well as selected transitions at critical points of the Brillouin zone together with experimental values which include those according to Chelikowsky and Cohen [13] are tabulated in Table 1. An extensive summary of the critical point transitions of CdTe from reflectivity experiments and pseudopotential calculations has been compiled by Zanio [2].

Obviously all LDA and *GW* calculated energy levels in the conduction band are about 1.3 and 0.3 eV too low, respectively, whereas the Cd 4d states are about 2.1 and 1.1 eV too high, respectively, but they are shifted by nearly the same magnitude. As mentioned above, these two effects are related and these incorrect energies remain an hereto unsolved problem in *ab initio* theories. Nevertheless, the shape of the dispersion of both LDA and *GW* are in good agreement with the experimental angle-resolved photoelectron spectroscopic (ARPES) results of Niles and Höchst [14]. Obviously agreement for the Γ_7 band is better, whereas, it must be pointed out that the light and heavy hole bands in the ARPES results shown in Fig. 2 are unresolved peaks whose energies depend on the experimental fitting procedure. Furthermore, the experimental photoelectrons correspond to a distribution of $\boldsymbol{k}$ values and the resulting $\boldsymbol{k}$ dispersion depends on the fitting procedure.

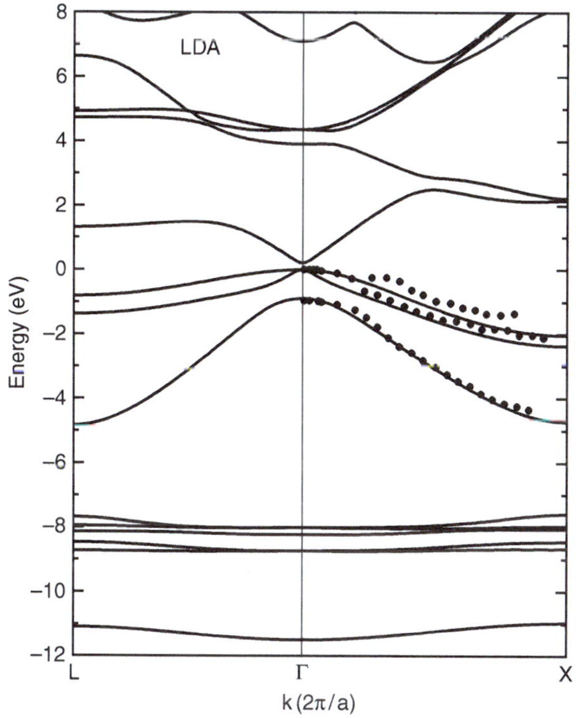

Figure 1 The calculated LDA band structure (lines) of CdTe at the Γ, L, and X points in the Brillouin zone and its energy dispersion between these points at 0 K. The ARPES results of Niles and Höchst [14] are reproduced as filled circles.

Table 1 The energy gap of CdTe and 4d states of Cd as well as selected transitions at critical points of the Brillouin zone of CdTe in eV

	$\Gamma_8^v - \Gamma_6^c$	Cd 4d states	$L_{4,5}^v - L_6^c$	$L_6^v - L_6^c$
Theory; LDA	0.21	−8.37	2.13	2.78
Theory; *GW-RPA*	1.26	−9.43	3.14	3.70
Experiment	1.60	−10.50	3.46[a]	4.03[a]

[a]Experimental values according to Chelikowsky and Cohen [13].

1.1.3 Effective masses

Experimental values of the electron, light hole, and heavy hole effective masses have been determined primarily from cyclotron resonances as well as from Faraday rotation. The former results are reproduced in Tables 2 and 3 and the latter result for the electron effective mass according to Marple [15] is $m_e^*/m_o = 0.11 \pm 0.01$ and is independent of carrier concentration.

Table 2 Electron effective mass (m_e^*/m_o) measured at the cyclotron resonance energy ($\hbar\omega$), the band effective mass (m_b^*/m_o), and the corresponding Fröhlich constant (α)

($\hbar\omega$) (meV)	m_e^*/m_o	m_b^*/m_o	α	References
0.29	0.096 ± 0.005			[16]
1.28	0.0963 ± 0.0008			[17]
3.68	0.0979	0.09	0.4	[18]
15.808	0.1124			[18]
0.29	0.096 ± 0.003			[19]
0.29	0.094 ± 0.004	0.088 ± 0.004	0.4	[20]
		0.0900 ± 0.0005	0.35	[21]
		0.0909	0.4	[22]
		0.0920	0.3	[23]
		0.0898	0.4	[23]

Table 3 Light hole and heavy hole effective masses in CdTe for conduction in the plane normal to the indicated magnetic field orientation

$H_{\|\|}$	m_{lh}^*/m_o	m_{hh}^*/m_o	References
$\langle 110 \rangle$	0.12 ± 0.02	0.81 ± 0.05	[19]
$\langle 111 \rangle$	0.12 ± 0.01	0.84 ± 0.02	[20]
$\langle 100 \rangle$	0.13 ± 0.01	0.72 ± 0.01	[20]

1.1.4. Density of states

The density of states based on the LDA and *GW* calculations both with spin-orbit interaction are shown in Fig. 3 as dotted and solid lines, respectively [12], because as is well-known spin-orbit interaction in CdTe is very important as indeed it is in all II^B-VI semiconductors.

1.2. Complex loss function

In a simple metal or some semiconductors, the excitation of electrons against a background of positively charged ionic cores or semicores normally results in a single damped plasmon [24]. In contrast, a number of loss structures of comparable strength have been observed in CdTe in the vicinity of the classical plasmon energy [25]. By means of high-resolution electron energy loss spectroscopy (ELLS), Dröge et al. [25] observed three main peaks and a few less pronounced peaks between 11.9 and 16.3 eV as shown in Fig. 4. Using *ab initio* calculations of the bulk dielectric function and the loss function of CdTe, the origin of these features were investigated and shown to be due to the presence of two interband transitions from the occupied Cd 4d

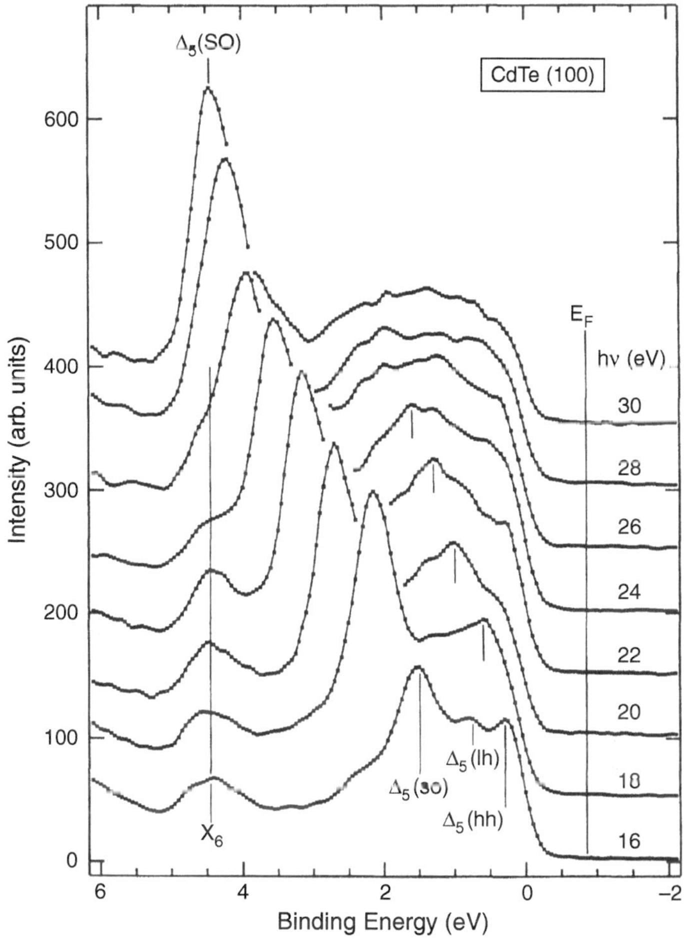

Figure 2 Valence band spectra of normal photoemission of CdTe(001) after Ref [14]. Dispersion over the entire Brillouin zone is apparent for excitation energies ranging from $h\nu = 16\text{-}30$ eV. Reprinted figure with permission from Niles and Höchst [14]. Copyright (1991) by the American Physical Society.

states into the high energy regime of unoccupied states. Because these energies are very close to that of the *free electron* plasmon in CdTe, a strong interference between the interband transitions and plasmon takes place.

The complex loss function;

$$W(\mathbf{q}, \omega) = -\mathrm{Im}\, \varepsilon^{-1}(\mathbf{q}, \omega) \tag{3}$$

was calculated from the LDA bands. The results of Dröge et al. [25] for CdTe(111) without crystal local fields as well as with crystal local fields

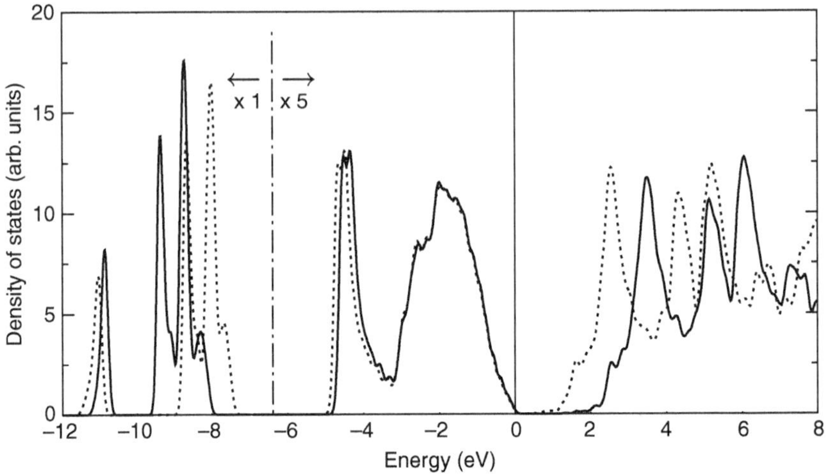

Figure 3 The calculated density of states of CdTe by the LDA method (dotted line) and the *GW-RPA* method (solid line) with spin-orbit interaction.

and using a time dependent local density approximation (TDLDA) are compared with experimental EELS results in Fig. 4. Agreement of the full matrix LDA approximation without the inclusion of crystal local fields is apparently better. This may be due to the known fact that crystal local fields and many body local fields, that is, correlation effects, nearly cancel one another. According to Hanke and Sham [26] both factors contribute to the strength of the response function almost equally and the inclusion of only one can lead to a deterioration of the results. Apparently the TDLDA approximation underestimates the correlation effect strength in the screening of CdTe.

Obviously the experimental and theoretical peak positions are in good agreement, however, to achieve this agreement the experimental Cd 4d energy level (see Table 1) was used in the LDA and *GW* calculations. Agreement was significantly worse when the LDA and *GW* calculated Cd 4d energy levels were employed. The fine structure in peak A of the EELS spectrum is probably due to the spin-orbit splitting of the Cd 4d states [27], but is not reproduced in the theoretical spectrum because of the finite *k* mesh and the resulting insufficient resolution.

1.3. The CdTe(001) surface

Knowledge of the electronic and atomic structure of the CdTe surface is often a prerequisite for epitaxial growth and modern technological applications of CdTe-based heterostructures. This can be gained with the *ab initio* methods described above and, where applicable, compared with

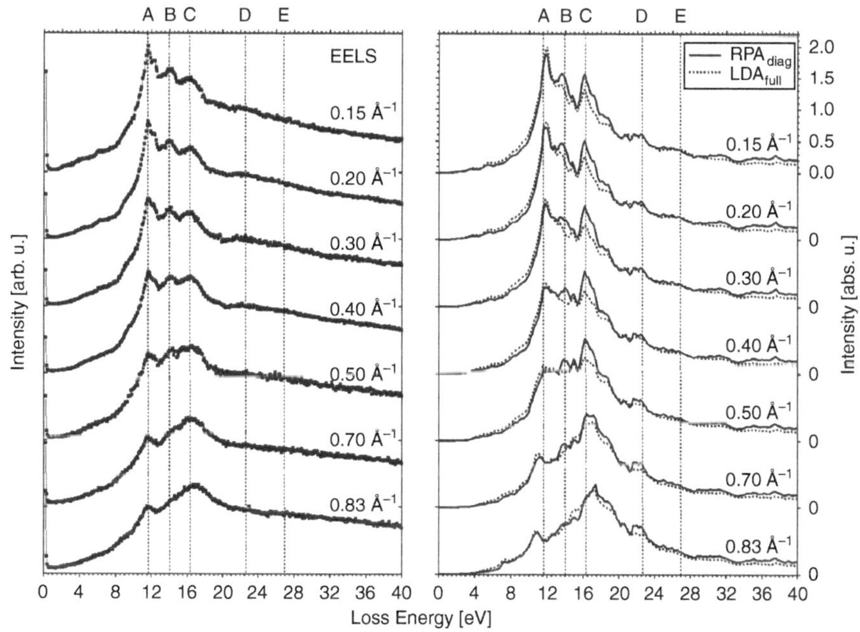

Figure 4 Comparison of experimental ELLS spectra along the (111) direction in the left panel and the *ab initio* calculated loss function of CdTe in the right panel after Ref. [25]. In the right panel, *GW-RPA* calculations without crystal local fields are shown as solid lines and LDA calculations with crystal local fields and many-body corrections are showed as dotted lines. Reprinted figure with permission from Dröge et al. [25]. Copyright (1999) by the American Physical Society.

experimental results such as angle resolved photoemission measurements; The CdTe(001) surface was recently investigated theoretically by means of LDA and many body *GW* methods [28].

A number of possible surface reconstructions were considered by Gundel et al. [28], which are schematically represented as the ball and stick models shown in Fig. 5. For the calculations of the surface energy, a slab geometry was employed which consisted of six or seven monoatomic layers of CdTe, depending on the surface termination, and five or six monoatomic layers of vacuum between adjacent monoatomic layers of CdTe. Their results for the energies of the surface reconstructions are plotted against the chemical potential of Te ($\Delta\mu$) in Fig. 6. They predict that the Cd-terminated half-covered surface is the most stable over a greater portion of the range of $\Delta\mu$. Of the two potentially stable reconstructions, $c(2 \times 2)$ is slightly more stable than (2×1). This results are consistent with the patterns of reflected high energy electron refraction (RHEED) during epitaxial growth, which is either dominated by a $c(2 \times 2)$ structure or a mixture of $c(2 \times 2)$ and (2×1) structures.

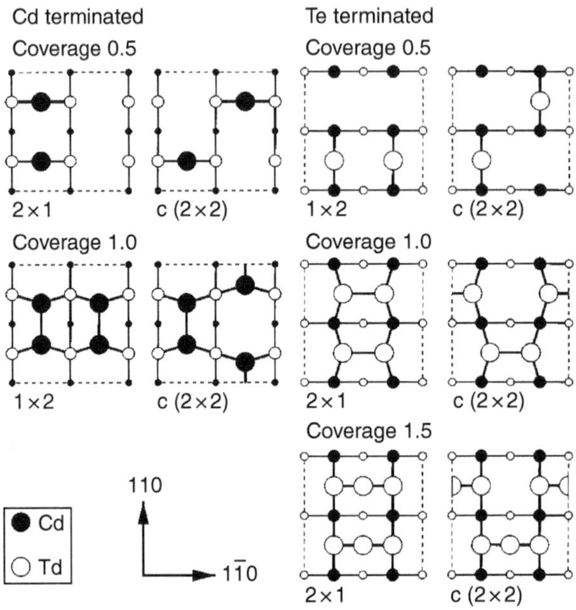

Figure 5 Schematic ball and stick models of CdTe surface reconstructions considered in Ref. [28]. Reprinted figure with permission from Gundel et al. [28]. Copyright (1999) by the American Physical Society.

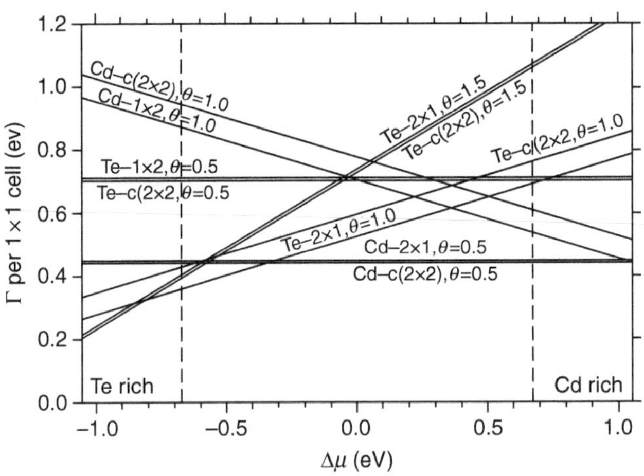

Figure 6 The calculated energies of the CdTe surface reconstructions shown in Fig. 5 and considered in Ref. [28]. Reprinted figure with permission from Gundel et al. [28]. Copyright (1999) by the American Physical Society.

1.4. Donors and acceptors

A number of donors and acceptors have been identified in both bulk CdTe and epitaxially grown CdTe primarily by means of photoluminescence (PL) studies and the dependence of electrical properties on dopant concentration. In most cases, but not all, ionization energies have been determined.

1.4.1. Shallow donors

The identification of shallow donors has not been an easy task due to the low ionization energies on the order of 14 meV for these hydrogenic donors. Nevertheless, the identity of most donors in CdTe is now well established [29, 30], see Table 4. Assignment of photoluminescence peaks to specific impurities is conclusive if their intensities are correlated with doping concentration, which is the case for iodine [30], chlorine [31], and indium [32]. However, the required higher concentrations leads to broader features in the photoluminescence spectra and is a hindrance as far as the detection and detailed analyses of donor bound exciton peaks is concerned.

Francou et al. [29] investigated the PL and high-resolution selective excitation of luminescence of unintentionally doped CdTe. They found six native donors, five of which they assigned or tentatively assigned to chemical species, see Table 4. The effective mass theory, modified to take the polaron effect into account, was employed to analyze the data. The donor energy levels $-R_0/n$ [2] are shifted by a correction of ΔE such that;

Table 4 $D_{n=1,2,3}$ are the energies of donor bound exciton peaks, E_{bind} is the binding energy of the exciton, $E(1s) - E(2s)$ is the 1s-2s transition energy, R_0 is the modified Rydberg, and $E_\alpha(1s)$ and $E_\beta(1s)$ are calculated ionization energies as described below or extrapolated from the plot of $(E(ns) - E(1s))$ versus n^2 shown in Fig. 7 according to Ref. [29]

Chemical species	D_1	D_2	D_3	E_{bind}	$E(1s)-E(2s)$	R_0	$E_\alpha(1s)$	$E_\beta(1s)$	
	$\pm 2 \times 10^{-5}$ eV			meV	meV	meV	meV	meV	
F	1.59339	1.58286	1.58097	3.36	10.28	12.85	13.71	13.67	I
Ga*	1.59309	1.58272	1.58076	3.41	10.37	12.96	13.93	13.88	A
Al*	1.59305	1.58252	1.58056	3.46	10.53	13.16	14.04	14.05	B
In	1.59302	1.58246	1.58044	3.48	10.56	13.20	14.08	14.15	C
Cl	1.59296	1.58230	1.58006	3.54	10.86	13.58	14.48	14.48	E
	1.59284	1.58175	1.57979	3.66	11.09	13.86	14.79	14.60	F
I[a]	1.593						14		

The letters I, A, B, C, E, and F refer to lines in Fig. 7. Asterisks denote tentative identification. I[a] indicates results for iodine after Ref. [30].

$$E(n) = -\frac{R_0}{n^2} + \Delta E(n), \tag{4}$$

where

$$-\frac{\Delta E(n)}{\alpha} = \frac{R_0}{6n^2}. \tag{5}$$

R_0 is the modified Rydberg and α is the polaron coupling constant. Using $\alpha = 0.40 \pm 0.03$ according to Litton et al. [18] results in

$$E(1\text{s}) = 1.067 R_0, \tag{6}$$

$$E(2\text{s}) = 0.267 R_0, \tag{7}$$

$$E(3\text{s}) = 0.119 R_0, \tag{8}$$

etc.

To determine the ionization energy $E(1\text{s})$, a more precise value of R_0 is required than is normally available. Hence the authors employed a direct empirical approach in which they compared the experimental shift $E(1\text{s}) - E(2\text{s})$ with the difference deduced from Eqs. (6)–(8);

$$E(1\text{s}) - E(2\text{s}) = 0.800\, R_0. \tag{9}$$

The resulting R_0 values and ionization energies $E_\alpha(1\text{s})$ for all donors are listed in Table 4. The authors also determined the ionization energy in a more direct manner by assuming a pseudoacceptor model for the (D^0X) complex. The experimental shift $E(n\text{s}) - E(1\text{s})$ was plotted versus n^2 and by extrapolating to $n \to \infty$ as shown in Fig. 7, $E_\beta(1\text{s})$ was obtained. Obviously agreement between $E_\alpha(1\text{s})$ and $E_\beta(1\text{s})$ is excellent.

Ionization energies for a number of donors from Hall effect measurements agree reasonably well within their experimental uncertainties, for example, 14.8 meV for iodine according to Brun-Le-Cunff et al. [33] compared to 14 meV from PL results [30] and 14.5 meV for indium according to Hwang et al. [34] compared to 14.1 meV from PL results [29]. Of the six identified donors, three are due to impurities which are substitutionally incorporated on a Cd site (Ga, Al, and In) and three on a Te site (F, Cl, and I).

1.4.2. Acceptors

A number of acceptors in CdTe have been identified whose ionization energies, E_A, are in the range of 56 meV $< E_A <$ 273 meV. They have been investigated at low temperatures by high-resolution PL, PL excitation spectroscopy, infrared absorption, and magneto-optical experiments, the results of which have been supplemented and reviewed by Molva et al. [37]. These results are reproduced in Table 5, and the 1s, 2s, and 3s energy levels of these acceptors are displayed in Fig. 8.

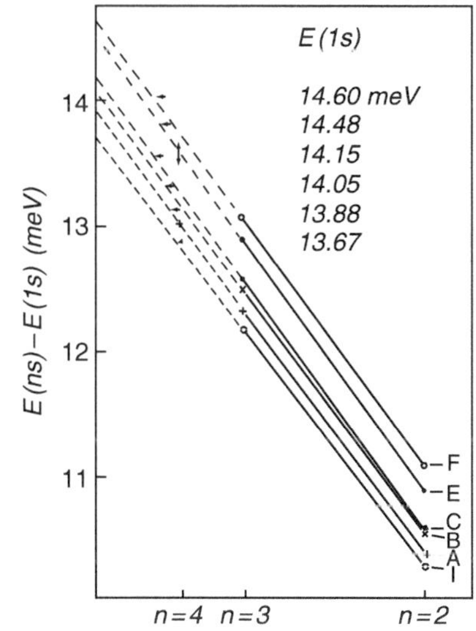

Figure 7 The experimental shift $E(ns) - E(1s)$ versus n^2 of donors in CdTe. $E_\beta(1s)$ was obtained by extrapolating, that is., $n \rightarrow \infty$. The letters I, A, B, C, E, and F represent the chemical species F, Ga, Al, In, Cl and an unknown donor, respectively. Reprinted figure with permission from Francou et al. [29]. Copyright (1990) by the American Physical Society.

Table 5 Acceptor states in CdTe above the valence band

Chemical species	A_1	$1s_{3/2}$	$2p_{3/2}$	$2s_{3/2}$	$2p_{5/2}$ (Γ_8)	$2p_{5/2}$ (Γ_7)	$3s_{3/2}$	$4s_{3/2}$	$5s_{3/2}$	$6s_{3/2}$	
	eV	meV	meV	meV	meV	meV	meV	meV	meV	meV	References
EM		56.8	23.8	15.8	15.3	11.7	8.6				[36]
Li	1.58923	58.0	23.9	15.1	13.6	11.2	8.7	5.7	3.4		[37, 38]
Na	1.58916	58.7	23.8	15.4	13.7	11.2	8.8	5.8	3.4		[37, 38]
Cu	1.58956	146.0		21.6	15.1	11.4	10.0	6.1	4.3	3.1	[39]
Ag	1.58848	107.5		19.6	15.0	11.3	9.6	5.9	4.0	3.0	[39]
Au	1.57606	263		28	(16)		(12)	(9)			[40]
N	(1.5892)	56.0									[41]
P	1.58897	68.2	23.3	17.4	15.1	11.6	9.4	5.7	3.7	3.1	[37, 41]
As	1.58970	92.0		18.8			9.7	5.9	3.8		[41]

The excited energies were obtained from two hole transition peaks, donor acceptor pair excitation spectroscopy, free to bound exciton transitions, and infrared absorption spectroscopy. The overall precision is ±0.5 meV and the values in parentheses are tentative assignments. EM corresponds to the effective mass acceptor calculated within the theory of Baldereschi and Lipari [35] with the parameter values of $R_0 = 30$ meV, $\mu = 0.69$ and $\delta = 0.12$. A_1 is the principal bound exciton peak. The Li, Na, Ag, Cu, and Au impurities are on Cd sites and N, P, and As are on Te sites.

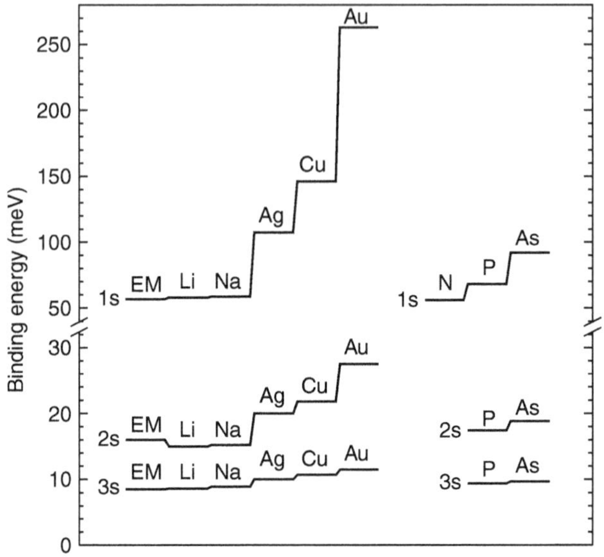

Figure 8 Binding energies of 1s, 2s, and 3s levels of acceptors in CdTe after Ref. [37]. Theoretical effective mass, energy states are indicated with EM. Acceptors which are formed by substitution of impurities on Cd sites are shown on the left and those on Te sites on the right.

A portion of the PL spectrum of P-doped CdTe according to Molva et al. [37] is reproduced in Fig. 9. The temperature dependence of the donor-acceptor pair (DAP) band and the free electron neutral acceptor transition $(e, A_{\mathrm{P}}^{\mathrm{o}})$ for low excitation energies is shown. As can be seen, the intensity of the $e, A_{\mathrm{P}}^{\mathrm{o}}$ peak increases with temperature. Its relative position is related to the ionization energy of the acceptor, $E_{\mathrm{A}} = 68.2 \pm 0.5$ meV, see Table 5.

A value for E_{A} of 56.0 ± 0.5 meV was determined for N by Molva et al. [41], see Table 5. More recently in an investigation combining *C-V* and PL measurements [42] the ionization energy of N was found to be 57 ± 1 meV for a N-doped CdTe sample with a hole concentration of 2.75×10^{17} cm^{-3}, in good agreement with the value according to Molva et al. [41].

2. ELECTRICAL PROPERTIES

Technological applications of semiconductors depend on their electrical properties which will be reviewed here. The ability to extrinsically dope CdTe is crucial to most applications as well as sufficiently large carrier mobilities and diffusion lengths.

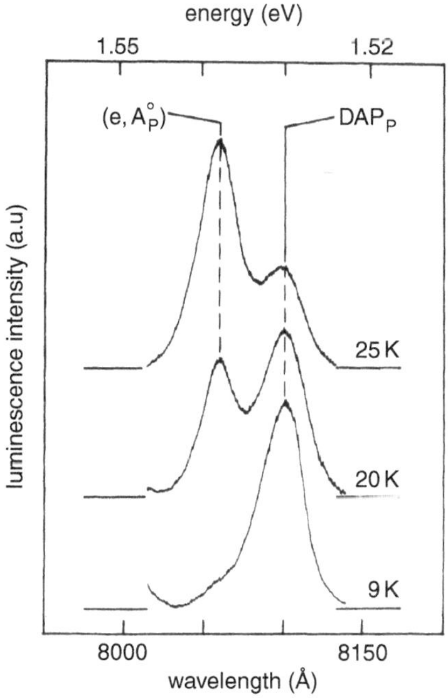

Figure 9 The donor-accepter pair (DAP) band and the free electron neutral acceptor transition (e, A°_{P}) related to P in the photoluminescence spectrum of CdTe at 1.8 K with a low excitation power after Ref. [37]. Upon heating, the intensity of the free to bound transition increases and its relative peak position corresponds to the ionization energy of the acceptor P, that is, 68.2 ± 0.5 meV. Reprinted figure with permission from Molva et al. [37]. Copyright (1984) by the American Physical Society.

2.1. Carrier concentration limits

We begin with extrinsic n-type and p-type doping of CdTe. Potential donors and acceptors are discussed in Section 1.4, however, only I, Cl, Br, In, and Al have proved to be practical impurities for n-type extrinsic doping of CdTe and only N and As for p-type extrinsic doping. The maximum experimental electron and hole concentrations in CdTe as well as other II-VI materials are shown in Fig. 10. As can be seen, CdTe is a better n-type conductor than p-type, albeit not as extreme as is the case for ZnO or ZnS. The maximum electron concentration of 5.3×10^{18} cm^{-3} at room temperature was achieved with I in the form of ZnI_2 by Fischer et al. [43]. The same authors recorded somewhat lower values of about 1.8 and 1.2×10^{18} cm^{-3} for Br and Cl. In all three cases compensation set in at higher doping levels.

The difference in maximum doping levels for In, Cl, Br, and I is particularly pronounced for ternary compounds based on CdTe, such as

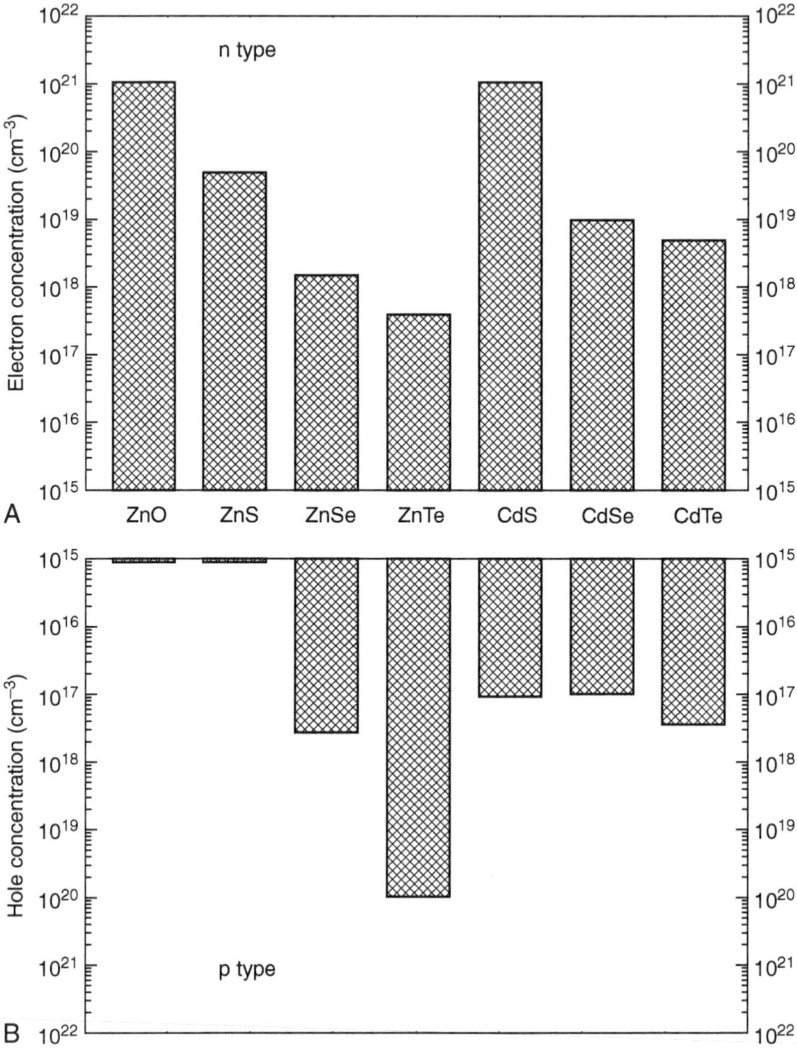

Figure 10 Experimental maximum carrier concentrations in various II-VI semiconductors. Note that the data employed in this plot may have been determined at different temperatures, which can be found in a summary of the data in Ref. [46] and references therein.

CdMgTe. With increasing Mg concentration, the conduction band is shifted upward until it exceeds the energy of the donor related defect state and the Fermi energy. At this point substantial compensation sets in, leading to a rapid decrease in carrier concentrations for larger Mg concentrations [44]. Iodine is by far the best donor for $Cd_{1-x}Mg_x$ Te, being efficient up to $x = 0.35$. This corroborates the results which have been established for pure CdTe. Furthermore, only with iodine donors, has the fabrication of modulation doped Cd(Mn)Te-CdMgTe heterostructures been possible, which have potential applications in spintronics.

Maximum values for hole concentrations are more than an order of magnitude lower, that is, approximately 2.8×10^{17} cm^{-3} for diffused P according to Hall and Woodbury [45] as well as for N by means of cyclotron resonance excited N plasma by Oehling et al. [42].

Zhang et al. [46] have developed a phenomenological model of carrier concentration limits based on the "doping pinning rule" [47]; In essence, doping limits exist because extrinsic doping with donors (acceptors) moves the Fermi energy E_F toward the conduction band minimum, CBM (valence band maximum, VBM), thus lowering the formation energy of spontaneously formed acceptors (donors) which then compensate some of the extrinsic donors (acceptors). This is a consequence of the formation enthalpy of defect α with charge q;

$$\Delta H(\alpha,q) = \text{Constant} + qE_F, \tag{10}$$

where the constant is usually on the order of a few eV and depends only on growth parameters such as atomic chemical potentials but not on the Fermi energy. When donors are introduced, E_F moves toward the CBM and consequently according to Eq. (10), the formation enthalpy for $\alpha =$ acceptors ($q < 0$) is lowered thus leading to spontaneously formed acceptors. This results in a pinning of the Fermi energy near the CBM, at an energy which the authors call E^n_{pin}. Similarly when acceptors are introduced, the Fermi energy is pinned near the VBM at E^p_{pin}. These processes form the upper and lower bounds of the Fermi energy;

$$E_{pin}(p) \le E_F \le E_{pin}(n). \tag{11}$$

Hence the maximum doping limits are determined by the value of E_F at which there are sufficient spontaneously generated defects (acceptors or donors) to compensate the intentional dopants. The net free carrier concentration $N^{(n/p)}(T, E_F^{(n/p)})$ in the single, parabolic band approximation is determined by means of the Fermi-Dirac integral;

$$N^{(n/p)}\left(T, E_F^{(n/p)}\right) = \frac{1}{2\pi^2}(2m^{*,(n/p)})^{3/2} \times \int \frac{E^{1/2}dE}{\exp\left[\left(E - E_F^{(n/p)}\right)/(kT)\right] + 1} \tag{12}$$

Therefore, if the experimental maximum electron and hole concentrations, $N_{max}^{(n/p)}$, are known, the upper and lower bounds for E_F, $E^n_{pin,exp}$, and $E^p_{pin,exp}$, can be determined simply by inverting Eq. (12). The values according to Zhang et al. [46] for CdTe and several other II-VI semiconductors are shown in Fig. 11 as short lines. The short line for $E^n_{pin,exp}$ in

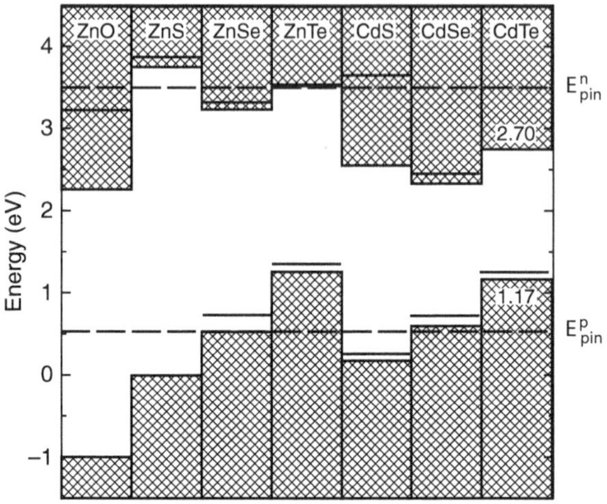

Figure 11 Band diagram for II-VI compounds. Numerical values indicate the energies of the VBM and CBM of CdTe in eV. Short solid lines represent experimental pinning energies, $E^{(n/p)}_{pin,exp}$ calculated via Eq. (12) and the energies of the long-dashed lines, E^n_{pin} and E^p_{pin}, are arithmetic averages of $E^{(n/p)}_{pin,exp}$ according to Ref. [46]. Note that the short line for $E^n_{pin,exp}$ in CdTe has been omitted because the maximum electron concentration for CdTe according to Ref. [46] is a factor of 10 too large.

CdTe has been omitted because the maximum electron concentration for CdTe according to Ref. [46] is a factor of 10 too large.

The pinning energies $E^n_{pin, exp}$ and $E^p_{pin, exp}$ have to be calculated for each material with Eq. (12) and consequently no information is gained for other materials. Caldas et al. [48] noted that if the band edges of different materials are aligned with respect to their band offsets, then the positions of a given deep impurity level is nearly constant, which is termed the "vacuum pinning rule." Walukiewicz [47] has suggested that E^n_{pin} and E^p_{pin} can similarly be related to the vacuum level rather than the band edges. This approach was first employed by Fashinger et al. [49] for several II-VI compounds and later by Zhang et al. [46].

To line up $E^{(n/p)}_{pin}$ in the manner described above, the corresponding band offsets between semiconductors are required. Zhang et al. [46] argue that experimental band offsets are not appropriate because they include the effects of a rough impure interface or interfacial strains as well as the intrinsic or "natural" band line up. Therefore, they calculated these "natural" band offsets using LDA as implemented by the general potential, linearized augmented plane wave (LAPW) method [50]. The resulting band alignment is shown in Fig. 11.

As can be seen in Fig. 11, the vacuum related $E^{(n/p)}_{pin,exp}$ values tend to line up. Arithmetic averages of the values for these II-VI materials, denoted

$E_{\text{pin}}^{(n/p)}$ in Ref. [46], are shown as dashed lines in Fig. 11. Zhang et al. [46] have formulated the following doping limit rule: Compounds whose CBM is much higher or lower than E_{pin}^{n} are difficult or easy to dope n-type. Similarly compounds whose VBM is much lower or higher than E_{pin}^{p} are difficult or easy to dope p-type. This is consistent with the fact that CdTe can be heavily n-type doped and only moderately p-type doped. It also explains why ZnO and ZnS can be heavily n-type doped, contrary to conventional wisdom that wide bandgap materials are difficult to dope.

2.2. Mobilities

The transport properties of bulk CdTe and the corresponding scattering mechanisms have been extensively reviewed by Zanio [2] in 1978. An electron mobility of $5.7 \times 10^4\ \text{cm}^2/(\text{V s})$ was measured at 30 K for multiply zone refined high-purity CdTe by Segall et al. [51]. According to the authors the mobility at low temperatures is limited by impurity scattering. Values as high as $1.4 \times 10^5\ \text{cm}^2/(\text{V s})$ were later reported by Triboulet and Marfaing [52] and Woodbury [53]. However these mobilities are higher than predicted by simple charge scattering models [2]. This anomaly is associated with the pairing of charged defects and the subsequent reduction in the density of scattering centers. Hole mobilities up to about $500\ \text{cm}^2/(\text{V s})$ at low temperatures have been reported by Yamada [54]. These values are supplemented by results of Smith [55] at temperatures between 600 and 900 K and have been extensively analyzed by Zanio [2].

Currently interest is centered on solar cells (see chapter on solar cells), X-ray and gamma ray detection, and spintronic applications involving heterostructures based on CdTe. In the case of X-ray and gamma ray detection, the largest possible resistivity is the main goal as discussed in the chapter on radiation detectors. In the latter case, the transport properties of epitaxially grown thin CdTe layers are of significance.

The mobilities of epitaxially grown CdTe are significantly lower than those for multiply zone refined bulk CdTe; The largest reported low-temperature electron mobilities are 6600 and $5700\ \text{cm}^2/(\text{V s})$ according to Hwang et al. [34] and Bassani [56], respectively; Corresponding values for hole mobilities are 80 and $90\ \text{cm}^2/(\text{V s})$ according to Hwang et al. [34] and Moesslein [57], respectively. Even though reasons for this discrepancy have not been reported, the lower values must result from scattering due to point defects and extended defects, as well as scattering at interfaces and the surface of these thin layers whose thicknesses are on the order of 1 μm.

2.3. Carrier diffusion lengths and lifetimes

By definition, the carrier diffusion length is the average distance traveled due to diffusion during the lifetime of the carrier. Because of the required charge neutrality, both types of charge carriers must be involved in the diffusion. Nevertheless, in the case of extrinsic semiconductors, carrier diffusion is controlled by the minority carriers and the diffusion length L is given by

$$L = (D\tau)^{1/2}, \tag{13}$$

where D and τ are the diffusion constant and the lifetime, respectively. If D and τ are known then L can be calculated. Experimentally L is determined from the decay of excess minority carriers which are generated locally by either a photon flux or an electron beam, or by a more complex photocurrent analysis.

Experimental diffusion lengths for both conductivity types and various doping levels in both single crystal and polycrystalline CdTe have been tabulated by Marfaing [58]. The hole diffusion length [59–68] in these n-type CdTe decreases from 5 to about 0.18 μm when the electron concentration n increases from 2×10^{13} to 8×10^{17} cm^{-3};

$$\log(L_\mathrm{h}) \approx 4.9 - 0.315 \log(n), \tag{14}$$

where L_h is in μm and n is in cm^{-3}. In contrast, the electron diffusion length [60, 61, 63, 65, 68, 69] in p-type CdTe samples, which lies in the range of 0.4-2.0 μm for hole concentrations p between 1×10^{15} and 8×10^{16} cm^{-3}, displays almost no correlation with the hole concentration, This is consistent with the fact that the electron mobility in n-type CdTe is much larger than the hole mobility in p-type CdTe [58].

Minority carrier lifetimes have also been reported for various carrier concentrations; Similar to the behavior of the results for carrier lengths, hole lifetimes were found to decrease from 20 to 0.5 ns with increasing electron concentration in n-type CdTe [59, 62, 65]; Electron lifetimes in p-type CdTe between 10 and 30 ns have been reported [65].

ACKNOWLEDGMENTS

The author gratefully acknowledges helpful discussions with A. Fleszar.

REFERENCES

[1] S.-H. Wei, A. Zunger, Phys. Rev. B 37 (1988) 8958.
[2] K. Zanio, in: R. Willardson, A.C. Beer (Eds.), Semiconductors and Semimetals, vol. 13, Academic Press, New York, 1981, p. 77.
[3] D.T.F. Marple, Phys. Rev. 150 (1966) 728.
[4] M.L. Cohen, T.K. Bergstresser, Phys. Rev. 141 (1966) 789.
[5] P. Hohenberg, W. Kohn Phys, Rev. 136 (1964) B864.
[6] W. Kohn, L.J. Sham, Phys. Rev. 140 (1965) A1133.
[7] L. Hedin, Phys. Rev. 139 (1965) A796.

[8] L. Hedin, S. Lundqvist, in: H. Ehrenreich, F. Seitz, D. Turnbull (Eds.), Solid State Physics, vol 23, Academic, New York, 1969, p. 1.

[9] G. Strinati, H.J. Mattausch, W. Hanke, Phys. Rev. Lett. 45 (1980) 290; Phys. Rev. B 25 (1981) 2867; W. Hanke, M. Meskini, H. Weiler, in: J.T. Devreese, P. van Camp (Eds.), Electronic Structure, Dynamics and Quantum Structural Properties of Condensed Matter, Plenum, New York, 1985, p. 113.

[10] M.S. Hybertsen, S.G. Louie, Phys. Rev. Lett. 55 (1985) 1418; Phys. Rev. B 34 (1986) 5390.

[11] A. Fleszar, W. Hanke, Phys. Rev. B 71 (2005) 045207.

[12] A. Fleszar, private communication

[13] J. Chelikowsky, M.L. Cohen, Phys. Rev. B 14 (1976) 556.

[14] D.W. Niles, H. Höchst, Phys. Rev. B 43 (1991) 1492.

[15] D.T.F. Marple, Phys. Rev. 129 (1963) 2466.

[16] K.K. Kanazawa, F.C. Brown, Phys. Rev. 135 (1964) A1757.

[17] A.L. Mears, R.A. Stradling, Solid State Commun. 7 (1969) 1267.

[18] C.W. Litton, K.J. Button, J. Waldman, D.R. Cohn, B. Lax, Phys. Rev. B 13 (1976) 5392.

[19] R. Romestain, C. Weisbuch, Phys. Rev. Lett. 45 (1980) 2067.

[20] L.S. Dang, G. Neu, R. Romestain, Solid State Commun. 44 (1982) 1187.

[21] M. Helm, W. Knap, W. Seidenbusch, R. Lassnig, E. Gornik, Solid State Commun. 53 (1985) 547.

[22] H. Kobori, T. Ohyama, E. Otsuka, Solid State Commun. 84 (1992) 383.

[23] F.M. Peeters, J.T. Devreese, Physika B 127 (1984) 408.

[24] H. Raether, Excitation of Plasmons and Interband Transitions by Electrons, Springer Tracts in Modern Physics, vol. 88, Springer-Verlag, Berlin, 1980.

[25] H. Dröge, A. Fleszar, W. Hanke, M. Sing, M. Knupfer, J. Fink, F. Goschenhofer, C.R. Becker, R. Kargerbauer, H.P. Steinrück, Phys. Rev. B 59 (1999) 5544.

[26] W. Hanke, L.J. Sham, Phys. Rev. B 21 (1980) 4656.

[27] A. Wall, Y. Gao, A. Raisanen, A. Pranciosi, J.R. Chelikowsky, Phys. Rev. B 43 (1991) 4988.

[28] S. Gundel, A. Fleszar, W. Faschinger, W. Hanke, Phys. Rev. B 59 (1999) 15261.

[29] J.M. Francou, K. Saminadayar, J.L. Pautrat, Phys. Rev. B 41 (1990) 12035.

[30] N.C. Giles, Jaesun Lee, D. Rajavel, C.J. Summers, J. Appl. Phys 73 (1993) 4541.

[31] G. Neu, Y. Mayfaing, R. Lagos, R. Triboulet, L. Svob, J. Lumin. 2 (1980) 293.

[32] R.N. Bicknell, N.C. Giles, J.F. Schetzina, Appl. Phys. Lett. 49 (1986) 1095.

[33] D. Brun-Le-Cunff, T. Baron, B. Daudin, S. Tatarenko, B. Blanchard, Appl. Phys. Lett. 67 (1995) 965.

[34] S. Hwang, R.L. Harper, K.S. Harris, N.C. Giles, R.N. Bicknell, J.W. Cook Jr., J.F. Schetzina, J. Vac. Sci. Technol. A 6 (1988) 2821.

[35] A. Baldereschi, N.O. Lipari, Phys. Rev. B 9 (1974) 1525.

[36] G. Milchberg, Ph.D. thesis, University of Grenoble, France, 1983.

[37] E. Molva, J.L. Pautrat, K. Saminadayar, G. Milchberg, N. Magnea, Phys. Rev. B 30 (1984) 3344.

[38] E. Molva, J.P. Chamonal, J.L. Pautrat, Phys. Status Solidi B 109 (1982) 635.

[39] E. Molva, J.P. Chamonal, G. Milchberg, K. Saminadayar, B. Pajot, G. Neu, Solid State Commun. 44 (1982) 351.

[40] E. Molva, F.M. Francou, J.L. Pautrat, K. Saminadayar, Le Si Dang, J. Appl. Phys. 56 (1984) 2241.

[41] E. Molva, K. Saminadayar, J.L. Pautrat, E. Ligeon, Solid State Commun. 48 (1983) 955.

[42] S. Oehling, H.J. Lugauer, M. Schmitt, H. Heinke, U. Zehnder, A. Waag, C.R. Becker, G. Landwehr, J. Appl. Phys. 79 (1996) 2343.

[43] F. Fischer, A. Waag, G. Bilger, Th. litz, S. Scholl, M. Schmitt, G. Landwehr, J. Cryst. Growth 141 (1994) 93.

[44] A. Waag, F. Fischer, J. Gerschütz, S. Scholl, G. Landwehr, J. Appl. Phys. 75 (1994) 368.

[45] R.B. Hall, H.H. Woodbury, J. Appl. Phys. 39 (1968) 5361.
[46] S.B. Zhang, S.-H. Wei, A. Zunger, J. Appl. Phys. 83.
[47] W. Walukiewicz, J. Vac. Sci. Technol. b 5 1062 (1987) 3192.
[48] M. Caldas, A. Fazzio, A. Zunger, Appl. Phys. Lett. 45 (1984) 671.
[49] W. Fashinger, S. Ferreira, H. Sitter, J. Cryst. Growth 151 (1995) 267.
[50] S.-H. Wei, H. Krakauer, Phys. Rev. Lett. 63 (1993) 2549.
[51] B. Segall, M.R. Lorenz, R.E. Halsted, Phys. Rev. 129 (1963) 2471.
[52] R. Triboulet, Y. Marfaing, J. Electrochem. Soc. 120 (1973) 1260.
[53] H.H. Woodbury, Phys. Rev. B 9 (1974) 5188.
[54] S. Yamada, J. Phys, Soc. Jpn. 15 (1960) 1940.
[55] F.T.J. Smith, Metall. Trans. 1 (1970) 617.
[56] F. Bassani, S. Tatarenko, K. Saminadayar, N. Magnea, R.T. Cox, A. Tardot, J. Appl. Phys. 72 (1992) 2927.
[57] J. Moesslein, A. Lopez-Otero, A.L. Fahrenbruch, D. Kim, R.H. Bube, J. Appl. Phys. 73 (1993) 8359.
[58] Y. Marfaing, in: P. Capper (Ed.), Properties of Narrow Gap Cadmium-Based Compounds, INSPEC, the Institution of Electrical Engineers, London, 1994, p. 542.
[59] D.A. Cusano, M.R. Lorenz, Solid State Commun. 2 (1964) 125.
[60] K. Yamaguchi, N. Makayama, H. Matsumoto, S. Ikegami, Jpn. J. Appl. Phys. 16 (1977) 1203.
[61] S.N. Maximovski, I.P. Revocatova, V.M. Salman, M.A. Selezneva, P.N. Lebedeu, Rev. Phys. Appl. 12 (1977) 161.
[62] M. Chu, A.L. Fahrenbruch, R.H. Bube, J.F. Gibbons, J. Appl. Phys. 49 (1978) 322.
[63] J. Mimila-Arroyo, Y. Marfaing, G. Cohen-Solal, R. Triboulet, Sol. Energy Mater. 1 (1979) 171.
[64] A. Lastras-Martinez, P.M. Raccah, R. Triboulet, Appl. Phys. Lett. 36 (1980) 469.
[65] D.R. Wight, D. Bradley, G. Williams, M. Astles, S.J.C. Irvine, J. Cryst. Growth 59 (1982) 323.
[66] J. Gautron, P. Lemasson, J. Cryst. Growth 59 (1982) 332.
[67] T.L. Chu, S.S. Chu, Y. Pauleau, K. Murthy, E.D. Stokes, P.E. Russel, J. Appl. Phys. 54 (1983) 398.
[68] S.P. Albright, V.P. Singh, J.F. Jordan, Sol. Cells 24 (1988) 43.
[69] K. Mitchell, A.L. Fahrenbruch, R.H. Bube, J. Appl. Phys. 49 (1977) 829.

CHAPTER IID

Optical Properties of CdTe

V. Consonni[1]

As CdTe-based materials are direct bandgap semiconductors, belonging to the II-VI type group, the maximum of the valence band is located at same wave vector as the minimum of the conduction band. When photons of sufficient energies—typically higher than the bandgap energy in the case of photoluminescence (PL) measurements—, or when an electron beam—in the case of cathodoluminescence (CL) measurements—, encounter the CdTe surface, their absorption induces electronic excitations, from the ground state in the valence band to excited states in the conduction band. Subsequently, such electronic excitations are relaxed by the return of electrons to their ground state. If radiative relaxations proceed, photons with specific wavelengths are emitted from the CdTe surface and are characteristic of the energy levels in the bandgap. As such energy levels are directly related to the presence of native point defects, dopants, or even extended defects in CdTe, carrying out the study of such optical properties is of primary importance and opens the way to a real defect engineering. Its better understanding is essential for the control of the electronic and transport properties in optoelectronic devices.

1. GENERAL FEATURES

1.1. Different radiative recombination processes

A luminescence spectrum typically gathers a multitude of radiative recombination processes, as described in Fig. 1. The near band edge emission region, corresponding to a luminescence spectrum in CdTe, can be divided into two main parts: one excitonic emission band, which

CEA-LETI, Minatec, 17 rue des Martyrs, 38054 Grenoble Cedex 9, France
[1] Current address: Paul-Drude-Institut für Festkörperelektronik, Hausvogteiplatz 5-7, D-10117 Berlin, Germany.

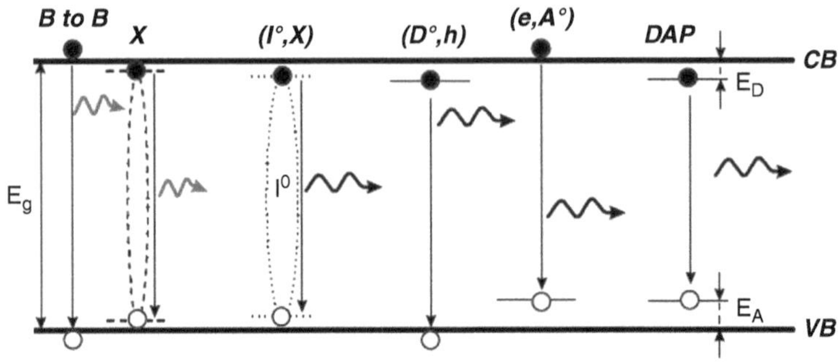

Figure 1 The different processes of radiative recombinations.

can be attributed to X and (I°, X) type radiative transitions, and one shallow emission band at lower energies, which corresponds to DAP, (D°, h), and (e, A°) type radiative recombinations.

1.1.1. Excitons: Definition and properties

The excitonic emission band brings a remarkable quasiparticle into play, namely the so-called exciton: an exciton consists in a free electron in the conduction band (CB) bound to a free hole in the valence band (VB) by coulombian interaction. Electrically neutral excitons form a mobile pair of opposite charge carriers, which is able to freely move in CdTe and thus act as a real probe of the crystalline quality [1]. The evident similarities of the exciton configuration with the structure of the standard hydrogen atom permit the use of the hydrogenoid model, in a first approximation, so as to determine its binding energy and orbital radius by application of the effective-mass approach [2]:

$$E_{\mathrm{B,X}} = \frac{m_{\mathrm{X}}^{*} \cdot e^{4}}{8h^{2}\varepsilon_{0}^{2}\varepsilon_{\mathrm{r}}^{2}} \quad \text{and} \quad a_{\mathrm{X}} = \frac{\varepsilon_{0}\varepsilon_{\mathrm{r}}h^{2}}{\pi \cdot m_{\mathrm{X}}^{*}e^{2}} \tag{1}$$

in which ε_{r} is the dielectric constant and m_{X}^{*} is the reduced mass defined by:

$$m_{\mathrm{X}}^{*} = \frac{m_{\mathrm{e}}^{*} \cdot m_{\mathrm{h}}^{*}}{m_{\mathrm{e}}^{*} + m_{\mathrm{h}}^{*}}. \tag{2}$$

The binding energy of excitons is 10.6 meV in CdTe while its orbital radius equals 72.2 Å, by taking $\varepsilon_{\mathrm{r}} = 10.16$ [3]. Such a value of the binding energy is in relatively good agreement with the experimental value of 10 meV as deduced from optical absorption edge measurements [4]. Furthermore, in order to develop a physically more realistic approach, Baldereschi et al. considered the VB degeneracy, leading to a theoretical value closer to the experimental one [5].

Typically, excitons are liable to be into two different states. On the one hand, excitons can be free and lead to X type radiative transitions, as represented in Fig. 1. On the other hand, excitons can also be bound to neutral or ionized impurity centers acting as donors or acceptors. Indeed, a free exciton can feel in a short distance from the impurity center either its coulombian field when ionized or its exchange interaction when neutral. The main consequence is to associate a localization energy E_{LOC} of excitons bound to an impurity center, which is typically 3 meV for donors and 7 meV for acceptors in CdTe [6, 7]. Interestingly, Haynes showed that the localization and ionization energies E_D (E_A) of a same donor (acceptor)—defined as the energy necessary to excite an electron from the donor level (VB) to the CB (acceptor level)—are related by a linear dependence in silicon [8]. The linear coefficient for a simple donor (acceptor) is approximately 0.2 (respectively, 0.1) in CdTe, as given by Halsted et al. [9]. However, such a linear relation is valid provided that the given impurity center follows a quasihydrogenoid behavior. Radiative recombinations of bound excitons to donors and acceptors are, respectively, referred to as (D°, X) (or D^+, X) and (A°, X) type transitions, as described in Fig. 1. In CdTe, the electron effective-mass (i.e., $0.0963m_0$) is very low as compared to the hole effective-mass (i.e., $0.35m_0$), indicating that donors can be either in a neutral state or in an ionized state after radiative recombinations. On the contrary, acceptors are inevitably neutral, as shown by Hopfield [2, 10, 11]. Furthermore, excitation of PL induces the existence of two-electron satellite transitions (TES) and two-hole satellite transitions (THS): some of the energy corresponding to TES or THS permits to achieve internal electronic transitions, which provides excited states and thus a signing of the nature of the impurity center at work [6, 7]. Such an excitonic emission band typically lies in the energy range of 1.6-1.56 eV in CdTe.

1.1.2. The shallow emission band

Below 1.56 eV in CdTe, radiative recombinations of free electrons in the CB to acceptors are referred to as (e, A°) type transitions in Fig. 1. The energy position $E_{(e,A^\circ)}$ of such transitions, which is given by $E_{(e,A^\circ)} = E_G - E_A + \frac{1}{2}k_B T$, is of great interest since directly dependent upon the ionization energy E_A of the acceptors involved: such transitions thus represent a powerful means to determine the ionization energy E_A of acceptors [12]. Radiative recombinations of free holes in the VB to donors are also expected and referred to as (D°, h) type transitions in Fig. 1. By analogy with (e, A°) type transitions, such transitions yield the ionization energy E_D of the involved donors.

Furthermore, radiative recombinations of electrons from donors with holes from acceptors contribute to the formation of donor-acceptor pairs (DAP) and are referred to as DAP type transitions in Fig. 1. By neglecting the van der Waals type interaction, the energy position E_{DAP} directly depends on the DAP separation R and the ionization energies E_D of the donor and E_A of the acceptor, forming the pair, according to the following relation [13, 14]:

$$E_{DAP} = E_G - (E_D + E_A) + \frac{e^2}{4\pi\varepsilon_0\varepsilon_r R}. \tag{3}$$

The determination of the DAP separation R supplies a direct assessment of the acceptor density N_A through the simple relation: $N_A \approx (3/4\pi)R^{-3}$. Furthermore, selective excitation of DAP can also probe excited states of the impurity center at work, as in the case for TES and THS transitions. Consequently, such radiative transitions yield valuable information concerning the compensation mechanisms in CdTe through a thorough description of the defect complexes involved. Such a shallow emission band typically lies in the energy range of 1.56-1.35 eV in CdTe.

1.2. Theoretical determination of ionization energy and orbital radius

The ionization energy and orbital radius of an impurity center are two important specific features representing a kind of signing. These are essential to determine in order to study the nature of point defects and dopants involved.

1.2.1. Simple hydrogenoid model

The configuration of a simple donor is relatively close to that of a standard hydrogen atom: the single difference arises from its localization in the solid instead of the vacuum. Therefore, the periodic potential of the solid directly influences the electron propagation while the ionized donor also interacts by polarization with the solid. By application of the effective-mass approach, the ionization energy of a simple donor E_D or of a simple acceptor E_A and their orbital radius are given by the following relations [2]:

$$E_{D(A)} = \frac{m^*_{e(h)} \cdot e^4}{8h^2\varepsilon_0^2\varepsilon_r^2} \quad \text{and} \quad a_{D(A)} = \frac{\varepsilon_0\varepsilon_r h^2}{\pi \cdot m^*_{e(h)} e^2} \tag{4}$$

in which $m^*_{e(h)}$ are the effective-masses of electrons and holes, respectively.

The ionization energies of a simple donor E_D and a simple acceptor E_A in CdTe are, respectively, 12.7 and 46.1 meV while their corresponding orbital radius are 55.8 and 15.4 Å, by taking $\varepsilon_r = 10.16$ [3]. The simple hydrogenoid model supplies some correct assessments of these physical features in CdTe; nevertheless, it is based on several simple approximations that are not physically realistic.

1.2.2. Refined hydrogenoid model

The hydrogenoid model is approximate for an acceptor, as regards especially the specific features of the VB. Baldereschi et al. developed a thorough approach by taking into account the degeneracy of the VB. The ionization energy E_A of a simple acceptor is found to be 56.8 meV with the following set of VB parameters: $R_0 = 30$ meV, $\mu = 0.69$ and $\delta = 0.12$ [7, 15, 16]. In order to refine further the hydrogenoid model, it is also crucial to take into account both the chemical nature of the impurity center and the different interactions with the crystalline lattice [17]. On the one hand, the coulombian potential felt by charge carriers is strictly dependent upon the chemical nature of the impurity center in a very short distance from its core. Soltani et al. introduced a central cell correction term to consider the core of the impurity center, which brings two parameters λ and K into play: the former accounts for the larger attractive potential due to the replacement of the host atom whereas the latter describes changes in screening effects and equals 1.353 Å^{-1} [11]. On the contrary, the approach followed by the quantum defect model consists in directly introducing a parameter υ into the trial wave function to consider these physical effects [18, 19]. On the other hand, the respective interactions of the impurity center and the bound charge carriers with the longitudinal optical (LO) phonons induce slight lattice distortions. The latter interactions give rise to the consideration of the so-called Huang-Rhys coupling constant S, which directly assesses the coupling strength [20]:

$$S = \sum_q \frac{|V_q|^2}{(\hbar \cdot \omega_{\mathrm{LO}})^2} \cdot |\rho_q|^2 \tag{5}$$

in which q is the wave vector, V_q is the coupling strength, and ρ_q is the charge distribution around the core of the impurity involved [21].

Hopfield showed that such a coupling constant can experimentally be determined by fitting the envelop of the band, consisting of a zero phonon line (ZPL) and n multiple LO phonon replicas, with a Poisson distribution combined with a Gaussian function for each line [22]:

$$I_{\mathrm{PL}}(E) = \Delta \times E^2 \exp(-S) \times \sum_{n=0}^{n} \left[\frac{S^n}{n!} \exp\left(-\frac{(E - E_0 - nE_{\mathrm{LO}})^2}{2\sigma^2} \right) \right], \tag{6}$$

in which Δ is a numerical constant, σ is the average width of each line, E_0 is the energy position of the ZPL, and E_{LO} is the phonon energy.

These approaches are commonly applied to the case of CdTe and give accurate values for all these physical features and in particular for the ionization energies of donors and acceptors [11].

1.3. Evolution of photoluminescence with experimental parameters

The nature of radiative recombination processes at work for each line in PL spectra can basically be identified from both its energy position and its shape. For instance, a sharp line is commonly referred to as excitonic radiative recombinations. Nevertheless, such a single identification process is usually not sufficient enough to unravel the problem and can induce some great difficulties since most of the lines peak at very close energy positions. The needs for high spectral resolution setup as well as for high crystalline quality are essential. Therefore, studying the PL variation with several experimental parameters is particularly relevant to bring out sound arguments for the identification of each line in PL spectra.

1.3.1. Variation of photoluminescence with excitation power density

It was early stated that the PL intensity I_{PL} and the excitation power density L are both related by a power-type law [23]:

$$I_{PL} \approx L^k \tag{7}$$

The plot of $\log(I_{PL}) = f(\log(L))$ is a straight line, whose the slope is the power coefficient k. The k value depends on the nature of the radiative recombinations process at work and is comprised between 0 and 2. By assuming that the concentrations of neutral donors and acceptors are dependent upon the excitation power density L, Schmidt et al. proposed several analytical relations in order to assess the power coefficient k [24]. According to this approach and several experimental studies, k values around 1.2 are associated with X type radiative transitions whereas (A°, X) and (D°, X) radiative transitions are involved for higher k values, typically between 1.2 and 2. On the contrary, k values smaller than 1 are attributed to the presence of (e, A°) or DAP radiative recombinations [1, 23–25].

1.3.2. Variation of photoluminescence with temperature

When the temperature raises, the stored thermal energy implies, in a first time, the delocalization of bound excitons from impurity centers: such delocalizations induce the decrease in the intensities of (I°, X) type radiative transitions until they vanish while the intensity of X type radiative transitions increases. If the temperature keeps on raising, the stored thermal energy ends up dissociating, in a second time, free excitons themselves, leading to the disappearance of X type radiative transitions. Bimberg et al. proposed the following relation to describe the dependence of the intensity of (I°, X) type radiative transitions on the temperature [26]:

$$I_{PL}(T) = \frac{I_{PL}(0)}{\left[1 + C_1 \exp\left(\frac{-E_1}{k_B T}\right) + C_2 \exp\left(\frac{-(E_1+E_2)}{k_B T}\right)\right]} \tag{8}$$

in which E_1 represents the activation energy of the dissociation resulting in one free hole (although no strong argument prevents from choosing the localization energy of bound excitons in CdTe instead, as suggested by Taguchi et al.) and E_2 is the sum of the localization energy of bound excitons with the binding energy of free excitons [23, 27]. C_1 and C_2 constants directly depend on the energy density of states involving both the delocalization of bound excitons and then the dissociation of free excitons [27]. Interestingly, the extension of such an approach to DAP radiative recombinations is conceivable by considering a distinct dissociation path from that of bound excitons: the ionization energies of donors E_D and acceptors E_A are thus introduced in Eq. (8), instead. Furthermore, an increase in the temperature can also ionize donors involved in the pairs, resulting in the increase in the intensity of (e, A°) radiative recombinations with respect to that of DAP radiative recombinations: such a trend constitutes a powerful means so as to distinguish these two types of radiative transitions in PL spectra.

2. UNDOPED AND DOPED CADMIUM TELLURIDE

2.1. Undoped CdTe

Very few research works really deal with the optical properties of undoped CdTe. First, its overall properties are usually more interesting for electronic or optoelectronic devices when CdTe is doped: the control of acceptor and donor densities is easier by means of doping and ensures a better reliability of such devices. Second, the high purity 5N CdTe sources—that are commonly employed for a wide variety of growth methods—are seldom pure enough to grow perfectly intrinsic CdTe: despite the relatively effective purification at work during the epitaxial, film, or bulk growth, the incorporation of residual impurities is inevitably achieved. Very low doping levels are thus observed in the majority of as-grown CdTe and referenced as unintentional doping.

2.1.1. The intrinsic native point defects

The equilibrium phase diagram of CdTe indicates a more pronounced deviation around the stoichiometry in the Te side, giving rise to Te overpressure during the growth [28]: Te-rich conditions for the growth of either monocrystalline or polycrystalline CdTe variants are thus favored, unless a Cd overpressure is balanced. The main intrinsic point defects are expected to be the majority Cd vacancies (V_{Cd}) and the minority Te antisites (Te_{Cd}) [29, 30]. Most of the corresponding energy levels were determined by electron paramagnetic resonance (EPR) measurements. V_{Cd} is a doubly ionized deep acceptor with an ionization energy of 470 meV for the 2−/− level, which accounts for the p-type conductivity of undoped CdTe [31]. It is worth noticing that some research works reported shallower levels for V_{Cd} around 50 meV above the VB corresponding to its first ionization degree (i.e., for the $-/0$ level) [32]. On the contrary, Te_{Cd}, which is in smaller concentration, is a doubly ionized deep donor with an ionization energy of 750 meV for the $+/2+$ level [33]. Furthermore, the very minority Te vacancy (V_{Te}) is a single deep donor lying 1.4 eV below the CB for the $0/+$ level [34]. As regards the interstitial defect, most of the research works agree with their actual depth even if the situation has not completely been elucidated yet [35]. Therefore, PL measurements, in which the energy range is often limited between 1.606 and 1.35 eV according to the setup used, cannot generally detect the presence of intrinsic native point defects. Actually, PL measurements can only detect combinations of native point defects between them or with dopants, which give rise to shallower defect complexes, as widely represented by the so-called A centers in the following.

2.1.2. Typical spectra of undoped CdTe

Dal'Bo et al. reported typical spectra of epitaxial undoped CdTe layers, which are completely dominated by excitonic radiative transitions with no other type of transitions, as shown in Fig. 2 [36]. More recently, Song et al. also obtained typical spectra with only excitonic radiative recombinations but in very high purity bulk CdTe [37]. Thus, most of the research works concerning the optical properties in epitaxial or bulk undoped CdTe reported the existence of strong luminescence lines around 1.589 eV. Such lines were attributed to (A°, X) radiative recombinations involving either Cu or Ag residual impurities, V_{Cd}, (V_{Cd}, D), or (V_{Cd}, 2D) defect complexes,

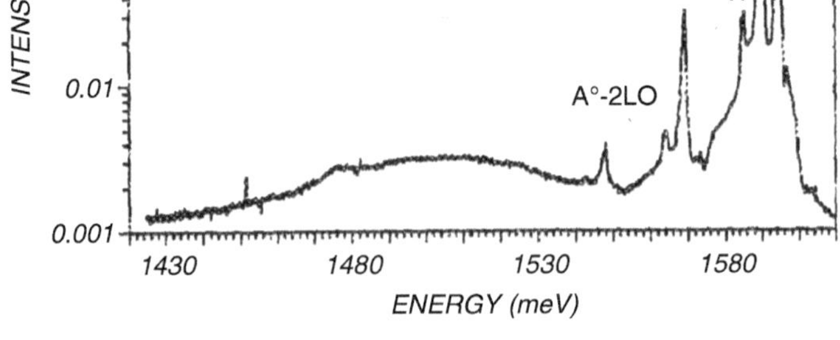

Figure 2 PL spectrum of high purely epitaxial undoped CdTe from Ref. [36].

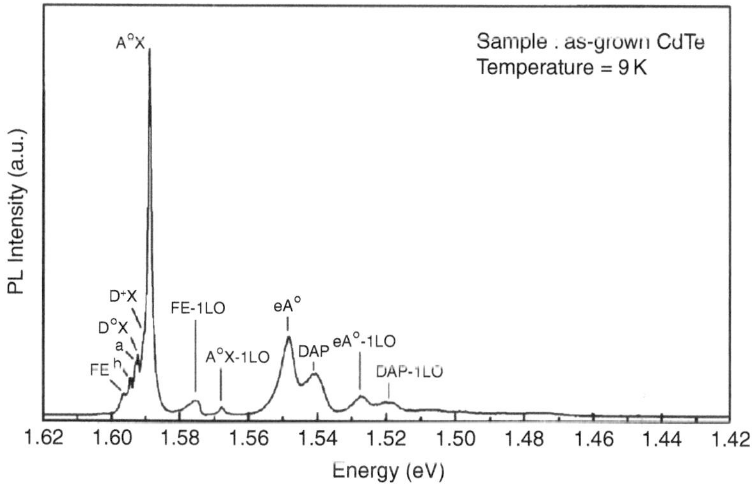

Figure 3 PL spectrum of bulk undoped CdTe from Ref. [44].

from annealing under Cd atmosphere or irradiation experiments [7, 38–43]. Interestingly, Shin et al. identified the presence of several native point defects, as represented in Fig. 3 [44]. The 1.596 eV line is associated with X type radiative transitions, for which the energy position E_X yields an experimental value of 10 meV for the binding energy of free excitons $E_{B,X}$, through $E_{B,X} = E_G - E_X$, in agreement with the referenced one [4]. The excitonic band is again dominated by the 1.589 eV line, which is related here to (A°, X) radiative transitions involving α acceptor complexes. Such complexes also lead to the presence of a doublet at energies of 1.549 and 1.541 eV, respectively, referred to as $(e, A°)_\alpha$ and DAP_α radiative recombinations, respectively. An ionization energy of 56.4 meV was deduced for α acceptor complexes but their nature has not completely been determined yet: Shin et al. suggested assigning it to intrinsic compensating acceptor complexes, as previously proposed by Seto et al., but in the form of (V_{Cd}, 2D) with an unknown donor D [42, 44]. As several research works proposed to consider the formation of (V_{Cd}, Te_{Cd}) defect complexes in order to account for the relatively high resistivity of undoped CdTe, the unknown donor D could be attributed to Te_{Cd} [29, 33]. Furthermore, Shin et al. identified, from annealing under Te atmosphere, (e, A°) and DAP radiative recombinations involving V_{Cd} at energies of 1.554 and 1.547 eV, respectively. An ionization energy of 50.4 meV was deduced for V_{Cd} [44].

2.1.3. The extended defect band

For both monocrystalline and polycrystalline variants of CdTe, it is worth noticing that the optical properties can be related to the dislocation density in PL spectra [45, 46]. Several research works showed the strong

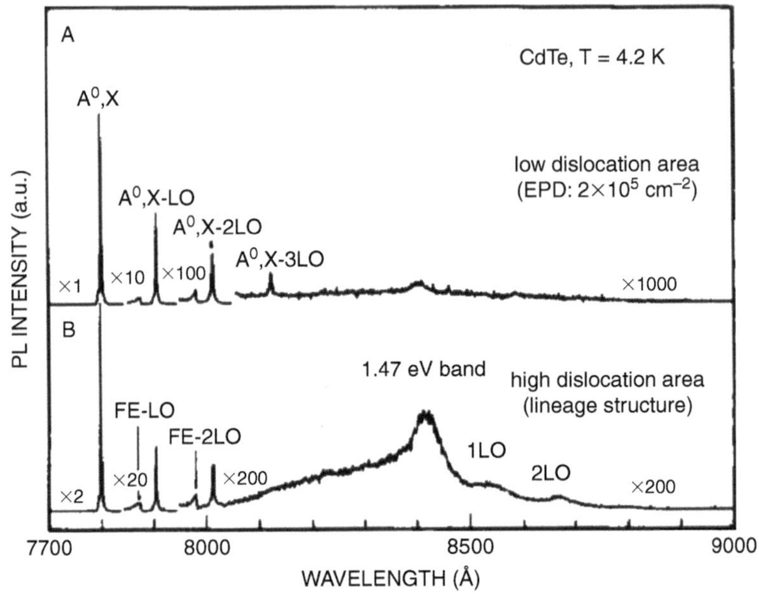

Figure 4 PL spectrum of undoped CdTe on area of low and high dislocation densities from Ref. [47].

correlation between the existence of a band located around 1.47 eV (i.e., the so-called Y band) and the presence of dislocations previously revealed by chemical etching, as represented in Fig. 4 [47, 48]. The Y band consists of a ZPL located at energy of 1.475 eV, which is associated with radiative recombinations of excitons bound to extended defects, followed by several LO phonon replicas, each separated by a phonon energy of 21.3 meV: the extended defects were found to be Te(g) type glide dislocations [48]. Interestingly, the typical value of the Huang-Rhys coupling constant S for the Y band is smaller than 0.5 and thus rather low: the charge distribution around extended defects is partially delocalized, inducing a weak coupling strength with the crystalline lattice, as shown in Eq. (5). This specific feature is a powerful means to distinguish the Y band from DAP_A radiative recombinations involving A centers for instance, for which the energy positions of the corresponding lines are very close, as can be seen in the following.

2.2. p-Type doped CdTe

Despite the p-type conductivity of undoped CdTe, the control of p-type doping is tricky due to its relatively high intrinsic resistivity. Typically, the dopants of the I and metal alkali groups were early used for the p-type doping in CdTe, by substituting for Cd sites, and widely studied through the related optical properties. Nevertheless, their relative instability, which was evidenced through aging phenomena, constitutes a major obstacle for their efficient use. More recently, the dopants of the VI groups have been consequently more and more employed for the p-type doping in CdTe, by substituting for Te sites.

2.2.1. Substitutional Cd sites: Alkali metal, Cu, Ag, Au doping

Molva et al. and Chamonal et al. early determined the typical PL spectra of Li-, Na-, Cu-, Ag-, and Au-doped CdTe. They observed (A°, X) and DAP radiative recombinations involving Li_{Cd}, Na_{Cd}, Cu_{Cd}, Ag_{Cd}, and Au_{Cd} acceptors, and systematically determined their ionization energy, as given in Table 1 [7, 49, 51, 54–56]. The latter were assessed either from the energy position of (e, A°) radiative transitions or from THS lines and typically varies in the energy range between 58.0 and 263 meV, corresponding to Li_{Cd} and Au_{Cd} acceptors, respectively. The excited state energies of such acceptors were also systematically determined from THT lines and selective excitation of DAP [7, 55].

Nevertheless, it was early stated that Ag and Cu dopants are thermally unstable on substitutional Cd sites in CdTe due to their large diffusion coefficient: after a storage of several weeks at room temperature, the amount fraction of Cu_{Cd} and Ag_{Cd} acceptors drastically decreases and such acceptors are preferentially involved in the formation of defect complexes.

In the case of Ag-doped CdTe, a prominent line associated with (A°, X) radiative recombinations is revealed at energy of 1.5815 eV after a storage at room temperature, as represented in Fig. 5 [57]. The exact mechanism, which can give rise to such an aging process, has not clearly been evidenced yet. It was suggested that new defect complexes involving Ag_{Cd} acceptors on distinct sites are formed: the nature of such defect complexes could be (Ag_{Cd}, Ag_i), as proposed by Monemar et al. [58]. More recently, Zelaya-Angel et al. showed that the features of PL spectra are dependent upon the way to incorporate Ag in CdTe, namely either by *ex-situ*

Table 1 Energy positions of (A°, X) and DAP radiative recombinations, and ionization energies E_A for metal alkali and group I dopants on Cd sites

$Acceptor_{Cd}$	(A°, X) (eV)	DAP (eV)	E_A (meV)	References
Li_{Cd}	1.58923	1.540	58.0	[7, 49, 50]
Na_{Cd}	1.58916	1.540	58.7	[7, 49]
Cu_{Cd}	1.58956	1.453	146.0	[49, 51–53]
Ag_{Cd}	1.58848	1.491	107.5	[49, 51]
Au_{Cd}	1.57606	1.335	263.0	[54]

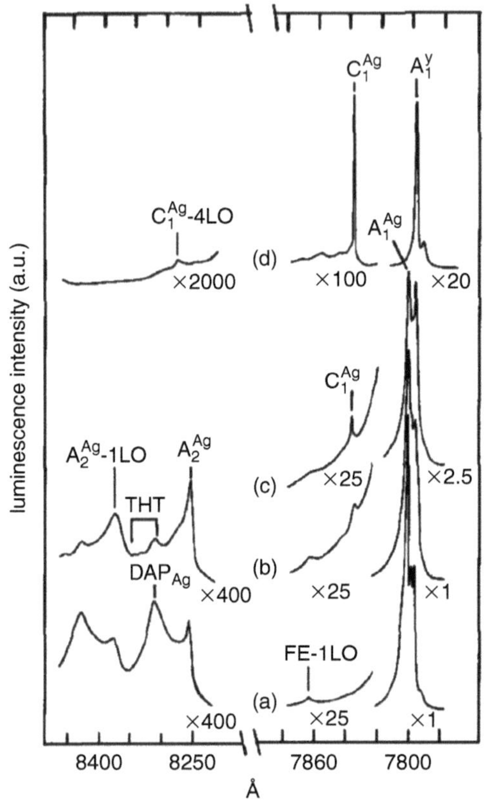

Figure 5 Variation of PL spectra versus time in Ag-doped CdTe after re-annealed treatment at 800 °C under Te atmosphere: (A) 2 days, (B) and (C) 40 days, (d) 75 days, from Ref. [57].

diffusion processes or by *in-situ* doping during the growth: (V_{Cd}, Ag_i) defect complexes were proposed to be involved in such an aging process [59]. Furthermore, other research works reported Ag related DAP radiative recombinations between 1.52 and 1.55 eV [59–61].

Recently, there has been increasing efforts as regards the case of Cu-doped CdTe, since Cu is commonly used in CdTe-based solar cells as an ohmic back contact and is thus likely to be incorporated in CdTe. Grecu et al. identified, just after the diffusion process of Cu, a new 1.555 eV line attributed to ($A°$, X) radiative transitions: the proposed acceptor could be (V_{Cd}, Cu_i) defect complexes by analogy with Ag-doped CdTe, which was later confirmed by Aguilar-Hernández et al. [62–64]. Interestingly, such a 1.555 eV line also vanishes after a few days, evidencing the formation of nonradiative recombination centers [62, 63]. Furthermore, the appearance of the 1.456 eV line after annealing under an oxidizing

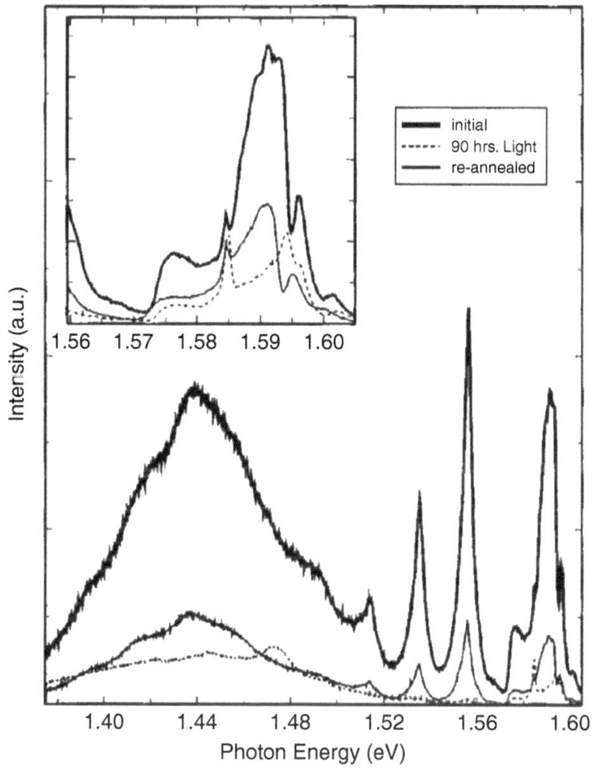

Figure 6 Aging and restoration effects in Cu-doped CdTe from Ref. [63].

atmosphere was recently reported and associated with (D°, h) radiative recombinations involving (Cu_i, O_{Te}) defect complexes [65].

It is also worth noticing that the aging process is reversible in the sense that the restoration of the optical properties of Ag- and Cu-doped CdTe is possible after specific treatments, as shown in Fig. 6 [63]. However, the instability of both Cu_{Cd} and Ag_{Cd} acceptors is a severe drawback to reach permanent, efficient p-type doping in CdTe. Unfortunately, a similar aging behavior was also revealed concerning Li_{Cd} and Na_{Cd} acceptors [66]. This opens the way to the use of group V dopants, for which the diffusivity is limited in CdTe.

2.2.3. Substitutional Te sites: N, P, As, Sb, Bi doping

Molva et al. early determined the typical PL spectra of N-, P-, and As-doped CdTe, as represented in Fig. 7 [67]. They observed (A°, X) and DAP radiative recombinations involving N_{Te}, P_{Te}, and As_{Te} acceptors and systematically assessed their ionization energy, as presented in Table 2 [7, 67]. The latter was determined from the energy position of (e, A°)

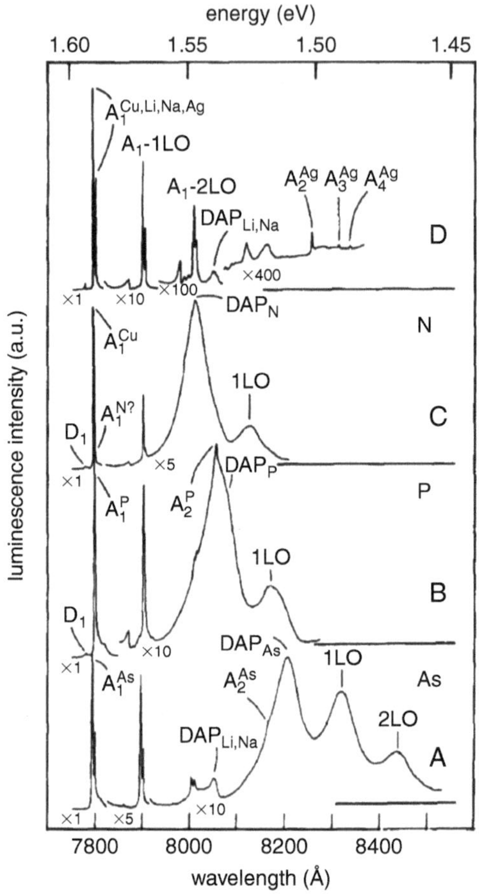

Figure 7 Typical spectra of N-, P-, and As-doped CdTe from Ref. [67].

Table 2 Energy positions of (A°, X) and DAP radiative recombinations, and ionization energies E_A for group V dopants on Te sites

$Acceptor_{Te}$	(A°, X) (eV)	DAP (eV)	E_A (meV)	References
N_{Te}	1.5892	1.547	56.0	[7, 67–71]
P_{Te}	1.58897	1.537	68.2	[7, 67, 72]
As_{Te}	1.58970/1.5903	1.510/1.542	92.0/58-60	[67, 74, 75/ 11, 76–78]
Sb_{Te}	1.5908	1.541-1.548	51-61	[11, 70, 71]

radiative recombinations. N_{Te} acceptors are the shallower acceptors with an ionization energy of 56.0 meV, which is very close to the hydrogenic acceptor like ionization energy [7, 15, 67–71]. The excited state energies of P_{Te} and As_{Te} acceptors were also given from THT lines [7, 67]. Furthermore, Oehling et al. mentioned more recently (A°, X) radiative recombinations at energy of 1.582 eV in N-doped CdTe, which are tentatively assigned to (N_{Cd}, N_i) defect complexes [73].

In the case of As-doped CdTe, the situation remains controversial and quite unclear, as suggested by Table 2, highlighting two different types of results. The first group supports an ionization energy of 92.0 meV for As_{Te} acceptors, which was determined both from the energy position of (e, A°) radiative recombinations and from THT lines [67, 74, 75]. On the contrary, the second groups mentioned more recently an ionization energy around 58-60 meV from the energy position of (e, A°) radiative recombinations located at a shallower energy [11, 76–78]. Besides, Soltani et al. and Dhese et al. determined the general features of Sb_{Te} acceptors in Sb-doped CdTe, for which the properties may be close to those of As_{Te} acceptors in the second case [11, 70, 71].

More recently, an increasing interest in Bi-doped CdTe arises from its unexpected ability to reach very high resistivity whereas Bi should occupy Te sites and act as acceptors by strengthening the p-type conductivity of CdTe. Saucedo et al. widely carried out the study of the optical properties of Bi-doped CdTe and identified some lines related to Bi around 1.45 and 1.55 eV, as well as deeper in the bandgap [79, 80].

2.3. n-Type doped CdTe

The control of n-type doping in CdTe is of great interest as regards the infrared detectors but is quite tricky to control due to the p-type conductivity of undoped CdTe and to the predominant role of V_{Cd} as compensating agent in Te-rich growth conditions. However, such compensation mechanisms are not always detrimental and even required concerning γ- and X-ray detectors, in which a very high resistivity of up to 10^{10} Ω cm is expected to reduce their dark current. Consequently, PL spectra exhibit general features that are governed by the formation of defect complexes between V_{Cd} and the dopants, such as the so-called A centers for instance. Dopants of the III and VII groups are commonly employed for the n-type doping in CdTe, by substituting for Cd and Te sites, respectively.

2.3.1. Substitutional Cd sites: Al, Ga, In doping

Francou et al. early determined the typical PL spectra of Al-, Ga-, and In-doped CdTe. They observed (D°, X) radiative recombinations involving Al_{Cd}, Ga_{Cd}, and In_{Cd} donors and determined their ionization energy from TET lines, as presented in Table 3 [6]. The corresponding localization

Table 3 Energy positions of (D°, X) and DAP radiative recombinations, localization and ionization energies E_{LOC} and E_D for group III dopants on Cd sites

$Donor_{Cd}$	(D°, X) (eV)	E_{LOC} (meV)	E_D (meV)	References
Al_{Cd}	1.59305	3.46	14.04	[6, 81]
Ga_{Cd}	1.59309/1.5932/1.594	3.41	13.83/18	[6/87/88]
In_{Cd}	1.59302	3.48	14.08	[6, 82–86]

energies of excitons bound to such donors were also assessed and are in agreement with the Hayne's rule with a slope of 0.246 [6]. On the contrary, Ekawa et al. and Gold et al. rather proposed (D°, X) radiative recombinations involving Ga_{Cd} donors at slightly different energies of 1.5932 and 1.594 eV, respectively [87, 88].

Al_{Cd}, Ga_{Cd}, and In_{Cd} donors usually form shallow acceptor complexes with V_{Cd}, namely the so-called A centers. (A°, X) and DAP radiative recombinations were widely reported for such centers as well as the corresponding Huang-Rhys coupling constant S and ionization energies E_A, as given in Table 4. Nevertheless, in the case of In-doped CdTe, Worschech et al. attributed, from magneto-luminescence spectroscopy, the 1.584 eV line (i.e., the so-called C line) to (A°, X) radiative recombinations involving (V_{Cd}, $2In_{Cd}$) defect complexes rather than (V_{Cd}, In_{Cd}) A centers, as commonly suggested [82–86, 89]. A line at 1.581 eV (i.e., the so-called W line) was also considered to be due to transversal acoustical (TA) phonon sideband [85, 86]. In the case of Ga-doped CdTe, the correlation between the 1.581 eV line related to (A°, X) radiative recombinations and Ga doping was also observed [90]. Furthermore, Babentsov et al. suggested the role of (Cd_{Te}, Ga_{Cd}) defect complexes instead of (V_{Cd}, Ga_{Cd}) A centers for (A°, X) radiative recombinations at energy of 1.5841 eV [93]. Recently, Song et al. reported a specific feature concerning the case of Al-doped CdTe, as represented in Fig. 8: the existence of the 1.5906 eV line is attributed to (A°, X) radiative recombinations involving possible (V_{Cd}, $2Al_{Cd}$) defect complexes by analogy with Cl-doped CdTe [94]. Such complexes also lead to DAP radiative recombinations at energy of 1.553 eV and present an ionization energy of 52.1 meV [94].

Table 4 Energy positions of (A°, X) and DAP radiative recombinations, ionization energies E_A and Huang-Rhys coupling constants S for A centers from group III dopants on Cd sites

A center	(A°, X) (eV)	DAP (eV)	E_A (meV)	S	References
(V_{Cd}, Al_{Cd})	1.5848/1.5905	1.453/1.451	145		[94/81]
(V_{Cd}, Ga_{Cd})	1.5841	1.475	131	1.7	[89–92]
(V_{Cd}, In_{Cd})	1.5840	1.454	142	1.8/2.6	[82–85, 89]

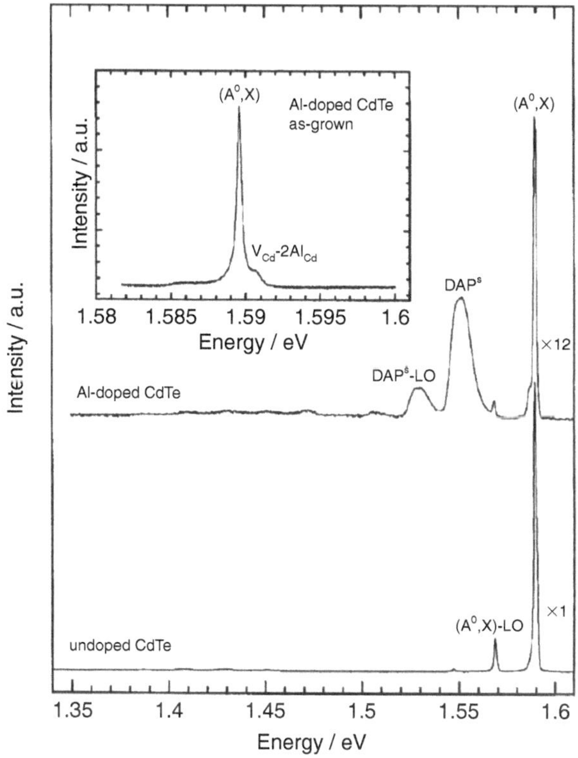

Figure 8 Typical spectra of Al-doped CdTe from Ref. [94].

2.3.2. Substitutional Te sites: F, Br, I doping

The PL spectra of F-, Br-, and I-doped CdTe are seldom mentioned in the literature, which is probably due to the popular success of Cl doping in CdTe, for which a special section is dedicated in the following. Francou et al. tentatively assigned the 1.59314 eV line to (D°, X) radiative recombinations involving F_{Te} donors: the ionization energy E_D of such donors is found to be 13.71 meV [6]. Giles et al. reported similar radiative recombinations involving I_{Te} donors at energy of 1.593 eV and obtained a related ionization energy E_D of 14-15 meV [95, 96]. Interestingly, F_{Te}, Br_{Te}, and I_{Te} donors also form A centers with V_{Cd}, for which Stadler et al. reported the energy positions of DAP radiative recombinations as well as the Huang-Rhys coupling constant S and ionization energies E_A, as shown in Table 5 [89]. Moreover, in the case of I-doped CdTe, a sharp line at 1.49 eV is revealed, but its nature remains quite unclear, even if (Na_{Cd}, I_{Te}) defect complexes were suggested [98].

Table 5 Energy positions of DAP radiative recombinations, ionization energies E_A and Huang-Rhys coupling constant S for A centers from group VII dopants on Te sites

A center	DAP (eV)	E_A (meV)	S	References
(V_{Cd}, F_{Te})		116	3.2	[89]
(V_{Cd}, Br_{Te})	1.478	119	2.6	[89, 97]
(V_{Cd}, I_{Te})	1.470	128	1.5	[89, 95, 96]

3. THE SPECIAL CASE OF CHLORINE DOPING

As Cl-doped CdTe is used as a p-type semiconductor in the manufacturing of solar cells in photovoltaics and also very promising as a photoconductor for γ- or X-ray detectors in the nuclear or medical field, the study of both its monocrystalline and polycrystalline variants have been widely carried out.

3.1. Cl-doped monocrystalline CdTe

The mechanisms of Cl doping in monocrystalline CdTe have received an increasing interest for several decades. They represent a kind of model system so as to identify the compensation processes at work and to find direct relationships with the optical properties.

3.1.1. Predominance of A centers

The optical properties of Cl-doped monocrystalline CdTe are well known and governed by the presence of the so-called chlorine A centers, as represented in Fig. 9 [99]. The atomic structure of chlorine A centers was elucidated by optically detected magnetic resonance measurements: it consists of a V_{Cd} on a tetrahedral site related to a Cl_{Te} donor on a nearest Te site, as (V_{Cd}, Cl_{Te}) acceptor complexes [100, 101]. The 1.586 eV line is thus commonly referred to as (A°, X) radiative recombinations involving chlorine A centers [102, 103]. Furthermore, Cl atoms act as hydrogenic donors by substituting for Te sites and leads to the presence of the 1.59296 eV line associated with (D°, X) radiative recombinations [6, 102]. Francou et al. early determined the exact ionization energy E_D of Cl_{Te} donors, which equals 14.48 meV [6]. Below 1.5 eV, DAP_A radiative recombinations involving chlorine A centers present a ZPL located at energy of about 1.47 eV according to its atomic concentration [99–102]. Such a ZPL is commonly followed by several LO phonon replicas, leading to a value of the Huang-Rhys coupling constant S of about 2.2 [89, 100]. Cl_{Te} donors

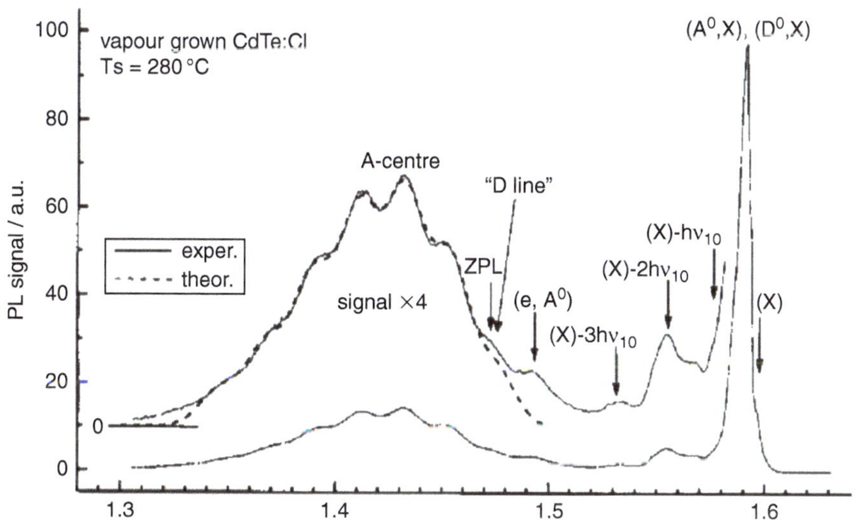

Figure 9 Typical spectra of Cl-doped monocrystalline CdTe from Ref. [99].

are involved in DAP_A radiative recombinations. The ionization energy E_A of chlorine A centers still remains controversial: some values between 120 and 170 meV are commonly reported, even if several more recent research works tend to propose a rough value of about 120-135 meV [99–111]. Such experimental uncertainties partially originate from the variation of the energy position of DAP_A radiative recombinations with the atomic concentrations of Cl_{Te} donors and A centers forming the pair.

3.2.1. Specific observations for high Cl atomic concentration

Some research works reported the appearance of new radiative transitions around 1.55 eV in monocrystalline CdTe when the Cl atomic concentration is typically higher than about 10^{18} at/cm^3, as represented in Fig. 10 [102, 112]. Shin et al. identified (e, A°) and DAP radiative recombinations involving β acceptors at energies of 1.559 and 1.553 eV, respectively. An ionization energy of 45 meV for such complexes was deduced [102]. The 1.590 eV line is also clearly attributed to (A°, X) radiative recombinations involving β acceptors [102, 113]. Furthermore, Ossau et al. earlier showed that β acceptor has a structure of defect complex magneto-optical measurements [112, 114]. Its nature is expected to be (V_{Cd}, $2Cl_{Te}$) three defect complex (i.e., the so-called β acceptor complexes), for which Bell early predicted theoretically its existence [115].

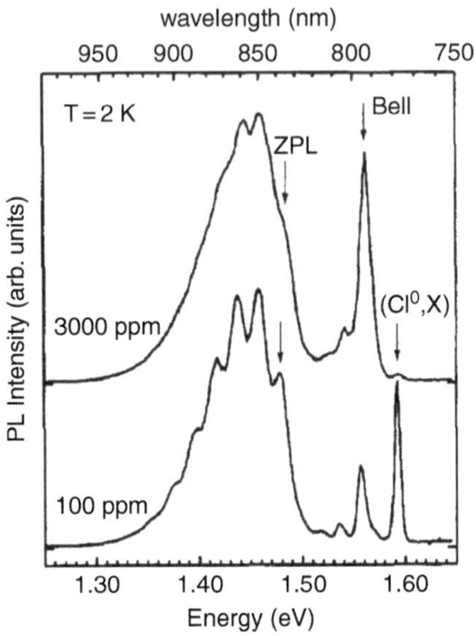

Figure 10 Typical spectra of lightly and highly Cl-doped monocrystalline CdTe from Ref. [112].

3.2. Cl-doped polycrystalline CdTe

Most of the current optoelectronic devices, such as solar cells or X-ray detectors, require very large dimensions of CdTe-based materials, especially imposing the use of its polycrystalline variant for cost and technological reasons.

3.2.1. Predominance of β acceptor complexes

The optical properties of Cl-doped polycrystalline CdTe are much less known as compared to its monocrystalline variant. It was recently shown that such a variant is governed by the presence of the so-called β acceptor complexes, even if the formation of chlorine A centers still occurs [116, 117]. The 1.590 eV line, which is attributed to (A°, X) radiative recombinations involving β acceptor complexes, usually dominates the excitonic band. (D°, X) radiative recombinations involving Cl_{Te} donors are also reported around 1.593 eV . As in its highly doped monocrystalline variant, β acceptor complexes contribute to the presence of a doublet at energies of 1.557 and 1.552 eV, which is respectively associated with (e, A°) and DAP radiative recombinations, as evidenced by Consonni et al. [117]. Below 1.5 eV, a mixed band composed of both DAP_A radiative recombinations involving chlorine A centers and radiative transitions of excitons bound to extended defects is evidenced: it consists of a ZPL located at energy of about 1.45-1.48 eV followed by several LO phonon replicas, each separated by a phonon energy of 21.3 meV [118]. An intermediate value for the Huang-Rhys coupling constant S of about 1.3-1.5 was deduced [118].

3.2.2. Strong relationship between the structural and optical properties

In order to determine the origin of the optical properties in Cl-doped polycrystalline CdTe, Consonni et al. collected several localized CL spectra as growth proceeds [118]. They focused in particular on the effects of the stage of island coalescence, which is related to the formation of the so-called grain boundaries, as shown in Fig. 11 [118]. An evolution of the nature of acceptor complexes was evidenced as a function of the crystalline structure considered (i.e., single or coalesced islands, polycrystalline films) and ends up leading to the predominance of β acceptor complexes. Interestingly, the existence of β acceptor complexes in the polycrystalline variant is directly associated with segregation phenomena of chlorine around grain boundaries, leading to the presence of highly Cl concentrated regions [118]. β acceptor complexes dissociate when the Cl atomic concentration is drastically reduced, following annealing under nitrogen atmosphere for instance [111]. By observing the simultaneous formation of β acceptor complexes with the shift of DAP_A radiative recombinations toward higher energies on the luminescence spectra, an atomic structure was proposed concerning such three defect complexes: it could consist in a V_{Cd} on a tetrahedral site related to two Cl_{Te} donors on nearest Te sites [118]. The correlation of the existence of β acceptor complexes with highly concentrated regions of chlorine is consistent with the fact that such complexes need more Cl atoms to be formed than A centers. Consequently, the polycrystalline variant of Cl-doped CdTe is characterized by an inhomogeneous spatial distribution of the optical properties, which were also confirmed by several monochromatic CL images at different energies in Ref. [117] on a same crystalline structure.

4. PROSPECTS

Carrying out the study of the optical properties in CdTe is a powerful means to determine the nature of defects, playing a role in most of the physical processes responsible for the device properties. Currently, the most recent research works aim at better understanding the properties of Cl and Cu doping within the polycrystalline variant of CdTe, which is promising for large dimension devices such as X-ray detectors and solar cells. Alternative solutions to reach very high-resistivity are also proposed: there are increasing interests as regards the control of Bi doping for instance. Similarly, p-type conductivity of CdTe by means of As doping is also widely studied for infrared detectors. However, as it has been stated in this entire chapter, identifying the nature of the lines in PL

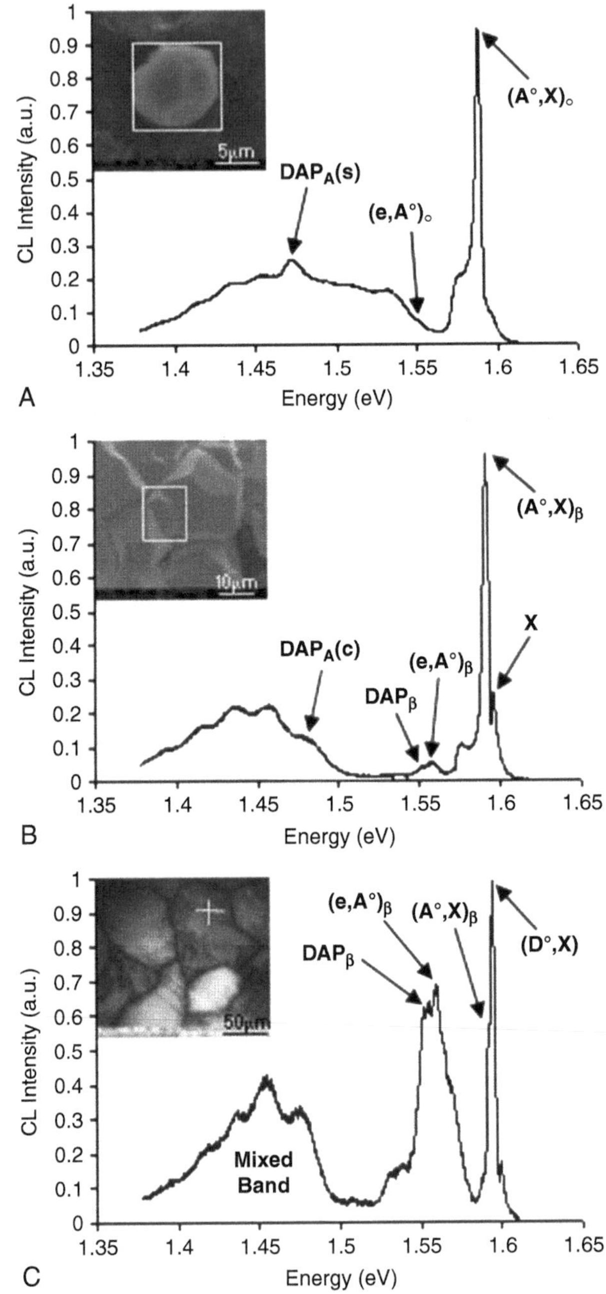

Figure 11 CL spectra of Cl-doped CdTe on (A) single and (B) coalesced islands, and on (C) thick polycrystalline films, from Ref. [118].

or CL spectra and correlating them with the defects involved is not straightforward: most of the dopants lead to the presence of lines with similar characteristics at very close energy positions. Therefore, coupling the study of the optical properties with chemical analysis, such as secondary ion mass spectroscopy (SIMS) or time-of-flight SIMS for instance, should systematically be carried out. Furthermore, increasing efforts concerning the study of the optical properties in the polycrystalline variant of CdTe raises new problems, which are related to their inhomogeneous spatial distribution, mainly related to the presence of grain boundaries [117–119]. The use of spatially resolved luminescence measurements or CL imaging thus becomes crucial to elucidate the exact physical phenomena at work in such a polycrystalline variant and may still require many efforts and interests in a near future [117, 118, 120].

ACKNOWLEDGMENTS

The author would like to thank all the research groups who took part in the study of the optical properties in CdTe and, in particular, the teams from the CEA and the CNRS in Grenoble, France.

REFERENCES

[1] D.E. Cooper, J. Bajaj, P.R. Newman, J. Cryst. Growth 86 (1988) 544–551.
[2] K. Zanio, Semiconductors and Semimetals, vol. 13, Academic Press, New York, 1978.
[3] A.N. Pikhtin, A.D. Yas'kov, Sov. Phys. Semicond. 22 (1988) 613–626.
[4] D.T.F. Marple, Phys. Rev. 150 (1966) 728–734.
[5] A. Baldereschi, N.O. Lipari, Phys. Rev. B 3 (1971) 439–451.
[6] J.M. Francou, K. Saminadayar, J.L. Pautrat, Phys. Rev. B 41 (1990) 12035–12046.
[7] E. Molva, J.L. Pautrat, K. Saminadayar, G. Milchberg, N. Magnea, Phys. Rev. B 30 (1984) 3344–3354.
[8] J.R. Haynes, Phys. Rev. Lett. 4 (1960) 361–363.
[9] R.E. Halsted, M. Aven, Phys. Rev. Lett. 14 (1965) 64–67.
[10] L.S. Dang, G. Neu, R. Romestain, Sold State Commun. 44 (1982) 1187–1190.
[11] M. Soltani, M. Certier, R. Evrard, E. Kartheuser, J. Appl. Phys. 78 (1995) 5626–5632.
[12] D.M. Eagles, J. Phys, Chem. Solids 1 (1960) 76–83.
[13] J.J. Hopfield, Phys. Rev. Lett. 10 (1963) 162–164.
[14] L. Mehrkam, F. Williams, Phys. Rev. B 6 (1972) 3753–3756.
[15] A. Baldereschi, N.O. Lipari, Phys. Rev. B 8 (1973) 2697–2709.
[16] A. Baldereschi, N.O. Lipari, Phys. Rev. B 9 (1974) 1525–1539.
[17] S. Munnix, E. Kartheuser, Phys. Rev. B 26 (1982) 6776–6787.
[18] H.B. Bebb, Phys. Rev. 185 (1969) 1116–1126.
[19] M. Germain, E. Kartheuser, A.L. Gurskii, E.V. Lutsenko, I.P. Marko, V.N. Pavlovskii, G.P. Yablonskii, K. Heime, M. Heuken, B. Schineller, J. Appl. Phys. 91 (2002) 9827–9834.
[20] K. Huang, A. Rhys, Proc. Roy. Soc. A 204 (1950) 406–423.
[21] H. Fröhlich, H. Pelzer, S. Zienau, Philos. Mag. 41 (1950) 221–242.

[22] J.J. Hopfield, J. Phys, Chem. Solids 10 (1959) 110–119.
[23] T. Taguchi, J. Shirafuji, Y. Inuishi, Phys. State Solidi B 68 (1975) 727–738.
[24] T. Schmidt, K. Lischka, W. Zulehner, Phys. Rev. B 45 (1992) 8989–8994.
[25] Z.C. Feng, A. Mascarenhas, W.J. Choyke, J. Lumin. 35 (1986) 329–341.
[26] D. Bimberg, M. Sondergeld, E. Grobe, Phys. Rev. B 4 (1970) 3451–3455.
[27] H. Zimmermann, R. Boyn, K. Piel, J. Phys. Condens. Matter 4 (1992) 859–868.
[28] M. Aven, J.S. Prener, Physics and Chemistry of II-VI Compounds, North-Holland Publishing Company, Amsterdam, 1967.
[29] M.A. Berding, Phys. Rev. B 60 (1999) 8943–8950.
[30] S.H. Wei, S.B. Zhang, Phys. Rev. B 66 (2002) 155211.
[31] P. Emanuelsson, P. Omling, H.G. Grimmeiss, B.K. Meyer, M. Wienecke, M. Schenk, Phys. Rev. B 47 (1993) 15578–15580.
[32] F.J. Bryant, D.H.J. Totterdell, W.E. Hagston, J. Phys. C: Solid State Phys. 4 (1971) 641–653.
[33] M. Fiederle, C. Eiche, M. Salk, R. Schwarz, K.W. Benz, W. Stadler, D.M. Hofmann, B.K. Meyer, J. Appl. Phys. 84 (1998) 6689–6692.
[34] B.K. Meyer, P. Omling, E. Weigel, G. Müller-Vogt, Phys. Rev. B 46 (1992) 15135–15138.
[35] X. Mathew, Sol. Energy Mater. Sol. Cells 76 (2003) 225–242.
[36] F. Dal'Bo, G. Lentz, N. Magnea, H. Mariette, L.S. Dang, J.L. Pautrat, J. Appl. Phys. 66 (1989) 1338–1346.
[37] S.H. Song, J.F. Wang, M. Isshiki, J. Cryst. Growth 236 (2002) 165–170.
[38] F.J. Bryant, H.J. Totterdell, Radiat. Effects 9 (1971) 115.
[39] C.E. Barnes, C. Kikuchi, Radiat. Effects 26 (1975) 105.
[40] B. Yang, Y. Ishikawa, T. Miki, Y. Doumae, T. Tomizono, M. Isshiki, J. Cryst. Growth 159 (1996) 171–174.
[41] S.H. Song, J.F. Wang, M. Isshiki, J. Cryst. Growth 257 (2003) 231–236.
[42] S. Seto, A. Tanaka, Y. Masa, S. Dairaku, M. Kawashima, Appl. Phys. Lett. 53 (1988) 1532–1534.
[43] T.A. Kuhn, W. Ossau, R.N. Bicknell-Tassius, G. Landwehr, Appl. Phys. Lett. 55 (1989) 2637–2639.
[44] H.Y. Shin, C.Y. Sun, Mater. Sci. Eng. B 52 (1998) 78–83.
[45] C. Onodera, T. Taguchi, J. Cryst. Growth 101 (1990) 502–506.
[46] M. Cardenas-Garcia, J. Aguilar-Hernandez, G. Contreras-Puente, Thin Solid Films 480-481 (2005) 269–272.
[47] S. Seto, A. Tanaka, F. Takeda, K. Matsuura, J. Cryst. Growth 138 (1994) 346–351.
[48] S. Hildebrandt, H. Uniewski, J. Schreiber, H.S. Leipner, J. Phys. III 7 (1997) 1505–1514.
[49] E. Molva, J.P. Chamonal, J.L. Pautrat, Phys. State Solidi (b) 109 (1982) 635–644.
[50] G. Neu, Y. Marfaing, R. Legros, R. Triboulet, L. Svob, J. Lumin. 21 (1980) 293–304.
[51] J.P. Chamonal, E. Molva, J.L. Pautrat, Solid State Commun. 43 (1982) 801–805.
[52] K.M. James, J.L. Merz, C.E. Jones, J. Vac. Sci. Technol. A 6 (1988) 2664–2669.
[53] J.P. Laurenti, G. Bastide, M. Rouzeyre, R. Triboulet, Solid State Commun. 67 (1988) 1127–1130.
[54] E. Molva, J.M. Francou, J.L. Pautrat, K. Saminadayar, L.S. Dang, J. Appl. Phys. 56 (1984) 2241–2249.
[55] E. Molva, J.P. Chamonal, G. Milchberg, K. Saminadayar, B. Pajot, G. Neu, Solid State Commun. 44 (1982) 351–355.
[56] J.P. Chamonal, E. Molva, M. Dupuy, R. Accomo, J.L. Pautrat, Phys. B 116 (1983) 519–526.
[57] J.P. Chamonal, E. Molva, J.L. Pautrat, L. Revoil, J. Cryst. Growth 59 (1982) 297–300.
[58] B. Monemar, E. Molva, L.S. Dang, Phys. Rev. B 33 (1986) 1134–1145.
[59] O. Zelaya-Angel, M. Garcia-Rocha, J.G. Mendoza-Alvarez, M. Cardenas, J. Aguilar-Hernandez, J. Appl. Phys. 94 (2003) 2284–2288.

[60] M. Cárdenas, J.G. Mendoza-Alvarez, F. Sánchez-Sinencio, O. Zelaya, C. Menezes, J. Appl. Phys. 56 (1984) 2977–2980.
[61] J. Hamann, A. Burchard, M. Deicher, T. Filz, V. Ostheimer, C. Schmitz, H. Wolf, Th. Wichert, the ISOLDE collaboration, Appl. Phys. Lett. 72 (1998) 3029–3031.
[62] D. Grecu, A.D. Compaan, Appl. Phys. Lett. 75 (1999) 361–363.
[63] D. Grecu, A.D. Compaan, D. Young, U. Jayamaha, D.H. Rose, J. Appl. Phys. 88 (2000) 2490–2496.
[64] J. Aguilar-Hernández, M. Cárdenas-Garcıá, G. Contreras-Puente, J. Vidal-Larramendi, Mater. Sci. Eng. B 102 (2003) 203–206.
[65] C.R. Corwine, J.R. Sites, T.A. Gessert, W.K. Metzger, P. Dippo, J. Li, A. Duda, G. Teeter, Appl. Phys. Lett. 86 (2005) 221909.
[66] B. Yang, Y. Ishikawa, T. Miki, Y. Doumae, M. Isshiki, J. Cryst. Growth 179 (1997) 410–414.
[67] E. Molva, K. Saminadayar, J.L. Pautrat, E. Ligeon, Solid State Commun. 48 (1983) 955–960.
[68] T. Baron, S. Tatarenko, K. Saminadayar, N. Magnea, J. Fontenille, Appl. Phys. Lett. 65 (1995) 1284–1286.
[69] T. Baron, K. Saminadayar, N. Magnea, J. Appl. Phys. 83 (1998) 1354–1370.
[70] K.A. Dhese, P. Devine, D.E. Ashenford, J.E. Nicholls, C.G. Scott, D. Sands, B. Lunn, J. Appl. Phys. 76 (1994) 5423–5428.
[71] K.A. Dhese, D.E. Ashenford, J.E. Nicholls, P. Devine, B. Lunn, C.G. Scott, J. Jaroszyfiski, J. Cryst. Growth 138 (1994) 443–447.
[72] C. Uzan, R. Legros, Y. Marfaing, J. Cryst. Growth 72 (1985) 252–257.
[73] S. Oehling, H.J. Lugauer, M. Schmitt, H. Heinke, U. Zehnder, A. Waag, C.R. Becker, G. Landwehr, J. Appl. Phys. 79 (1995) 2343–2346.
[74] J.M. Arias, S.H. Shin, D.E. Cooper, M. Zandian, J.G. Pasko, E.R. Gertner, R. E. DeWames, J. Singh, J. Vac. Sci. Technol. A 8 (1990) 1025–1033.
[75] M. Ekawa, K. Yasuda, T. Ferid, M. Saji, A. Tanaka, J. Appl. Phys. Lett. 71 (1992) 2669–2671.
[76] S.K. Gandhi, N.R. Taskar, I.B. Bhat, Appl. Phys. Lett. 50 (1987) 900–902.
[77] R.L. Harper, S. Hwang, N.C. Giles, J.F. Schetzina, D.L. Dreifus, T.H. Myers, Appl. Phys. Lett. 54 (1989) 170–172.
[78] N.C. Giles, K.A. Bowers, R.L. Harper, S. Hwang, J.F. Schetzina, J. Cryst. Growth 101 (1990) 67–72.
[79] E. Saucedo, C.M. Ruiz, V. Bermúdez, E. Dieguez, E. Gombia, A. Zappettini, A. Baraldi, N.V. Sochinskii, J. Appl. Phys. 100 (2006) 104901.
[80] E. Saucedo, J. Franc, H. Elhadidy, P. Horodysky, C.M. Ruiz, V. Bermúdez, N.V. Sochinskii, J. Appl. Phys. 103 (2008) 094901.
[81] L. Worschech, W. Ossau, F. Fischer, A. Waag, G. Landwehr, J. Cryst. Growth 161 (1996) 134–138.
[82] C.E. Barnes, K. Zanio, J. Appl. Phys. 46 (1975) 3959–3964.
[83] N.C. Giles, S. Hwang, J.F. Schetzina, S. McDevitt, C.J. Johnson, J. Appl. Phys. 64 (1988) 2656–2665.
[84] F. Bassani, S. Tatarenko, K. Saminadayar, J. Bleuse, N. Magnea, J.L. Pautrat, J. Appl. Phys. 58 (1991) 2651–2653.
[85] S. Seto, K. Suzuki, V.N. Abastillas, K. Inabe, J. Cryst. Growth 214-215 (2000) 974–978.
[86] L. Worschech, W. Ossau, G. Landwehr, Phys. Rev. B 52 (1995) 13965–13974.
[87] M. Ekawa, K. Yasuda, T. Ferid, M. Saji, A. Tanaka, J. Appl. Phys. 72 (1992) 3406–3409.
[88] J.S. Gold, T.H. Myers, N.C. Giles, K.A. Harris, L.M. Mohnkern, R.W. Yanka, J. Appl. Phys. 74 (1993) 6866–6871.
[89] W. Stadler, D.M. Hofmann, H.C. Alt, T. Muschik, B.K. Meyer, E. Weigel, G. Müller-Vogt, M. Salk, E. Rupp, K.W. Benz, Phys. Rev. B 51 (1995) 10619–10630.

[90] J.M. Wrobel, J.J. Dubowski, P. Becla, J. Vac. Sci. Technol. A 7 (1989) 338–342.
[91] S. Seto, A. Tanaka, K. Suzuki, M. Kawashima, J. Cryst. Growth 101 (1990) 430–434.
[92] W. Joerger, M. Laasch, T. Kunz, M. Fiederle, J. Meinhardt, K.W. Benz, K. Scholz, W. Wendl, G. Müller-Vogt, Cryst. Res. Technol. 32 (1997) 1103–1113.
[93] V. Babentsov, V. Corregidor, J.L. Castaño, M. Fiederle, T. Feltgen, K.W. Benz, E. Diéguez, Cryst. Res. Technol. 36 (2001) 535–542.
[94] S.H. Song, J. Wang, Y. Ishikawa, S. Seto, M. Isshiki, J. Cryst. Growth 237-239 (2002) 1726–1730.
[95] N.C. Giles, J. Lee, D. Rajavel, C.J. Summers, J. Appl. Phys. 73 (1993) 4541–4545.
[96] J. Lee, N.C. Giles, D. Rajavel, C.J. Summers, J. Appl. Phys. 78 (1995) 5669–5674.
[97] H. Zimmermann, R. Boyn, C. Albers, K.W. Benz, D. Sinerius, C. Eiche, B.K. Meyer, D.M. Hoffmann, J. Cryst. Growth 128 (1993) 593–598.
[98] J. Lee, N.C. Giles, C.J. Summers, J. Appl. Phys. 77 (1995) 4544–4547.
[99] T. Kunz, M. Laasch, J. Meinhardt, K.W. Benz, J. Cryst. Growth 184-185 (1998) 1005–1009.
[100] D.M. Hofmann, P. Omling, H.G. Grimmeiss, B.K. Meyer, K.W. Benz, D. Sinerius, Phys. Rev. B 45 (1992) 6247–6250.
[101] S. Biernacki, U. Scherz, B.K. Meyer, Phys. Rev. B 48 (1993) 11726–11731.
[102] H.Y. Shin, C.Y. Sun, J. Cryst. Growth 186 (1998) 354–361.
[103] R. Triboulet, Y. Marfaing, A. Cornet, P. Siffert, J. Appl. Phys. 45 (1974) 2759–2765.
[104] B. Furgolle, M. Hoclet, M. Vandevyver, Y. Marfaing, R. Triboulet, Solid State Commun. 14 (1974) 1237–1240.
[105] A. Castaldini, A. Cavallini, B. Fraboni, Appl. Phys. Lett. 69 (1996) 3510–3512.
[106] S. Seto, A. Tanaka, Y. Masa, M. Kawashima, J. Cryst. Growth 117 (1992) 271–275.
[107] D.M. Hofmann, W. Stadler, P. Christmann, B.K. Meyer, Nucl. Instrum. Meth. A 380 (1996) 117–120.
[108] K. Suzuki, S. Seto, A. Tanaka, M. Kawashima, J. Cryst. Growth 101 (1990) 859–863.
[109] D.P. Halliday, J.M. Eggleston, K. Durose, J. Cryst. Growth 186 (1998) 543–549.
[110] D.P. Halliday, J.M. Eggleston, K. Durose, Thin Solid Films 322 (1998) 314–318.
[111] V. Consonni, G. Feuillet, J. Appl. Phys. 105 (2009) 083535.
[112] B.K. Meyer, W. Stadler, J. Cryst. Growth 161 (1996) 119–127.
[113] W. Ossau, T.A. Kuhn, R.N. Bicknell-Tassius, J. Cryst. Growth 101 (1990) 135–140.
[114] T.A. Kuhn, W. Ossau, R.N. Bicknell-Tassius, G. Landwehr, Appl. Phys. Lett. 55 (1989) 2637–2639.
[115] R.O. Bell, Sol. State Commun. 16 (1975) 913–916.
[116] O. Ka, G. Neu, J. Appl. Phys. 68 (1990) 3386–3389.
[117] V. Consonni, G. Feuillet, S. Renet, J. Appl. Phys. 99 (2006) 053502.
[118] V. Consonni, G. Feuillet, F. Donatini, J. Bleuse, J. Appl. Phys. 101 (2007) 063522.
[119] A. Gukasyan, A. Kvit, Y. Klevkov, S. Oktyarbrsky, Solid State Commun. 97 (1996) 897–902.
[120] M.J. Romero, T.A. Gessert, M.M. Al-Jassim, Appl. Phys. Lett. 81 (2002) 3161–3163.

CHAPTER IIE

Mechanical Properties

J.C. Moosbrugger

1. ELASTICITY PROPERTIES

Table 1 lists four sets of independent experimental results for the elastic constants of CdTe single crystals [1–4]. Each of these was obtained using ultrasonic or resonance frequency techniques, measured in the range 77–300 K. While specific techniques employed differ among the results cited, the methods are all based on the relationship between the longitudinal and shear acoustic wave speeds in a crystal and its elastic properties (e.g., [5]). Maheswaranathan et al. [6] reported crystal elastic constant values for $Cd_{0.52}Zn_{0.48}Te$ that were comparable to, though slightly higher than, those in Table 1, indicating that substitution of Zn does not have a major impact on the elastic constants. Substitution of Mn in $Cd_{(1-x)}Mn_xTe$ was found to lower the elastic constants slightly.

Three elastic constants are required for the cubic symmetry, reported here as elsewhere (e.g., [7–9]) as the components of the elastic stiffness matrix (e.g., [5]) for a Cartesian coordinate system aligned with $\langle 100 \rangle$ axes. Young's modulus E_{ijk} associated with uniaxial tension or compression along any crystallographic axis $[ijk]$ can then be computed (e.g., [10, 11]), as

$$\frac{1}{E_{ijk}} = S_{11} - 2\left(S_{11} - S_{12} - \frac{1}{2}S_{44}\right)\left(l_{i1}^2 l_{j2}^2 + l_{j2}^2 l_{k3}^2 + l_{k3}^2 l_{i1}^2\right), \quad (1)$$

$$C_{11} = \frac{S_{11} + S_{12}}{(S_{11} + 2S_{12})(S_{11} - S_{12})}, \quad (2)$$

$$C_{12} = \frac{-S_{12}}{(S_{11} + 2S_{12})(S_{11} - S_{12})}, \quad (3)$$

Department of Mechanical and Aeronautical Engineering, Center for Advanced Materials Processing, Clarkson University, Potsdam, NY 13699-5725, USA

Table 1 Elastic constants obtained experimentally (using ultrasonic or resonance frequency methods)

References	c_{11} (GPa)	c_{12} (GPa)	c_{24} (GPa)	Temperature (K)
[1]	53.51	36.81	19.94	298
[2]	61.5	43.0	19.6	77
[3]	53.3	36.5	20.44	300
	55.7	38.4	20.95	77
[4]	53.8	37.4	20.18	298
	56.2	38.3	20.61	77

$$C_{44} = \frac{1}{S_{44}}, \tag{4}$$

where S_{ij} are the components of the stiffness matrix inverse (i.e., compliance matrix) and l_{i1}, l_{j2}, l_{k3} are direction cosines between the [*ijk*] direction and the crystallographic axes 1, 2, and 3, respectively (i.e., [100], [010], and [001]).

Tables 2 and 3 list experimental parameters of quasistatic mechanical tests reported for single crystals of CdTe and related compounds. Most of these tests were performed in compression to finite strains. It should be noted that these tests were apparently performed without direct measurement of, or control of, local specimen displacements. This is evidenced by large discrepancies in elastic properties, both among some of the results inferred from the data corresponding to Tables 2 and 3 and, particularly, between the quasistatic test results and those listed in Table 1 at the same temperatures. Equations (1)–(4) and average values from Table 1 were used to compute Young's modulus for the crystallographic orientations reported in Table 3 and it was also computed from the initial linear portion of reported stress-strain curves. Those derived from ultrasonic or resonance frequency techniques and those computed from the stress-strain curves differ by an order of magnitude or more, in some cases. A logical conclusion is that the quasistatic test results are influenced by testing machine compliance effects, compression specimen surface roughness, nonparallel specimen-platen contact surfaces, and so on.

Support for the aforementioned conclusion is provided by computational results as well as microindentation test results. Recently reported molecular dynamics computations using empirical interatomic potentials [12], band structure calculations [13] and *ab initio* calculations [14, 15] are in general agreement with the experimental results obtained using ultrasonic and resonance frequency techniques. Likewise, microindention tests results reported in [16] yield Young's modulus values in the neighborhood of 45 GPa, while those in Table 1 yield Young's moduli for

Table 2 Mechanical tests and experimental parameters reported in the literature

References	Temperature range (K)	Nominal strain rate range (s^{-1})	Alloy tested $Cd_{(1-x)}El_xTe$	Type of test
[17]	298–573	2×10^{-5}–2×10^{-4}	$x = 0$ (n and p types)	Compression
[18]	298	–[a]	$x = 0$ (n and p types)	Compression
[19]	170–300	10^{-4}	$x = 0$ (n types)	Compression
[20]	298–773	2×10^{-4}	$x = 0$	Compression
[21]	77–500	10^{-5}–10^{-4}	$x = 0$ (n and p types)	Compression
[22, 23]	77–500	10^{-5}	$x = 0$ (n and p types)	Compression
[24]	303–363	10^{-5}	El = Hg, $0.70 \leq x \leq 0.82$	Three-point bending
[25]	262–468	10^{-4}	El = Hg; $x = 1$, $0.70 \leq x \leq 0.82$	Four-point bending
[26]	298	1.5×10^{-5}	El = Zn, Mn; $x = 0$, 0.04(Zn), 0.10(Mn)	Compression
[27]	77–300	4×10^{-5}	El = Zn, $x = 0$, 0.025	Compression
[28]	473–873	2×10^{-4}	El = Zn, $x = 0$, 0.045	Compression
[29]	195–1100	3×10^{-5}	EL = Zn, $x = 0$, 0.04	Compression
[30]	738–1137	–[b]	EL = Zn, $x = 0$, 0.04	Tension
[31]	300–1353	10^{-4}	$x = 0$	Compression

[a] Crosshead speed: 0.83×10^{-4} cm/s.
[b] Crosshead speed: 0.85×10^{-3} cm/s.

various orientations ranging from 23 to 52 GPa. One can reasonably conclude then that the elastic constants listed in Table 1 are an accurate representation of CdTe elasticity properties. Elastic properties (Young's modulus) from stress-strain curves obtained or inferred from the quasi-static tests reported in Tables 2 and 3 will not be accurate.

2. INELASTIC BEHAVIOR

It is common to ascribe the transition from elastic to inelastic deformation in single crystals to a single quantity referred to as the critical-resolved shear stress (CRSS). This quantity denotes the shear stress on the slip system at which inelastic deformation will initiate, a slip system being comprised a slip plane and a slip direction. The shear stress on the slip system (resolved shear stress) for any state of stress is the component of the traction on the slip plane that is acting in the slip direction. For a uniaxial state of stress (a single tensile or compressive normal stress acting along some axis), the resolved shear stress is obtained as the product of the axial stress and the Schmid factor, the latter being the

Table 3 Mechanical tests and experimental parameters reported in the literature

References	Initial dislocation density (cm^{-2})	Light conditions	Test axis orientation
[17]	10^4–10^5	Light, dark	[110]
[18]	–[a]	Light, dark	–[a]
[19]	–[a]	Dark	$[\bar{1}23]$
[20]	–[a]	–[a]	Approx $[\bar{2}35]$
[21]	–[a]	–[a]	[111]
[22, 23]	–[a]	Light, dark	Polycrystal, large grain size
[24]	10^6–10^7	Dark	Multiple orientations tested
[25]	2×10^6	–[a]	[123] beam span axis
[26]	5×10^5 ($x = 0$, El = Mn) 5×10^4 (El = Zn)	–[a]	[132]
[27]	3×10^5	Dark	[100], [110], [111] ($x = 0$)
[28]	3×10^5 10^5 ($x = 0.045$)	–[a]	$[\bar{1}23]$
[29]	10^5-10^6 ($x = 0$) 10^4-10^5 (El = Zn)	Dark up to 600 K	[132]
[30]	–[a]	–[a]	Multiple orientations tested
[31]	10^4-10^5	–[a]	[132]

[a]Not reported.

product of the direction cosine between the stress axis and the slip plane normal and the direction cosine between the stress axis and the slip direction. With the sphalerite/zinc blende structure, CdTe and its alloys possess 12 close-packed slip plane-slip direction pairs (the $\{111\}\langle 1\bar{1}0\rangle$ slip systems as in fcc crystals). All available indications are that the mechanism of inelastic deformation in CdTe and its alloys is dislocation glide on these slip systems [20, 32]. When a uniaxial test is performed with a stress axis oriented such that the stereographic projection of that axis lies inside the standard stereographic triangle, there is a single slip system with the highest Schmid factor. Slip typically occurs only on that system, at least up to some level of axial strain (i.e., within the stage I work hardening regime, e.g., [33]), and such orientations are referred to as single slip orientations. For axial stress axes aligned with $\langle 100\rangle$, $\langle 110\rangle$, or $\langle 111\rangle$ directions, or any direction with a stereographic projection lying on the boundary of the standard stereographic triangle, there are two or more slip systems with the highest Schmid factor. Such orientations are often referred to as multiple slip orientations.

Figure 1 shows what will be termed CRSS τ_c (using either back extrapolation of the stress versus strain responses reported to zero plastic shear

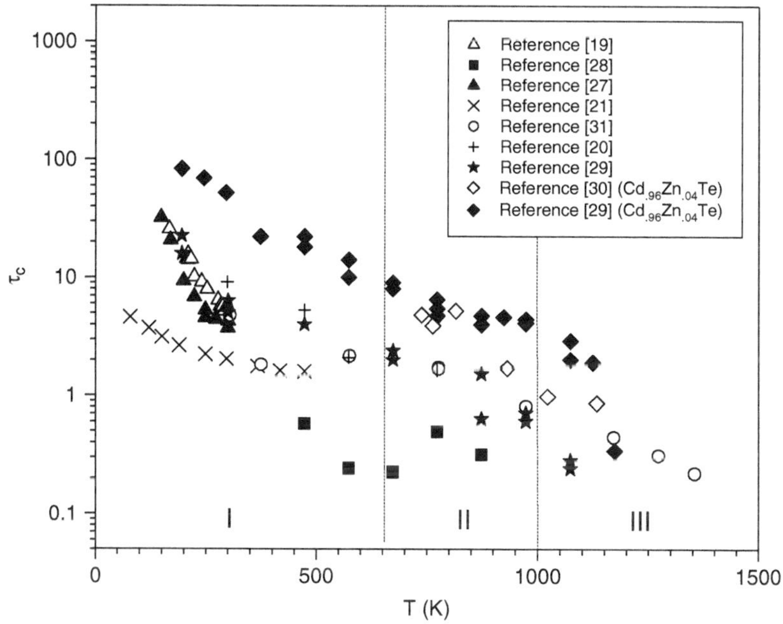

Figure 1 Critical-resolved shear stress versus temperature from results reported in the literature. Vertical-dashed lines delineate transition temperatures between regimes I, II, and III as discussed in Ref. [29].

strain or as the lower yield stress for tests which showed a distinct yield drop) versus temperature T for the single crystal CdTe and $Cd_{0.96}Zn_{0.04}Te$ references in Tables 2 and 3 which reported such results. The scatter among the different sets of experimental results is apparent.

Nonlinear dependence of τ_c on plastic strain rate (or dislocation velocity) can account for some of this scatter and, since some of the tests were performed on crystals oriented for multiple slip, some may be accounted for by virtue of the fact that the deformation was accommodated by slip on more than one slip system. Additionally, initial dislocation density variation among the materials tested can contribute to the scatter.

To examine some of these factors, we may employ Haasen's model [34] which incorporates both the temperature and dislocation density dependence (as well as strain rate dependence) of the flow stress, viz.

$$v = B(T)\tau_{\text{eff}}^{1/m}, \quad \tau_{\text{eff}} = \tau - A\sqrt{N}, \quad A = Gb/2\pi(1-\mu), \tag{5}$$

where v is the average dislocation velocity, N is the dislocation density, τ is the resolved shear stress on an active slip system, $B(T)$ is a temperature-dependent function, G is the shear modulus, b is the Burger's vector magnitude, and μ is Poisson's ratio. Combining this with Orowan's equation, that is, $\dot{\gamma}_{\mathrm{p}} = bNv$, we arrive at the following expression:

$$\dot{\gamma}_{\mathrm{p}} = bNB(T)\tau_{\mathrm{eff}}^{1/m}, \tag{6}$$

where $\dot{\gamma}_{\mathrm{p}}$ is the plastic strain rate. The preceding tacitly assumes single slip conditions and that the mobile and total dislocation densities are effectively equal. If we accept the latter assumption, but note that for symmetric orientations (i.e., uniaxial compression/tension axes aligned with $\langle 100 \rangle$, $\langle 110 \rangle$, or $\langle 111 \rangle$ directions, or any direction with a stereographic projection lying on the boundary of the standard stereographic triangle) there are n slip systems for which the magnitude of the resolved shear stress is equal, then the axial plastic strain rate is $\dot{\varepsilon}_{\mathrm{p}} = nf_{\mathrm{s}}\dot{\gamma}_{\mathrm{p}}$ so that

$$\dot{\varepsilon}_{\mathrm{p}} = nf_{\mathrm{s}}bNB(T)\tau_{\mathrm{eff}}^{1/m}, \tag{7}$$

where f_{S} is the magnitude of the Schmid factor. Equation (7) then assumes that for $n > 1$, each nominally equal stressed slip system contributes equally to the deformation; for $\langle 110 \rangle$, $\langle 100 \rangle$, and $\langle 111 \rangle$, $n = 4$, 6, and 8, respectively. Among the results shown in Fig. 1, those reported in Refs. [21, 27, 30] tested multiple slip orientations (in the latter case, as deduced from the orientation angles reported). Normalization for the dislocation density is somewhat problematic, since it was not reported for all of the tests in Fig. 1. However, the quantity $A\sqrt{N}$, representing the athermal component of the flow stress (or internal stress due to an array of dislocations) is generally somewhat small. Using the shear modulus correlation suggested in Ref. [29], Burgers vector magnitude $b = a/2$ with $a = 6.48$ Å [7] and Poisson's ratio $\nu = 0.3$, it is found that $0.032\ \mathrm{MPa} \leq A\sqrt{N} \leq 0.36\ \mathrm{MPa}$ for the temperature range $300\ \mathrm{K} \leq T \leq 1323\ \mathrm{K}$ and for a range of dislocation densities, $10^5\ \mathrm{cm}^{-2} \leq N \leq 10^7\ \mathrm{cm}^{-2}$. Thus, we can attempt to normalize for strain rate and slip system multiplicity by neglecting the athermal component of stress ($A\sqrt{N} \cong 0$) and noting then that

$$\frac{\tau^{1/m} nf_{\mathrm{s}}}{\dot{\varepsilon}_{\mathrm{p}}} = \frac{1}{bNB(T)}. \tag{8}$$

In Fig. 2 is plotted $\tau^{1/m} nf_{\mathrm{s}}/\dot{\varepsilon}$, where $\dot{\varepsilon}$ is the total nominal axial strain rate estimated for a particular data set as listed in Table 2 (i.e., the right-hand side of Eq. (8), except using total axial strain rate as an estimate for plastic axial strain rate) versus temperature using $1/m = 3$. This ($1/m = 3$) is somewhat representative of the strain rate sensitivity over the entire temperature range [8, 35], though there is likely no strain rate sensitivity (or even possibly negative strain rate sensitivity) in the temperature range $550\ \mathrm{K} \leq T \leq 800\ \mathrm{K}$ where strain aging phenomena are exhibited [29]. Figure 2 demonstrates that the results reported in Refs. [21, 27] exhibit

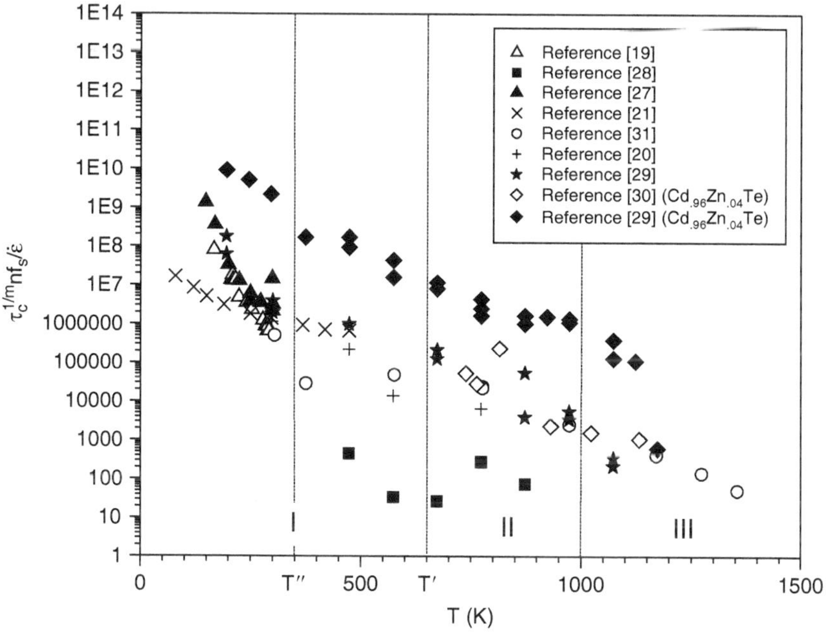

Figure 2 Normalization of reported critical-resolved shear stress data of Fig. 1 for strain rate and slip system multiplicity (see Eq. (8)).

improved congruence with the other data for CdTe in the lower temperature range ($T \leq 500$ K), when the data are normalized for strain rate and slip multiplicity. However, the results of Parfeniuk et al. [30] for $Cd_{0.96}Zn_{0.04}Te$ show poorer congruence with those of [29], but it is incontrovertible that the substitution of 2% Zn for Cd elevates τ_c considerably. Imhoff et al. [29] attributed this hardening effect of the Zn substitution to a size effect, primarily. The results in Ref. [28] remain as clear outliers.

In addition to strain rate and slip system multiplicity, there are several other possibilities for the scatter in τ_c versus T exhibited in Fig. 1. For example, CdTe, and other II-VI and III-V compound semiconductors exhibit flow stress sensitivity to light at low to intermediate temperatures [17, 22]. In many instances, the conditions under which the tests were performed (i.e., illuminated or dark) are not reported. Some of the scatter in the lower temperature range might be attributable to this photoplastic effect (PPE). Another source of uncertainty is stoichiometry. Buch and Ahlquist [18] reported a near 100% variation in resolved shear yield stress from specimens which are strongly p-type (Cd deficient) to those which are strongly n-type (Cd rich), with a minimum occurring for intrinsic CdTe. Other sources of uncertainty are, of course, initial dislocation density, initial dislocation substructure and density distribution within a specimen, specimen size effects as arise due to inhomogeneous deformation [34] impurity concentrations, sample preparation, test temperature control, and testing machine stiffness variation.

The temperature variation of τ_c has been addressed comprehensively for both CdTe and $Cd_{0.96}Zn_{0.04}Te$ by Imhoff et al. [29]. They determined activation parameters for both materials using three independent methods and noted the three regimes demarcated on Figs. 1 and 2 with the transition temperatures at 650 and 1000 K. They noted that for $T \leq 650$ K (regime I) and $T \geq 1000$ K (regime III) plastic deformation was highly thermally activated, these two regimes being separated by a plateau regime II similar to behavior observed in other semiconductor materials as well as fcc metals. Using activation energy variation with temperature, they identified two "athermal temperatures" at $T'' = 350$ K and $T' = 650$ K corresponding also to slope variations in τ_c versus T. They surmised that the former corresponded to a transition in dominant deformation mechanism from one controlled by screw dislocations at low temperatures to one controlled by edge dislocations, supported by the transmission electron microscopy (TEM) observations of Hall and Vander Sande [20]; they asserted that the transition at $T' = 650$ K represented the conventional athermal temperature.

Imhoff et al. [29] also reported serrated stress-strain behavior and performed static strain aging tests on $Cd_{0.96}Zn_{0.04}Te$ single crystals. The results of these strain aging experiments supported the hypothesis that the hardening effect of the Zn substitution was primarily a size effect and confirmed the Portevin Le Chatelier effect between 770 and 920 K. Balasubramanian [36] also reported serrated stress-strain curves for CdTe tested at intermediate temperatures (373 K $\leq T \leq$ 773 K); such serrated stress-strain curves are often associated with strain aging phenomena and inverse (or negative) strain rate sensitivity (e.g., [37]).

Observation of yield drop phenomena at the onset of inelastic deformation in quasistatic tests, as are typically associated with rapidly increasing mobile dislocation densities in the small strain regime, sometimes are reported and sometimes are not. It is not clear whether this is due to machine stiffness variations among experiments or due to differences in initial dislocation density, the latter being a key technological variable. Machine stiffness will influence initial yield point behavior as will initial mobile dislocation density, as both affect inelastic strain rate in the initial yielding regime for a displacement-controlled test. Yield drops are also associated with the Portevin Le Chatellier effect at higher temperatures [29].

Analyses indicate that for CdTe the rate limiting step for inelastic deformation in the low temperature-to-room temperature regime is the formation and/or motion of double kinks [20, 21, 23, 27]. Activation energies and activation volumes determined for the low temperature regime are consistent with such a mechanism. Using *in situ* TEM observations at room temperature as well as cathodoluminescence using scanning electron microscopy (SEM-CL) and surface etch pits, Nakagawa et al. [38] concluded that moving dislocations are likely to be in the extended state (separated into partial dislocations bounded a stacking fault) as they were observed to be when at rest by Hall and Vander Sande [20]. To reconcile TEM observations with SEM-CL and etch pit patterns, they surmised that rate of multiplication of mobile dislocations and their mean free path control the yield and flow stress, primarily. This would indicate an expression $\dot{\gamma}_p = b\dot{N}\bar{l}$, where $\bar{l}$ is the mean free path length, in contrast to the traditional use of Orowan's equation in rate form, that is, $\dot{\gamma}_p = bNv$. They observed frequent cross-slips, indicating that the mean free path is determined by jog dragging, the density of jogs increasing by cross-slips as dislocations advance. At intermediate and high temperatures other mechanisms seem to become important [29, 39–41]. In the intermediate temperature range, relatively athermal, long-range internal stresses would appear to dominate and at the higher temperatures, diffusion-controlled recovery mechanisms such as dislocation climb will play a significant role and cross-slip may also be important. Such an interpretation is supported by the work of Fissel et al. [39–41] who observed time-varying hardness on various compounds and documented deformation characteristics around the indents. The energetic analysis in [29] indicates a transition in rate controlling mechanisms occurring between 600 and 800 K. Stevens et al. [35] performed constant load tensile creep experiments on CdZnTe (3.5 and 4.5 at % Zn) at high homologous temperatures ($T/T_m = 0.79$ and 0.86) and found that the creep exponent and activation energy was consistent with diffusion assisted dislocation creep (combination of climb and glide). Finally, it is noted that twinning has generally not been observed as a mechanical deformation mechanism [32, 42, 43] though microtwinning has been observed around indents at low temperature in CdTe [44] and (Cd, Zn)(Te, Se) solid solutions [41].

3. FRACTURE PROPERTIES

The fracture resistance of CdTe has been quantified experimentally and using theoretical approaches. Using notched, three-point bending, Wermke and Petzold [45] estimated the room temperature fracture toughness for cleavage of CdTe single crystals along {111} planes, obtaining $K_{IC} = 0.158 \pm 0.016$ MPa m$^{1/2}$. They also estimated the fracture toughness for cleavage along {110} planes using an indentation fracture technique, obtaining $K_{IC} = 0.177 \pm 0.016$ MPa m$^{1/2}$. Based on the assumption of isotropic elastic behavior, with estimates of Young's modulus and Poisson's ration of $E = 40{,}000$ MPa and $\nu = 0.3$, respectively, these can be converted to cleavage energies using the plane strain relation $w_f = K_I^2/2E$ (e.g., [46]). This results in $w_f = 0.312\,\mathrm{J/m^2}$ and $w_f = 0.774\,\mathrm{J/m^2}$ for the three-point bending fracture along {111} and the indentation fracture along {110}, respectively. Berding et al. [47] computed cleavage energies for {111} and {110} cleavage using a Green's function theory, employing second-neighbor tight-binding Hamiltonian's. They obtained fracture energies of $w_{f,\{111\}} = 0.580\,\mathrm{J/m^2}$ and $w_{f,\{110\}} = 180\,\mathrm{J/m^2}$. They note that neglect of surface relaxation in their calculations would tend to make these computed values higher than the actual values, but also irreversible contributions (e.g., dislocation inelasticity) might tend to make experimental values higher. Both the experimentally obtained and the calculated values are lower than those obtained for GaAs and Si, as would be expected based on bond strength, length, and density arguments.

4. OPTOELECTRONIC-MECHANICAL COUPLINGS

4.1. Photoplastic effect

As noted earlier, CdTe exhibits a flow stress sensitivity to light at lower temperatures, as do other II-VI and III-V compound semiconductors. Gutmanas et al. [22] observed both positive (increase in flow stress with illumination compared to deformation in darkness) and negative PPEs in both n- and p-type CdTe samples, depending on the plastic strain, illumination power intensity and heat treatment. They proposed two different models for the PPE. Only a positive PPE for both n- and p-type samples was observed in [17, 48]. The authors noted a minimum in τ_c for samples annealed in a Cd vapor pressure corresponding to that which would yield minimum carrier concentration (near stoichiometric composition). There also appeared to be a minimum in the observed PPE at that composition. They proposed a model based on light-induced reduction of mobile dislocation density due to pinning of charged dislocations by oppositely charged native defects. Buch and Ahlquist [18] also studied the effect of plastic strain on electrical conductivity and observed that conductivity decreased with plastic strain in p-type material rather significantly, while n-type material showed a modest increase in conductivity with plastic strain. Both cases correspond to a Fermi energy shift toward the conduction band. They produced an energy band diagram for the native point-defect and dislocation states based on the observed optical, electrical, and mechanical behaviors. Based on their *in situ* TEM observations, Nakagawa et al. [38, 49] ascribed the PPE to a decrease in the mean free path of multiplied dislocations. Enhancement of the immobilization rate by jog formation of glide dislocations, by their interaction with charged point defects that trap minority carriers generated by dislocation jogs formed by cross-slip, was asserted to be the mechanism for the decrease in mean free path.

4.2. Piezoelectric constant and stress/strain-dependence of band characteristics

The piezoelectric constant referred to the cubic crystallographic axes for CdTe was measured directly at 77 K by Berlincourt et al. [2] and found to be $d_{14} = 1.68\mathrm{C/N}$. They extrapolate this value to $d_{14} = 1.54\mathrm{C/N}$ at 298 K using similar measurements on CdS. Dependence of band energy characteristics on stress has been characterized experimentally by [50–52]; it was also characterized theoretically/computationally as reported in Refs. [15, 53–55]. Dunstan et al. [52] found that the hydrostatic pressure dependence of the direct bandgap energy $\mathrm{d}E_g/\mathrm{d}P = 65$ meV/GPa, considerably lower than the value of 110 meV/GPa found by Thomas [50], but in better agreement with that found by Babonas et al. [51] and estimated in Refs. [53–55]. Deligoz et al.'s [15] recent computational result yielded $\mathrm{d}E_g/\mathrm{d}P =$ 84 meV/GPa, closer to the value obtained in [51, 53]. As a parallel result, Deligoz et al. [15] also computed the pressure dependence of the elastic constants. In addition to shifting of the band energy due to hydrostatic pressure, the dependence of valence band splitting on uniaxial stress (or, more precisely, principal stress difference) was also characterized in Refs. [50, 52]. The latter concluded that, using an isotropic characterization, the valence band splitting was approximately $\Delta_{001} = \Delta_{111} = 87$ meV/GPa, lower than the 139 meV/GPa found by Thomas [50].

The valence band splitting gives rise to birefringence induced by stress in CdTe. This was investigated as a function of wavelength by Wardzyńoski [56]. In this study, it was found that the birefringence changes sign near the absorption edge, with the position of this inversion point depending on both the stress and light directions relative to the crystallographic axes. The components of the piezo-optic tensor were estimated for wavelengths from 20,000 Å to the absorption edge. Kloess et al. [57] and Laasch et al. [58] used this effect with infrared polarizing microscopy to map long-range internal stress fields due to defects and other sources arising from crystal growth processes in CdTe.

5. SUMMARY

Elastic stiffness tensor components for CdTe single crystals measured between 77 and 300 K using ultrasonic or resonance frequency methods are fairly consistent and these are consistent with values obtained using computational means and from microindentation measurements. Young's modulus computed from stress-strain curves taken from or inferred from data reported for quasistatic compression or tension tests reported in the literature are not consistent.

CRSS versus temperature for CdTe single crystals shows considerable scatter among results reported in the literature. Some of this scatter can be reduced by accounting for strain rate and multiple slip. Alloying of CdTe with Zn elevates the CRSS considerably. Apparent transitions in rate controlling mechanisms occur at temperatures of approximately 350, 650, and 1000 K, though dislocation slip is the inelastic deformation mechanism throughout. Twinning does not appear to be a significant factor in mechanical deformation.

Fracture toughness of CdTe single crystals is lower than for Si and GaAs, with measured and predicted fracture energies in the range of 0.15–0.78 J/m^2.

Optoelectronic-mechanical couplings include: dependence of flow stress on illumination intensity (photoplastic effect); direct bandgap energy dependence on hydrostatic pressure; valence band splitting dependence on principal stress difference. The latter gives rise to stress-dependent birefringence.

REFERENCES

[1] H.J. McSkimin, D.G. Thomas, J. Appl. Phys. 33 (1962) 56–59.
[2] D. Berlincourt, H. Jaffe, L.R. Shiozawa, Phys. Rev. 129 (1963) 1009–1017.
[3] Y.K. Vekilov, A.P. Rusakov, Sov. Phys. Solid State 13 (1971) 956–960.
[4] R.D. Greenough, S.B. Palmer, J. Phys. D Appl. Phys. 6 (1973) 587–592.
[5] R.F.S. Hearmon, An Introduction to Applied Anisotropic Elasticity, Oxford University Press, London, 1961.
[6] P. Maheswaranathan, R.J. Sladek, U. Debska, Phys. Rev. B 31 (1985) 5212–5216.
[7] K. Zanio, Semiconductors and Semimetals: Volume 13, Cadmium Telluride, Academic Press, New York, 1978.
[8] J.C. Moosbrugger, A. Levy, Met. Mater. Trans. A 26 (1995) 2687–2697.
[9] J.C. Moosbrugger, Int. J. Plasticity 11 (1995) 799–826.
[10] R.W. Hertzberg, Deformation and Fracture Mechanics of Engineering Materials, second ed., John Wiley and Sons, New York, 1983.
[11] T.H. Courtney, Mechanical Behavior of Materials, second ed., McGraw-Hill, New York, 2000.
[12] M.B. Kanoun, W. Sekkal, A. Aourag, G. Merad, Phys. Lett. A 272 (2000) 113–118.
[13] M. Kitamura, S. Muramutsu, W. Harrison, Phys. Rev. B 46 (1992) 1351–1357.
[14] B.K. Agrawal, S. Agrawal, Phys. Rev. B 45 (1992) 8321–8327.
[15] E. Deligoz, K. Colakglu, Y. Ciftci, Physica B 373 (2006) 124–130.
[16] I.V. Kurilo, V.P. Alekhin, I.O. Rudyi, S.I. Bulychev, L.I. Osypyshin, Phys. State Sol. (a) 163 (1997) 47–58.
[17] L. Carlsson, C.N. Ahlquist, J. Appl. Phys. 43 (1972) 2529–2536.
[18] F. Buch, C.N. Ahlquist, J. Appl. Phys. 45 (1974) 1756–1761.
[19] K. Maeda, K. Nakagawa, S. Takeuchi, Phys. State Solidi (a) 48 (1978) 587–591.
[20] E.L. Hall, J.B. Vander Sande, J. Am. Ceram Soc. 61 (1978) 417–425.
[21] E.Y. Gutmanas, N. Travitzky, U. Plitt, P. Haasen, Scripta Metall. 13 (1979) 293–297.

[22] E.Y. Gutmanas, N. Travitzky, P. Haasen, Phys. State Sol. (a) 51 (1979) 435–444.
[23] E.Y. Gutmanas, P. Haasen, Phys. State Sol. (a) 63 (1981) 193–202.
[24] S. Cole, J. Mater. Sci. 15 (1980) 2591–2596.
[25] S. Cole, A.F.W. Willoughby, M. Brown, J. Mater. Sci. 20 (1985) 274–288.
[26] K. Guergouri, R. Triboulet, A. Tromson-Carli, Y. Marfaing, J. Cryst. Growth 86 (1988) 61–65.
[27] S.V. Lubenets, L.S. Fomenko, Sov. Phys. Solid State 31 (1989) 256–260.
[28] R.S. Rai, S. Mahajan, D.J. Michel, H.H. Smith, S. McDevitt, C.J. Johnson, Mater. Sci. Eng. B10 (1991) 219–225.
[29] D. Imhoff, A. Zozime, R. Triboulet, J. Phys. III France 1 (1991) 1841–1853.
[30] C. Parfeniuk, F. Weinberg, I.V. Samarasekera, C. Schvezov, L. Li, J. Cryst. Growth 119 (1992) 261–270.
[31] R. Balasubramanian, W.R. Wilcox, Mater. Sci. Eng. B 16 (1993) 1–7.
[32] J. Shen, R. Balasubramanian, D.K. Aidun, L.L. Regel, W.R. Wilcox, J. Mater. Eng. Perf. 7 (1998) 555–563.
[33] S.J. Basinski, Z.S. Basinski, Dislocations in metallurgy—Chapter 16, in: F.R.N. Nabarro (Ed.), Dislocations in Solids, vol. 4, North-Holland, New York, 1979, 263 pp.
[34] H. Alexander, P. Haasen, in: F. Seitz, D. Turnbull, H. Ehrenreich (Eds.), Advances in Solid State Physics, vol. 22, Academic Press, New York, 1968, 27 pp.
[35] T.E. Stevens, J.C. Moosbrugger, F.M. Carlson, J. Mater. Res. 14 (1999) 3864–3869.
[36] R. Balasubramanian, Ph.D. Thesis, Mechanical Behavior of CdTe and Real-Time, In-Situ Observation of Stress-Induced Dislocation Motion in Single Crystals, Clarkson University, Potsdam, NY, USA, 1992.
[37] R.B. Schwarz, Scripta Metall. 16 (1982) 385–390.
[38] K. Nakagawa, M. Maeda, S. Takeuchi, J. Phys. Soc. Jpn. 49 (1980) 1909–1915.
[39] A. Fissel, M. Schenk, A. Engel, Cryst. Res. Technol. 24 (1989) 557–565.
[40] A. Fissel, M. Schenk, Cryst. Res. Technol. 25 (1990) 89–95.
[41] A. Fissel, M. Schenk, J. Mater. Sci.: Mater. Electr. 3 (1992) 147–156.
[42] A.W. Vere, S. Cole, D.J. Williams, J. Electron. Mater. 12 (1983) 551–561.
[43] A. Orlova, B. Sieber, Acta Metall. 32 (1984) 1045–1052.
[44] W. Winkler, M. Schenk, I. Hähnert, Cryst. Res. Technol. 27 (1992) 1047–1051.
[45] B. Wermke, M. Petzold, Cryst. Res. Technol. 25 (1990) K121–K124.
[46] T.L. Anderson, Fracture Mechanics: Fundamentals and Applications, CRC Press, Boca Raton, FL, 1991.
[47] M.A. Berding, Krishnamurthy, A. Sher, A.B. Chen, J. Appl. Phys. 67 (1990) 6175–6178.
[48] L. Carlsson, J. Appl. Phys. 42 (1971) 676–680.
[49] K. Nakagawa, M. Maeda, S. Takeuchi, J. Phys. Soc. Jap. 50 (1981) 3040–3046.
[50] D.G. Thomas, J. Appl. Phys. 32 (1961) 2298–2304.
[51] G.A. Babonas, R.A. Bendoryas, A.Y. Shileika, Sov. Phys. Semicond. 5 (1971) 392–397.
[52] D.J. Dunstan, B. Gill, K.P. Homewod, Phys. Rev. B 38 (1988) 7862–7865.
[53] D.L. Camphausen, G.A.N. Connel, W. Paul, Phys. Rev. Lett. 26 (1971) 184–188.
[54] Y.F. Tsay, S.S. Mitra, B. Bendow, Phys. Rev. B 10 (1974) 1476–1481.
[55] M. Cardona, N.E. Christensen, Phys. Rev. B 35 (1987) 6182–6194.
[56] W. Wardzyńoski, J. Phys. C: Sol. St. Phys. 3 (1970) 1251–1263.
[57] G. Kloess, M. Laasch, R. Schwarz, K.W. Benz, J. Cryst. Growth 146 (1995) 130–135.
[58] M. Laasch, G. Kloess, Th. Kunz, R. Schwarz, K. Grasza, C. Eiche, K.W. Benz, J. Cryst. Growth 161 (1996) 34–39.

CHAPTER III

CdTe-Based Nanostructures

Henri Mariette

Contents

The development of molecular beam epitaxy (MBE) in the 1970s for arsenic III-V semiconductors has allowed to produce high-quality epitaxial layers with very abrupt interfaces, good control of thickness, doping, and composition. All these achievements are exactly what are required to perform semiconductor heterostructures to demonstrate and develop new concepts of sophisticated electronic and optoelectronic devices. These bottom-up approaches started in the 1980s for low-dimensional structures based on CdTe: due to the strong exciton binding energy in II-VI as compared to arsenic III-V semiconductors, it was possible, with these CdTe nanostructures, to reveal various new optical effects which are

Equipe CEA – CNRS "Nanophysique et Semiconducteurs," Institut Néel/CNRS and CEA/INAC/SP2M, 25 avenue des martyrs, 38054 Grenoble, France

CDTE and Related Compounds
DOI: 10.1016/B978-0-08-046409-1.00003-4

briefly presented here. In the first part of this chapter, we will illustrate the up-to-date know-how in the MBE growth of CdTe-based heterostructures. Especially, the ultimate control of the structure dimensionality will be presented by using various approaches. In the second part, we will show few quantum effects which were discovered in CdTe-based heterostructures due to the specificity of II-VI compound semiconductors as compared to the III-V one. Moreover the possibility to dope CdTe with magnetic atoms (such as manganese), makes this system fascinating for basic studies in spintronics: especially the magnetic quantum dots open new ways for the coherent control of a single spin.

1. GROWTH

1.1. Atomic layer epitaxy: The ultimate control of II-VI nanostructures growth

In the last two decades, the II-VI heterostructures based on CdTe have been the subject of intensive work. Growth process such as molecular beam epitaxy was developed for the II-VI heterostructures in the late 1980s: first started with high-quality thick CdHgTe epilayers grown on either GaAs [1] or CdZnTe [2] substrates, the layer-by-layer growth process required to achieve heterostructures was demonstrated with the observation of RHEED oscillations during the growth of CdTe homo-[3] and hetero-[4] epitaxy (Fig. 1). Indeed the periodic variation of the reflection high-energy electron diffraction (RHEED) intensity corresponds to the surface step distribution [5] and provides a precise determination of monolayer incorporation in the case of two-dimensional growth procedure. A protocol involving an excess of Cd (Fig. 1) and growth interruptions at the well-barrier interfaces has been found necessary to observe a 2D growth during the epitaxy of CdTe/CdZnTe quantum wells and superlattices [4]. This method gives accurate *in situ* thickness measurements of all the layers during the super lattice growth, in agreement with *ex situ* X-ray diffraction data. The sharpness of X-ray diffraction satellites confirms the high crystalline quality with a period fluctuation of less than one monolayer. Later on, by extending this know-how to both, ultrathin layers with subnanometer length scale [6], and heterostructures containing either Mn or Mg [7, 8], an ultimate growth control technique was developed, namely the atomic layer epitaxy (ALE).

The ALE consists in sending *alternatively* the cations (Cd, Zn, Mg, or Mn), and the anions (Te), coming from elemental sources onto the surface, leaving dead times (denoting in the following as growth interruptions or GI) between each vapor pulse in order to stabilize the surface and to reevaporate the possible excess material. The sequence of these stages

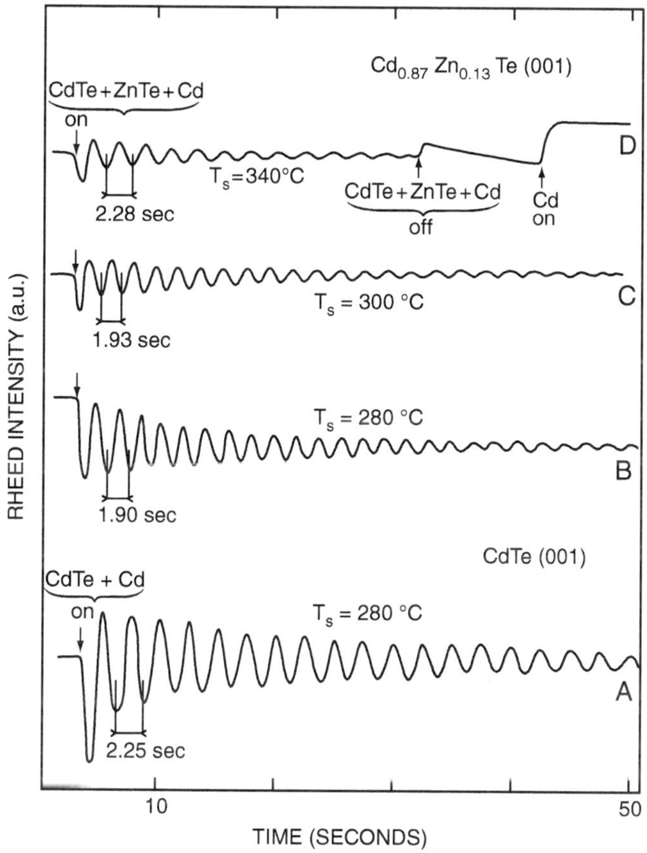

Figure 1 RHEED specular beam intensity oscillations during growth of (A) CdTe (001) at substrate temperature $T_s = 280\ °C$, of $Cd_{0.87}Zn_{0.13}Te$ (001) at (B) $T_s = 280\ °C$, (C) $T_s = 300\ °C$, and (D) $T_s = 340\ °C$. From Lentz et al. [4].

(exposure to cations, GI, exposure to Te, CI) corresponds to one ALE cycle. Compared to conventional MBE, in which the two constituent elements are brought simultaneously onto the surface, the mobility of adsorbed species on the growing surface is expected to be greatly enhanced, as it was demonstrated for III-V semiconductor compounds such as GaAs [9].

By contrast to standard III-V materials, a CdTe surface can be stabilized under a flux of both its cation or anion without any degradation. For typical substrate temperatures T_s ($\approx 280\ °C$), the stable surface reconstruction obtained is (2×1) for the Te stabilized surface, with a full Te surface coverage. For the Cd stabilized surface, a mixture of $c(2 \times 2)$ and (2×1) reconstructions is expected, with a half surface coverage [10].

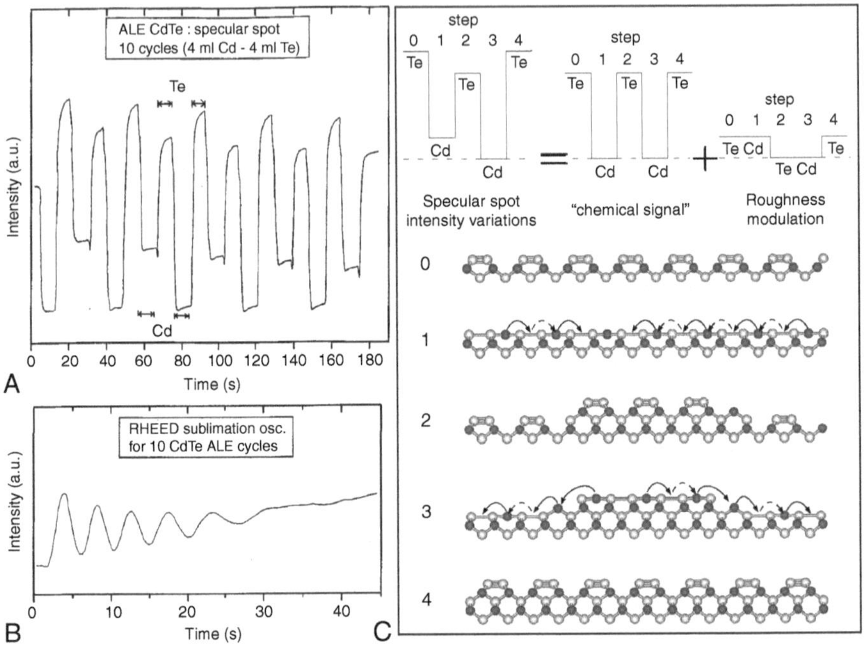

Figure 2 (A) RHEED specular spot intensity variations versus time during a 10 cycle CdTe ALE growth at a substrate temperature of 280 °C. ALE cycle: 8 s Cd and Te pulses; 1 s dead time; Cd and Te fluxes: 0.5 ML/s. (B) CdTe RHEED sublimation oscillations corresponding to the evaporation of the ALE grown CdTe layer shown in Fig. 1A. Direct evidence of the autoregulated growth rate at 0.5 ML/cycle. 10 ALE cycles, five oscillations ($T_{sublimation} = 400$ °C). (C) Atomic model explaining the autoregulated growth rate of 0.5 ML/cycle (white balls: Te atoms, black balls: Cd atoms). From Hartmann et al. [7].

Figure 2A shows the RHEED specular spot intensity variations observed during the CdTe ALE growth [7]. The strong periodic intensity variations with a period equal to that of the ALE cycle correspond to the "chemical signal," that is to the presence of, respectively, Cd and Te atoms and to their position on the surface. Superimposed on this, a bi-cycle periodicity of the envelope signal is clearly observed: it has been attributed to a periodic roughness variation which corresponds to a deposition of half a monolayer of CdTe per ALE cycle. This has been evidenced by observing the RHEED oscillations during CdTe sublimation: for the 10 CdTe ALE cycles shown in Fig. 2, one observes five sublimation oscillations (Fig. 2B), directly demonstrating an ALE self-regulating growth rate at 0.5 ML/cycle [7].

A simple atomic model shown in Fig. 2C, which relies on the mobility of Cd atoms chemisorbed on the surface, can account for this self-regulating regime and for the periodic roughness modulation. Starting from a flat (2 × 1) Te-rich surface [step (0)], a $c(2 \times 2) + (2 \times 1)$ Cd-rich half-filled surface is formed upon exposure to the Cd flux [(step (1)]. At this point, the roughness of the surface is still minimum as revealed by the STM picture [11] at this stage (Fig. 3). When completing the first CdTe ALE cycle with Te exposure, a (2 × 1)Te-rich surface is reformed (step (2)); small islands of CdTe are present on the surface, with a half-monolayer surface coverage (see the STM picture of Fig. 3). The surface roughness is then maximum. The strong nucleation of islands observed at this step can be explained by the weak mobility of the Cd atoms chimisorbed on the surface. When beginning the second CdTe ALE cycle, a $c(2 \times 2) + (2 \times 1)$ Cd-rich half-filled surface is formed (step (3)), still with a maximum roughness. Upon completion of the second CdTe ALE cycle, a flat (2 × 1) Te-rich surface, similar to the starting surface, is again reformed (step (4)). Some holes on the flat surface appear, however, on the STM picture (step (4) in Fig. 3), indicating that a full CdTe monolayer has not been completely grown after two ALE cycles. This small difference is probably due to a partial sublimation of the Cd atoms when they have to go down the steps.

To summarize, a complete description of the ALE growth process of (001) CdTe has been proposed, based on the different surface structures and stoechiometry. A self-regulation at about half a monolayer per ALE

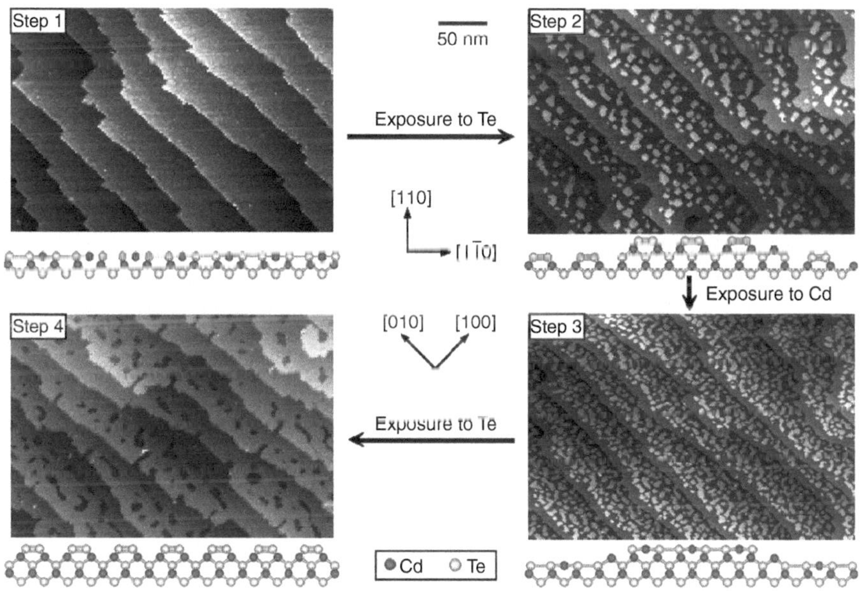

Figure 3 Surface morphology after each steps of two ALE cycles of CdTe growth leading to the deposition of 1 ML of CdTe. From Martrou et al. [11, 20].

cycle is demonstrated. The same results have been obtained for the ALE growth of ZnSe [12]. Moreover, the CdTe growth proceeds by the formation of small and isolated islands [11], which is very different from what is obtained with a MBE growth process. Such self-regulated growth process is not valid for Mn and Mg rich ternary alloys: in that case the exact amount of Mn of Mg material should be sent during the corresponding step of the ALE cycle.

1.2. Control of thickness and interfaces

Interfaces play a crucial role in the electron confinement and, as a consequence, in the optical or magneto-optical properties of the heterostructures. Chemical gradient as well as roughness affect the efficiency of this confinement. Therefore, it is important to measure both quantities as a function of the different growth parameters. Direct methods usually employed are X-ray diffraction on multilayers, X-ray reflectivity, or high resolution transmission electron microscopy (HRTEM). Each has its own advantage. In particular, HRTEM in cross section provides local information at a scale close to atomic distances but averaged over 10-20 nm while X-ray measurements give an average on several micrometers. The averaging along the atomic column in HRTEM is particularly relevant as the radius of the exciton is of the same order of magnitude (7 nm in CdTe). More indirect determinations of the concentration profiles are also available: for instance Mn profile into CdTe can be studied by magneto-optic measurements. With the next figures, we will illustrate these three methods which allow deducing some quantitative information on the concentration profile at the interfaces.

As far as the HRTEM studies are concerned [6, 13], one can measure local distortions within the layers directly from analyzing the images (Fig. 4). These distortions are due to lattice parameters differences between the materials (i.e., CdTe and ZnTe) and elastic deformations of the strained layers. Assuming linear elasticity to be valid (even for such thin layer), quantitative information on the chemical profile at the interfaces can be deduced. The method provides then the location of Cd in each (002) plane and a profile of composition along the growth axis. For CdTe/CdZnTe heterostructures grown in optimal conditions ($T_{substrate} < 300\,°C$), the interfaces are very abrupt and symmetric, whereas for CdTe/CdMnTe heterostructures, one can clearly evidence a direct interface (CdMnTe on CdTe) from an indirect one (CdTe on CdMnTe) which is much broader.

In the first case (CdTe/CdZnTe superlattices), we evidence the mechanism which can be at the origin of a deviation from a perfect planar interface: chemical interdiffusion may occur toward the ternary alloy, the most stable state of two binary compounds with no demixtion.

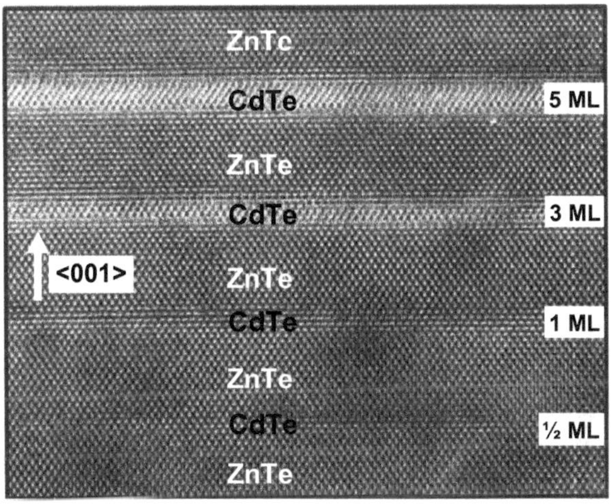

Figure 4 High-resolution TEM image of ultrathin CdTe layers embedded in ZnTe observed in the ⟨110⟩ direction. From Ph.D. thesis of P.H. Jouneau, Univ. J. Fourier, Grenoble.

The interdiffusion coefficient at CdTe/CdZnTe interfaces has been measured [14] by following the evolution of the X-ray diffraction pattern of superlattices annealed at various temperatures under a Cd overpressure (Fig. 5): the period, the width of the satellites, together with the splitting between the substrate peak and the zero-order satellite, remain constants, meaning the absence of strain relaxation during the treatment. The only observable effect is the decrease of the integrated intensities of the ± 1 and ± 2 peaks normalized to the zero-order one versus annealing time (Fig. 5). These experimental results are well described by a diffusion process in bulk material and at CdTe/CdZnTe interfaces. However, the deduced interdiffusion coefficient is such that, at growth temperatures generally used for the MBE of CdTe-based heterostructures ($T_{substrate} < 300$ °C), the contribution of interdiffusion to the interface morphology is strongly limited [14].

As far as the CdTe/CdMnTe heterostructures are concerned, the asymmetry between the two interfaces has been clearly confirmed by magneto-optical measurements [15]: by studying a CdTe well inserted between two barrier layers, a magnetic one with CdMnTe and a nonmagnetic one with CdZnTe grown in different order (see Fig. 6), a different Zeeman energy splitting was obtained depending on the growth order of the two barriers. The magneto-optical method is sensitive to the presence of diluted magnetic atoms and probes the low concentration part of the Mn distribution. The conclusions of these results are the following: (i) a low-concentration tail of Mn extends at the inverse interface (CdTe on CdMnTe) but not at the direct one, (ii) the results are compatible with a

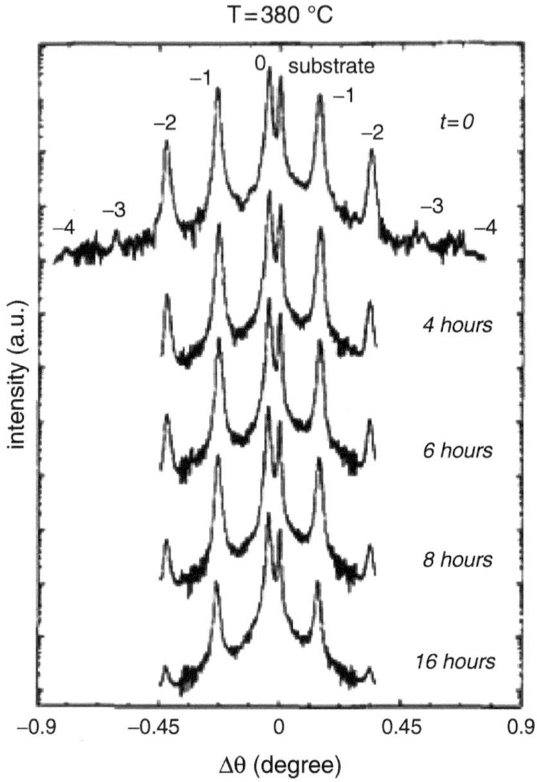

Figure 5 X-ray diffraction patterns recorded near (004) of a CdTe(39 monolayers)/ $Cd_{0.92}Zn_{0.08}Te$ (40 monolayers) superlattice as-grown ($t = 0$) or annealed for various durations at 380 °C under a saturated Cd overpressure. The substrate peak is indicated on the figure, and so is the integer index of the superlattice peaks. From Tardot and Magnea [14].

Mn composition profile decreasing quasi-exponentially. The inverse interface width is equal to 2.6 ML and corresponds exactly to a complete intermixing of the two top monolayers during deposition. In contrast to the previous mechanism (interdiffusion), the intermixing involves the presence of a surface during the growth process. It is based on an atomic exchange at the growing surface as it was first demonstrated for III-V heterostructures [16]. It corresponds to a surface segregation due to electronic or size effects of one species on a material of given composition. The complete intermixing between the top two monolayers during deposition, as deduced from the magneto-optical results for Mn atoms, is clearly compatible with the observed width of both the direct and indirect interfaces in high resolution TEM images.

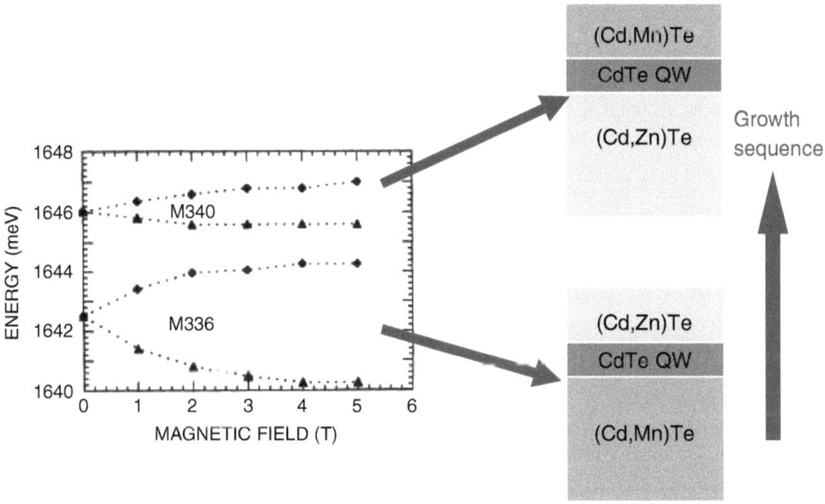

Figure 6 Energy of the e_1h_1 exciton lines for a pair of structures, samples M340 and M336, plotted versus magnetic field. The only difference between the two samples is related to the growth order: in M340, the magnetic barrier was grown after the CdTe QW (normal interface), whereas in M336 the order is opposite (inverted interface). Triangles symbols are for σ^+ polarization, diamond symbols for σ^-. The dots curves are guide for the eyes. From Grieshaber et al. [15].

Finally, let us mention two points concerning the study of the interfaces morphology:

(i) The Mn segregation revealed by the magneto-optical study was successfully used to control a very low concentration of Mn onto the growth front to dope CdTe quantum dots with only one Mn atom (on average).

(ii) When doping the heterostructures with nitrogen to have a p-type conductivity, a strong interdiffusion is observed, especially in the presence of magnesium (CdTe/CdMgZnTe heterostructures) [17]. This phenomenon can be overcome by using low temperature substrate and limited nitrogen doping.

1.3. Control of morphology: 2D/1D/0D

To decrease the dimensionality of CdTe nanostructures, namely from quantum wells (2D) to quantum wires (1D) and quantum dots (0D), two types of approaches have been demonstrated. One is based on the peculiar aspects of the CdTe growth by ALE as presented above: the amount of material deposited by ALE cycle, namely 0.5 ML/cycle for substrate temperature around 260 °C, and the geometry of the deposited CdTe which tends to form preferentially square islands with ⟨100⟩ edges (see Fig. 3). We used these results to prepare specific patterns on vicinal surfaces and demonstrate the possibility of growing both tilted 1D superlattices made out of CdTe-rich and MnTe-rich wires, and CdTe islands artificially organized onto these surfaces.

The other approach is a self-organized one on "standard" substrates, given rise to Stranski-Krastanow quantum dots and nanowires. This approach corresponds to a natural tendency, in given growth conditions, to proceed either in a one- or zero-dimensional regime.

1.3.1. Artificial organization of CdTe nanostructures

STM studies of (001) surfaces of these II-VI semiconductors have shown that the deposited CdTe tends to form preferentially square 2D islands with edges aligned along the ⟨100⟩ crystallographic directions [11] (see Fig. 3). These features which contrast with what is observed for (001) Si and GaAs surfaces, can be explained by the strong ionicity of the atomic bonding in these materials which induces a large electrostatic interaction between charged atoms on the step edges [18]. As a consequence these experimental results have been used to find out and to prepare the vicinal surfaces suitable for growing self-organized quantum structures. If the vicinal surface corresponds to steps aligned along the the ⟨100⟩ direction (C-type misoriented surface), then the initial misorientation and the preferential orientation of the steps edges during the CdTe growth tend both to maintain the step alignment along the ⟨100⟩ direction. This misoriented surface is then appropriated to obtain a regular staircase with straight and equally spaced steps that can be used as a template for the fabrication of quantum wires. Such 1D nanostructures arrays have been demonstrated for CdTe/CdMnTe and CdTe/CdMgTe [19]. However, the lateral ordering created by this structure is strongly limited by two effects, the Cd/Mn atomic exchange and the step array disorder which induce a tilted superlattice (TSL) (see Fig. 7).

By contrast, if the vicinal surface corresponds to steps aligned along the the ⟨110⟩ directions (A-type for steps running along [1–10] and B type for steps running along [110]), the miscut axis do not correspond to the energetically most favorable steps: they gives rise to steps at 45° from the preferential (100) growth. In that case, for the A-type vicinal surface, the step-ordering gives rise to a checkerboard structure (Fig. 8): the terraces are anticorrelated and they have a regular square shape with lateral dimensions of 25×25 nm, in good agreement with the value expected from the miscut angle of 1°.

The checkerboard structure which appears on a A-type vicinal surface is an appropriate template to grow quantum dots and to force them to have an in-plane ordering [18, 20]. Such results are presented in Fig. 8: the ALE growth of half a monolayer of CdTe gives rise to the formation of

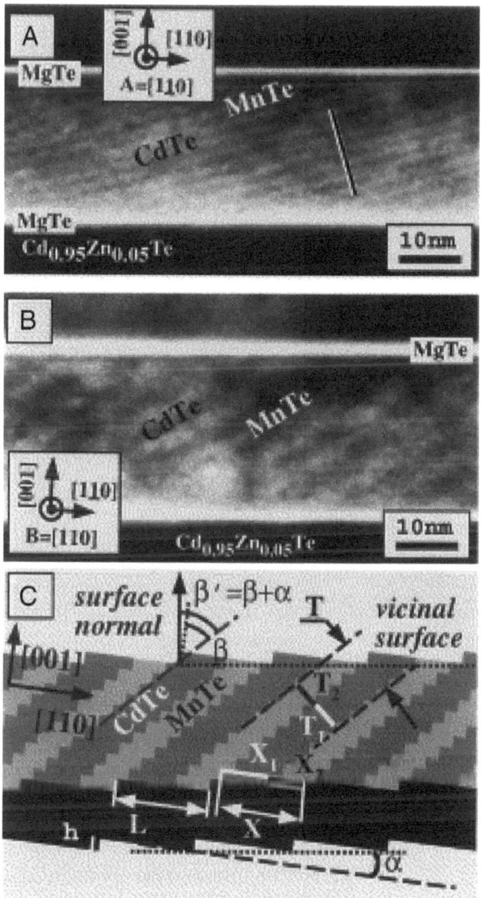

Figure 7 (A) Dark field images $g = (002)$ of a TSL grown on a 2°A (001) vicinal surface. MnTe rich layers are white. CdTe regions are dark. $T = 2.2$ nm, $\beta = 75°$, $\alpha = 2.08°$, $p = 0.878$ with $n = 0.409$, and $m = 0.468$. The inclined line indicates the direction, perpendicular to the TSL, along which the profile of Cd/Mn was realized; (B) dark field images $g = (002)$ of a TSL grown on a 2°B (001) vicinal surface. $T = 5.1$ mn, $\beta = 57°$, $\alpha = 1.94°$, $p = 0.97$; (C) scheme illustrating the structure and the geometric parameters of a $(CdTe)_m(MnTe)_n$ TSL: $m = X_1/L$, $n = X_2/L$, $p = X/L$, $m/n = X_1/X_2 = T_1/T_2$, $h = a/2$. T and X are the periods of the TSL measured, respectively, perpendicularly to the TSL and in the $\langle 110 \rangle$ direction From Marsal et al. [19].

islands where the size and the number per square strongly depend on the miscut angle: for 1.1° miscut angle (Fig. 8B), the average number of islands per square is equal to two, whereas for 1.5° miscut angle (Fig. 8A), most of the squares have only one CdTe island. The size and the density of the islands can be adjusted also with the substrate

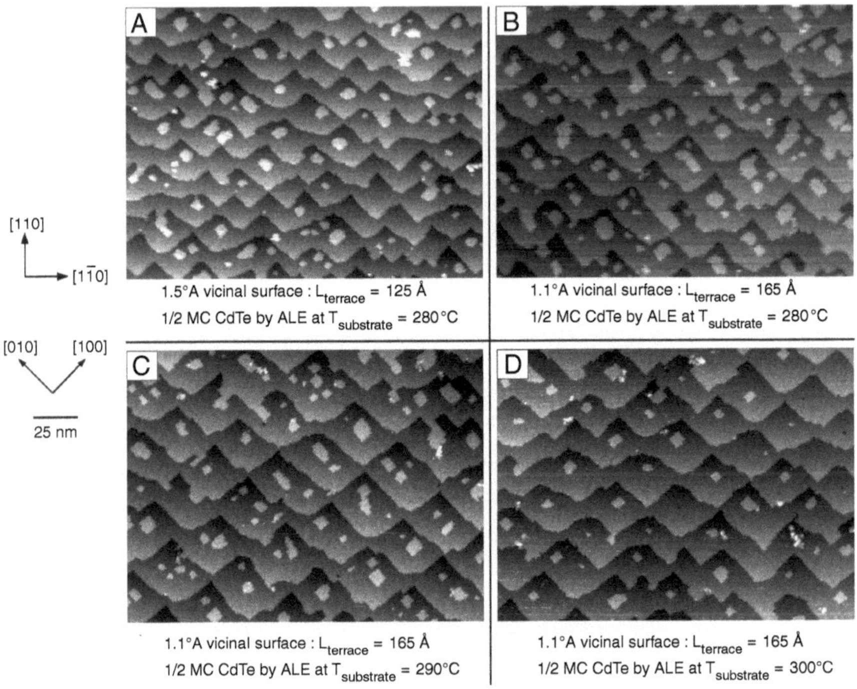

Figure 8 STM images of A-type vicinal surfaces organized in checkerboard array with the growth of half a monolayer of CdTe by atomic layer epitaxy. The CdTe islands appear as the white squares in the center of each terrace. From Martrou and Magnea [18].

temperature (Fig. 8C and D): the larger the substrate temperature, the smaller the island number per square.

All these results illustrate the possibility to induce an in-plane organization of the CdTe islands by using these well-controlled template structures. The limitation of this approach, however, is due to the disappearance of this artificial-ordered structure after the deposition of few monolayers.

1.3.2. Self-organization of CdTe nanostructures

Some combinations of lattice-mismatched semiconductors can exhibit, under specific epitaxial growth conditions, a sharp transition from a layer-by-layer 2D growth to the formation of 3D islands. This Stranski-Krastanow (SK) growth mode [21] allows the relaxation of highly strained 2D layers through the stress-free facets of 3D islands instead of generating misfit dislocations (MDs) [22]. These islands are expected to be dislocation-free and are thus of high structural quality. Usually their typical sizes are on the scale of a few nanometers, so that these self-assembled quantum dots (QDs) are attractive nanostructures for the study of zero-dimensional effects. In particular, the growth of these QDs, including the ability to tune their dimensions, surface density, and positions, is nowadays a topic of intense research effort to control their optical properties for optoelectronic applications.

The formation, above a critical film thickness, of such QDs by molecular beam epitaxy has been first discovered by Goldstein et al. [23] by observing *in situ* the 2D-3D morphology change of an InAs layer grown on GaAs; the process is now well established for this III-V semiconductors system [24]. The large lattice mismatch ($\Delta a/a \approx 7\%$) between these two semiconductors is seen as the driving force which induces the 2D-3D change of the surface morphology with the formation of SK islands. However, in the case of II-VI systems, which can exhibit mismatch as large as 6% for CdTe/ZnTe or CdSe/ZnSe, the 2D-3D transition is much less obvious: no clear 3D RHEED pattern has been reported during growth although zero-dimensional behavior was obtained [25–27]. In II-VIs indeed, above a critical thickness, MDs form easier than in III-Vs as clearly observed for CdTe/ZnTe by Cibert et al. [28], which corresponds to a plastic relaxation as first considered by Frank and Van der Merwe [22]. To account for the different behavior between these III-V and II-VI systems which have the same lattice mismatch, other parameters have to be considered in order to induce the SK transition.

We have developed [29] a simple equilibrium model taking into account not only the lattice mismatch, but also the dislocation formation energy and the surface energy. It demonstrates the importance of these parameters especially for II-VI systems such as CdTe/ZnTe and CdSe/ZnSe. For II-VIs indeed, as MDs are easier to form than in III-Vs (such as InAs/GaAs) or IV systems (Ge/Si), the 3D elastic transition is short circuited by the plastic one. This appears in Fig. 9A and C by comparing the total free energy in various growth modes, namely: (i) a 2D-coherent growth mode, (ii) a SK-coherent mode with the formation of coherent SK islands only, (iii) a 2D-MD mode with only the formation of MDs, and (iv) a SK-MD mode with both SK islands and MD. For a given epilayer thickness h, the equilibrium growth mode is the one exhibiting the minimum energy. The comparison between the energies deduced from this model [29] enables us therefore to predict which growth mode behavior is expected. Nevertheless, by lowering the surface energy cost (Fig. 9B and D), telluride and selenide quantum dots can also be grown as predicted with this calculation. By exposing our surface to amorphous anions (Te or Se) and then reevaporating this amorphous layer [30], we think that we realize this surface energy variation: indeed the 2D-3D transition occurs as evidenced experimentally by the strong change of the RHEED pattern, by atomic force microscopy and TEM images (see Fig. 10), and optical measurements.

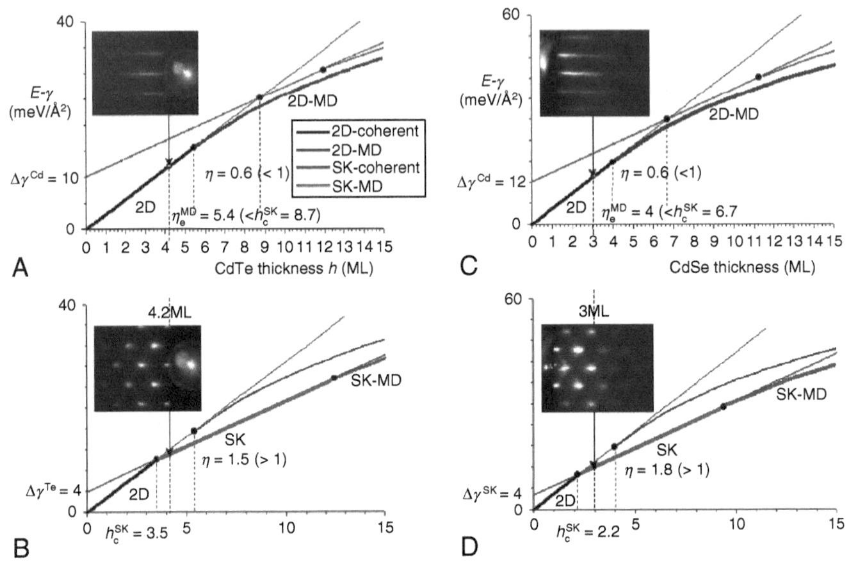

Figure 9 CdTe/ZnTe (respectively, CdSe/ZnSe)—(001) film's free energy as a function of CdTe (CdSe) thickness *h*. The experimental RHEED pattern in the inserts correspond, for (A) and (C), to Cd (Se)-rich growth conditions, and for (B) and (D), to a postgrowth treatment, namely the deposition and desorption of a thin amorphous Te (Se) layer. The spotty RHEED after such a treatment is a direct fingerprint of the SK transition. It can be understood as a decrease of the surface energy cost as illustrated with the equilibrium model applied for both CdTe/ZnTe and CdSe/ZnSe systems. The bold lines correspond to the growth modes expected with our model assuming a decrease of the surface energy cost by a factor 3 (respectively, 4) when going from cation-rich surfaces to anion-rich ones. From Tinjod et al. [29].

In summary, we have induced for the II-VI QDs a 2D-3D transition, which occurs *after and not during the growth* by the rearrangement of a strained 2D layer. This morphology transition under a fixed amount of deposited material (vertical grey line in Fig. 11) can be understood as a surface energy variation obtained by exposing the II-VI layer to a large quantity of the group-VI element. This method allows us to avoid a plastic relaxation and to get coherent grow QDs according to deposit an amount of material less than the one predicted by the plastic critical thickness for a standard SK transition occurring during the growth (gray path in Fig. 11). Moreover it opens the way toward the growth of QDs with two materials having even a smaller mismatch $\Delta a/a$ provided that it is possible to decrease sufficiently the surface energy cost.

This model, however, describes the growth behavior at the equilibrium. In fact kinetics effects (mobility of adatoms, of dislocations) can also play an important role during the heteroepitaxy of such mismatched

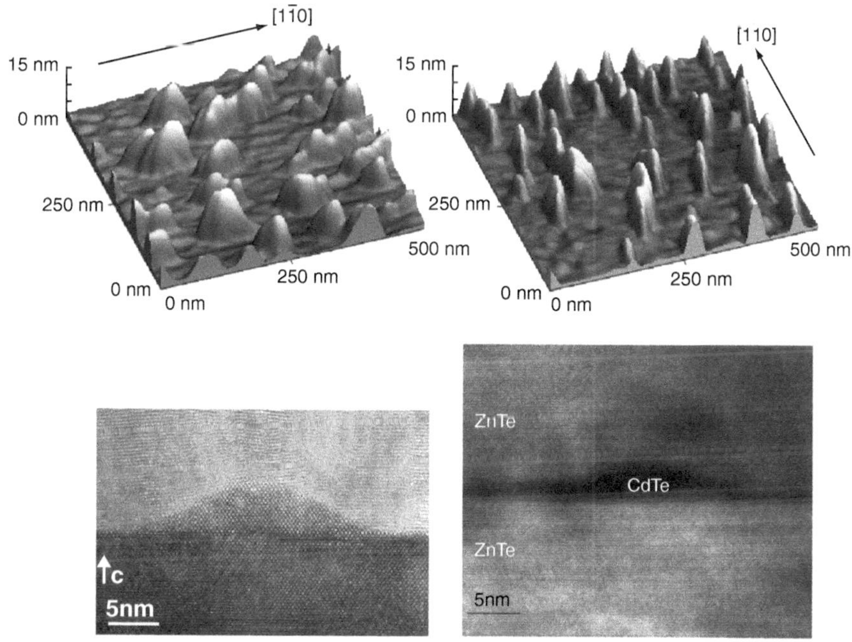

Figure 10 UHV-AFM images of CdTe QDs showing for the two scan directions a preferential orientation along the [1 −1 0] axis. The two dark field TEM images (C. Bougerol) correspond to both CdSe and CdTe 0Ds embedded in ZnSe (respectively, ZnTe) performed by changing the surface energy as described in the text.

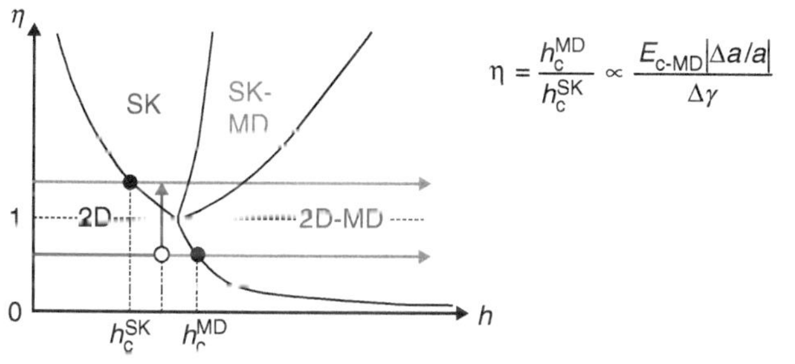

Figure 11 Schematic diagram illustrating the growth modes as a function of the deposited material and the ratio η of both critical thicknesses, the plastic one h_c^{MD} and the elastic one h_c^{SK}. This ratio depends on the lattice mismatch $\Delta a/a$, the dislocation formation energy E_{MD} and the surface energy variation $\Delta\gamma$. The grey horizontal line (with $\eta > 1$) corresponds to the case of InAs/GaAs, the grey one (with $\eta < 1$) to the case of II-VIs. The vertical path which induces the 2D-3D transition corresponds to a variation of $\Delta\gamma$ obtained by saturation with the VI element. From Tinjod et al. [29].

systems. This influences the QDs shape, their density, and even their formation. For example, the substrate temperature can activate or not adatoms' mobility to form (or not) the QDs. Finally, the challenge nowadays is to control precisely both the position and the size of the dot onto the surface which is a prerequirement to ultimately use these dots in quantum devices.

To conclude this section about the know-how to control the dimensionality of CdTe-based nanostructures, let us mention that, nowadays, CdTe and CdSe nanowires are also performed by MBE using metal droplets as a catalyst to initiate and control the one-dimensional growth mode [31].

2. ELECTRONIC PROPERTIES

2.1. Quantum effects discovered in 2D CdTe heterostructures

Low-dimensional structures based on II-VI semiconductors show strong excitonic effects and large optical nonlinearities. Among II-VI systems, the CdTe/CdZnTe quantum well one has been intensively studied due to its interesting fundamental properties such as, for example, a large value of the excitonic Rydberg and large strain splitting of the valence band. This latter effect induces that the lowest energy transitions are spatially direct (type I) for heavy-hole excitons and spatially indirect (type II) for light-hole excitons [32–35]. The robustness of the exciton Rydberg in II-VIs ($R =$ 10.5 meV in bulk CdTe) explains why it was possible to evidence, in 2D CdTe heterostructures, original quantum effects, such as the exciton centre-of-mass quantization, the strong coupling in a 2D microcavity leading to the Bose-Einstein condensation of polaritons, the first observation of the trion (X^- and X^+), and the tunneling of excitons as a whole.

2.1.1. Centre-of-mass quantization

By confining the excitons in quantum wells (QWs), we can study excitons of specific translational wave vectors and therefore of specific kinetic energy. The essential feature of our work was the use of wells that are sufficiently wide (relative to the exciton Bohr radius) for the two particles motion to be considered within the "adiabatic" approximation [36], in which the exciton is treated as a composite particle formed by the electron and hole mutually orbiting each other (the internal motion) plus a translational motion of their centre of mass. In the simplest description [37], the internal motion of the electron and hole is totally unaffected by the exciton confinement. However, Davies et al. [38] provided evidence recently that the internal structure of excitons becomes significantly changed as a result of their translational motion.

The quantization of the exciton centre of mass is revealed by the presence of satellites peaks on the high-energy side of the free exciton

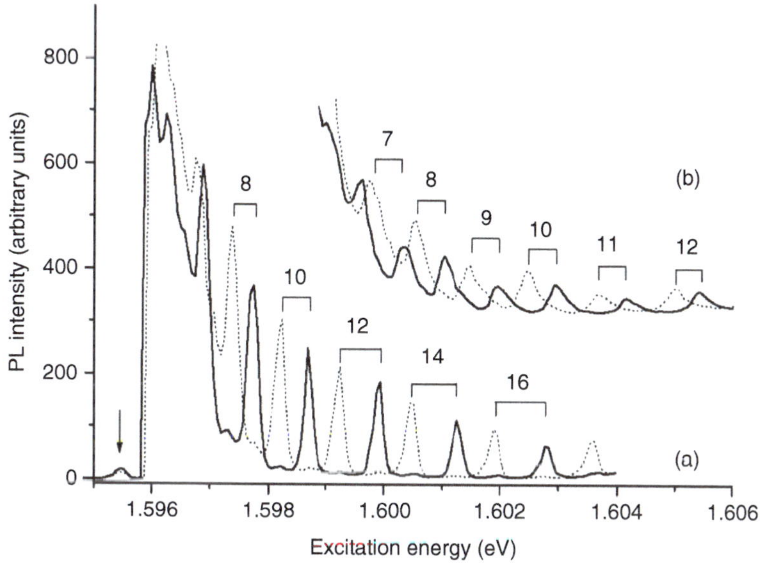

Figure 12 (A) The σ^+ (solid lines) and σ^- (broken lines) PLE spectra at 3 K from a 144.2 nm CdTe (001) QW embedded in $Cd_{0:94}Zn_{0:06}Te$ barriers in a magnetic field of 4.0 T. The increase of Zeeman splitting with quantization index N is apparent. The arrow indicates the detection energy. For energies below 1.5959 eV, an attenuator was inserted. (B) Corresponding data at 6.5 T for the 66 nm (110) CdTe QW. From Davies et al. [38].

line (Fig. 12), with spacing and relative intensities that depend on the thickness of the CdTe layer [37]. The situation studied here is rather different from the extensively studied case of much thinner GaAs wells, where exciton states E_iH_i are formed by binding an electron from an electron subband E_i and a hole from a hole subband H_i [39]. In the present case (thicker layers, smaller potential, larger electron mass), the confinement energies for an electron and a hole (approximately 1.3 and 0.2 meV, respectively for a 500 Å layer) are small compared to the exciton binding energy of about 13 meV. The electron-hole Coulomb interaction then completely mixes the subbands and we have to directly consider the quantification of the exciton itself. In the first approximation, the component K_z of the exciton wave vector perpendicular to the plane of the well (i.e., in the growth direction, taken to be the z axis) is quantized according to $K_z = N\pi/L$, where N is a nonzero integer and L is the well width [37]. More strictly, at least for energies near E_L and E_T (the energies of longitudinal and transverse excitons with $k = 0$), we must consider the quantification of the exciton-polariton.

The nature of the quantized exciton states observed (centre-of-mass quantization versus separate-carrier quantization) was easy to obtain in II-VI QWs as compared to GaAs ones due to larger exciton binding energies in II-VI's as compared to III-V's. However, clear-cut evidence for the concept of centre-of-mass quantization of excitons was then obtained for GaAs layers between GaAlAs barriers [40, 41].

2.1.2. Strong coupling and Bose-Einstein condensation

When quantum wells are embedded in a microcavity, the interaction between discrete excitonic levels and a single photon mode of a Fabry-Perot cavity is enhanced and could give rise to the so-called strong coupling regime [42]. This regime is characterized by an anticrossing behavior between exciton and photon states, giving rise to coupled modes called cavity polaritons. This effect is very similar to the Rabi oscillations which occur between atom and photon states in atomic physics and by analogy the polariton lines splitting, at the anticrossing point, is usually called (vacuum field) Rabi splitting. Cavity polaritons and strong coupling regime have been observed in CdTe-based QW microcavity sample [43]. The mirrors are Bragg reflectors consisting of periodic stacks of CdMnTe and CdMgTe $\lambda/4$ layers. The extremely steep dispersion of the cavity polariton modes, due to the optical confinement along the z direction, results in a typical polariton effective mass of 10^{-4} times the free electron mass. Thus, temperature and density criteria for Bose-Einstein condensation of polaritons in their ground state should be satisfied much more easily, especially for II-VIs which allow high density of excitons-polaritons [44].

The first indication of spontaneous quantum degeneracy of polaritons was the observation of stimulated emission under nonresonant pumping in CdTe microcavities [45]: above some excitation power threshold, the polariton emission exhibits a strong nonlinearity, while the linewidth shows significant narrowing (Fig. 13). Besides the observation of such massive occupation of the ground state developing from a polariton gas at thermal equilibrium, an increase of temporal coherence, and the buildup

Figure 13 Far-field emission measured at 5 K for three excitation intensities: $0.55P_{thr}$ (left panels), P_{thr} (centre panels), and $1.14P_{thr}$ (right panels); $P_{thr} = 1.67$ kW/cm^2 is the threshold power of condensation. (A) Pseudo-3D images of the far-field emission within the angular cone of $\pm23°$, with the emission intensity displayed on the vertical axis (in arbitrary units). With increasing excitation power, a sharp and intense peak is formed in the center of the emission distribution ($\theta_x = \theta_y = 0°$), corresponding to the lowest momentum state $k_{//} = 0$. (B) Same data as in (A) but resolved in energy. For such a measurement, a slice of the far-field emission corresponding to $\theta_x = 0°$ is dispersed by a spectrometer and imaged on a CCD-camera. The horizontal axis displays the emission angle (top) and the in-plane momentum (bottom); the vertical axis displays the emission energy in a false color scale (different for each panel). Below threshold (left panel), the emission is broadly distributed in momentum and energy. Above threshold, the emission comes almost exclusively from the $k_{//} = 0$ lowest energy state (right panel). From Kasprzak et al. [44].

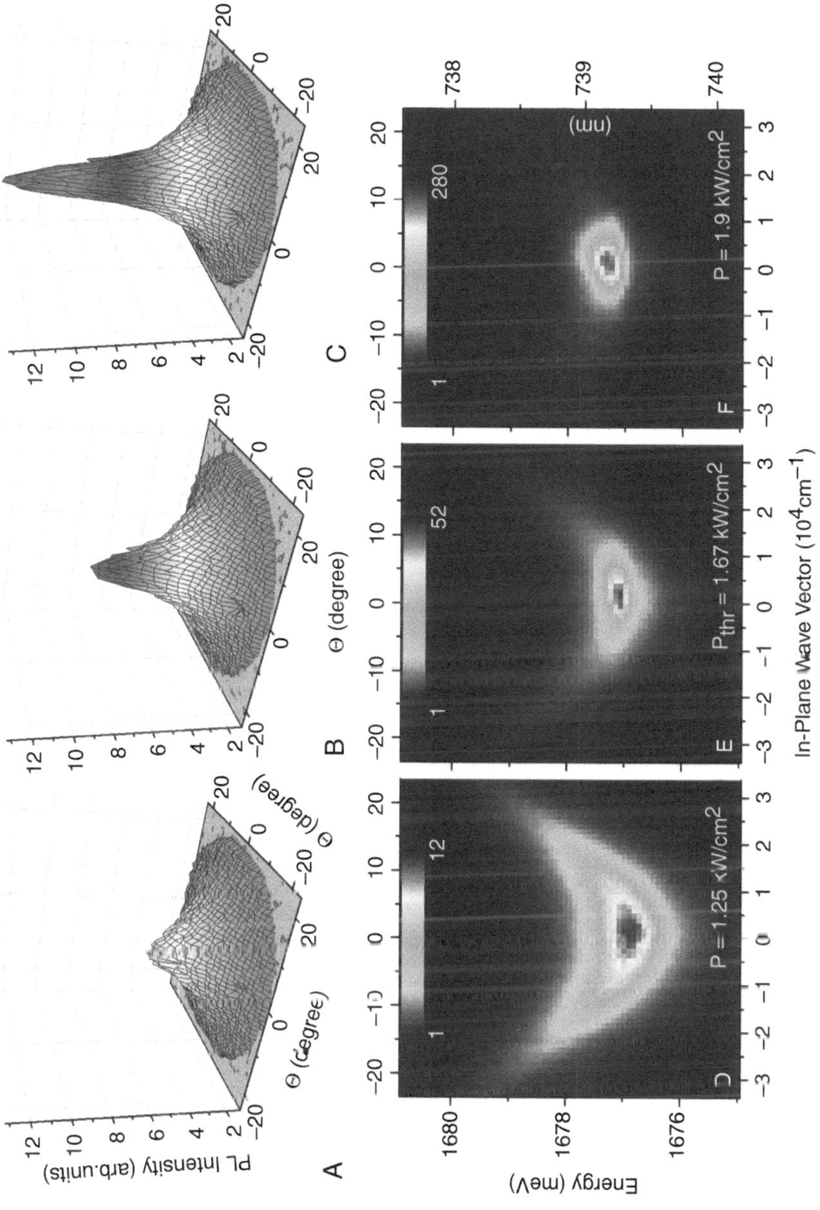
A
B
C
D
E
F
PL Intensity (arb.units)
Θ (degree)
Energy (meV)
(nm)
P = 1.25 kW/cm2
Pthr = 1.67 kW/cm2
P = 1.9 kW/cm2
In-Plane Wave Vector (10^4cm^-1)

of long-range spatial coherence and linear polarization, have been recently reported [44]. All these features indicate the spontaneous onset of a macroscopic quantum phase.

2.1.3. Existence of the trions

The exciton in a semiconductor is the analogy of the hydrogen atom in vacuum. The exciton binding energy is typically 1-50 meV ($R = 10.5$ meV for bulk CdTe). The existence of a related species, the negative charged exciton (X^-), that is two electrons and one hole, analogous to H^-, was predicted 50 years ago [46]. The energy of binding the second electron in X^-, E_{b2}, was expected to be $0.055R$, by analogy with the dissociation energy of H^- into H^0 and a free electron (R is the effective Rydberg). More precisely, calculations for bulk semiconductors taking into account the finite ratio of electron and hole masses predict somewhat lower values, namely $E_{b2} = 0.030R$ for $m_e/m_h = 0.5$ [47]. This means binding energies E_{b2} in the meV range or smaller, so resolved spectra of X^- are difficult to obtain. However, in a QW, E_{b2} was expected to increase dramatically: for example, Stébé and Ainane [48] calculated, for $m_e/m_h = 0.5$, a factor of 10 increase, to $E_{b2} = 0.30R$ in the ultimate 2D limit.

By using CdTe QWs, that is, a semiconductor compound with a large Rydberg together with a 2D system, it was possible to optimize the conditions to identify for the first time [49] the negatively charged exciton X^-. It was done with CdTe/$Cd_{0.84}Zn_{0.16}Te$ modulated doped MQW with electron concentration of few 10^{10} cm^{-2}. The identification of the species X^- was performed by its creation and annihilation transition in magneto-optical spectra (in Fig. 14 these transitions are called Y). Also the fairly large electron g-factor for CdTe, $g_e = -1.6$, increases circular polarization effects, which helps us to attribute the Y line to the X^- transition. The binding energy of the second electron of X^- has been found to be $E_{b2} = 2.65$ meV, that is $0.20R$ for a 100 Å well width [49]. Since then, X^- (and X^+) has been observed in other QW systems [50] and is the object of area of intensive research.

2.1.4. Exciton tunneling

Tunneling of carriers from one QW to an adjacent one through a thin potential barrier in semiconductor heterostructures has been intensively investigated because of its basic quantum mechanical aspect and its importance for tunneling devices [51]. A basic question for the transfer of an optical excitation is: Do the carriers—electrons or holes—tunnel independently or is it a bound electron-hole pair, that is, an exciton, which transits between the wells? In a first approach, experiments results on transfer mechanisms between adjacent QWs made of III-V semiconductors were interpreted as a tunneling of free electrons (free holes)

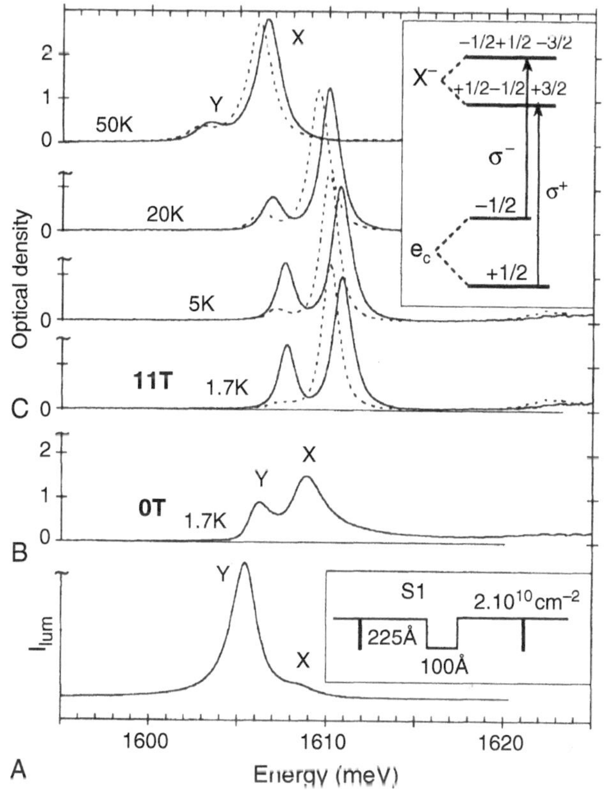

Figure 14 Optical spectra for sample S1: a 100 Å CdTe-450 Å $Cd_{0.84}Zn_{0.16}Te$ MQW (10 periods) doped nominally 2×10^{10} indium cm^{-2} at the barrier centers. (A) Luminescence and (B) absorption at 0 T, 1.7 K. The line labeled Y is the negatively charged exciton X^-. (C) absorption in a field $B = 11$ T applied perpendicular to the QW planes for various temperatures. Full lines are σ^+ polarization; dotted lines are σ^-. The optical density scale units at left are ln(1/transmission). The inset shows the allowed ($\Delta M = \pm 1$) transitions $e + h\nu \rightarrow X^-$. From Kheng et al. [49].

whose energy levels were brought into resonance by an electric field [52]. More detailed results have demonstrated that the correct description is a transfer from a direct to an indirect exciton: in this mechanism the electron effectively transfers from a bound state within a direct exciton (electron and hole wave functions localized in the same well), to a bound state within an indirect exciton (electron and hole localized in different wells), whereas "the hole only participate as a spectator" [53]. All these studies concluded then, that although excitonic effects in carrier tunneling must be considered, the simultaneous transfer of electrons and holes between direct excitons was not the dominant mechanism.

In II-VI based heterostructures, the tunneling properties are much more affected by excitonic effects, especially because the exciton binding energies can be larger than the hole confinement potential. Indeed we have found strong evidence for tunneling of excitons as a whole in CdTe/CdZnTe [54] and CdTe/CdMnTe [55] asymmetric double quantum wells (ADQW). Resonances between excitons were induced by a magnetic field in the two different types of II-VI ADQWs, using the enhanced Zeeman splitting of an exciton in the first type of system, and the strong diamagnetic shift of excited exciton states in the second one. Indeed, in CdTe/CdMnTe ADQWs, where only the wider QW contains some Mn, the magnetic field allows a continuous tuning of the coupling between the magnetic QW and the nonmagnetic one, due to the giant Zeeman effect. The tunneling dynamics were investigated by time-resolved and steady-state photoluminescence (PL) spectroscopy under such magnetic field [54, 55]. Very efficient tunneling was found when the transfer of a spatial direct exciton was possible, either with the emission of LO phonons or via the resonance with the 2s state of the low-energy QW (see Fig. 15). Such results have revealed the importance of excitons rather than free-carrier states for the tunneling mechanism in II-VI heterostructures.

2.2. Quantum effects discovered in 0D CdTe nanostructures

2.2.1. Fine structure and multiplet excitons

Most of the work concerning single-dot spectroscopy has been done on materials systems such as InAs/GaAs and is mostly concentrated on luminescence measurements (see e.g., [24, 56, 57]). The II-VI compounds on the one hand and excited states in single QDs on the other hand have been much less explored. As concerns materials, II-VI QDs such as CdTe/ZnTe [58, 59] or CdSe/ZnSe [27, 60, 61] are interesting due to their large excitonic binding energies. The relatively large Coulomb interaction makes such systems well adapted for the realization of a single photon emitter that might operate up to room temperature [62].

Figure 16 shows the PL and PL excitation (PLE) spectra of a single QD obtained by microspectroscopic techniques [63] Such μPLE measurements of a single QD provide quantitative information on the confinement potential and give the electronic structure of these nanostructures. Due to the unavoidable dispersion of size, shape and composition, different excited-state structures can be observed for various CdTe QDs. For the one presented in Fig. 16, two groups of sharp resonant peaks appear about 25 and 50 meV above the ground state and dominate the continuum absorption background. They are attributed to the first two excited states of the QD. They reveal the discrete atomic-like density of states of semiconductor QDs. The nearly equal spacing between the optical transitions

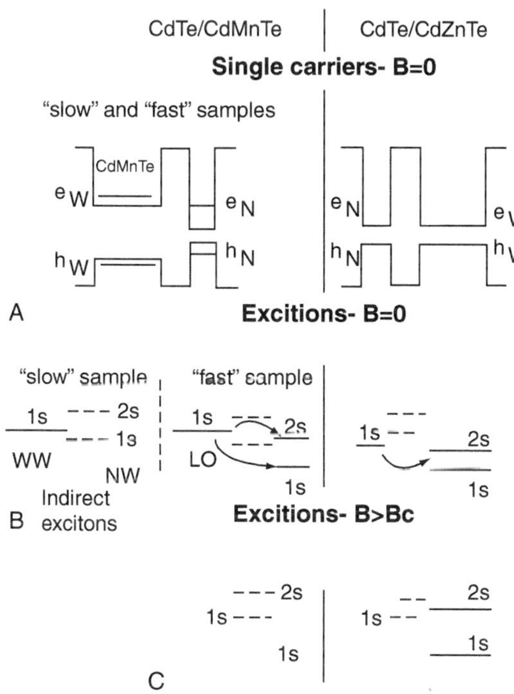

Figure 15 Scheme of the levels in II-VI asymmetric quantum wells: on left column, two CdTe/CdMnTe samples having a "fast" and "slow" tunneling time, and differing in the Mn composition of CdMnTe QW; on the right a CdTe/CdZnTe ADQW as described in Ref. [54]; (A) Schematic representation of the carrier energy levels in the conduction and valence band potential wells in the two studied systems. (B) and (C) show direct exciton states (full levels) and indirect exciton states (dotted levels) in the two systems. At $B = 0$, high efficiency transfer between the two QWs is possible through emission of LO phonon or via resonance with the 2 s state (curved arrows in (B)). At a magnetic field above a critical value B_c (C), these transfer are inhibited. From Lawrence et al. [55].

in this QD suggests that a simple effective parabolic potential can be used to describe the QD confinement [64]. A two-dimensional harmonic-potential model is particularly appropriate when interdiffusion smoothes the lateral confinement while the confinement remains strong along the growth direction. Interdiffusion has been detected in our QDs by a compositional analysis of transmission electron microscopy images. This reveals that the QD structure is formed by CdTe-rich islands (with about 20% Zn) in the ZnTe matrix [58].

In the framework of this model, the electronic levels (single particle states) in the dot can be labeled by quantum numbers $n = 1, 2, 3, \ldots$ corresponding to s, p, d,... shells, by analogy to atomic-like symmetries.

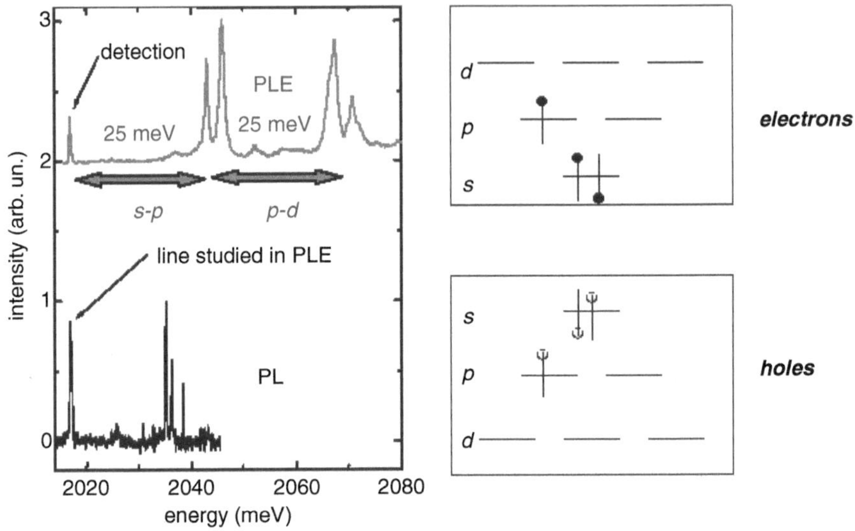

Figure 16 Microphotoluminescence spectrum (PL) and microphotoluminescence excitation spectrum (PLE) of a single CdTe QD carried out at 4 K. The two excited states which appear in this dot, p and d-like, are illustrated in the scheme of the left hand side. From Besombes et al. [63].

Optical interband transitions are allowed between shells with equal quantum number n. The two groups of absorption lines in Fig. 16 are then attributed to p-shell and d-shell transitions, respectively (the ground state corresponding to the s-shell).These p-shell and d-shell absorption lines are split: Hawrylak [65] has shown that this kind of splitting can result from Coulomb-interaction-induced mixing of the excited p and d shells.

These μPLE spectra reveal that up to two excited shells are clearly observed in some of the QDs. With such excited states, accumulation of carriers in the QDs under high excitation density is possible and allows to observe multiexciton in the emission spectra [63]. These two types of experiments, absorption and emission, provide complementary information about the structure of the excited states in individual QDs.

Multiexciton complexes corresponding to the occupation of the first-excited state appear at high excitation densities. A simplification of the structure of the emission spectra is found for a symmetric dot and is attributed to the fulfilling of the high-symmetry condition that leads to cancellation of interparticle interactions in the multiple-particle Hamiltonian. This allows to observe clearly up to four excitons (X_4) in the QD [63], that is, two excitons in the s-shell and two in the p-shell. Comparison of X_4 with X_2 is interesting because X_4 is the equivalent of the biexciton in the p-shell (with the s-shell fully occupied by two excitons). It was observed that adding a second exciton in the p-shell leads to a larger binding energy than adding a second exciton in the s-shell. Let us mention that by increasing the excitation power, an other few-particle species was identified in single QDs [66], namely the negative charged exciton, which appears at an energy between the neutral and the biexciton (see Fig. 18).

2.2.2. Exciton-phonon coupling

The linewidth of an optical transition is well known to be inversely proportional to the lifetime of the radiative state. In solid crystals, with increasing temperature, *inelastic* scattering of the exciton by optical or acoustic phonons reduces the exciton lifetime and broadens the excitonic line [67]. In such a mechanism, the corresponding loss of population of the radiative state induces a loss of phase coherence (dephasing) that can be measured by four-wave-mixing experiments. QD systems were thought to be very insensitive to *inelastic* scattering by low-energy acoustic phonons because of the absence of suited states between the QD discrete energy levels (bottleneck effect). However, previous studies of single QD lines have shown the persistence of some significant dephasing despite this bottleneck effect [68]. On the other hand, simultaneous measurements of the dephasing and the population decay for localized excitons in narrow GaAs quantum wells show that, with increasing temperature, *elastic* interaction with acoustic phonons also contributes to dephasing [69]. Such a loss of phase coherence that is not related to a population relaxation is called *pure dephasing*. Usually, pure dephasing is treated using an additional phenomenological phase damping linearly proportional to time [69] that straightforwardly leads to a Lorentzian line shape for the homogeneously broadened transition.

For CdTe QDs, we have shown [70] that this *elastic* exciton-phonon interaction in the low temperature range can no longer be described by a single dephasing rate (or a simple full width at half maximum—FWHM—as usual) [69, 71]. This is evidenced by the special temperature dependence behavior of the line shape of CdTe QD's emission (Fig. 17): the zero-phonon line and its acoustic phonon sidebands are distinctly observed due to a suited phonon coupling strength. These two components of the emission line are well described by a theoretical model that considers recombination from stationary eigenstates formed by the mixing of the discrete excitonic states with each acoustic phonon mode. This nonperturbative coupling creates a discrete set of polaron states which can recombine radiatively but with different probabilities depending on the phonon part of each exciton-acoustic phonon state. This allows us to give new insights into the acoustic phonon broadening mechanism which controls the exciton dephasing and imposes the real limits to the optical properties of single QD's emission.

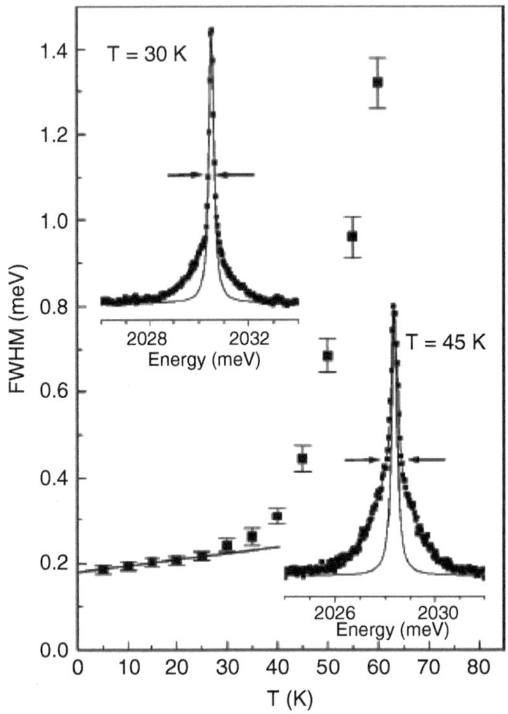

Figure 17 Temperature dependence of the exciton PL line FWHM. The linear fit at low temperature (solid line) gives a slope of 1.5 meV/K. Insets show PL lines measured at $T = 30$ and 45 K. The line shape strongly deviates from a Lorentzian profile (solid line) and the sidebands which appear around the central zero-phonon line progressively control the FWHM. From Besombes et al. [70].

This elastic exciton-acoustic phonon interaction was confirmed with four-wave mixing experiments by measuring, as a function of temperature, two dephasing times resulting in a non-Lorentzian line shape [72]. The observation of phonon wings on single QD spectra was then reported for QDs made of InAs [73, 74], GaAs [75], and GaN [76].

2.2.3. Coupling with a single magnetic atom

Finally, we want to show that CdTe nanostructures doped with magnetic atoms are the model system to study the coupling between the carriers and the local spins. There are three main reasons for that (i) Mn atoms are isoelectronic to the Cd ones which they replace in the matrix. As a consequence they do not introduce any carriers into the layer, only some local spins due to their d-electrons, by contrast to the III-V for which Mn is an acceptor. Then it is possible with II-VI based-heterostructures to tune independently the electrical and magnetic doping which allows to perform magnetic nanostructures not perturbed by ionized acceptors, (ii) the diluted magnetic semiconductor (DMS) CdMnTe can be achieved at optimal growth temperature and with any Mn compositions. This allows to obtain well-controlled DMS nanostructures with optimized optical properties by contrast to GaMnAs, and (iii) the strong sp-d exchange interactions between the band carriers and the transition metal ions give rise to large magneto-optical effects [77] which can be measured directly by magneto-optical experiments. Having in hand these advantages of II-VI magnetic nanostructures, three types of magneto-optical effects have been demonstrated:

- In a magnetic CdMnTe QW, it was possible to change the magnetic phase (from paramagnetic to ferromagnetic one above the critical temperature $T_c = 3$ K) by changing the density of a two-dimensional hole gas [78]. Moreover, Boukari et al. [78] reported that both photon beam and electric field can isothermally drive the system between the ferromagnetic and paramagnetic phases, in a direction which can be selected by an appropriate design of the structure. This offers new tools for patterning magnetic nanostructures as well as for information writing and processing, beyond the heating effects of light exploited in the existing magneto-optical memories. Obviously, however, practical applications of the tuning capabilities put forward here have to be preceded by progress in the synthesis of functional room temperature ferromagnetic heterostructures.
- In a magnetic QD, the sp-d interaction takes place with a single carrier or a single electron-hole pair. Recently, the formation of quasi-zero-dimensional magnetic polarons (i.e., regions with correlated carrier and magnetic ion spins) has been demonstrated [79] in individual QDs. In such work on diluted magnetic QDs, all the experimental studies were focused on the interaction of a *single* carrier spin with its paramagnetic environment (*large number* of magnetic atoms).
- Finally, in the case of a quantum dot incorporating a single magnetic atom (spin S) and a single confined exciton, the exchange interaction between the exciton and the magnetic atom reveals the various spin states of the magnetic atom. Instead of a single sharp peak usually recorded for a single QD, one observes six sharp equidistant lines (Fig. 18). This dramatic change is easy to understand if one considers the excited state in the photoluminescence process. The anisotropy of the hole, in such rather flat QD, is a key feature: due to the strong spin-orbit coupling in the valence band, the hole moment is normal to the plane of the sample (say z). A hole can have $J_z = 3/2$ (respectively, $J_z = -3/2$), then it will recombine with an electron of $s = -1/2$ ($s = 1/2$) and emit a σ^- (σ^+) polarized photon. In absence of magnetic field both

> transitions take place at the same energy. If the QD contains a Mn atom with a spin 5/2, the strong spin-hole interaction splits the initial multiplet into six equidistant sublevels (Fig. 18). The degeneracy of the Mn system is restored in the final states, so that the "standard" QD's single line emission splits into a comb of six lines.

In other words, the exciton acts as an effective magnetic field, so that the atom's spin levels are split even in the absence of any applied magnetic field [80]. The set of $(2S + 1)$ discrete emission lines provide a direct view of the atom's spin state at the instant when the exciton annihilates (Fig. 18). Then the created exciton appears as a probe of the spin state of the Mn atom: knowing the energy position of the emission together with its polarization, one can determine the Mn spin projection along the quantized axis (growth axis) in the QD. One should notice that the exciton is more than a probe, it also changes the spin distribution of the magnetic atom. This was illustrated by applying a magnetic field and by recording the magnetic field dependence of the emission of a Mn-doped QD [80]. More generally, the interaction between the magnetic atom and the carriers (or exciton) in the QD could be exploited to manipulate the quantum state of an individual spin by optical or electrical injection of polarized carriers. Coherent manipulation of the spin state of a single magnetic atom could also be performed under pulsed resonant optical excitation, suggesting implementation of controlled spin-qubit operations.

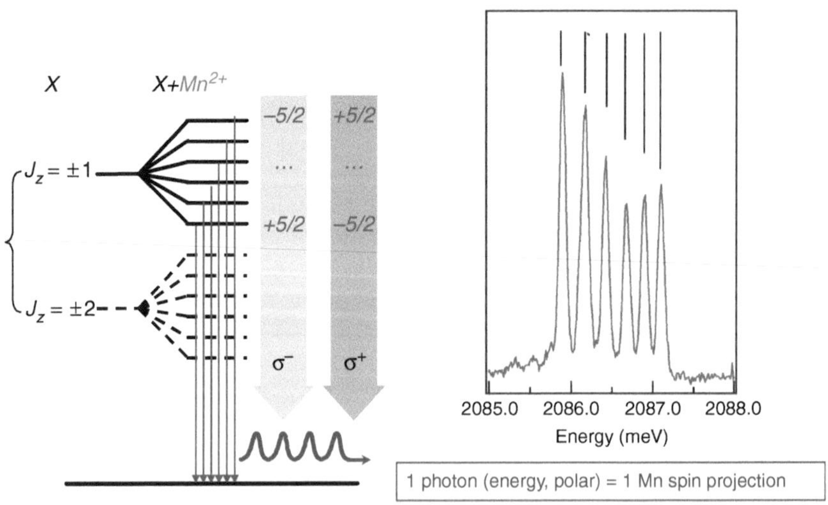

Figure 18 Photoluminescence spectrum at 2 K of a single CdTe quantum dot containing a single Mn impurity. The left scheme shows the initial (top) and final (bottom) states involved in the emission of a σ^+ (σ^-) photon. From Besombes et al. [80].

Finally, we reported on the reversible electrical control of a single Mn atom in an individual QD. Our device allows to prepare the dot in states with three different electric charges, 0, +1e and −1e by applying an external bias voltage V on an aluminum Schottky gate with respect to a back contact on the p-type substrate [81]. The PL emission pattern of the charged excitons differs strongly from that of the neutral exciton (Fig. 19). This difference reflects the fact that the Mn spin is very sensitive to the number of electrons and holes in both the excited and the ground states of the optical transitions [82, 83]. Figure 19 shows a detail of the recombination spectrum obtained for X^- coupled with a single Mn atom. Eleven emission lines are clearly observed with intensity decreasing from the outer to the inner part of the emission structure. A simple model taking into account the interaction between the spins of the Mn, the hole and the two electrons was proposed [81] to account quantitatively for this emission spectrum.

The emitting state in the X^- transition has two conduction band electrons and one hole coupled to the Mn. The effect of the two spin-paired electrons on the Mn is strictly zero. Thereby, the spin structure of the X^- state is governed by the interaction of the hole with the Mn and

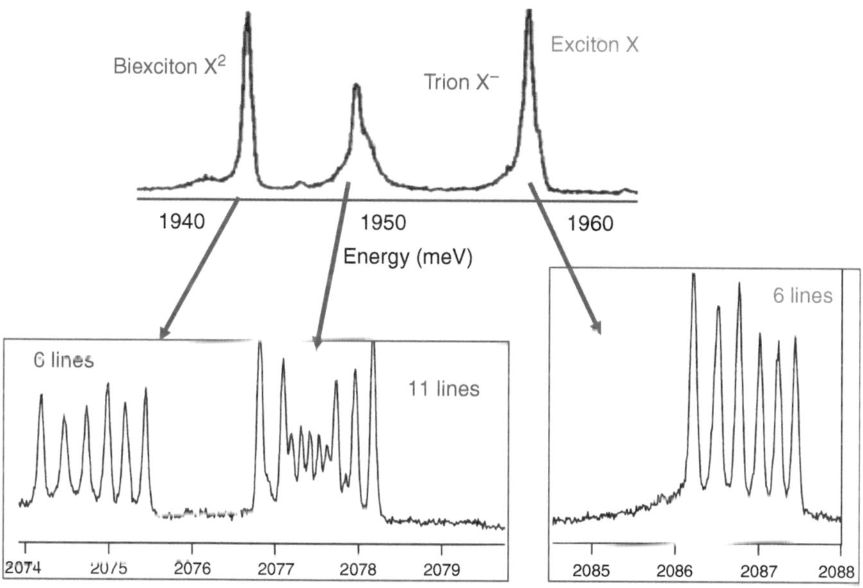

Figure 19 Comparison of the photoluminescence spectra of a pure single CdTe QD and a CdTe QD having one Mn atom. Besides the neutral exciton lines X and X^2 biexciton, the X^-emission appears when applying appropriate bias. In the presence of Mn, the emission of the trion X^- gives rise to 11 distinct lines which can be account quantitatively in energy and relative intensities. From Léger et al. [81].

composed of six doublets as seen before. After the electron-hole pair recombination, there are two possible final states, coming from the two isotrope spins 1/2 and 5/2, and giving rise to spin states with either $J = 2$ or $J = 3$. From this consideration alone, we expect 12 spectrally resolved lines. Their relative weight is given by both optical and spin conservation rules.

More generally, whereas in the neutral configuration the quantum dot is paramagnetic, the electron-doped dot behaves like a spin rotational invariant nanomagnet with $S = 3$ and the hole-doped dot behaves like a magnet with a well-defined easy axis parallel to the growth direction [81].

3. PERSPECTIVES

CdTe-based nanostructures can be controlled nowadays at the ultimate limit like other semiconductors using ALE in ultra-high vacuum chamber. As far as the dots are concerned, the control of the QD's position together with the possibility to insert them in a field effect structure, are the two next technological steps to put forward this system. In the future, the development of nanowires by initiating the growth with a catalyst will probably be a promising way to perform nanostructures, suppose that the CdTe surface can be stable in air. For the optical properties, the II-VI low-dimensional systems are very powerful to elucidate some new optical effects due to their strong light-matter interaction and the robustness of their excitons. Let us mention that we did not review in this chapter, the II-VI colloidal dots or nanocrystals which are very efficient light emitters even at room temperature. With an individual II-VI QD, we can hope to perform single photon source emitter at room temperature by contrast to InAs/GaAs-based system which operates only at low temperature. Talking about QDs, the possibility to dope these dots with magnetic atoms add a new degree of freedom with fascinating aspects due to the exchange interaction of the charge carriers with the magnetic moment. Since these two types of spins (magnetic atom and charge carrier spins) can be distinguished by their different g-factors, they form an ideal system to study the coherent interaction of spins in zero-dimensional object.

In the field of spin-based quantum computing in the solid state, II-VI quantum dots permit to expect much longer spin coherence time than their III-V counterpart. Indeed, it appears in recent studies of spin coherence time at very low temperature (100 mK) carried out in electrostatic QDs, that the main mechanism of spin decoherence is the hyperfine coupling with the nuclear spins [84, 85]. Regarding this limitation, the advantage of II-VI materials is the low proportion of isotope carrying a nuclear spin. For instance, for CdTe, only 24% of the cadmium and 7% of the tellurium atoms carry a nuclear spin $I = 1/2$ compared to 100% of spin $I = 9/2$ for the indium and 100% of spin $I = 3/2$ for the arsenic in InAs QDs. This low density of nuclear spins allows expecting long enough spin coherence times to control coherently a single spin by using microwave magnetic resonance techniques.

ACKNOWLEDGMENTS

We would like to warmly thank all our colleagues from the CEA-CNRS group "Nanophysique et Semiconducteurs" in Grenoble, who are at the origin of all these results. Without their strong motivation and the good synergy which exists in this group for many years, such review on CdTe nanostructures results would have been impossible.

REFERENCES

[1] J.P. Faurie, S. Sivananthan, D. Boukerche, J. Reno, Appl. Phys. Lett. 45 (1984) 1307–1309.
[2] A. Million, L. Di Cioccio, J.P. Gailliard, J. Piaguet, J. Vacuum Sci. Technol A6 (1988) 2813–2820.
[3] J.M. Arias, G. Sullivan, J. Vac, Sci. Technol. A5 (1987) 3143–3146.
[4] G. Lentz, A. Ponchet, N. Magnea, H. Mariette, Appl. Phys. Lett. 55 (1989) 2733–2735; L. Ulmer, N. Magnea, H. Mariette, P. Gentil, J. Cryst. Growth 111 (1991) 711–714.
[5] J.H. Neave, P.J. Dobson, B.A. Joyce, J. Zhang, Appl. Phys. Lett. 47 (1985) 100–102.
[6] P.H. Jouneau, A. Tardot, G. Feuillet, J. Cibert, H. Mariette, J. Appl. Phys. 75 (1994) 7310–7316.
[7] J.M. Hartmann, G. Feuillet, M. Charleux, H. Mariette, J. Appl. Phys. 79 (1996) 3035–3041.
[8] J.M. Hartmann, J. Cibert, F. Kany, H. Mariette, M. Charleux, P. Alleysson, R. Langer, G. Feuillet, J. Appl. Phys. 80 (1996) 6257–6265.
[9] J. Kim, M.C. Gallagher, R.F. Willis, J. Fu, D.J. Miller, J. Vac. Sci. Technol. B11 (1993) 1370–1373.
[10] B. Daudin, S. Tatarenko, D. Brun-Le Cunff, Phys.Rev. B 52 (1995) 7822–7825.
[11] D. Martrou, J. Eymery, N. Magnea, Phys. Rev. Lett. 83 (1999) 2366–2369.
[12] A. Ohtake, T. Hanada, T. Yasuda, K. Arai, T. Yao, Phys. Rev. B 60 (1999) 8326–8332.
[13] M. Charleux, J.L. Rouvière, J.M. Hartmann, A. Bourret, J. Appl. Phys. 84 (1998) 756–764.
[14] A. Tardot, N. Magnea, J. Cryst. Growth 148 (1995) 236–240.
[15] W. Grieshaber, J. Cibert, J.A. Gaj, Y. Merle d'Aubigné, A. Wasiela, Phys. Rev. B 50 (1994) 2011–2014, idem Phys. Rev. B 50, 5512-5527, 1994.
[16] J.M. Moison, C. Guille, F. Houzay, F. Barthe, M.V. Rompay, Phys. Rev. B 40 (1989) 6149–6162.
[17] T. Baron, F. Kany, K. Saminadayar, N. Magnea, R. Cox, Appl. Phys. Lett. 70 (1997) 2963–2965.
[18] D. Martrou, N. Magnea, Thin Solid Films 367 (2000) 48–57.
[19] L. Marsal, A. Wasiela, G. Fishman, H. Mariette, F. Michelini, S. Nagahara, T. Kita, Phys. Rev. B 63 (2001) 165304–165313.
[20] H. Mariette, M. Charleux, J.M. Hartmann, F. Kany, D. Martrou, L. Marsal, N. Magnea, J.L. Rouvière, Microelectron. J. 30 (1999) 329–334.

[21] I.N. Stranski, L. Krastanow, Ber. Akad. Wiss. Wien, Math.-Naturwiss. Kl. Abt. IIb 146 (1938) 797.

[22] F.C. Frank, J.H. Van der Merwe, Proc. Roy. Soc. (Lond.) A 198 (1949) 216.

[23] L. Goldstein, F. Glas, J.Y. Marzin, M.N. Charasse, G. Le Roux, Appl. Phys. Lett. 47 (1985) 1099–1101.

[24] J.-Y. Marzin, J.-M. Gérard, A. Izrael, D. Barrier, G. Bastard, Phys. Rev. Lett. 73 (1994) 716–719; D. Leonard, K. Pond, P.M. Petroff, Appl. Phys. Lett. 50 (1994) 11687–11689; R. Nötzel, Semicond. Sci. Technol 11 (1996) 1365–1380.

[25] Y. Terai, S. Kuroda, K. Takita, T. Okuno, Y. Masumoto, Appl. Phys. Lett. 73 (1998) 3757–3759.

[26] G. Karczewski, S. Mackowski, M. Kutrowski, T. Wojtowicz, J. Kossut, Appl. Phys. Lett. 74 (1999) 3011–3013.

[27] S.H. Xin, P.D. Wang, A. Yin, C. Kim, M. Dobrowolska, J.L. Merz, J.K. Furdyna, Appl. Phys. Lett. 69 (1996) 3884–3886.

[28] J. Cibert, Y. Gobil, L.S. Dang, S. Tatarenko, G. Feuillet, P.H. Jouneau, K. Saminadayar, Appl. Phys. Lett. 56 (1990) 292–294.

[29] F. Tinjod, I.-C. Robin, R. Andre, K. Kheng, H. Mariette, J. Alloys Compd. 371 (2004) 63; H. Mariette, CR Acad. Sci. Paris. 6 (2005) 23.

[30] F. Tinjod, B. Gilles, S. Moehl, K. Kheng, H. Mariette, Appl. Phys. Lett. 82 (2003) 4340–4342.

[31] E. Janik, J. Sadowski, P. Dluzewski, S. Kret, A. Presz, L.T. Baczewski, E. Lusakowska, J. Wrobel, G. Karczewski, T. Wojtowicz, in: W. Jantsch, F. Schäfler (Eds.), Proc of 28th Int. Conf. Phys. Semicond., AIP Conf., vol. 893 (2007) 103; G. Karczewski, S. Mahapatra, T. Borzzenko, P. Dluzewski, S. Kret, L. Klopotowski, C. Schumacher, K. Brunner, L.W. Molenkamp, T. Wojtowicz, idem. AIP Conf. 893 (2007) 65.

[32] H. Tuffigo, N. Magnea, H. Mariette, A. Wasiela, Y. Merle d'Aubigné, Phys. Rev. B 43 (1991) 14629–14637.

[33] E. Deleporte, J.M. Berroir, C. Delalande, N. Magnea, H. Mariette, J. Allegre, J. Calatayud, Phys. Rev. B 45 (1992) 6305–6308.

[34] P. Peyla, Y. Merle d'Aubigné, A. Wasiela, R. Romestain, H. Mariette, M.D. Sturge, N. Magnea, H. Tuffigo, Phys. Rev. B 46 (1992) 1557–1563.

[35] A. D'Andrea, N. Tomassini, Phys. Rev. B 47 (1993) 7176–7181.

[36] E.L. Ivchenko, G. Pikus, Superlattices and other Microstructures, Springer Verlag, Berlin, 1995.

[37] Y. Merle d'Aubigné, L.S. Dang, A. Wasiela, N. Magnea, F. Dalbo, A. Million, J. Phys. (Paris) 48 (1987) C5-363-366; H. Tuffigo, R.T. Cox, N. Magnea, Y. Merle d'Aubigné, A. Million, Phys. Rev. B 37 (1988) 4310–4313.

[38] J.J. Davies, D. Wolverson, V. Kochereshko, A.V. Platonov, R.T. Cox, J. Cibert, H. Mariette, I.V. Ignatiev, E.V. Ubyivovk, Yu P. Ekimov, Phys. Rev. Lett. 97 (2006) 187403–187406.

[39] R. Dingle, W. Wiegmann, C.H. Henry, Phys. Rev. Lett. 33 (1974) 827–830.

[40] J. Kusano, Y. Segawa, M. Mihara, Y. Aoyagi, S. Namba, Solid State Commun. 72 (1989).

[41] K. Cho, A. D'Andrea, R. Del Sole, H. Ishibara, J. Phys. Soc. Jpn. 59 (1990) 1853–1856.

[42] C. Weisbuch, M. Nishioka, A. Ishikawa, Y. Arakawa, Phys. Rev. Lett. 69 (1992) 3314–3317.

[43] R. André, D. Heger, L.S. Dang, Y. Merle d'Aubigné, J. Cryst. Growth 184 (1998) 758.

[44] J. Kasprzak, M. Richard, S. Kundermann, A. Baas, P. Jeambrun, J.M.J. Keeling, F.M. Marchetti, M.H. Szymanska, R. André, J.L. Staehli, V. Savona, P.B. Littlewood, B. Deveaud, L.S. Dang, Nature 443 (2006) 409–414.

[45] L.S. Dang, D. Heger, R. André, F. Bœuf, R. Romestain, Phys. Rev. Lett. 81 (1998) 3920–3923.

[46] M.A. Lampert, Phys. Rev. Lett. 1 (1958) 450–453.

[47] G. Munschy, B. Stébé, Phys. State Solidi (b) 64 (1977) 213–222.
[48] B. Stébé, A. Ainane, Superlattices Microstruct. 5 (1989) 545.
[49] K. Kheng, R.T. Cox, Y. Merle d'Aubigné, F. Bassani, K. Saminadayar, S. Tatarenko, Phys. Rev. Lett. 71 (1993) 1752–1755.
[50] G. Finkelstein, H. Shtrikmana, I. Bar-Joseph, Phys. Rev. Lett. 74 (1995) 976–979.
[51] F. Capasso, K. Mohamed, A.Y. Cho, IEEE J. Q. Electron. 22 (1986) 1853–1869.
[52] M. Nido, M.G.W. Alexander, W.W. Rühle, T. Schweizer, K. Köhler, Appl. Phys. Lett. 56 (1990) 355–357; D.Y. Oberli, J. Shah, T.C. Damen, J.M. Kuo, J.E. Henry, J. Lary, S.M. Goodnick, Appl. Phys. Lett. 56 (1990) 1239–1241.
[53] R. Ferreira, P. Rolland, Ph. Roussignol, C. Delalande, A. Vinattieri, L. Carraresi, M. Colocci, N. Roy, B. Sermage, J.F. Palmier, B. Etienne, Phys. Rev. B 45 (1992) 11782–11794.
[54] S. Haacke, N.T. Pelekanos, H. Mariette, M. Zigone, A.P. Heberle, W.W. Rühle, Phys. Rev. B 47 (1993) 16643–16646.
[55] I. Lawrence, S. Haacke, H. Mariette, W.W. Rühle, H. Ulmer-Tuffigo, J. Cibert, G. Feuillet, Phys. Rev. Lett. 73 (1994) 2131–2134.
[56] F. Findeis, A. Zrenner, G. Bohm, G. Abstreiter, Solid State Commun. 114 (2000) 227–230.
[57] M. Bayer, O. Stern, P. Hawrylak, S. Fafard, A. Forchel, Nature 405 (2000) 923–926.
[58] L. Marsal, L. Besombes, F. Tinjod, K. Kheng, A. Wasiela, B. Gilles, J.-L. Rouviere, H. Mariette, J. Appl. Phys. 91 (2002) 4936–4943.
[59] S. Makowski, J. Wrobel, K. Fronc, J. Kossut, F. Pulizzi, P. Christianen, J.C. Maan, G. Karczewski, Phys. Status Solidi B 229 (2002) 493–496.
[60] V.D. Kulakovskii, G. Bacher, R. Weigand, T. Kümmell, A. Forchel, E. Borovitskaya, K. Leomardi, D. Hommel, Phys. Rev. Lett. 82 (1999) 1780–1783.
[61] M. Lowisch, M. Rabe, B. Stegermann, F. Henneberger, M. Grundmann, V. Türck, D. Bimberg, Phys. Rev. B 54 (1996) R11074–11077; F. Flack, V. Nikitin, P.A. Crowell, J. Shi, J. Levy, N. Samarth, D.D. Awschalom, Phys. Rev. B 54 (1996) R17312–17315.
[62] K. Sebald, P. Michler, T. Passow, D. Hommel, G. Bacher, A. Forchel, Appl. Phys. Lett. 81 (2002) 2920–2922.
[63] L. Besombes, K. Kheng, L. Marsal, H. Mariette, Euro. Phys. Lett. 65 (1994) 144–147.
[64] A. Wojs, P. Hawrilack, S. Fafard, L. Jacak, Phys. Rev. B 54 (1996) 5604–5608.
[65] P. Hawrylak, Phys. Rev. B 60 (1999) 5597–5608.
[66] L. Besombes, K. Kheng, L. Marsal, H. Mariette, Phys. Rev. B 65 (2002) R121314–121317.
[67] D. Gammon, E.S. Snow, B.V. Shanabrook, D.S. Katzer, D. Park, Science 273 (1996) 87–90.
[68] K. Ota, N. Usami, Y. Shiraki, Physica E (Amsterdam) 2 (1998) 573–577.
[69] X. Fan, T. Takagahara, J.E. Cunningham, H. Wang, Solid State Commun. 108 (1998) 857–861.
[70] L. Besombes, K. Kheng, L. Marsal, H. Mariette, Phys. Rev. B 63 (2001) 155307–155311.
[71] F. Gindele, K. Hild, W. Langbein, U. Woggon, Phys. Rev. B 60 (1999) R2157–R2160.
[72] P. Borri, W. Langbein, S. Schneider, U. Woggon, R.L. Sellin, D. Ouyang, D. Bimberg, Phys. Rev. Lett. 87 (2001) 157401–157404.
[73] M. Bayer, A. Forchel, Phys. Rev. B 65 (2002) 041308–041311.
[74] I. Favero, G. Cassabois, R. Ferreira, D. Darson, C. Voisin, J. Tignon, C. Delalande, G. Bastard, Ph. Roussignol, J.-M. Gérard, Phys. Rev. B 68 (2003) 233301–233304.
[75] E. Peter, J. Hours, P. Senellart, A. Vasanelli, A. Cavanna, J. Bloch, J.-M. Gérard, Phys. Rev. B 69 (2004) 041307–041310.
[76] F. Rol, S. Founta, H. Mariette, B. Daudin, L.S. Dang, J. Bleuse, D. Peyrade, J. M. Gérard, B. Gayral, Phys. Rev. B 75 (2007)125306-(1–7).
[77] J.K. Furdyna, J. Appl. Phys. 64 (1988) R29–R64.
[78] H. Boukari, P. Kossacki, M. Bertolini, D. Ferrand, J. Cibert, S. Tatarenko, A. Wasiela, J.A. Gaj, T. Dietl, Phys. Rev. Lett. 88 (2002) 207204–207207.

[79] G. Bacher, A.A. Maksimov, H. Schömig, V.D. Kulakovskii, M.K. Welsch, A. Forchel, P.S. Dorozhkin, A.V. Chernenko, S. Lee, M. Dobrowolska, J.K. Furdyna, Phys. Rev. Lett. 89 (2002) 127201–127204.
[80] L. Besombes, Y. Leger, L. Maingault, D. Ferrand, H. Mariette, J. Cibert, Phys. Rev. Lett. 93 (2004) 207403–207406.
[81] Y. Léger, L. Besombes, J. Fernandez-Rossier, L. Maingault, H. Mariette, Phys. Rev. Lett 97 (2006) 107401–107404.
[82] J. Fernandez-Rossier, L. Brey, Phys. Rev. Lett. 93 (2004) 117201–117204.
[83] A.O. Govorov, A.V. Kalameitsev, Phys. Rev. B 71 (2005) 035338-1–5; A.O. Govorov, ibid 72 (2005) 075359-1–8.
[84] J.R. Petta, A.C. Johnson, J.M. Taylor, E.A. Laird, A. Yacoby, M.D. Lukin, C.M. Marcus, M.P. Hanson, A.C. Gossard, Science 309 (2005) 2180–2184.
[85] F.H.L. Koppens, J.A. Folk, J.M. Elzerman, R. Hanson, L.H. Willems van Beveren, I.T. Vink, H.P. Tranitz, W. Wegscheider, L.P. Kouwenhoven, L.M.K. Vandersypen, Science 309 (2005) 1346–1350.

CHAPTER IV

CdTe-Based Semimagnetic Semiconductors

Robert R. Gałązka and **Tomasz Wojtowicz**

Contents

1. INTRODUCTION

CdTe is a component of the best known and widely investigated semimagnetic semiconductor (SMS) $Cd_{1-x}Mn_xTe$. There exist also other CdTe based SMSs such as CdHgMnTe, CdZnMnTe, CdMnTeSe, CdCoTe, CdCrTe, CdFeTe, not so well known as CdMnTe but also investigated and showing interesting properties. Numerical data and functional relationships for

Institute of Physics, Polish Academy of Sciences, Al. Lotników 32/46, 02-668 Warsaw, Poland

CDTE and Related Compounds
DOI: 10.1016/B978-0-08-046409-1.00004-6

these materials are collected in review papers, books and chapters of handbooks [1–11]. In this chapter only the most important features and results characteristic for this group of solids will be presented.

SMSs are the group of solids at the interface between semiconductors and magnetic materials. SMSs (also referred to as diluted magnetic semiconductors) are semiconductor-based solid solutions where a part of cations are replaced by transition metals or rare earth elements. Crystallographic structure of semiconductor is conserved; the lattice constant is a function of composition. From a magnetic point of view SMS is a disordered magnetic material, since magnetic atoms are randomly distributed in the cation sublattice of semiconductor compound.

Generally speaking, in SMSs two interrelated and interacting subsystems coexist: mobile delocalized charge carriers and localized magnetic moments connected with paramagnetic ions. Electronic properties of SMSs are the subject of intensive studies since early seventies years. Due to strong spin exchange interaction between mobile carriers and localized magnetic moment (exchange constant of sp-d interaction is of the order of 1 eV for II-VI SMSs) significant changes of the band structure and behavior of carriers were observed. A number of new physical phenomena were discovered, such as giant Faraday rotation, magnetic field induced metal-insulator transition, bound magnetic polaron (BMP). Magnetic properties of SMSs have been a subject of studies since about 1980.

In 1987 the first paper devoted to layered structure and magnetic properties of low-dimensional (LD) SMS was published starting intensive studies of superlattices and other LD structures made of SMS.

2. CRYSTAL GROWTH TECHNOLOGY OF CdMnTe

CdMnTe crystal growth procedure is very similar to the methods applied for CdTe. Commonly used is the Bridman technique with some modifications. Source materials are usually CdTe, Mn, and Te. Sometimes prereacted MnTe is added to CdTe. In principle, all components Cd, Mn, and Te can be added separately. However, because of very different melting point of Mn in respect to Cd and Te, and also exothermic chemical reaction between Cd and Te at high temperature this method is less applicable. Melting point of alloys decreases with Mn concentration from 1090 °C of CdTe to about 1060 °C for 70% of Mn molar fraction. Bulk CdMnTe crystals can be obtained in CdTe zinc blende crystallographic structure up to 77% of Mn. Lattice constant decreases linearly with Mn concentration. Phase diagram [12] shows a rather unique property of CdMnTe solid solution—liquidus and solidus curves are very close to each other, so segregation coefficient is very close to 1 in the whole range of Mn concentration. Thus the whole crystal is very homogenous independently of the chemical composition.

Crystallographic investigations reveal cation vacancies, Te inclusions, and twins. All of these defects are also present in CdTe and can be, at least partially, removed by appropriate annealing. As grown crystals are usually p-type with hole concentration about 10^{16} cm^{-3}. Ionicity of chemical bond in CdMnTe slightly increased in respect to CdTe, reaches maximum for about 30% of Mn than decreases to the value close to CdTe. Vickers microhardness increases from 40 kg mm^{-2} for CdTe to 60 kg mm^{-2} for 25% of Mn and remains at this level until 70% of Mn [13].

Influence of impurities on electronic properties of CdMnTe was not so widely investigated as in CdTe. Doping with Cu, Ag, P, and N can increase the hole concentration originally present due to Cd vacancies. Doping with typical for CdTe donors such as In, Ga, or Cl is less effective. For Mn concentration higher than 5% doping for n-type results in high resistivity, highly compensated material containing probably neutral complexes of Cd vacancies and shallow donors.

3. PHYSICAL PROPERTIES OF BULK CdMnTe

3.1. Energy band structure of SMS

Replacing cations such as Cd with a paramagnetic ion such as Mn in the same crystallographic structure does not markedly disturb the semiconductor properties of the material. The energy gap changes (increases) but the conduction and valence bands conserve their symmetry and character as in nonmagnetic semiconductor mixed crystals. The spin momentum of paramagnetic ions is connected with the 3d or 4f shell: for transition metals or rare earth elements, respectively. The energy level of 3d or 4f electrons lies below the top of the valence band and has thus negligible influence on its shape or the shape of the bottom of the conduction band. Thus all basic semiconductor properties connected with the band symmetry and topology are the same as for a typical semiconductor and the only influence of the magnetic subsystems on electrons comes from the spin exchange interaction between mobile carriers and localized magnetic ions. This interaction can be represented by a Heisenberg term

$$H_{\mathrm{ex}} = \bar{\sigma}\sum_{i} \bar{S}_i J(\vec{r} - \vec{R}_i), \qquad (1)$$

where $\bar{\sigma}$ and $\bar{S}$ are spin operators of band electron and magnetic ion, respectively, summation is over all lattice sites occupied by magnetic ions, and J is an exchange constant.

Assuming mean field and virtual crystal approximations this term can be rewritten in the form periodic with a lattice constant:

$$H_{ex} = x\langle\bar{S}\rangle_{av}\bar{\sigma}\sum_{R} J(\vec{r} - \vec{R}), \tag{2}$$

where x is a molar fraction of magnetic ions, $\langle\bar{S}\rangle_{av}$ is the average over all magnetic ions and directly related to magnetization of sample by the equation:

$$M = N_0 g\mu\langle\bar{S}\rangle_{av}, \tag{3}$$

where N_0 is the number of unit cells in unit volume, g is the Lande factor for magnetic ions, and μ is the Bohr magneton.

The Heisenberg term must be added to the effective mass Hamiltonian to solve the energy eigen problem for SMS. Realizing this was the turning point in understanding their properties and provided a basis for this group of alloys to be distinguished from other semiconductor mixed crystals. Since that time SMS has become the subject of intensive studies in many laboratories worldwide [14].

Strictly speaking the Hamiltonian should also contain the term representing paramagnetic ion-ion exchange interaction responsible for magnetic properties of SMS. However, this term is about three orders of magnitude smaller than term (2). For this reason for energy band structure determination one can neglect paramagnetic ion interaction term, what significantly simplifies further calculations.

It is worth to notice that the experimentally measured magnetization, which can serve to evaluate $\langle\bar{S}\rangle_{av}$ in Eq. (3), reflects all magnetic interactions present in the crystal, therefore, also the ion-ion exchange interaction is, in fact, included in the band structure determination although not in a straightforward way.

The exchange interaction depends on the value of ion spin and the exchange constant. The spin of Mn is the highest for transition metals and equals 5 Bohr magnetons, whereas a typical value for the exchange coupling between magnetic ions $J_{d\text{-}d}$ is 10^{-3} eV, and the exchange constant for electrons and holes with magnetic ions $J_{sp\text{-}d}$ is about 1 eV in SMS.

The exchange interaction is also strongly temperature and magnetic field dependent. As was said earlier the macroscopic magnetization of a sample is proportional to the thermodynamic average value of the magnetic ion spin, and therefore to a good approximation we can replace the ion spin operators in the Hamiltonian by their average values, calculated or taken from measurements of the magnetization. From typical magnetic functions we can thus obtain information on the electronic behavior.

In semiconductors, an external magnetic field acts on both the orbital motion and the spin of electrons producing Landau quantization and spin splitting, to an extent dependent on the effective mass of the carriers. The exchange interaction acts on their spin only and is, in fact, mass independent. Because of this and also because of the large value of $J_{sp\text{-}d}$, the band

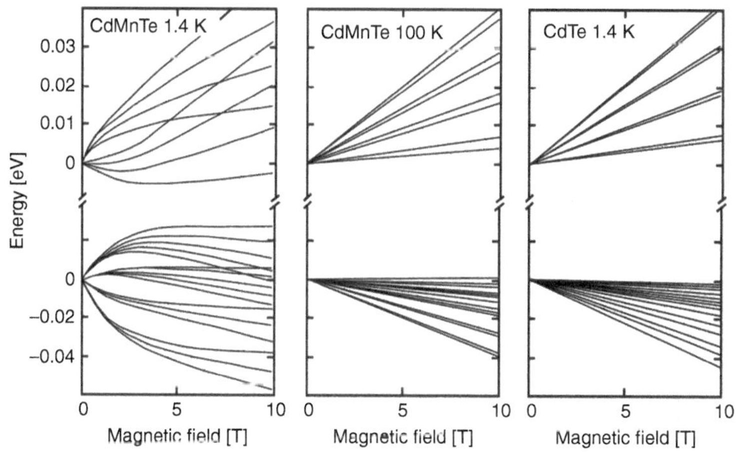

Figure 1 Conduction and valence band quantization in an external magnetic field for $Cd_{0.95}Mn_{005}Te$ and CdTe. Spin splitting, which is very weak in CdTe, is the main effect in CdMnTe. Notice the almost equal splitting of the heavy- and light-hole bands in CdMnTe, and crossing of spin levels. The influence of exchange interaction is visible even at 100 K. The picture of CdTe over this range of temperature is practically unchanged. After Gałązka [15].

structure in turn changes drastically under the influence of an external magnetic field and depends strongly on temperature. From Fig. 1 we can see that the very typical structure of the degeneracy of the light- and heavy-hole valence bands disappears in SMS when a magnetic field is applied. The change is accompanied by a drastic increase of hole mobility like that observed in low-dimensional structures [15].

Exchange interaction not only splits the bands but also introduces substantial band anisotropy. Calculations of the band structure of SMS have to take into account the symmetry of the crystal, appropriate form of Hamiltonian and proper set of wave functions. To illustrate band structure calculation, let us consider a case that is simpler than this presented in Fig. 1. For zinc blende structure semiconductor with open energy bandgap to determine the band structure in the presence of exchange interaction, one has to take into account the exchange splitting of the Γ_6, Γ_8, and Γ_7 levels when performing the $\vec{k} \cdot p$ band calculation [16]. This was done by keeping only terms of the order of k^2 and neglecting the inversion asymmetry and warping, with the following results. The splitting of Γ_6 band changes only slowly with $\vec{k}$ and can be described by an effective g-factor. This g-factor, for typically used x, is very high and positive even in narrow-gap SMS, in spite of the fact that in narrow-gap nonmagnetic semiconductors the g-factor of conduction electrons is negative.

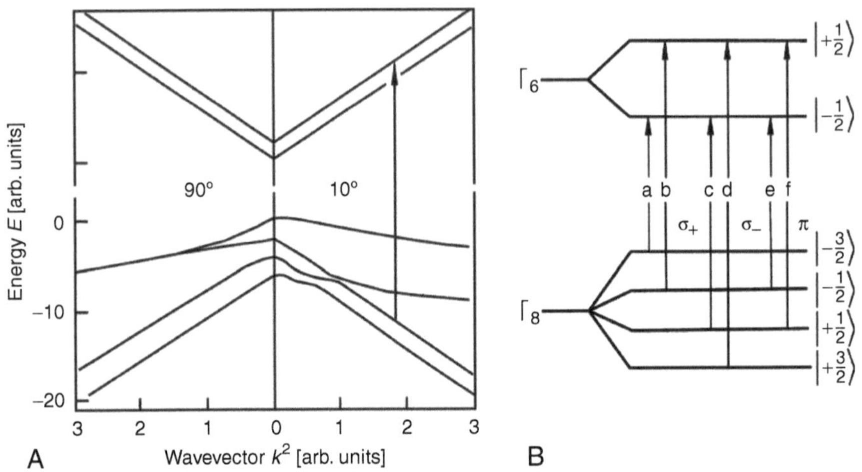

Figure 2 (A) Energy band structure of CdMnTe, *E*(*k*). *E* versus *k* is calculated for two values of the angle between *k* and external magnetic field 90° and 10°. (B) Schematic representation of bands splitting, transitions allowed in the Faraday (σ^+ and σ^- polarizations) and Voigt (π polarization) configurations are marked by arrows. After Gaj et al. [16].

For the Γ_8 bands, the splitting depends dramatically on both the direction and the absolute value of $\vec{k}$. This is shown schematically for an open-gap band structure in Fig. 2. The reduced variables used here are $\varepsilon = E/B_{\mathrm{exch}}$, $\vec{\kappa} = \hbar\vec{k}/(2B_{\mathrm{exch}}m_{\mathrm{hh}})^{1/2}$, where E is the electron energy and m_{hh} is the heavy-hole mass. Of particular interest are the dispersion relations close to the Γ point, that is, for

$$\hbar^2k^2/2m_{\mathrm{lh}} \ll |B_{\mathrm{exch}}|, \tag{4}$$

where B_{exch} is the value connected with H_{ex}, Eq. (2), and equals $B_{\mathrm{exch}} = -1/6\ N_0\beta\ \langle S_Z\rangle\ x$, where β is the exchange integral for valence band and m_{lh} is the light-hole mass. The four nondegenerate valence bands are (the warping of the valence band is neglected)

$$E_{\pm 3/2}(\vec{k}) = -\frac{1}{2}\hbar^2\left[\left(\frac{3}{4m_{lh}} + \frac{1}{4m_{hh}}\right)k_\perp^2 + \frac{k_z^2}{m_{hh}}\right] \pm 3B_{\mathrm{exch}}, \tag{5}$$

$$E_{\pm 1/2}(\vec{k}) = -\frac{1}{2}\hbar^2\left[\left(\frac{1}{4m_{lh}} + \frac{3}{4m_{hh}}\right)k_\perp^2 + \frac{k_z^2}{m_{lh}}\right] \pm B_{\mathrm{exch}}, \tag{6}$$

where $\vec{k}_\perp$ is the wave-vector component perpendicular to the magnetic field. Thus, the constant energy surfaces are rotational ellipsoids, cucumber-like for the highest and the lowest bands (±3/2) and disc-like for the intermediate bands (±1/2). For $m_{\mathrm{hh}} \gg m_{\mathrm{lh}}$, the mass anisotropy is particularly strong for the bands ±3/2, being equal to $3m_{\mathrm{hh}}/4m_{\mathrm{lh}}$. As $J_{\text{p-d}} > 0$ for know materials, that is, $B_{\mathrm{exch}} < 0$, $E_{-3/2}(\vec{k})$ is the highest valence band.

For $\vec{k}$ not fulfilling Eq. (4) there is a substantial mixing of the wave functions and the dependence of energy on $\vec{k}$ is more complicated.

So far we have neglected the direct effect of the external magnetic field on the orbital motion of the carrier, as well as the intrinsic spin splitting, taking into account only the ordering of manganese spins by the field and the exchange interaction of carriers with these spins. The direct effect may be neglected, in fact, at rather weak magnetic fields in wide-gap SMS, like $Cd_{1-x}Mn_xTe$ with relatively large x, because of high effective masses of carriers (i.e., small Landau splittings) and low intrinsic g-factors.

Impurity levels are also influenced by exchange interactions: the ionization energy of an acceptor decreases and its wave function becomes anisotropic under the influence of a magnetic field. Both effects produce a giant negative magnetoresistance and a field-induced insulator-metal transition. Indeed all the properties susceptible of being changed by a magnetic field are very different for CdMnTe in comparison with ordinary semiconductor—the Faraday effect, magnetooptical properties, luminescence; these are all different and are also strongly temperature and magnetic field dependent.

In the absence of a magnetic field, the average value of the ion spins is zero, the magnetization is zero too, and SMS should behave as typical nonmagnetic semiconductors. Whereas this is true for delocalized band electrons of low density (see farther about carrier induced ferromagnetism), when an electron is localized around an impurity, a BMP can be observed. This concept was first introduced to explain the behavior of magnetic semiconductors (Eu chalcogenides) and then later was applied also to SMS. A localized electron can produce a spontaneous magnetization within the range of its wave function via the exchange interaction. Local ferromagnetic ordering in that range produces a Stoke's shift in spin-flip Raman scattering even in the absence of an external magnetic field and other related phenomena. Local fluctuations of magnetization and a detailed knowledge of electronic states are essential for a theoretical description of this effect. The BMP is a subtle example of the feedback between electronic and magnetic subsystems present in SMS.

3.2. Optical properties

Optical properties of CdMnTe as well as of the other SMSs with Mn in the absence of external magnetic field exhibit some extra features in respect to typical semiconductor mixed crystal, for example, CdZnTe. Mn 3d level lies 3.5 eV below the top of the valence band and does not influence the energy gap (E_g) behavior. $Cd_{1-x}Mn_xTe$ energy gap increases with Mn concentration x according to experimentally established rule: $E_g(x) = (1.595 + 1.598x)$ eV at 10 K. Absorption edge corresponds to energy gap shift up to about 2.2 eV at 4.2 K and is pinned at this energy because the

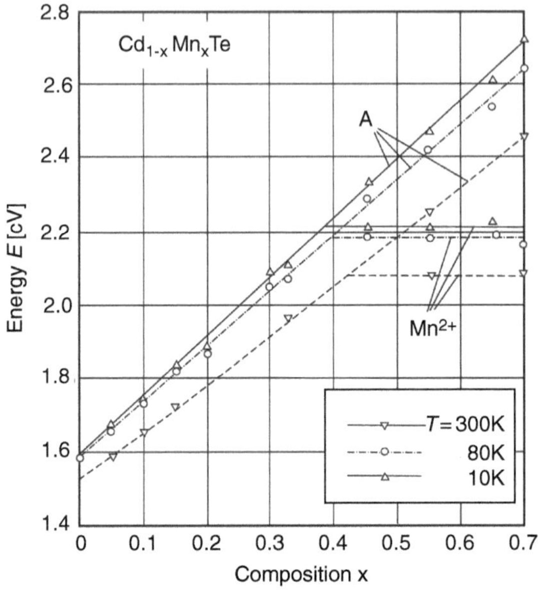

Figure 3 Variation of the energy gap and Mn transition with Mn concentration in $Cd_{1-x}Mn_xTe$ for 10, 80, and 300 K. In the figure, the peak labeled A is identified as the free exciton. The concentration-independent feature Mn^{2+} is associated with the leading edge of the Mn $^6A_1(^6S) \rightarrow {}^4T_1(^4G)$ absorption band. After Lee and Ramdas [17].

strong intra-Mn transitions stop the transmission for a higher energy—Fig. 3. E_g decreases approximately linearly with temperature above liquid nitrogen temperature with average coefficient -3.5×10^{-4} eV/K. However, departure from this behavior—extra shift to higher energies—below the spin freezing temperature has been observed [1, 2, 4].

In spite of fundamental absorption edge, the intra-Mn transitions – broad excitation and emission features – have been observed for $x \geq 0.6$ when fundamental absorption edge is shifted to the higher energy opening the window for these transitions. The energy associated with intra-Mn transitions is usually given as a function of crystal field parameter D_q and Racah parameters B and C (Tanabe-Sugano model) [4]. Because all three parameters are a function of the distance between the transition ion and ligand ion, intra-Mn emission and excitation spectra are function of Mn concentration x, hydrostatic pressure, and temperature.

External magnetic field, ordering Mn spins, causes significant changes in the energy band structure—Figs. 1 and 2. In turn both absorption and reflectivity spectra are now strongly magnetic field dependent and depend also on the light polarization due to selection rules. The exciton ground state splits to six components: four visible in the σ polarization and two visible in the π polarization. In Faraday configuration four

σ-components exhibit two σ^+ and two σ^- circular polarizations. The whole splitting pattern is approximately symmetric to the zero-field position. Components a and d are significantly stronger then b and c (Fig. 2B). The splitting depends on $J_{sp\text{-}d}$ exchange constant values and is proportional to the magnetization, which in turn is temperature and magnetic field dependent [1, 2].

All the other magnetooptical effects in CdMnTe are also strongly and, differently than in typical semiconductors, magnetic field and temperature dependent. Faraday effect is a particularly spectacular example of the influence of exchange sp-d coupling on the magnetooptical effect in CdMnTe. The effect is orders of magnitude stronger than in CdTe with a sign opposite to that in nonmagnetic II-VI compounds [18, 19].

Luminescence studies and Raman spectroscopy supplied additional information concerning impurity levels, localized and collective excitations being vibrational, electronic, or magnetic in character. Free and bound to impurity luminescence spectra are strongly magnetic field dependent. Many luminescence measurements have been performed to study BMP effects. In excitonic spectra acceptor BMP has been observed and analyzed by many authors.

Detailed studies, both experimental and theoretical have given insight into the behavior of BMP. At high temperatures magnetic fluctuations determine the behavior of BMP. As temperature is lowered, the BMPs continuously evolve from fluctuation-dominated regime to a collective regime in which the carrier and magnetic ion spins are strongly correlated with each other. Experimentally it can be seen in magnetooptical studies as a deviation from strict proportionality of Zeeman splitting to magnetization. Theoretical approach must go beyond the molecular field approximation and include mutual interaction between the ions and electron spins at microscopic scale [4].

Vibrational Raman spectra for CdMnTe exhibit a two-mode behavior. In other words in mixed crystals TO and LO phonons structures characteristic for CdTe-like and MnTe-like vibrational modes are observed. Spin-flip Raman scattering in Ga doped CdMnTe and finite Raman shift observed in zero magnetic field was attributed to donor BMP. For high Mn concentration, $x > 0.60$, partial ordering in antiferromagnetic phase takes place. A distinct magnon feature was observed in CdMnTe for $0.40 \leq x \leq 0.70$ in Raman scattering. From polarization characteristics and temperature behavior of Raman line the one-magnon excitation was deduced [4].

Optical transitions away from the center of Brillouin zone exhibit the same structure as for CdTe although observed maxima for transitions corresponding to the off center points of Brillouin zone are broader with increasing Mn concentration. Commonly denoted peak positions E_1 and $E_1 + \Delta_1$ only slightly shift with x to higher energy indicating that the energy band structure in the whole Brillouin zone remains nearly the same as for CdTe although some broadening of the bands is observed due to disorder characteristic for mixed crystals [1, 20].

3.3. Transport properties of CdMnTe

As grown CdMnTe is always p-type with typical hole concentration $p = 10^{15}$-10^{17} cm^{-3} and mobility about 40 cm^2/Vs at room temperature. At low temperatures hopping transport is observed. Resistivity measurement as a function of temperature for samples with different Mn concentration exhibits substantially higher resistivity activation energy than in CdTe [1].

In CdMnTe doped for n-type with In, Ga, and Al persistent photoconductivity effect is observed. Illuminations of In and Ga doped CdMnTe with white light at temperature below 120 K produce an increase of n-concentration of occupied shallow donors of above three orders of magnitude. The decay rate of n after illumination was of the order of a few hours at 77 K, and at 4.2 K it was too slow to measure. This photomemory effect was used in doped CdMnTe to study a number of physical phenomena such as metal-insulator transition, electron localization effect, electron-electron interaction, influence of donor BMP on the scattering of electrons, thermodynamics properties of BMP, identification of DX centers in CdMnTe with negative correlation energy (negative U) [21–24]. These investigations shed light not only on the properties of CdMnTe but also on the more general problems of phase transitions, scaling theory, nature of DX centers, dynamics of photoionization process, excitonic magnetic polarons effect and others. Semiinsulating CdMnTe is believed to be good candidate for X-ray and γ-ray large area detector. Promising results have been obtained for vanadium doped CdMnTe where high resistivity of sample is accompanied with relatively high value of mobility-lifetime product, both required for good detectors [25].

3.4. Magnetic properties of CdMnTe

Magnetic properties of CdMnTe and also other SMSs are connected with exchange interaction between paramagnetic ions. Some influences of semiconductor properties, such as band structure and carrier concentration, on magnetic properties are also visible. The fact that bulk CdMnTe is available in a wide range of Mn concentrations (up to $x = 0.77$) enabled one to trace the development of specific features in various magnetic properties with increasing number of magnetic atoms incorporated into the host lattice. The continuous transition from diamagnetic behavior (for $x \leq 0.001$) of CdTe, to paramagnetism, spin-glass (for $x \geq 0.05$) and finally antiferromagnetic ordering (for $x > 0.6$) has been observed.

Superexchange is the dominant mechanism leading to the coupling of Mn magnetic moments. Type three antiferromagnetic ordering is observed for $x > 0.6$ indicating that the magnetic elementary cell contains two elementary cells of the cation sublattice. From topological point of view the magnetic subsystem is disordered, the distribution of the magnetic atoms is random in the cation sublattice [6].

Interaction between Mn ions becomes evident for x as small as 0.002, EPR spectra becomes structureless and single resonance line is observed [4, 26, 27].

A spin-glass like state is observed in CdMnTe and other SMSs in the broadest range of paramagnetic ion concentrations ($0.02 \leq x \leq 0.60$) below and above the percolation threshold, Fig. 4. Indeed the temperature dependence of the magnetic susceptibility below and above the transition temperature and its dynamic properties at low and high magnetic fields suggest rather the existence of two spin-glass phases arising from the competition between spin-glass and clustering behavior. Below the percolation threshold ($x \leq 0.17$ for fcc lattice) the mechanism responsible for a spin-glass phase is even less clear, although for such a dilute system it must be long-range interaction to produce freezing of the spins.

The antiferromagnetic order has been inferred from specific heat and magnetic susceptibility measurements. Additional information concerning this phase has come from neutron diffraction studies: only a certain fraction of the total number of magnetic ions (e.g., about 50% for

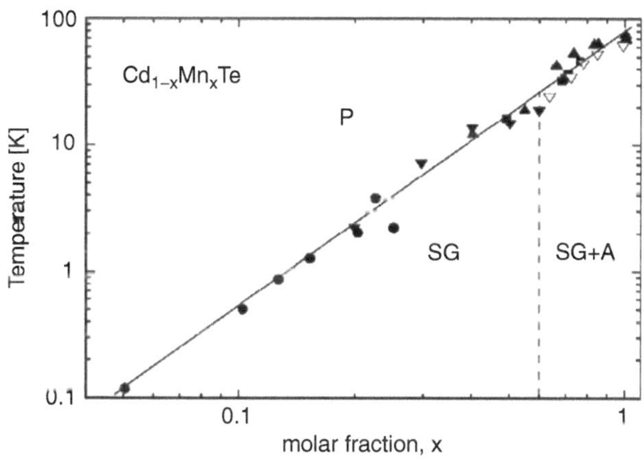

Figure 4 Magnetic phase diagram for CdMnTe in double logarithmic scale. P, SG, A, denotes paramagnetic, spin-glass and antiferromagnetic phases, respectively. For $x \geq 0.8$ long-range antiferromagnetic phase is observed. Experimental points: •, Novak et al. [28]; ▼, Gałązka et al. [29]; ▲, Pietruczanis et al. [30]; Δ, Ando and Akinaga [31]; • and ▼, bulk crystal samples; ▲ and ∇, epilayer samples.

CdMnTe, $x = 0.65$) is well ordered, the rest remaining in a disordered spin-glass phase. Thus a mixed phase (antiferromagnetic and spin-glass coexisting together) rather than truly antiferromagnetic phase is observed.

Many magnetic properties of CdMnTe are common to the whole group of SMS:

- Magnetic phase diagrams are very similar
- Interaction between paramagnetic ions is antiferromagnetic for nearest neighbors
- Both short-range and long-range magnetic interactions are present and are important
- Different magnetic phases coexist

The influence of basic electronic properties such as band structure and carrier concentration on the magnetic properties of SMS is also significant.

In highly doped for p-type CdMnTe thin epitaxial layers (modulation doped quantum well structures) at low temperatures ferromagnetic behavior was observed confirming theoretical predictions [9] (see further).

4. OTHER CdTe-BASED SEMIMAGNETIC SEMICONDUCTORS

As mentioned in Section 1 there exist also other SMSs based on CdTe. In the case of SMSs containing Mn, two quaternary alloys have been particularly investigated: $Cd_xHg_yMn_zTe$ and $Cd_xZn_yMn_zTe$ ($x + y + z = 1$). Quaternary alloys offer the possibility of additional engineering of the band structure keeping magnetic interaction unchanged and *vice versa*.

$Cd_xHg_yMn_zTe$ single crystals and epitaxial layers are grown in wide range of z and y but with $x < 0.1$. Small amount of Mn improves the structure of the alloy. Room temperature energy gap obeys the formula E_g (eV) $= 1.46x - 1.62y + 1.33z$. Exchange constants $J_{\text{sp-d}}$ usually extrapolated from HgMnTe and CdMnTe give satisfactory description of magnetic parameters. High amount of Hg imposes narrow energy gap what makes these alloys suitable for infrared detectors and diodes. Due to exchange interactions photovoltaic spectra are magnetic field tunable [3, 32].

$Cd_xZn_yMn_zTe$ can be grown in wide range of x and y, and $z < 0.7$. Band structure parameters as well as exchange constants can be linearly extrapolated from the values of CdMnTe and ZnMnTe. All these quaternary alloys exhibit higher energy gap than CdTe.

CdTe-based SMSs with transition elements other than Mn, such as $Cd_{1-x}Fe_xTe$, $Cd_{1-x}Co_xTe$ and $Cd_{1-x}Cr_xTe$ can be grown only for $x < 0.06$ for CdFeTe and CdCoTe, and for $x < 0.03$ for CdCrTe [2, 6]. In contrary to Mn, the 3d levels of Fe, Co, and Cr are located in the fundamental energy gap. Ground state of these atoms is split in the crystal field of host CdTe crystal. Additional splittings due to spin-orbit interaction and Jahn-Teller effect are also observed. Giant spin splittings of conduction and valence bands due to sp-d exchange interaction are observed at Γ point of the Brillouin zone. Location of 3d level inside the energy gap produces impurity states and influences optical and transport properties of these alloys differently than in Mn containing SMSs where 3d level lies deep in the valence band. Also magnetic properties are different. In CdFeTe Van Vleck paramagnetism is observed and magnetization is anisotropic with respect to the crystallographic axes, what is also true in CdCrTe. Generally speaking transition elements other than Mn act not only via the spin exchange interaction but also dope CdTe, introducing impurity states into the energy gap.

5. EPITAXIAL LAYERS AND LOW-DIMENSIONAL STRUCTURES

A huge impact on the development of CdTe-based semimagnetic semiconductors was brought about by the application of molecular beam epitaxy (MBE). Many research groups all over the world, such as Purdue, Notre Dame in US, Tsukuba in Japan, Grenoble, Wuerzburg, Hull, and many others, and much later Warsaw, have contributed to the development of MBE technology and to studies of CdTe-based SMS layers and low-dimensional structures. In a short chapter it is just impossible to discuss all properties or to even briefly mention all the interesting results that have been obtained. Therefore, we decided to give only a few, relatively new examples from various areas, which we think are from one side the most representative and from the other prove that the contribution of CdTe-based SMS was crucial for the development of physics of low-dimensional semimagnetic semiconductor structures as a whole, and in most cases was unmatched by that of any other semimagnetic semiconductor. After short general overview and some words about the growth, we will first discuss chosen results obtained with the use of undoped structures, then describe those obtained on n- and p-type doped structures.

5.1. Introduction: Growth and general overview

The MBE technique was proved to be especially successful in the case of crystals containing Mn, although some attempts to introduce Cr into CdTe-based structures have also been undertaken [33]. This nonequilibrium growth technique allowed, in particular, for the extension of accessible range of Mn concentrations in $Cd_{1-x}Mn_xTe$ random-mixed crystal alloys to the molar fractions x above 0.7, the range of the so-called "weakly diluted magnetic semiconductors" [34] which was previously inaccessible. Also the originally "hypothetical" zinc blende endpoint of

$Cd_{1-x}Mn_xTe$ mixed crystals, the binary MnTe was grown [35, 36] and thoroughly studied. Thick $Cd_{1-x}Mn_xTe$ layers with $0.7 < x \leq 1$ were used to extend the magnetic phase diagram of this SMS to the full composition range. In addition to the common "disordered alloys" the new type of (Cd, Mn)Te semimagnetic semiconductor alloys, called digital alloys (DA), became available [37]. The concept of digital alloys is based on the digital growth technique brought about by atomic precision of MBE. During the digital growth some constituent of the structure, in the discussed case MnTe or $Cd_{1-x}Mn_xTe$, with a certain thickness, frequently submonolayer, is introduced into a base CdTe material, at strictly predefined positions. For example, this technique allows for engineering almost at will the shape of the confining potential in quantum structures in digital fashion (see further). If the constituents are built-in periodically with a very small period of the order of a single monolayer, the result is called a digital alloy. Digital alloying was also demonstrated to be very effective tool for independent tailoring of static and dynamic magnetic properties taking advantage of the fact that the later is strongly sensitive to the Mn clustering [38]. Finally MBE also allowed overcoming the difficulty of efficient n-type and p-type doping of CdTe-based semimagnetic semiconductors, taking advantage on the one hand of lower growth temperature and better control of stoichiometry and on the other of a remote doping technique.

Although semimagnetic semiconductors layers grown by MBE were important extension of bulk SMS crystals the real impact of the MBE technique is in the creation of a new field of low-dimensional semimagnetic semiconductor structures. This new field stands out from the general field of low-dimensional structures by the unique spin-splitting engineering (SSE) that is available in low-dimensional SMSs (see further) as well as by their interesting magnetic properties. CdTe-based SMS structures have their really strong contribution to the field of low-dimensional SMS structures. Let us start by stating that virtually all kinds of basic types of low-dimensional nanostructures known from the field of nonmagnetic structures have also been built of CdTe-based SMS by either direct MBE growth or by the combination with postgrowth structurization. In the area of quasi-two-dimensional structures nominally rectangular single and multiple quantum wells (QWs) as well as superlattices (SLs), graded potential QWs (parabolic, half-parabolic, triangular) [39, 40], QW structures with grading or step-like profiles of various layer thickness (of QW region, barrier, cap, or doped region) in the direction perpendicular to the growth axis (resulting, e.g., in "wedge" QWs) were grown [40]. Quasi-one-dimensional structures were produced from layered structures by electron-beam lithography followed by etching [41, 42], by focused ion beam lithography and subsequent thermal annealing [43], by the fractional monolayer growth on (001) vicinal substrates [44] and by cleaved edge overgrowth method, the last resulting in the T-shaped wires [45]. Finally quasi-zero-dimensional structures, CdTe-based SMS quantum dots (QDs), were either formed naturally due to the width fluctuation of narrow QWs [46, 47] or were formed by interdiffusion process in either as grown [48] or ion-beam prepatterned QW structures [43], or were produced with the use of Stranski-Krastanow growth mode on ZnTe-based lattice mismatched substrates by MBE. In the last method, various techniques of incorporating Mn ions into the QDs were used: depositing a submonolayer of Mn ions over a ZnTe surface prior to deposition of the CdTe dot layer [49], diffusion of Mn from a remote Mn-rich layer [50] and direct incorporation of Mn either into the whole [51, 52] or only central part [53] of the CdTe layer from which QD were formed.

The simplest examples of quasi-two-dimensional structures are nominally rectangular QW and superlattices, with Mn introduced either into the QW region or barrier region, or both. CdTe-rich material was used to made a QW region and barriers were typically made of ternary alloys (including digital): $Cd_{1-x}Mn_xTe$ ($E_g \leq 3.4$ eV), $Cd_{1-y}Mg_yTe$ ($E_g \leq 3.7$ eV, however, for y larger than about 0.6 this alloys is hygroscopic and unstable in the air), $Cd_{1-z}Zn_zTe$ ($E_g \leq 2.4$ eV), or even $Mg_{1-x}Mn_xTe$ ($E_g \approx 3.4$-3.7 eV relatively weakly dependent on x), and their mixtures resulting in quaternary alloys. The $Cd_{1-y}Mg_yTe$-based barriers [54] were particularly useful for the growth of high-quality quantum structures due to the relatively good lattice match to the CdTe ($\Delta a/a = 9.3 \times 10^{-3}$ for $y = 1$). Quantum well structures made of all above-listed combinations were type I, with CdTe-rich narrower gap material forming potential well for both electrons and holes (in structures having $Cd_{1-z}Zn_zTe$ barrier, the hole confinement is weak due to the small valence band offset between ZnTe and CdTe).

The growth of CdTe-based SMS layers and quantum structures by MBE is similar to that of nonmagnetic structures and can be performed from either elemental, Cd, Te, Zn, Mg, and Mn sources or binary sources (e.g., CdTe), in the latter case, however, additional elemental fluxes are required to obtain highest quality structures and efficient incorporation of dopants (e.g., additional Cd flux is required in the case of (001)-oriented growth of CdTe). As an n-type dopant: the In [55] and Al [56] substituting metal, and halogens: Br, Cl, and I substituting Te were used (the fluxes of last three delivered from Cd- and Zn-halogen compound sources, e.g., ZnI_2), and allowed to achieve room temperature free electron concentrations n above 10^{18} cm^{-3} in mixed $Cd_{1-x}Mn_xTe$ and $Cd_{1-y}Mg_yTe$ crystal layers having x and y below 0.05, and above 5×10^{17} cm^{-3} for x and y up to 0.1 [57]. Out of these dopants the iodine is probably the most effective for obtaining both the large concentrations (up to 10^{12} cm^{-2}) and highest mobilites of 2D electron gas in $Cd_{1-x}Mn_xTe$-based quantum structures

[58, 59], since electron concentrations of 2×10^{17} cm^{-3} can be achieved in $Cd_{1-y}Mg_yTe$ layers with y as high as 0.37 [60]. Very efficient p-type doping of $Cd_{1-x}Mn_xTe$-based QWs was reported for nitrogen acceptors introduced remotely into the CdZnMgTe barriers with the use of electron cyclotron resonance source [61, 62] and from the surface states [63, 64].

The most typically the (001)-oriented substrates are used for the growth of CdTe-based SMS structures, but successful growths were also reported on (111), (110), and (120) oriented substrates. Apart from CdTe, CdZnTe, and InSb (as well as ZnTe, and GaSb for CdTe/ZnTe quantum dot structures) well lattice matched substrates also strongly mismatched substrates, such as GaAs, were used. Before the growth of actual II-VI structure on GaAs substrate, the appropriate, thick II-VI buffer has to be grown first. The growth of this buffer can be done either in the same or in a separate growth run. In the latter case the growth of the buffer can be performed on the whole wafer, say (001)-oriented epi-ready GaAs, so as to produce the so-called "hybrid" II-VI/GaAs substrate. Hybrid substrate can then be taken out of MBE apparatus and used for the growth of CdTe-based SMS structures later on, either as a whole or after cleaving it into smaller pieces. The number of lattice mismatched induced dislocations in hybrid substrates can be reduced by growing thicker buffer, typically 4-5 μm, and by introducing additional superlattice. In order to stabilize (001) oriented growth of CdTe-rich buffers on (001)-oriented GaAs substrate a thin ZnTe layer (5 nm) is grown first, directly on the GaAs surface. Various types of hybrid II-VI/(001)-GaAs substrates, with II-VI being CdTe, CdMgTe, ZnTe, and ZnMgTe, were developed and proved to be very useful, especially in applications requiring large and flat surfaces, such as patterning of quantum structures with the use of electron beam lithography. It is enough to mention that the record high mobility of two-dimensional electron gas in CdTe and $Cd_{1-x}Mn_xTe$ QWs (recently exciding at helium temperatures values of 400,000 and 100,000 cm^2/Vs, for $x = 0$ and $x = 0.01$, respectively) was achieved in structures grown on such hybrid substrates in Warsaw.

5.2. Undoped structures

The most intriguing and interesting property of undoped quantum structures built of CdTe-based SMS is again, as in the case of bulk crystals, related to the strong sp-d exchange interaction resulting on the one hand in the giant spin splitting of the electronic states and on the other in the formation of magnetic polarons. Low dimensionality and lattice mismatch induced strain in heretostructures introduce, however, the anisotropy of the hole effective g-factors (see for 2D case, e.g., [65], and 1D, e.g., [66, 67])—and anisotropy and strong enhancement of the excitonic magnetic polarons effects (for review on 2D case see [68, 69]).

Moreover, whereas the spin splitting in bulk SMSs is tunable either by varying the molar fraction of paramagnetic ions or by an external magnetic field and/or temperature, MBE grown two-dimensional structures introduce further flexibility in such SSE [70]. First, by changing the quantum confinement, one can strongly influence the value of the exchange constant α of the conduction electrons. The possibility of modification of α is related to the fact that, while in 3D SMSs the value of α is determined only by direct or potential exchange mechanism with the kinetic exchange contribution vanishing exactly because of the symmetry of the conduction electron wave function, in the 2D configuration the latter mechanism does operate and may even dominate [71]. Second, by a laser illumination one can heat 2D electrons which then—via the exchange scattering—change the population of Mn spin sublevels, thus increasing Mn spin temperature T_{Mn}. The change of T_{Mn}, which now is also magnetic field dependent, modifies, in turn, the value of the spin splitting [72]. Third, by an atomic scale control of the spatial distribution of paramagnetic ions in the MBE growth direction, allowing to obtain a desired shape of the confining potential and its variation with the magnetic field, one can intentionally modify the field dependence of the spin splitting [73]. The last method of SSE is based on the fact that the spin splitting in SMSs is a nonlinear function of Mn concentration and, therefore, the application of an external field translates into a strong perturbation of the potential profile from its zero-field shape. These deviations are most pronounced in the valence band where, in the case of a parabolic potential shape in the absence of the field, for one of the spin species (namely, for $|3/2, -3/2\rangle$ state) one can expect even a "camel-back" shape of the confining potential profile after application of a magnetic field. The field-induced change of the potential shape for different spin components, which determines to a great degree the field dependence of confined energy levels and hence the spin splitting of different excitonic transitions, depends sensitively on the actual distribution of Mn ions (i.e., composition profile). Therefore, by properly modeling the spatial distribution of Mn, one can obtain a required field dependence of the spin splitting.

The most straightforward application of SSE in rectangular QW structures is the possibility of varying the alignment of the conduction and valence band edges of layers comprising the structure. These band offsets are one of the most important parameters of semiconductor heterostructures since they determine the quantum effects exhibited by the system. By introducing the Mn into the QW region or into the barrier and taking advantage of giant spin splitting, one can significantly vary positions of the bottoms of potential wells or of the tops of the barriers for a given spin component, respectively. This provides not only very strong variation of optical transition energies with magnetic field, but also enables quite precise measurement of band offsets at the SMS/non-SMS interface by analyzing Zeeman splitting of the excitonic transitions [74]. Even more precise determination of the band offset with this method can be performed with the use of graded potential QWs (e.g., parabolic QWs), in which the distances between energetic levels of confined carriers, contrary to the case of rectangular QWs, depend sensitively and explicitly on the actual value of the bandgap discontinuity between the materials used to produce these graded QWs, and for which, apart from diagonal $h_n \rightarrow e_n$ optical transitions also nondiagonal $h_n \rightarrow e_m$ optical transitions are allowed. Using SMS parabolic and half-parabolic QWs the values of the valence band offset Vb_{off}, defined as the ratio of valance bands discontinuity to the energy gap discontinuity between materials, was determined for two material systems: CdTe/MnTe and CdTe/MgTe, to be 0.4 ± 0.05 [73] and 0.45 ± 0.1 [40], respectively. It is worth mentioning that parabolic potential, apart from being useful for the band offset determination, introduces a new length scale to the exciton problem. This leads to a strong—stronger than in the case of rectangular QWs with the same width—enhancement of the exciton binding energy in parabolic CdMnTe QWs [33, 73].

Taking advantage of very close bandgap energy of $Cd_{1-x}Mn_xTe$ and $Cd_{1-y}Mg_yTe$ for $x \approx y$ and very similar values of the valence band offsets in CdTe/MnTe and CdTe/MgTe systems, just discussed, it is possible to realize with the use of CdTe-based semimagnetic semionductor a very interesting and unique for SMSs type of nanostructures called spin-superlattices (SSL) or spin-QWs. At zero magnetic field there is no confinement for the movement of either electrons or holes and the system behaves like 3D. Now application of external magnetic field in the direction normal to the surface induces the potential well in the CdMnTe layer for spin-down electrons ($|-1/2\rangle$) and spin-down holes ($|-1/2\rangle$ and $|-3/2\rangle$), and barrier for particles with opposite spin projections. Therefore, carriers with a particular spin orientation can become either fully spatially separated into 2D planes (in the case of multiple spin-QWs) or will have larger amplitudes of wave function in those planes (in the case of SSL). The formation of a multiple spin-QWs was realized in $Cd_{1-x}Mn_xTe/Cd_{1-y}Mg_yTe$ multi-QW structures with $x = y$ under an external magnetic field and revealed by the studies of magnetophotoluminescence excitation and PL [75, 76].

To correctly reproduce by theoretical modeling the experimentally observed giant spin splitting of excitonic states in nominally rectangular QWs and superlattices made of CdMnTe it is necessary to take into account two effects [77, 78]. First of all the interface between magnetic and nonmagnetic layers is not abrupt due to the intermixing (caused by segregation and diffusion). It is therefore necessary to use in calculations the real profile of the distribution of Mn^{2+} ions at the interfaces between the QW and the barrier regions. Second, and shown to be less important, the Mn^{2+} ions located at the interfaces between nonmagnetic and magnetic layers have smaller number of Mn^{2+} neighbors (since the number of Mn atoms on one side of the interface is much lower) and there is unavoidable interface roughness and, therefore, the influence of antiferromagnetic Mn-Mn interaction is reduced. Based on the above observations a very convenient and precise method of determining the real profile of Mn^{2+} distribution, utilizing the studies of spin splitting of excitonic state, called "spin tracing," was developed [77, 78] and used not only in the case of rectangular CdMnTe QWs but also for graded potential wells (parabolic, half-parabolic, and triangular) [40, 73]. Let us finally mention that the presence of Mn^{2+} ions inside the nominally pure CdTe QWs can sometimes be used to advantage. That was the case of CdTe/MnTe superlattice in which optically induced multispin entanglement was demonstrated in an ensemble of noninteracting electrons bound to donors and at least two Mn^{2+} ions inside CdTe QW [79, 80].

Another spectacular manifestation of the sp-d exchange interaction in CdTe-based SMS, additional to giant spin splitting, is already mentioned formation of magnetic polarons. While donor- and acceptor-BMPs are typically observed in bulk crystals of semimagnetic semiconductors, polarons associated with excitons are often found in CdMnTe-based heterostructures, such as QWs and superlattices. But although the reduction of dimensionality favors the formation of magnetic polarons in the experimentally available 2D systems initial localization is still necessary to start the exciton magnetic polaron (EMP) formation. The characteristic feature of EMP is that the process of its formation can be interrupted by exciton recombination before the MP reaches its equilibrium energy. Hence the EMP energy depends also on the interplay between formation time of EMP and exciton lifetime. There are already excellent reviews on the EMP in low-dimensional structures and we refer the reader to these reviews for a comprehensive treatment of the problem [68, 69]. The majority of the most important results concerning EMP in CdMnTe low-dimensional structures were obtained with the use of selective excitation spectroscopy [69] both in static- and time-resolved modes. Excitons were exited selectively in the band of localized states where the spectral diffusion due to phonon-assisted tunneling does not occur during the exciton lifetime. In such a case the Stokes shift between the luminescence line and the energy of selective excitation is determined by the magnetic polaron formation energy.

Here we would only like to underline that CdTe-based SMS quasi-two-dimensional structures have contributed immensely to the studies of EMP and to the current understanding of their formation and properties. CdTe QWs with CdMnTe barriers where actually one of the first, if not the first, quantum structures where 2D EMP were unambiguously identified

[81, 82]. We refer the reader to the reviews [68, 69] and papers cited therein for a comprehensive coverage of the very interesting results obtained on EMP formation in CdTe-based SMS. Here let us additionally mention two relatively new results that have not been included in these reviews. First, the extension of the previous studies of $CdTe/Cd_{1-x}Mn_xTe$ QWs to the structures with high Mn content, $0.4 \leq x \leq 0.8$, revealed unexpected increase of MP energy with increasing x and was attributed to the intermixing of interface [83]. Thus the measurements of MP energy can be used as a tool of interface profile characterization, especially useful in samples involving high-x materials, additional to "spin tracing" method. In these high-x structures, the large total bandgap discontinuity (up to 1.2 eV) caused large splitting between heavy- and light-holes states and led to a strong anisotropy in the suppression of the MP formation by magnetic fields applied parallel and perpendicular to the structure growth axis. Second, wide QWs made of (Cd, Mn)Te digital magnetic alloy grown on (120)-oriented substrates were used to show that the initial localization of exciton, which is required for the formation of EMP in 3D SMS and which predetermines its energy, can be enhanced not only by alloy potential fluctuations [84] but also by an increase of the heavy-hole effective mass in the [120] direction (as compared to hole mass in [100] direction) [85].

The number of interesting results obtained in the area of quasi-one-dimensional structures made of CdTe-based SMS, as opposite to the case of 2D or even 0D structures, is quite limited. This is related mainly to the difficulties in producing such structures. Apart from interesting experimental results concerning weak localization in the CdMnTe submicron wires, to be discussed later, unique properties of T-shaped CdMnTe quantum wires have been predicted theoretically [66, 86]. As a consequence of confinement in two spatial directions, the hole states in a quantum wire are known to be mixtures of heavy- and light-hole components. However, due to a strong p-d exchange interaction in semimagnetic semiconductors, the relative contribution of these components is strongly affected by an external magnetic field, a feature that is absent in nonmagnetic quantum wires. This leads, in turn, to a strong magnetic-field dependence of the probabilities of various optical dipole transitions in SMS quantum wires. Numerical calculations performed for the case of $Cd_{1-x}Mn_xTe/Cd_{1-x-y}Mn_xMg_yTe$ T-shaped quantum wires demonstrate the possibility to efficiently control the polarization characteristics of light emitted from such structures by means of an external magnetic field [66] as well as additional magnetic field dependence of the exciton binding energy [86].

Another consequence of valence-band mixing characteristic of the one-dimensional SMS is anisotropic Zeeman shift observed experimentally with the use of magnetophotoluminescence in the dense nanowires structures formed by the growth of fractional monolayer $(CdTe)_{0.5}(Cd_{0.75}Mn_{0.25}Te)_{0.5}$ superlattices on vicinal substrates [44]. These results were explained theoretically using multiband effective mass method and model of an ideal CdMnTe quantum wire surrounded by infinite potential barriers [67].

Further reduction of the dimensionality of the CdMnTe-based system, so as to produce quasi-zero-dimensional objects, leads to the effects, even at zero magnetic field, that have not been observed in nonmagnetic CdTe QDs as well as to the enhancement of some of sp-d exchange related phenomena observed also in SMS structures of higher dimensionality. Similarly to the case of 2D and 1D CdTe-based SMS structures in the QD structures the Mn^{2+} ions can be placed either inside the dots or close to their interface, where they interact with the tails of the excitonic wave function which penetrates into the barrier. The sizes of QDs produced by Stranski-Krastanow growth mode are not affected by the incorporation of Mn-ions during the growth. They are typically lens-like shaped, with an in-plane diameter of 20 nm and a height of 2 nm [51, 53]. Their sheet density is of the order of 10^{10}-10^{11} cm^{-2}, when the Mn content ranges from 0 to 10% and decreases abruptly for Mn content larger than 10% [51]. This is explained by a relatively large sticking coefficient of Mn ions which suppresses the migration of CdTe at the surface and inhibits the formation of the 3D islands. The number of Mn ions inside a QD is usually of the order of several hundreds per dot [46, 87, 88]. There are, however, some reports of CdMnTe QDs containing in average 10 Mn^{2+} ions per dot [53], or only one Mn^{2+} in a single QD [50].

One of the characteristic features of CdMnTe QDs, visible already at zero magnetic field and for QDs with different sizes and various number of Mn^{2+} ions inside them, is that the PL lines from individual QDs are by one or two orders of magnitude broader, as compared to their nonmagnetic counterparts [46, 47, 89, 90]. The reasons for this broadening are thermal fluctuations of the total spin of Mn^{2+} ions inside the QD. Moreover, the width of these PL-lines significantly decreases with an increasing external magnetic field applied in Faraday configuration, until it reaches values typical for nonmagnetic QDs. This effect is understood as a result of the dumping of magnetization fluctuations inside the dot which exhibits saturation at sufficiently high magnetic fields. The alignment of Mn ions is less efficient in QDs with relatively large Mn concentration due to the presence of the ion-ion interaction, which prevents Mn ions being aligned by the external magnetic field [91]. Simultaneously to the decrease of the spectral width, a giant Zeeman splitting of excitonic states is observed. However, there is a major difference between the spin splitting observed in the SMS structures of higher dimensionality and that of individual QDs. Each given quantum dot that has been chosen for the studies contains a very specific and quite limited number of Mn^{2+} that determines the exchange splitting, while in 2D and 3D structures one measures averaged splitting. Therefore, the spin splitting observed in various individual CdMnTe QDs was different, and to a great extent was determined by the specific number of Mn^{2+} ions inside the dot. This splitting was correctly reproduced by theoretical calculations based on muffin tin model assuming that the number of Mn^{2+} inside various QDs varied from 5 to 25 [53].

A careful study of the shapes and positions of PL-lines from individual QDs as a function of temperature and magnetic field reveals the formation of EMPs in $Cd_{1-x}Mn_xTe$ QDs [46]. The energy gain due to the MP effect is typically 10-17 meV at temperatures of few Kelvin and decreases with an increasing temperature and external magnetic field. The intrinsic magnetic fields related to a spontaneous alignment of Mn^{2+} ions in MP vary from 1 to 3 T at low temperatures, depending on the dot. They are expected to be strongly dependent on the size of the QDs [92]. Indeed, one observes some indications of an increased MP-effect in relatively small QDs, with MP energy reaching 30 meV [91, 93], significantly larger than the one typically observed in 2D and 3D CdMnTe with similar Mn concentrations [69].

A full spontaneous alignment of Mn^{2+} ions due to sp-d exchange interaction is possible only when the formation time of magnetic polaron is shorter than the recombination time of excitons, as already mentioned. The formation time of MPs in SMS QDs is expected to be shorter than in structures with a higher dimensionality, because it reflects purely the spin response time of the paramagnetic spin system on the exchange field [94, 95]. The transient changes of the wave function during the formation of the spin cloud, important in the case of 2D structures and bulk crystal, can be neglected in QDs because of the strong three-dimensional confinement of excitons. The MP formation time has not been measured directly in CdMnTe QDs, but it should not be much different from that in CdMnSe QDs, where it was determined to be 125-170 ps in the temperature range 2-25 K [94].

The excitonic lifetime, on the other hand, depends strongly on the energetic position of the PL emission with respect to the energy of internal Mn^{2+} transition. If the PL emission energy exceeds the energy of intra-Mn^{2+} transition, an Auger type nonradiative recombination channel results in a strong decrease of excitonic lifetime to the 10-20 ps range [96]. In the case of QDs with the emission energy below the intra-Mn^{2+} transition, the excitonic lifetime is considerably longer, of the order of 400 ps [88]. Therefore, one may conclude that the complete MP formation takes place only when the PL emission is below the intra-Mn transition. In the opposite case, the MP polaron formation is interrupted by the recombination of exciton.

The advantage of Te-based SMS QDs relies on the relatively small energy gap of ZnTe (2.4 eV) in comparison to that of ZnSe (2.7 eV). Therefore, the PL emission from CdMnTe QDs grown on ZnTe is very close or below the intra-Mn^{2+} transition. Even more, the PL emission can be shifted well below the intra-Mn^{2+} transition by adding some percentage of Cd into the ZnTe barrier [52, 53] or by forming relatively large CdMnTe QDs.

A direct consequence of the exciton MP formation is the possibility of an optical control of magnetization inside QDs. It is obtained by resonant excitation with circularly polarized light, which creates spin-polarized excitons that in turn form exciton MPs having Mn spins aligned in a given direction, even without any external magnetic field applied [93]. Moreover, the strong 3D spatial confinement significantly increases the stability of magnetic polarons so that the optically induced spin alignment persists to temperatures as high as 160 K.

Spin-lattice relaxation in CdMnTe QDs is found to be different than that in QWs and bulk structures [97]. The decay of nonequilibrium spins is nonexponential, what can be attributed to diffusive escape of spins into the wetting layer, where the spin phonon scattering is suppressed because of low density of Mn clusters.

5.3. Intentionally doped structures

The overall influence of free carriers on the properties of CdTe-based SMS in the absence of magnetic field, with the exception of hole mediated ferromagnetism to be discussed later, is quite similar to that in nonmagnetic CdTe structures. However, this influence differs substantially in the presence of external magnetic field due to the fact that carriers now posses gigantic g-factors that can additionally be engineered in quite a broad range (in the case of electrons from -1.6 to $+$ few hundreds), as already discussed. The SSE is quite useful in the studies of the influence of carriers by providing very convenient handle of the carriers' redistribution on the spin sublevels and by allowing unambiguous interpretation of the data. This was widely used for the studies of both QWs containing hole gas [98, 99] and QWs containing free electrons [98–100]. In particular the evolution of the optical properties of $Cd_{1-x}Mn_xTe$ QWs with the increasing concentration of the 2DEG in a wide range from nominally zero up to 5×10^{11} cm^{-2} were studied with the use of step-like modulation doped structures [100]. These structures were produced by moving the main shutter stepwise in front of a long substrate during the doping of the barrier, so as to produce various thickness of doped region and hence various 2DEG concentrations inside the different parts of the very same QW. By studying a series of such step-like samples having different concentration x of Mn inside the

$Cd_{1-x}Mn_xTe$ QW the g-factor dependence of evolution of magnetooptical spectra with the density of 2DEG have been also assessed. Let us consider, as an example, the magnetoreflectivity spectra of a typical SMS structure having $x = 0.01$ (low-temperature effective electron g-factor at $B = 0$ equals +55 and heavy-hole g-factor 220). In short, the spectra of undoped part of the structure were dominated by exciton line that was strongly splitting with B in accordance to excitonic g_{eff}. With increasing doping in addition to X line a new line due to negatively charged exciton X^- (following the spin splitting of X) and combined exciton cyclotron resonances (moving linearly with field at high B and corresponding to the creation of exciton with simultaneous excitation of electron from ground to upper Landau levels (LLs)) appeared, until finally—for the highest doping—transitions between the LLs were seen. Let us concentrate only on the brief discussion of the results concerning charged excitons, because we believe that in this area the contribution of CdTe-based SMS is particularly important. Charged excitons are three particles complexes, called therefore also trions, consisting of exciton bound to an electron (X^-) or to a hole (X^+) that were in fact for the first time unambiguously identified just in CdTe QWs [101]. Ground state of trion is a singlet, with spins of particles of the same type aligned antiparallel. The creation of X^-, as an example, by photon with a given helicity (σ^+ or σ^-), which predetermines the spin of the photo created electrons, depends therefore on the availability of 2D electrons with a proper spin projection (antiparallel to that of photon created electron), and thus on the electron spin-splitting leading to the redistribution of electrons among spin sublevels. This splitting, due to intrinsic negative value of electron g-factor and positive but saturating exchange contribution to effective g-factor, can be engineered to change its sign at some value of the magnetic field. Therefore, in CdTe-based SMS heterostructures one can realize a unique situation of a nonmonotonous field dependence of intensity of X^- line, as was demonstrated in $Cd_{0.998}Mn_{0.002}Te$ QW [100]. As can be seen in Fig. 5, with increasing magnetic field the X^- line first disappears in σ^+ circular polarization (since effective g-factor of electrons is large and positive at low B) as opposite to the case of CdTe (for which X^- line disappears for σ^- polarization—since g-factor is negative). However, the line can be recovered in σ^+ polarization at much higher magnetic fields, where the exchange contribution has already saturated and the spin splitting of electrons is governed again by the intrinsic g-factor.

Analogous and very interesting studies of the evolution of the optical spectra of CdMnTe QW were also performed as a function of the density of 2D hole gas, addressing additionally the issue of the dependence of X^+ binding energy on hole concentration (for a review see, e.g., [98, 99]). It these studies the 2DEG density was varied either by doping or, very conveniently since in one sample, by light illumination (see also

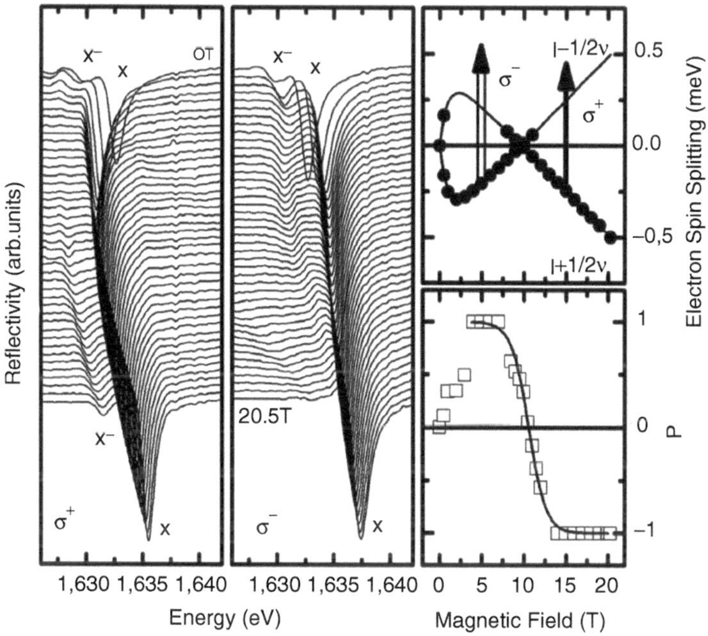

Figure 5 Magnetoreflectivity spectra at $T = 1.6$ K observed for σ^+ and σ^- circular polarizations in lightly n-type doped $Cd_{0.998}Mn_{0.002}Te$ QW structure. The magnetic field increases from 0 to 20.5 T with the step of 0.5 T between the curves. The upper panel at far right shows experimentally determined Zeeman shift of the electron levels which are the initial states for the trion creation. The single (double) arrow represents optical transition in σ^+ (σ^-) leading to creation of the singlet state of the trion. The lower panel at right shows the polarization degree P of the X^- line [100].

discussion concerning ferromagnetism). The polarization properties of positively charged excitons (X^+), although monotonous in magnetic field, are even stronger than those of X^-, since exchange constant β is four times larger than α, and total depopulation of higher spin sublevel of holes occurs at much lower magnetic fields. This property actually is a convenient way to distinguish between X^- and X^+ in weakly doped CdMnTe structures, since the binding energy of positively and negatively charged excitons is very similar and cannot therefore be used as a criterion of identification.

5.3.1. Transport properties of n-type doped structures and quantum Hall ferromagnetism

One of the most important applications of SSE in CdMnTe-based quantum structures and one of the most fascinating results obtained with the use of this engineering is the observation of the quantum Hall ferromagnetic (QHF) state [58, 102]. This state can be formed in 2DEG system if LLs corresponding to the opposite spin orientations of quasiparticles at Fermi level overlap because the spin degree of freedom is not frozen by the field so that a spontaneous spin order may appear at low temperatures [103]. Interestingly, the ground state is predicted to have the uniaxial anisotropy if the spin subbands involved originate from different LLs [104]. The level arrangement corresponding to such Ising QHF has been realized in various III-V 2DEG systems, mainly by tilting the magnetic field so as to independently vary spin- and Landau-splitting (see, e.g., [105]). The giant and strongly nonlinear spin splitting in SMS (due to the saturating exchange contribution to the effective g-factor of electrons) can be used to bring LLs having opposite real spins into the coincidence at selected magnetic fields B_C, even without application of any component of external magnetic field parallel to the sample plane. Moreover, the crossing of spin-up and spin-down sublevels in SMS can be engineered to occur either for spin sublevels with the same LLs index [100], as discussed before in connection with X^-, or for LLs having different indexes [58, 102]. This gives the unique opportunity to examine the quantum Hall ferromagnetism (QHF) at crossing of real-spin subbands in the perpendicular configuration, and to study how QHF evolves under in-plane magnetic field applied. This opportunity was fully exploited with the use of gated iodine modulation doped (Cd, Mn)Te/$Cd_{1-y}Mg_yTe$ quantum structures. In these studies CdTe-based SMS QWs were made either of regular $Cd_{1-x}Mn_xTe$ mixed crystal or of CdTe/$Cd_{1-x}Mn_xTe$ digital alloys.

The formation of Ising QHF is evidenced by anomalous magnetoresistance spikes that appear at various critical fields B_C corresponding to the crossing of $n\uparrow$ and $m\downarrow$ LLs, and that are distinct from usual Shubnikov-de Haas maxima. The spikes occur because in the region of LLs' coincidence, owing to differences in the local potential landscape, domains of different Ising ferromagnets coexist. Domain walls constitute one-dimensional conducting channels, which make scattering between edge channels possible [106]. Thus longitudinal resistance ρ_{xx} is no longer zero, as it should be in the quantum Hall effect (QHE) plateau regime.

In Fig. 6 five such QHF spikes at different LLs(m, n) crossing are visible in ρ_{xx} for various 2DEG concentrations n_S from 2.5 to 3.5×10^{11} cm^{-2}, as controlled by a gate voltage [107]. The experimental positions B_C of the spikes are substantially shifted toward higher B with respect to these predicted by the one electron model calculation, presented in the lower part of the figure. The observed shift stems mainly from the exchange interactions with frozen LLs$\downarrow$ lying well below Fermi energy. In CdMnTe there is a unique situation where a number of frozen LLs$\downarrow$ increases as B decreases.

In studied heterostructures the critical behavior of the spike resistance was also found, verifying the recent theoretical prediction [106] and

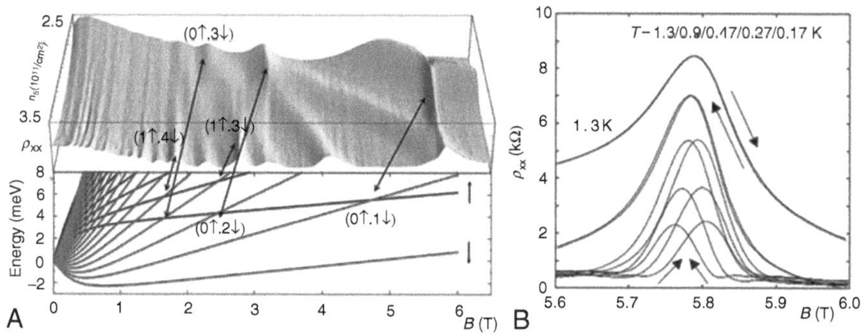

Figure 6 (A) Longitudinal resistance ρ_{xx} in the B-n_s plane and energy level diagram calculated within the independent electron model. Arrows show selected LLs crossings and corresponding QHF spikes [107], (B) Hysteresis loops in the region of (0↑,1↓) spike for sweeping the magnetic field in two directions [58].

making it possible to determine the phase diagram of a QHF as a function of the carrier density. At the same time, the Curie temperature T_C was found to reach 2 K, a value much higher than that observed and explained theoretically in the case of high electron mobility AlAs QW ($T_C \leq 0.5$ K) [105, 106]. This enhanced stability of the QHF phase is rather surprising in view of the significance of disorder in the CdMnTe material. Moreover, the position of the spikes at low T depends strongly on B sweep direction and rate leading to a history and time dependent resistance (see Fig. 6B). This metastability constitutes a clear indication of the fact that the peaks occur at the magnetic phase transition critical points. Additionally, Barkhausen-like noise observed in ρ_{xx} versus time exclusively in the region of QHF spikes [102] and only at temperatures below T_C clearly reflects the dynamics of ferromagnetic domains.

The spin-splitting engineering was also used to bring the energies of the spin excitations of free carriers to coincidence with energies of Mn spin excitations. At the magnetic field for which these two energies are almost identical an anomalously large Knight shift was observed in the resistively detected electron spin resonance experiments as well as spin-flip Raman scattering [108]. This suggests the existence of magnetic-field-induced ferromagnetic order at low temperatures.

The giant spin splitting of 2DEG states caused by the overlap of the electron wave function with the localized magnetic moments of Mn atoms located either in the SMS barrier (then weaker) [109] or inside the QW (e.g., in the planes of DA) [58, 59, 110] has dramatic consequences also for the more typical quantum transport phenomena like Shubnikov-de Haas (SdH) oscillations and integer quantum Hall effect. The difference between these phenomena observed in nonmagnetic QWs and magnetic

QWs can be conveniently traced by measuring the series of samples with increasing concentration of Mn ions, and hence increasing contribution to g-factor from s-d exchange [110]. While in CdTe QWs the minima in ρ_{xx} and plateaus of QHE are observed in low magnetic field region only for even filling factors, in the (Cd, Mn)Te QWs the plateaus are easily distinguishable for both odd and even filling factors (although not always simultaneously). Both the ρ_{xx} and ρ_{xy} field traces change strongly with the temperature, due to the strong temperature dependence of g_{exch}. Finally, the amplitude of the SdH oscillations in the superior quality CdMnTe QW with 2DEG mobility of 60,000 cm^2 V^{-1} s^{-1} and relatively low average Mn concentration $x = 0.003$ shows a distinct beating pattern with nodes corresponding to coincidences between the spin splitting and a half integer multiple of the cyclotron energy [59]. At low magnetic fields the well pronounced SdH oscillations with minima corresponding to *odd* filling factor as high as 53 were observed.

The giant and temperature dependent spin splitting in CdMnTe was also shown to have a profound impact on quantum phenomena in diffusive charge transport regime, where quantum interference between electron wave functions plays a crucial role. One of such phenomena is the occurrence of universal conductance fluctuations (UCFs). It was demonstrated with the use of submicron wires made of MBE grown CdMnTe films that giant spin splitting constitutes a novel mechanism by which the UCFs are generated as a function of magnetic field and temperature [41]. Giant spin splitting induces redistribution of carriers between spin subbands substantially changing electronic wavelength at Fermi level which in turn alters the quantum interference and generates UCFs when spin splitting changes (with B or T). These UCFs in SMS are therefore also sensitive to the occurrence of magnetization steps, which result from the field-induced change in the ground state of nearest neighbor Mn pairs. They were also used, together with the studies of $1/f$ noise, to extract information about the nature of spin-glass dynamics [42]. It is worth mentioning that mechanism of spin-splitting generation of UCFs is much less important in the case of wires made of 2DEG CdMnTe system since in modulation doped structures elastic mean free path becomes very long and hence the orbital generation of UCFs is a dominant one [41].

5.3.2. Ferromagnetism in p-type doped structures

One of the most fascinating and important results obtained with the use of p-type doped CdTe-based SMS quantum structures is the observation of ferromagnetism in the system of Mn ions [61]. The presence of high concentration of holes, by providing long-range indirect Zener/RKKY-type coupling channel between Mn^{2+} spins which overcomes the short-range antiferromagnetic super exchange, leads to the formation of a ferromagnetic state [111, 112]. And although the carrier induced ferromagnetism with higher Curie temperature was observed both in IV-Mn-VI, III-Mn-V (for a recent review see, e.g., Ref. [113]) and other II-Mn-VI [112] SMSs the advantage of CdMnTe quantum structures is that incorporation of the localized magnetic moment is independent of the introduction of mobile carriers (as opposite to III-Mn-Vs), and can also be relatively easy controlled by external factors other than a pressure (as opposite to IV-Mn-VIs). Therefore, the CdMnTe quantum structures provide one of the clearest demonstrations of the free carrier induced ferromagnetism. The disadvantage of this system is, however, that until now it was impossible to produce CdMnTe-based structures appropriate for electrical transport studies of ferromagnetism (e.g., through the anomalous Hall effect). Therefore, the magnetic properties of the system were assessed through magnetospectroscopy with the use of giant spin-splitting effect.

The ferromagnetic state was evidenced in both nitrogen modulation doped and surface doped $Cd_{1-x}Mn_xTe$ QW with $x \approx 0.02$-0.09, having width 8-15 nm, and containing 2D hole gas with concentration $p \approx 1.6$-3.2×10^{11} cm^{-2} [61, 64]. The manifestation of the ferromagnetic state is twofold. First, the magnetic susceptibility, as measured by the magnetooptical methods, is found to diverge at low temperatures and follow a Curie-law. Second, below a critical temperature T_C (of up to 6 K) the zero-field splitting of the photoluminescence line appears, indicating the onset of the local spontaneous magnetization. A simple mean-field model of carrier induced ferromagnetism [111] accounts well for the observed transition temperatures and the critical behavior of susceptibility.

One of the main interests in the carrier-induced ferromagnetism in semiconductors is related to the unique magnetism engineering offered by these systems [113]. There are two distinct types of such engineering possible. The first one can be performed at the stage of structure design/growth and the other, available in already existing structure, can be realized by application of external perturbations. Both types of magnetism engineering were convincingly demonstrated in the CdMnTe quantum structures.

The most straightforward engineering by structure design was realized via controlling the Mn^{2+}-Mn^{2+} indirect ferromagnetic coupling through the change of free hole density inside the QW. This hole density in modulation doped structures was varied by changing the width of undoped spacer separating QW from the nitrogen doped CdMgZnTe barrier [61] (but also could have been alternatively varied by the number of nitrogen acceptors introduced remotely), and by the difference between the valence band offset and the confinement energy [114]. In the "surface doped" structures the hole concentration was changed by varying the thickness of cap layer separating the QW from the acceptor-like surface states [63]. In all these cases, the CdMnTe QW without large concentration of free holes was paramagnetic and became ferromagnetic upon doping, with ferromagnetic transition temperature raising for increasing hole concentration (T_C changed with p since the system was not strictly 2D).

Since indirect ferromagnetic interaction among isotropic Mn spins is mediated by quasi-2D hole gas, the CdMnTe QW structures allow also for the engineering of magnetic anisotropy of the system by taking advantage of valence band engineering. Such engineering was realized through a strain and quantum confinement control of the light-hole/heavy-hole splitting. The uniaxial strain was imposed on the QW from substrates having various lattice constants, and on which the structure was pseudomorphically grown. It was demonstrated that the axis of easy magnetization can be turned from the growth direction, for the case of heavy-holes mediated coupling, to the in-plane direction in the case when light-holes mediate ferromagnetic coupling [64, 115]. The reason of the easy axis of magnetization being in the z growth direction is that due to the strong anisotropy of the hole wave function the exchange field seen by Mn spins in the presence of heavy holes is purely along the normal to the QW plane, and the spin-spin interaction takes the Ising form. On the other hand the exchange field seen by the Mn spins in the presence of light holes is larger within the QW plane than out of the plane, hence the spin-spin interaction is more xy-like and, therefore, spontaneous magnetization lies in the plane of QW.

Even more useful for potential spintronic devices built of semimagnetic semiconductor might be the control of magnetism in already existing device. This is also viable through the change of free hole concentration but now induced by some external factors. Two such factors were proved to be useful in CdMnTe-based structures for varying p and hence T_C, as well as the magnitude of low-T spontaneous magnetization: light illumination and electrical field. In the first case illumination of the structure with a light of energy exceeding bandgap energy of the barrier can either decrease or increase the free hole concentration, depending on the band alignment which determines spatial redistribution of the photo-generated carriers. Specifically in the p-i-p structure the light illumination, in the steady state situation, removes holes from the CdMnTe QW and thus weakens ferromagnetism. In the p-i-n structure on the other hand the light increases the 2D hole gas concentration and enhances ferromagnetism [61, 116] (see Fig. 7A). For the spintronic devices the electrical manipulation of ferromagnetism is probably more important than manipulation with light. Such a possibility was also demonstrated in p-i-n structures, where an application of a small negative voltage to the electrostatic gate (much smaller than that in the case of InMnAs field effect transistor [113]) depletes QW from holes and kills ferromagnetism [116] (see Fig. 7B).

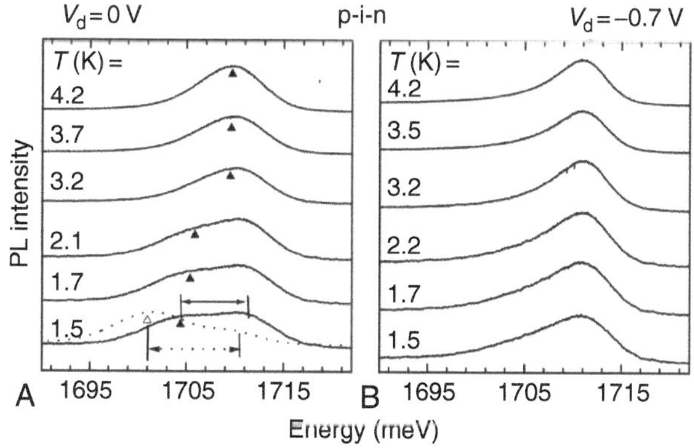

Figure 7 PL spectra for a modulation-doped p-type $Cd_{0.95}Mn_{0.05}Te$ QW located in a p-i-n diode: (A) without bias at various temperatures: the hole density is constant (solid lines) or increased (dotted line) by additional Ar-ion laser illumination; (B) with a −0.7 V bias (depleted QW) at various temperatures. Splitting and shift of the lines mark the transition to the ferromagnetic phase [116].

6. CONCLUSIONS AND PROSPECTS

In summary both bulk and low-dimensional structures made of CdTe-based SMSs have contributed immensely to the progress in physics and technology of semiconductors as a whole, and being one of the most important members of the family of semimagnetic semiconductors (or otherwise diluted magnetic semiconductors) gave strong input to the foundation of the recently vividly developing field of spin-based semiconductor electronics: "spintronics." It is worth to notice that SMSs were recently recognized by *Nature* as one of the 23 milestones of spin physics for the period 1806-1997 [117]. Probably one of the very first semiconductor spintronic devices, optical-separator based on giant Faraday rotation was built from CdMnTe crystal by Tokin Company. And although there is not too much hope to build the room temperature operating spintronics devices with the use of CdTe-based ferromagnetic semiconductor, we believe that in the years to come the CdTe-based SMSs will still be important materials and can be at least used for testing new ideas of such spintronic devices. To mention only one of the possible directions that has been practically unexplored so far is the area of hybrid structures combining SMS with superconductor or ferromagnetic metals. There one can make use of giant spin splitting of SMS combined with the very local magnetic fields, generated either by domains patterns of ferromagnet or by electron-lithographically defined nanomagnets [118, 119], or by vortices of superconductor, and allowing to produce spin and charge textures, that can even be controllable, for example, by manipulating vortices [120].

Another possibility, mentioned earlier, is the application of CdMnTe crystal in detection of nuclear radiation [25]. Additionally, as has been for many years by now, also in the future CdTe-based SMS can hopefully continue to be greatly appreciated object of many interesting studies in the area of basic research.

ACKNOWLEDGMENTS

The authors would like to thank Dr. Piotr Wojnar and Dr. Wojciech Szuszkiewicz for the discussions. T.W. acknowledges the financial support from the Foundation for Polish Science through subsidy 12/2007.

REFERENCES

[1] R.R. Gałązka, J. Kossut, Semimagnetic Semiconductors, in: O. Madelung, M. Schulz, H. Weiss (Eds.), Landolt-Bornstein Numerical Data and Functional Relationships in Science and Technology, New Series, Group III, vol. 17b, O. Madelung (volume Ed.), Springer, Berlin, 1982, pp. 302–318, 528–543.
[2] R.R. Gałązka, J. Kossut, T. Story, Semimagnetic Semiconductors, in: O. Madelung, U. Rossler, M. Schulz (Eds.), Landolt-Bornstein Numerical Data and Functional Relationships in Science and Technology (update to vol. 17b), Group III, vol. 41B, U. Rossler (volume Ed.), Springer, Berlin, 1999, pp. 650–721.
[3] M. Jain (Ed.), Diluted Magnetic Semiconductors, World Scientific, Singapore, 1991.
[4] J.K. Furdyna, J. Kossut (Eds.), Semiconductors and Semimetals, vol. 25, Diluted Magnetic Semiconductors, Academic Press, New York, 1988.
[5] J. Kossut, W. Dobrowolski, in: K.H.J. Buschow (Ed.), Handbook of Magnetic Materials, vol. 7, Elsevier North-Holland, Amsterdam, 1993, pp. 231–305.
[6] R.R. Gałązka, J. Magn. Magn. Mater. 140-144 (1995) 13–16.
[7] J.K. Furdyna, J. Appl. Phys. 64 (1988) R29-R64; M. Averous, M. Balkanski (Eds.), Semimagnetic Semiconductors and Diluted Magnetic Semiconductors, Plenum, New York, 1991.
[8] W. Dobrowolski, J. Kossut, T. Story, in: K.H.J. Buschow (Ed.), Handbook of Magnetic Materials, vol. 15, Elsevier North-Holland, Amsterdam, 2003, pp. 289–377.
[9] T. Dietl, Semicond. Sci. Technol. 17 (2002) 377–392.
[10] S. Adachi (Ed.), II-VI Compound Semiconductors, Handbook on Physical Properties of Semiconductors, vol. 3, Kluwer Academic, Boston, 2004.
[11] R.R. Gałązka, in: R.D. Guenther, L.P. Bayevel, D.G. Steel (Eds.), Encyclopedia of Modern Optics, vol. 4, Elsevier, Oxford, 2005, pp. 352–357.
[12] R. Triboulet, G. Didier, J. Cryst. Growth 52 (1981) 614–618.
[13] R. Triboulet, Semicond. Sci. Technol. 6 (1991) 18–23.
[14] R.R. Gałązka, in: B.L.H. Wilson (Ed.), Physics of Semiconductors, 1978, Institute of Physics Conference Series 43, The Institute of Physics, Bristol, 1979, pp. 133–140.
[15] R.R. Gałązka, Europhys. News 18 (1987) 90–92.
[16] J.A. Gaj, J. Ginter, R.R. Gałązka, Phys. Status Solidi B 89 (1978) 655–662.
[17] Y.R. Lee, A.K. Ramdas, Solid State Commun. 51 (1984) 861–863.
[18] J.A. Gaj, R.R. Gałązka, M. Nawrocki, Solid State Commun. 25 (1978) 193–195.
[19] D.U. Bartholomew, J.K. Furdyna, A.K. Ramdas, Phys. Rev. B: Condens. Matter. 34 (1986) 6943–6950.

[20] M. Zimnal-Starnawska, M. Podgorny, A. Kisiel, W. Giriat, M. Demianiuk, J. Zmija, J. Phys. C: Solid State Phys. 17 (1984) 615–621.

[21] N.G. Semaltianos, G. Karczewski, T. Wojtowicz, J.K. Furdyna, Phys. Rev. B: Condens. Matter 47 (1993) 12540–12549.

[22] T. Wojtowicz, S. Kolesnik, I. Miotkowski, J.K. Furdyna, Phys. Rev. Lett. 70 (1993) 2317–2320.

[23] C. Leighton, I. Terry, P. Becla, Phys. Rev. B: Condens. Matter 58 (1998) 9773–9782.

[24] C. Leighton, I. Terry, P. Becla, Phys. Rev. B: Condens. Matter 56 (1997) 6689–6697.

[25] M. Witkowska-Baran, A. Mycielski, A.J. Szadkowski, E. Lusakowska, V. Domukhovski, R. Jakiela, B. Witkowska, W. Kaliszek, Phys. Status Solidi B 244 (2007) 1706–1713.

[26] S. Oseroff, M. Mesa, M. Tovar, R. Arce, J. Appl. Phys. 53 (1982) 2208–2210.

[27] H.A. Sayad, S.M. Bhagat, Phys. Rev. B: Condens. Matter 31 (1985) 591–593.

[28] M.A. Novak, O.G. Symko, D.J. Zheng, S. Oseroff, J. Appl. Phys. 57 (1985) 3418–3420.

[29] R.R. Gałązka, S. Nagata, P.H. Keesom, Phys. Rev. B: Condens. Matter 22 (1980) 3344–3355.

[30] J. Pietruczanis, W. Mac, A. Twardowski, G. Karczewski, A. Zakrzewski, E. Janik, T. Wojtowicz, J. Kossut, Mater. Sci. Forum 182–184 (1995) 687–690.

[31] K. Ando, H. Akinaga, J. Magn. Magn. Mater. 140-144 (1995) 2029–2030.

[32] J. Pawlikowski, in: M. Jain (Ed.), Diluted Magnetic Semiconductors, World Scientific, Singapore, 1991, pp. 527–562.

[33] T. Wojtowicz, G. Karczewski, J. Kossut, Thin Solid Films 306 (1997) 271–282.

[34] T. Wojtowicz, G. Karczewski, J. Kossut, Acta Phys. Polon. A 88 (1995) 631–642.

[35] S.M. Durbin, J. Han, O. Sungki, M. Kobayashi, D.R. Menke, R.L. Gunshor, Q. Fu, N. Pelekanos, A.V. Nurmikko, D. Li, J. Gonsalves, N. Otsuka, Appl. Phys. Lett. 55 (1989) 2087–2089.

[36] E. Janik, E. Dynowska, J. Bak-Misiuk, J. Domagala, M. Kutrowski, T. Wojtowicz, A. Stachow, Appl. Phys. Lett. 68 (1996) 3796–3798.

[37] T. Wojtowicz, G. Karczewski, A. Zakrzewski, M. Kutrowski, E. Janik, E. Dynowska, K. Kopalko, S. Kret, J. Kossut, J.Y. Laval, Acta Phys. Polon. A 87 (1995) 165–168.

[38] M.K. Kneip, D.R. Yakovlev, M. Bayer, G. Karczewski, T. Wojtowicz, J. Kossut, Appl. Phys. Lett. 88 (2006) 1521051–1521053.

[39] W.E. Hagston, S.J. Weston, M. O'Neill, T. Stirner, P. Harrison, J.H.C. Hogg, D.E. Ashenford, B. Lunn, J. Vac. Sci. Technol. B (Microelectron. Nanometer. Struct.) 12 (1994) 1146–1149.

[40] T. Wojtowicz, M. Kutrowski, G. Karczewski, J. Kossut, Acta Phys. Polon. A 94 (1998) 199–217.

[41] J. Jaroszynski, J. Wrobel, M. Sawicki, E. Kaminska, T. Skoskiewicz, G. Karczewski, T. Wojtowicz, A. Piotrowska, J. Kossut, T. Dietl, Phys. Rev. Lett. 75 (1995) 3170–3173.

[42] J. Jaroszynski, J. Wrobel, G. Karczewski, T. Wojtowicz, T. Dietl, Phys. Rev. Lett. 80 (1998) 5635–5638.

[43] T. Kummell, G. Bacher, M.K. Welsch, D. Eisert, A. Forchel, B. Konig, C. Becker, W. Ossau, G. Landwehr, J. Cryst. Growth 214-215 (2000) 150–153.

[44] S. Nagahara, T. Kita, O. Wada, L. Marsal, H. Mariette, Phys. Rev. B: Condens. Matter Mater. Phys. 69 (2004) 2333081–2333084.

[45] G. Cywinski, T. Wojtowicz, G. Karczewski, J. Kossut, W. Gebicki, Acta Phys. Polon. A 94 (1998) 277–280.

[46] A.A. Maksimov, G. Bacher, A. McDonald, V.D. Kulakovskii, A. Forchel, C.R. Becker, G. Landwehr, L.W. Molenkamp, Phys. Rev. B: Condens. Matter 62 (2000) R7767–R7770.

[47] G. Bacher, H. Schomig, J. Seufert, M. Rambach, A. Forchel, A.A. Maksimov, V. D. Kulakovskii, T. Passow, D. Hommel, C.R. Becker, L.W. Molenkamp, Phys. Status Solidi B 229 (2002) 415–422.
[48] G. Prechtl, W. Heiss, S. Mackowski, E. Janik, Appl. Phys. Lett. 78 (2001) 2140–2142.
[49] S. Mackowski, H.E. Jackson, L.M. Smith, J. Kossut, G. Karczewski, W. Heiss, Appl. Phys. Lett. 83 (2003) 3575–3577.
[50] L. Besombes, Y. Leger, L. Maingault, D. Ferrand, H. Mariette, J. Cibert, Phys. Rev. Lett. 93 (2004) 2074031–2074034.
[51] S. Kuroda, Y. Terai, K. Takita, T. Takamasu, G. Kido, N. Hasegawa, T. Kuroda, F. Minami, J. Cryst. Growth 214-215 (2000) 140–149.
[52] S. Kuroda, N. Umakoshi, Y. Terai, K. Takita, Phys. E 10 (2001) 353–357.
[53] P. Wojnar, J. Suffczynski, K. Kowalik, A. Golnik, G. Karczewski, J. Kossut, Phys. Rev. B: Condens. Matter 75 (2007) 1553011–1553017.
[54] A. Waag, F. Fischer, T. Litz, B. Kuhn-Heinrich, U. Zehnder, W. Ossau, W. Spahn, H. Heinke, G. Landwehr, J. Cryst. Growth 138 (1994) 155–160.
[55] F. Bassani, S. Tatarenko, K. Saminadayar, N. Magnea, R.T. Cox, A. Tardot, C. Grattepain, J. Appl. Phys. 72 (1992) 2927–2940.
[56] A. Arnoult, D. Ferrand, V. Huard, J. Cibert, C. Grattepain, K. Saminadayar, C. Bourgognon, A. Wasiela, S. Tatarenko, J. Cryst. Growth 201-202 (1999) 715–718.
[57] A. Waag, T. Litz, F. Fischer, H. Heinke, S. Scholl, D. Hommel, G. Landwehr, G. Bilger, J. Cryst. Growth 138 (1994) 437–442.
[58] J. Jaroszynski, T. Andrearczyk, G. Karczewski, J. Wrobel, T. Wojtowicz, E. Papis, E. Kaminska, A. Piotrowska, D. Popovic, T. Dietl, Phys. Rev. Lett. 89 (2002) 2668021–2668024.
[59] F.J. Teran, M. Potemski, D.K. Maude, T. Andrearczyk, J. Jaroszynski, G. Karczewski, Phys. Rev. Lett. 88 (2002) 1868031–1868034.
[60] F. Fischer, A. Waag, G. Bilger, T. Litz, S. Scholl, M. Schmitt, G. Landwehr, J. Cryst. Growth 141 (1994) 93–97.
[61] A. Haury, A. Wasiela, A. Arnoult, J. Cibert, S. Tatarenko, T. Dietl, Y. Merle d'Aubigne, Phys. Rev. Lett. 79 (1997) 511–514.
[62] A. Arnoult, J. Cibert, S. Tatarenko, A. Wasiela, J. Appl. Phys. 87 (2000) 3777–3784.
[63] W. Maslana, P. Kossacki, M. Bertolini, H. Boukari, D. Ferrand, S. Tatarenko, J. Cibert, J.A. Gaj, Appl. Phys. Lett. 82 (2003) 1875–1877.
[64] P. Kossacki, W. Pacuski, W. Maslana, J.A. Gaj, M. Bertolini, D. Ferrand, S. Tatarenko, J. Cibert, Phys. E 21 (2004) 943–946.
[65] B. Kuhn-Heinrich, W. Ossau, E. Bangert, A. Waag, G. Landwehr, Solid State Commun. 91 (1994) 413–418.
[66] F.V. Kyrychenko, J. Kossut, Phys. Rev. B: Condens. Matter 61 (2000) 4449–4452.
[67] Y. Harada, T. Kita, O. Wada, L. Marsal, H. Mariette, H. Ando, Phys. Status Solidi C (2006) 667–670.
[68] D.R. Yakovlev, Acta Phys. Pol. A 90 (1996) 703–714.
[69] D.R. Yakovlev, K.V. Kavokin, Comment. Condens. Matter Phys. 18 (1996) 51–81.
[70] T. Wojtowicz, M. Kutrowski, G. Karczewski, J. Kossut, B. Konig, A. Keller, D.R. Yakovlev, A. Waag, J. Geurts, W. Ossau, G. Landwehr, I.A. Merkulov, F.J. Teran, M. Potemski, J. Cryst. Growth 214-215 (2000) 378–386.
[71] I.A. Merkulov, D.R. Yakovlev, A. Keller, W. Ossau, J. Geurts, A. Waag, G. Landwehr, G. Karczewski, T. Wojtowicz, J. Kossut, Phys. Rev. Lett. 83 (1999) 1431–1434.
[72] B. Konig, I.A. Merkulov, D.R. Yakovlev, W. Ossau, S.M. Ryabchenko, M. Kutrowski, T. Wojtowicz, G. Karczewski, J. Kossut, Phys. Rev. B: Condens. Matter 61 (2000) 16870–16882.
[73] T. Wojtowicz, M. Kutrowski, G. Cywinski, G. Karczewski, E. Janik, E. Dynowska, J. Kossut, R. Fiederling, A. Pfeuffer-Jeschke, W. Ossau, J. Cryst. Growth 184-185 (1998) 936–941.

[74] B. Kuhn-Heinrich, W. Ossau, T. Litz, A. Waag, G. Landwehr, J. Appl. Phys. 75 (1994) 8046–8052.
[75] W. Ossau, B. Kuhn-Heinrich, G. Mackh, A. Waag, G. Landwehr, J. Cryst. Growth 159 (1996) 1052–1056.
[76] M.C. Debnath, I. Souma, M. Takahashi, T. Sato, R. Pittini, F. Sato, M. Tanaka, Y. Oka, J. Cryst. Growth 229 (2001) 109–113.
[77] W. Grieshaber, A. Haury, J. Cibert, Y.M. D'Aubigne, A. Wasiela, J.A. Gaj, Phys. Rev. B: Condens. Matter 53 (1996) 4891–4904.
[78] J.A. Gaj, W. Grieshaber, C. Bodin-Deshayes, J. Cibert, G. Feuillet, Y. Merle d'Aubigne, A. Wasiela, Phys. Rev. B: Condens. Matter 50 (1994) 5512–5527.
[79] B. Jiming, A.V. Bragas, J.K. Furdyna, R. Merlin, Nat. Mater. 2 (2003) 175–179.
[80] J.M. Bao, A.V. Bragas, J.K. Furdyna, R. Merlin, Phys. Rev. B: Condens. Matter Mater. Phys. 71 (2005) 0453141–04531413.
[81] D.R. Yakovlev, W. Ossau, G. Landwehr, R.N. Bicknell-Tassius, A. Waag, S. Schmeusser, I.N. Uraltsev, Solid State Commun. 82 (1992) 29–32.
[82] D.R. Yakovlev, W. Ossau, G. Landwehr, R.N. Bicknell-Tassius, A. Waag, I.N. Uraltsev, Solid State Commun. 76 (1990) 325–329.
[83] G. Mackh, W. Ossau, D.R. Yakovlev, G. Landwehr, R. Hellmann, E.O. Gobel, T. Wojtowicz, G. Karczewski, J. Kossut, Solid State Commun. 96 (1995) 297–304.
[84] G. Mackh, M. Hilpert, D.R. Yakovlev, W. Ossau, H. Heinke, T. Litz, F. Fischer, A. Waag, G. Landwehr, R. Hellmann, E.O. Gobel, Phys. Rev. B: Condens. Matter 50 (1994) 14069–14076.
[85] R. Fiederling, D.R. Yakovlev, W. Ossau, G. Landwehr, I.A. Merkulov, K.V. Kavokin, T. Wojtowicz, M. Kutrowski, K. Grasza, G. Karczewski, J. Kossut, Phys. Rev. B: Condens. Matter 58 (1998) 4785–4792.
[86] F.V. Kyrychenko, J. Kossut, Phys. Status Solidi B 229 (2002) 557–561.
[87] G. Bacher, H. Schomig, J. Seufert, M. Rambach, A. Forchel, A.A. Maksimov, V. D. Kulakovskii, T. Passow, D. Hommel, C.R. Becker, L.W. Molenkamp, Phys. Status Solidi B 229 (2002) 415–422.
[88] S. Mackowski, S. Lee, J.K. Furdyna, M. Dobrowolska, G. Prechtl, W. Heiss, J. Kossut, G. Karczewski, Phys. Status Solidi B 229 (2002) 469–472.
[89] S. Mackowski, J. Wrobel, K. Fronc, J. Kossut, F. Pulizzi, P.C.M. Christianen, J.C. Maan, G. Karczewski, Phys. Status Solidi B 229 (2002) 493–496.
[90] R. Brazis, J. Kossut, Solid State Commun. 122 (2002) 73–77.
[91] P. Wojnar, J. Suffczynski, K. Kowalik, A. Golnik, G. Karczewski, J. Kossut, Nanotechnology 19 (2008) 2354031–2354035.
[92] A.K. Bhattacharjee, C. la Guillaume, Phys. Rev. B: Condens. Matter 55 (1997) 10613–10620.
[93] T. Gurung, S. Mackowski, H.E. Jackson, L.M. Smith, W. Heiss, J. Kossut, G. Karczewski, J. Appl. Phys. 96 (2004) 7407–7413.
[94] J. Seufert, G. Bacher, M. Scheibner, A. Forchel, S. Lee, M. Dobrowolska, J.K. Furdyna, Phys. Rev. Lett. 88 (2002) 0274021–0274024.
[95] G. Bacher, A.A. Maksimov, H. Schomig, V.D. Kulakovskii, M.K. Welsch, A. Forchel, P.S. Dorozhkin, A.V. Chernenko, S. Lee, M. Dobrowolska, J.K. Furdyna, Phys. Rev. Lett. 89 (2002) 1272011–1272014.
[96] T. Kuroda, N. Hasegawa, F. Minami, Y. Terai, S. Kuroda, K. Takita, J. Lumin. 83-84 (1999) 321–324.
[97] A.V. Scherbakov, A.V. Akimov, D.R. Yakovlev, W. Ossau, L.W. Molenkamp, Y. Terai, S. Kuroda, K. Takita, I. Souma, Y. Oka, Phys. Status Solidi B 241 (2004) 361–369.
[98] R.T. Cox, V. Huard, K. Kheng, S. Lovisa, R.B. Miller, K. Saminadayar, A. Arnoult, J. Cibert, S. Tatarenko, Acta Phys. Pol. A 94 (1998) 99–109.
[99] P. Kossacki, J. Phys.: Condens. Matter 15 (2003) 471–493.

[100] T. Wojtowicz, M. Kutrowski, J. Kossut, F.J. Teran, M. Potemski, Phys. Rev. B: Condens. Matter 59 (1999) 10437–10440.
[101] K. Kheng, R.T. Cox, Y. Merle d'Aubigne, F. Bassani, K. Saminadayar, S. Tatarenko, Phys. Rev. Lett. 71 (1993) 1752–1755.
[102] J. Jaroszynski, T. Andrearczyk, G. Karczewski, J. Wrobel, T. Wojtowicz, E. Papis, E. Kaminska, A. Piotrowska, D. Popovic, T. Dietl, Phys. Status Solidi B 241 (2004) 712–717.
[103] S.M. Girvin, Phys. Today 53 (2000) 39–45.
[104] T. Jungwirth, A.H. MacDonald, Phys. Rev. B: Condens. Matter 63 (2001) 0353051–03530510.
[105] E.P. De Poortere, E. Tutuc, S.J. Papadakis, M. Shayegan, Science 290 (2000) 1546–1549.
[106] T. Jungwirth, A.H. MacDonald, Phys. Rev. Lett. 87 (2001) 2168011–2168014.
[107] J. Jaroszynski, T. Andrearczyk, E.A. Stringer, G. Karczewski, T. Wojtowicz, J. Wrobel, D. Popovic, T. Dietl, AIP Conf. Proc. (2005) 1271–1272.
[108] F.J. Teran, M. Potemski, D.K. Maude, D. Plantier, A.K. Hassan, A. Sachrajda, Z. Wilamowski, J. Jaroszynski, T. Wojtowicz, G. Karczewski, Phys. Rev. Lett. 91 (2003) 0772011–0772014.
[109] S. Scholl, H. Schafer, A. Waag, D. Hommel, K. von Schierstedt, B. Kuhn-Heinrich, G. Landwehr, Appl. Phys. Lett. 62 (1993) 3010–3012.
[110] F. Takano, S. Kuroda, K. Takita, T. Takamasu, Y. Imanaka, G. Kido, Phys. B 298 (2001) 407–410.
[111] T. Dietl, A. Haury, Y.M. D'Aubigne, Phys. Rev. B: Condens. Matter 55 (1997) 3347–3350.
[112] T. Dietl, M. Sawicki, Le Van Khoi, J. Jaroszynski, P. Kossacki, J. Cibert, D. Ferrand, S. Tatarenko, A. Wasiela, Phys. Status Solidi B 229 (2002) 665–672.
[113] T. Dietl, H. Ohno, Mater. Today 9 (2006) 18–26.
[114] J. Cibert, P. Kossacki, A. Haury, D. Ferrand, A. Wasiela, Y. Merle d'Aubigne, A. Arnoult, S. Tatarenko, T. Dietl, J. Cryst. Growth 201-202 (1999) 670–673.
[115] J. Cibert, M. Bertolini, H. Boukari, W. Pacuski, D. Ferrand, S. Tatarenko, P. Kossacki, J.A. Gaj, Phys. Status Solidi B 241 (2004) 692–699.
[116] H. Boukari, P. Kossacki, M. Bertolini, D. Ferrand, J. Cibert, S. Tatarenko, A. Wasiela, J.A. Gaj, T. Dietl, Phys. Rev. Lett. 88 (2002) 2072041–2072044.
[117] F. Pulizzi, Nature Phys. 4 (Suppl.) (2008) S16–S17.
[118] J. Kossut, I. Yamakawa, A. Nakamura, G. Cywinski, K. Fronc, M. Czeczott, J. Wrobel, F. Kyrychenko, T. Wojtowicz, S. Takeyama, Appl. Phys. Lett. 79 (2001) 1789–1791.
[119] P. Redlinski, T. Wojtowicz, T.G. Rappoport, A. Libal, J.K. Furdyna, B. Janko, Phys. Rev. B: Condens. Matter Mater. Phys. 72 (2005) 0852091–0852099.
[120] M. Berciu, T.G. Rappoport, B. Janko, Nature 435 (2005) 71–75.

CHAPTER V

Defects

Contents

CHAPTER **VA**

Extended Defects in CdTe

Ken Durose

This chapter is dedicated to the late Prof. Jürgen Schreiber of Halle University, Germany.

1. INTRODUCTION

1.1. Importance of extended defects

Known types of extended defects in CdTe include dislocations, stacking faults, grain boundaries and inclusions of second phases. Generally these have a deleterious effect, both upon the efficient and stable working of optoelectronic devices and on the processibility of the crystals. Hence crystal defects are of profound importance to technology, for example for X-ray detection, infrared imaging and solar electricity generation. This is now exemplified:

1.1.1. Processibility

Many optoelectronic technologies, particularly the more demanding ones – for example infrared sensors – require uniform processing and demand monocrystalline wafers. Indeed, processes such as surface chemical treatment for contacting, the definition of structures by ion or chemical etching, and epitaxial overgrowth are all orientation dependent. Moreover, in epitaxy, not only do grain and twin orientations propagate into the overgrowth, but dislocations, dislocation arrays and inclusions [1] may also have a direct influence on the quality of the layers obtained.

Finally it should be noted that, for processes relying on cleaving the crystal, use of single crystals is a prerequisite – although cleaving is more commonly used in silicon and III–V technology than for CdTe.

Department of Physics, Durham University, South Road, Durham, UK

CDTE and Related Compounds
DOI: 10.1016/B978-0-08-046409-1.00005-8

1.1.2. Effect on devices

1.1.2.1. Diffusion and gettering Dislocations and grain boundaries are well known to attract impurities, and to promote diffusion (Willoughby [2] describes the main cases in a review). Such effects might be expected to lead to instability in devices, or else to have an influence on the thermal processing conditions chosen to fabricate certain devices. An example is the interdiffusion of CdTe and CdS in polycrystalline solar cells for which the grain boundary diffusion coefficient has been measured [3].

Grain boundary segregation is well known in metals (e.g. Cu in Pb [4], the driving force being strain reduction at the boundary plane. Decoration of grain and twin boundaries in CdTe with Te inclusions is widely reported – and is described in Section 2.4. Minor component impurities in CdTe have also been shown to segregate out to grain boundary regions [5] and to dislocation arrays [6].

1.1.2.2. Electrical activity The electrical states associated with grain boundaries and dislocations can have a number of adverse effects on the performance of optoelectronic devices.

Firstly, the deep states associated with extended defects can promote undesirable recombination. This is especially important where devices depend for their operation on minority carriers (e.g. photovoltaic infrared detectors and solar cells). This recombination is readily observed in CdTe using cathodoluminescence (CL) [7, 8] and electron beam induced current (EBIC) microscopies [9]. Sophisticated measurements of the density of grain boundary states in CdTe have also been reported, as has their temporary passivation with H and Li [10].

Secondly, grain boundaries act as resistive barriers. This is attributed to the grain boundary plane being a charged interface, causing it to present an electrical barrier to current transport. Such barriers have been observed directly for CdTe using the so-called 'remote' EBIC method [11] and are considered responsible for limiting effects in polycrystalline solar cells [12].

Thirdly, grain boundaries and dislocations may act as conduits for current transport rather than barriers. For the case of dislocations at the epitaxial interface of $Cd_xHg_{1-x}Te$/CdTe infrared detectors, this is considered to be a source of leakage current and it has been suggested that the upper limit to dislocation density acceptable is $\sim 10^5$ cm^{-2} [13]. On the contrary, for the case of polycrystalline CdTe solar cells, it has been suggested by one laboratory that grain boundary conduction is beneficial to efficient operation [14]. However, for solar cells, the condition of the grain boundaries is likely to depend on their position in the layer; near surface grain boundaries are likely to be Te-rich [15] (i.e. conducting) on account of the etching used to prepare contacts, whereas those remote from the free surface may nevertheless act as recombination centres [16]. There may therefore be no universal expectation that grain boundaries in

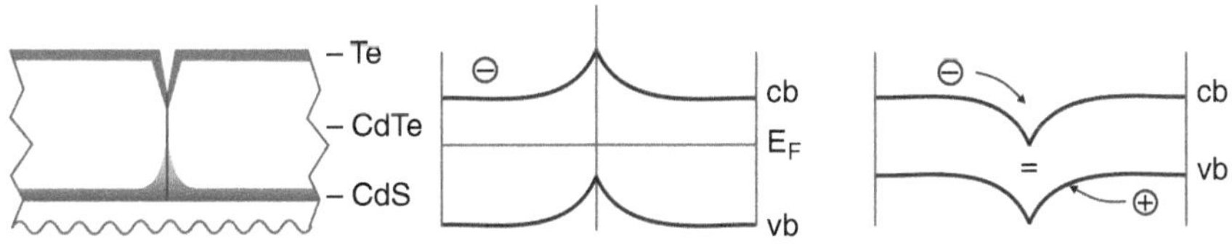

Figure 1 Some possible effects of grain boundaries on optoelectronic devices (schematic diagrams). Left: Segregation and diffusion at grain boundaries in CdTe/CdS solar cells. Interdiffusion of CdTe and CdS is promoted at the grain boundaries [3], while chemical etching to form an electrical contact to the CdTe causes enrichment of the boundaries with Te [15]. Middle: A grain boundary as a potential barrier to minority carrier transport. Such boundaries contribute to the resistance of the material [12]. Right: Enhanced recombination at deep states associated with a grain boundary [16].

CdTe behave in the same sense after different kinds of processing, from one device type to another, and even in the same device (Fig. 1).

1.2. Factors predisposing CdTe to dislocation and twinning

There are a number of factors that make CdTe particularly susceptible to the formation of extended defects.

Firstly, its thermal conductivity is relatively low (see Table 1), and compared to InP and Si is lower by factors of 10 and 20, respectively. Secondly, CdTe also has a low critical-resolved shear stress (CRSS) for slip, a factor of ~18 lower than that in Si. Hence CdTe is exposed to greater thermal stress during crystal growth, and this stress has a proportionately greater effect than in many other materials. High dislocation densities are therefore expected and are found in practice.

A third factor is the low stability of the sphalerite form of CdTe with respect to the wurtzite form. The stacking fault energy is correspondingly low. Whereas the wurtzite (hexagonal) form is not known for bulk CdTe, there are reports of heavily faulted and wurtzite phases in thin films (see Section 2.1.1). Rather than full polymorphism, CdTe is very strongly prone to twinning, with the low energy $\Sigma = 3$ coherent twin boundary being the most common (Section 2.3.2). Twinning occurs on all scales and may be found in all forms of CdTe. Stacking faults are common, and its dislocations are usually, but not always, dissociated (Section 2.2.1).

The fourth factor is that the solubility of excess Te in the lattice decreases with decreasing temperature, this being apparent from the T–x phase diagram. This 'retrograde solid solubility' ensures that for crystals grown and cooled from high temperatures with excess Te, then there shall be precipitation – or else inclusions of Te shall form (Section 2.4).

Finally, the crystallography of the sphalerite structure is of high symmetry, giving ample opportunities for the formation, multiplication and propagation of defects. In particular, the cubic symmetry affords four {111} type planes (or eight polar variants), each of which may participate

Table 1 Comparison of some factors which make semiconductors susceptible to the formation of extended defects

Materials	Melting point (°C)	Thermal conductivity (W/cm K)	CRSS (MPa)	SFE (mJ/m^2 or erg/cm^2)	Sphalerite stabilization $E_{ZB\text{-}W}$ (meV/atom) [19]	Thermal expansion coefficient (10^{-6} K^{-1})	Vickers hardness value (HV) at 300 K (kg/mm^{-2})
Si	1420	0.21	1.85	70	n/a	2.57 (300 K) [20]	
Ge	980	0.17	0.7	63	n/a	5.90 (300 K)	
GaAs	1238 1237 [21]	0.07 0.5 [21]	0.4	48	~8.5	6.7	
InP	1070	0.1	0.36	20	~3.3	4.5 [22]	
$Cd_{1-x}Zn_xTe$ ($x = 0.04$)	~1095 [23]				~9.5 (ZnTe)		
CdTe	1092	0.01 Liquid 0.3 [24] Solid 0.015 [24] At mpt 0.013 [25]	0.11	10	~8.8	5.31 4.99 [26]	40 50 [27]

CRSS, critical-resolved shear stress; SFE, stacking fault energy. All data from Thomas et al. [28] unless stated. CRSS is a strong function of temperature (see Section 2.2.4), and Thomas et al. reported values extrapolated to the melting point. Thomas's data for thermal conductivity shown in the table are for high temperatures. The value for CdTe varies with temperature in the range 300–700 K as $11.7477 - 15.22207T + 0.306916T^2$ [29]. Similarly thermal expansion coefficients vary with temperature, and that for CdTe varies as $4 \times 10^{-6} + 3.3 \times 10^{-9}T$ K^{-1} [26]. Variation of critical parameters with temperature may contribute to defect formation mechanisms.

in dislocation slip, or become the composition plane of a twin. By comparison, the wurtzite lattice with just one set of {0001} planes has more constrained opportunities for defect formation. This makes CdS for example easier to grow in bulk single crystal form than CdTe.

The strong tendency of CdTe to high levels of extended defects led to the development of (Cd, Zn)Te as an alternative [17]. Rai et al. [18] demonstrated that the yield strength of $Cd_{0.96}Zn_{0.04}Te$ is a factor of approx. three times greater than for CdTe in the range 200–600 °C. They point out that since the Zn–Te bond is shorter, then the alloying causes local strain centres which act as obstacles to dislocation motion. (Cd, Zn) Te is therefore often used as a substrate of choice for $Cd_xHg_{1-x}Te$ epitaxy on account of its improved dislocation densities. It also has the advantage of allowing for lattice matching by varying the composition.

1.3. Methods of investigating extended defects

This section provides detail on experimental methods of investigating extended defects in CdTe – but is restricted to those methods for which there are special issues or application to CdTe itself. The discussion therefore includes: (a) X-ray rocking curves – where there is uncertainty in their ability to determine dislocation density, (b) transmission electron microscopy – for which special sample thinning methods must be used for CdTe, (c) use of selective etches for the investigation of dislocation distributions – these are particular to a given substance.

1.3.1. X-ray rocking curves

The perfection of both bulk crystals and of epitaxial films is routinely evaluated by X-ray rocking curve measurements. Of the methods available, the most popular is double crystal X-ray rocking curve (DCXRC or DCRC). In DCRC measurement, the X-ray beam is first Bragg scattered from a reference crystal having a similar lattice parameter to the sample. For CdTe, InSb is often used as the reference crystal. Then the radiation is scattered from the sample and the intensity monitored with an 'open' (i.e. large) detector while the sample is rocked through a Bragg peak. The method has the advantage of being insensitive to the wavelength spread of the radiation used. A disadvantage of DCRC is that it is incapable of distinguishing between *tilts* arising from dislocations, and local lattice parameter variations (*dilatations*), arising from strain or composition, for example. Distinguishing between tilts and dilatations requires the more sophisticated high-resolution X-ray diffraction (HRXRD), or triple axis diffractometry, in which the beam is monochromated *and* collimated before reaching the sample, and the scattering angle measured with high precision. (Fewster provides a review of the methods [30].) HRXRD is most usually applied to ternary solid solutions, for example $Cd_xZn_{1-x}Te$ for which lattice parameter variation is expected.

Since DCRC measurements are often reported for CdTe, the factors contributing to the full width at half maximum (FWHM) of the rocking curves are now discussed. Contributions to the breadth of the curve may be added in quadrature as in Eq. (1) [31]:

$$\beta^2 = \beta_S^2 + \beta_B^2 + \beta_D^2, \tag{1}$$

where β is the total rocking curve width, and has contributions from strain (β_S), bending (β_B) and dislocations (β_D). An instrumental contribution might also be expected. Gay, Hirsch and Kelly gave the following relation between dislocation density and rocking curve width [32, 33]:

$$\mathrm{dd} = \frac{(\beta^2 - B^2)}{9b^2}, \tag{2}$$

where dd is the dislocation density (m^{-2}), b is the tilt component of the Burgers vector (m), β the total rocking curve width and B the non-dislocation contribution.

The Gay model is frequently cited by authors who are attempting to infer dislocation density from DCXC FWHM. However, for bulk CdTe at least, the agreement is not good. Figure 2 shows dislocation density versus FWHM for Bridgman-grown CdTe and is plotted using data from Refs. [32] and [33]. Gay's model plotted using $b = 1/2\langle 110 \rangle$, and attributing the broadening solely to dislocations, gives an order of magnitude overestimate of the experimentally observed maximum dislocation densities. Use of a single non-zero value of B improved the fit for

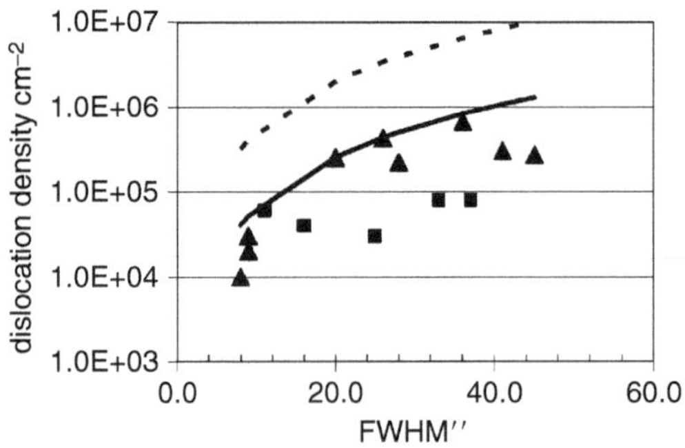

Figure 2 DCRC FWHM (arc sec) versus Nakagawa etch pit density for Bridgman-grown CdTe plotted using data from Johnson [35] (triangles) and Lay et al. [36] (squares). Gay's model for $b = 1/2\langle 110 \rangle$, shown as a dotted line, gives an overestimate for dislocation density. There are either fewer dislocations, or greater cooperative alignment than Gay's model suggests, or, alternatively there is significant and variable non-dislocation (strain) broadening. The solid line shows an envelope delineating the upper extent of the data (see Section 1.3.1).

low FWHM, but not overall. However, the experimental data did exhibit an upper envelope – see Fig. 2. Fitting of the envelope to Gay's equation, using $B = 0$ and using $9b^2$ as an adjustable parameter gave the solid line in the figure. For this fit, $9b^2$ was approximately 10 times its expected value. The behaviour may be explained as due to (a) there being fewer dislocations revealed than are actually present or (b) there being greater cooperative tilting effects than are expected or (c) there being a variable non-dislocation contribution to the FWHM. In comparing samples with dislocation densities of 9.6×10^5 and 3.3×10^4 cm^{-2}, Capper et al. [34] demonstrated (using triple axis diffractometry) that the dislocations influence the tilt contribution to FWHM very relatively weakly (17″–22″) compared to their effect on strain FWHM which reduced from 40″ to 26″. That study therefore implicates strain broadening as being significant – and likely to result in an underestimate of dislocation density if it is not taken into account.

Overall it may be concluded that DCRC FWHM is a widely used method of wafer quality control, but it convolves tilt, strain and instrumental broadening. Direct use of the Gay, Hirsch and Kelly formula to compute dislocation densities gives overestimates of at least a factor of 10. Both tilt and strain contribute to the FWHMs in CdTe.

1.3.2. Specimen preparation for transmission electron microscopy

Typically specimens are thinned to electron transparency by a two-stage process: Firstly, mechanical polishing is used to reduce the sample thickness to as little as 100 μm. This may be done with or without forming a dimple in the centre of the 3 mm disc. Secondly, final stage thinning is used to perforate the sample, the nearby regions being electron transparent. Since CdTe is highly optically absorbing, optical termination is generally used. Chemical and ion beam methods of final thinning are now briefly reviewed:

1.3.2.1. Chemical methods The most common is to use solutions of bromine in methanol, which has a smooth chemical polishing action on CdTe. The rate of dissolution may be adjusted by either increasing the viscosity of the solution (by adding ethylene glycol), or by changing the concentration of bromine. Use of highly concentrated solutions should be avoided since there can be a violent exothermic reaction upon mixing [37]. To get good results, care must be taken to wash the sample quickly in methanol – exposure to air and/or moisture causes a white deposit to form. Neutralisation of the waste solutions with aqueous sodium thiosulphate is recommended.

Chlorine in methanol is less popular, but gives a cleaner reaction. The solution may be prepared by bubbling the gas through the liquid until it becomes coloured.

1.3.2.2. Ion beam thinning Ion milling with argon is not suitable for CdTe since it causes damage in the form of a high density of dislocation loops lying on {111}. The accepted procedure is therefore to thin the material to near-transparency using argon ion milling, but to complete the operation using iodine [38]. This may be conveniently supplied to a milling instrument from a solid source, the vapour pressure at room temperature being sufficient, although it should be moderated by a PTFE valve. Cooling may be advantageous. Iodine is corrosive and hence its use is reserved for final stage thinning, often in a dedicated instrument.

1.3.3. Dislocation etchants

There is an extensive literature on dislocation etching for CdTe, and the review given here is selective (see Table 2). The key requirements for a dislocation etchant are that it should (i) reliably reveal the locations of grown-in and fresh dislocations, (ii) be easy to apply, (iii) leave the samples free from contamination, (iv) be effective on technologically important orientations.

While Inoue's E-Ag series etchants [39] were influential and were used to determine a great deal of crystallographic information, they are not considered to be reliable for dislocation counting. Nakagawa's reagent [7] however showed a 1:1 correlation between etch pits on {111}Cd and dark spots in cathodoluminescence microscopy. It became the industry standard. Nevertheless, it is reliance on fresh H_2O_2 and its non-revelation on $\{\bar{1}\bar{1}\bar{1}\}$, lead to the development of other chemistries. Hähnert's is an alternative [40] for the Cd-face, while $FeCl_3$ works on many orientations. However, while the former generates clear pits, the latter forms particularly small pits, which are only clear on well prepared surfaces that have been thoroughly washed after etching. Both contain metal ions. Everson's etch [41] was designed to overcome these limitations and the pits developed by it are correlated to dislocations.

Crystallographic polarity, and the methods used to identify it (including chemical etching) are described in Section 2.1.3.

1.3.4. Spatial distribution of dislocations in CdTe

Since there is a strong tendency for dislocation polygonisation in CdTe (Section 2.2.3) it is of interest to have a quantitative measure of the distribution of the dislocations. DCRC analysis (Section 1.3.1) gives an indication of mosaic tilt, but not of dislocation distribution itself.

The spatial distributions of dislocations may be recorded by digitising the coordinates of etch pits – this may be semi-automated. The spatial distribution may then be analysed by computer using a number of statistical tests, for example by comparison with a Poisson distribution, autocorrelation or nearest neighbour analysis [48]. Dislocations have been studied in CdTe grown by a variety of means using such methods

Table 2 Properties of some commonly used dislocation revealing etchants for CdTe

Etchant	Notes on action and verification of correlation with dislocations
Inoue, 1962/3 [39, 42] E-Ag-1 and 2	Effective on all surfaces, pits have crystallographic shapes with facets being $\{\bar{1}\bar{1}\bar{1}\}$Te planes [43, 44] Sometimes gives patchy results, especially on {111}Cd Useful in crystallographic analysis including polarity Results sensitive to stirring and other factors *Not favoured as a reliable dislocation etchant*
Nakagawa, 1979 [7]	Effective on {111}Cd. Rounded pits, most noticeable on Cd-face
H_2O: H_2O_2:HF 2:2:3 v/v (use fresh H_2O_2)	1:1 Correspondence with CL [7] Good correspondence with sub-grain boundaries seen in X-ray topography [45] and repeated in Ref. [46, 47]
Hähnert, 1990 [40, 47]	Effective on {111}Cd. Clear pits [40] Etch pits correlate with sites of Nakagawa etch pits [47]
H_2O (60), [50 wt% CrO_3 in H_2O](1), conc. HNO_3 (1), conc. HF(1) – brackets indicate parts by volume	Mosaicity revealed in XRT correlates to etching [47] Effective on {111}Cd for (Cd,Zn)Te
Watson, 1993 [8]	Effective on {111}Cd, Te$\{\bar{1}\bar{1}\bar{1}\}$B and other surfaces. Small pits
35 g of $FeCl_3 \cdot 6H_2O$ dissolved in 10 ml water. Solution aged for several weeks works best; aggressive washing needed	1:1 Correspondence with CL
Everson, 1995 [41]	Effective on $\{\bar{1}\bar{1}\bar{1}\}$ Te. Pits easily visible Etch pits correlated to dislocations by TEM of near-surface regions [41]
48%HF/HNO_3/lactic acid 1:4:25 v/v	Correlation with Nakagawa pitting (Cd surface) with defects revealed on Te surface by this etchant [41]

[48–51]. There is always some clustering, but the dislocations in CdTe grown using modern vapour methods have been shown to be have more randomised distributions than those in Bridgman-grown samples [48].

The methods of pattern analysis are general and have been applied to twinning in Cd(S,Te) [52] and v-pits in GaN [53]. Nevertheless such methods are not widely used in the characterisation of defects in crystals at the time of writing.

2. CRYSTALLOGRAPHY, DESCRIPTIONS OF THE DEFECT TYPES AND DEFECT PHENOMENA

2.1. Stable phases of CdTe, the sphalerite structure, crystallographic polarity, cleavage planes

2.1.1. Stable phases

Sphalerite is the usual crystal structure adopted by CdTe: other phases are rarely encountered and seldom studied. At high pressures, sphalerite CdTe transforms to the rocksalt (30 kbar, 3000 MPa) and then the white tin structure (90 kbar, 9000 MPa), with corresponding volume reductions of 24 and 3%, respectively. This is reviewed in Refs. [54, 55] and is summarised in Table 3.

The author is not aware of any reports of bulk CdTe having the wurtzite structure – the stable phase is generally considered to be sphalerite. However, there are reports of wurtzite phase CdTe in thin layers. Although some of these are convincing, the reader is advised to exercise caution: (i) CdTe is prone to twinning, and grains with twin orientations (rotated by 180° about ⟨111⟩ may contribute to diffraction intensities, (ii) CdTe is prone to stacking faults – heavy random faulting can give diffraction intensities that resemble unstable phases (wurtzite CdS when ball-milled has a sphalerite-like diffraction pattern [59, 60]).

Table 3 Stable phases of CdTe [54, 55]

Structure	Transition pressure	Lattice parameter $a - c$ (nm)	Volume reduction (%)	References
Sphalerite	Usual phase at 1 atm	0.648–	–	–
NaCl	30 kbar, 3000 MPa	0.592–	24	[56, 57]
		0.581–	28	[58]
White tin?	90–100 kbar, 9000–10,000 MPa	0.586–0.294	3	[56, 57]

Weinstein et al. [61] reported wurtzite CdTe grown on single crystal basal plane CdS, as evaluated by back-reflection Laue patterns. However, later work with evaporated and MOVPE deposition, and evaluated by TEM and RHEED revealed only the sphalerite phase, although twinning was common [62]. Spinulescu-Carnaru [63] reports that hexagonal character is encouraged by co-evaporation of CdTe with excess Cd, but not with Te. In a high-resolution TEM study of CdTe smoke particles, Fujita et al. [64] indicate that excess Cd may similarly encourage the formation of the wurtzite phase: they found that single particles could contain regions of both the sphalerite and wurtzite phases, and also contain stacking faults. Impurities may act to destabilise the lattice: Barrioz and colleagues [65] made attempts to nucleate CdTe nano-wires on gold dots: the resulting films had $\theta - 2\theta$ X-ray patterns characteristic of the wurtzite phase. Chandramohan et al. [66] report mixed phase CdTe films on stainless steel substrates.

2.1.2. Sphalerite structure

A conventional unit cell of the sphalerite (or zinc blende) structure is shown in Fig. 3. The structure has a face-centred cubic (fcc)Bravais lattice, with each lattice point populated by a basis comprising a Cd–Te pair; the separation between them being ¼ [111], i.e. ¼ of the body diagonal. An equivalent description is of two fcc sub-lattices, that for Cd being displaced from that for Te by a vector of ¼ [111]. The atom positions are the same as in diamond, but the Cd and Te sites are distributed over alternate close-packed (i.e. {111}) planes.

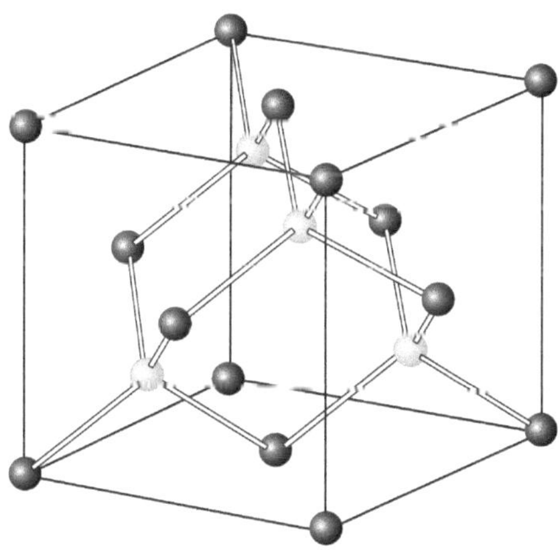

Figure 3 Conventional unit cell of the sphalerite structure. Cd sites at: 0, 0, 0 ½, ½, 0 ½, 0, ½ 0, ½,½ . Te sites at: ¼, ¼, ¼ ¾, ¾, ¼ ¾, ¼, ¾ ¼, ¾, ¾.

A convenient description of the fcc structure is in terms of the stacking sequence of successive close-packed planes, denoted by A, B and C. Atoms in all layers with the same designation lie vertically above one another. The stacking sequences corresponding to hexagonal close-packed (hcp) and fcc structures are as follows:

Hcp ABABAB...
Fcc ABCABCABC....

In the case of binary compounds, Roman letters are used to denote metal atoms (Cd) and Greek, non-metals (Te). The stacking sequences of wurtzite and sphalerite, derived from hcp and fcc, are therefore:

Wurtzite AαBβ AαBβ AαBβ...
Sphalerite AαBβCγ AαBβCγ AαBβCγ...

In sphalerite, these close-packed planes are {111} oriented and are arranged in tetrahedra, the angles between adjacent similar planes being 70°32′. Bonding in the sphalerite structure is tetrahedral, with there being 4:4 coordination. This, and the arrangement of the basis set of atoms A–B (Cd–Te in this case) gives rise to the non-centrosymmetric character of the sphalerite structure which has space group $F\bar{4}3m$. The structure therefore exhibits crystallographic polarity, as is described in Section 2.1.3.

2.1.3. Crystallographic polarity

The non-centrosymmetric symmetry of the sphalerite structure gives rise to the phenomenon of crystallographic polarity in CdTe, i.e. the structure and hence chemical properties of some pairs of opposing crystal planes – (hkl) and $(\bar{h}\bar{k}\bar{l})$ – are different. This is important for crystal growth and chemical etching, both of which have relevance to device fabrication.

Polar differences may be most clearly seen on the {111} faces of the sphalerite structure, as shown in Fig. 4. Each surface is terminated by a single atom type (Cd or Te), each atom being triply bonded to the atoms beneath. Since the bonding is tetrahedral, addition of a Te (Cd) atom to the Cd-(Te-)terminated surface, will be by the formation of a single bond – such adatoms are evidently less stable than those connected to the crystal by three bonds. Hence, the stable {111} surfaces are always terminated with either Cd or Te. By convention, the {111} surface is designated as A-terminated (Cd) and $\{\bar{1}\bar{1}\bar{1}\}$ as B-terminated (Te) – the surfaces being described as {111}Cd or {111}A and $\{\bar{1}\bar{1}\bar{1}\}$Te or $\{\bar{1}\bar{1}\bar{1}\}$B, respectively. Tetrahedra formed by the complete sets of both Cd- and Te-terminated {111} planes are shown in Fig. 5.

While this nomenclature (due to Gatos and Lavine [67]) is now universal, readers of the early literature (1950s) should be aware that an opposite convention was sometimes used. For ball and stick diagrams of

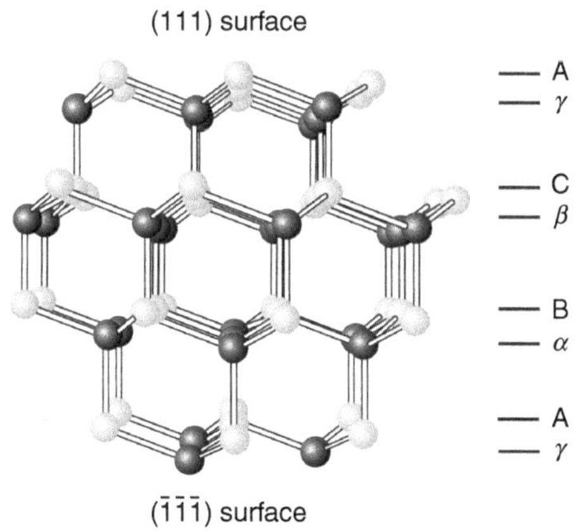

Figure 4 Model of the sphalerite structure showing the Cd- and Te-terminated close paced surfaces. These are labelled (111)Cd and (111)Te by convention.

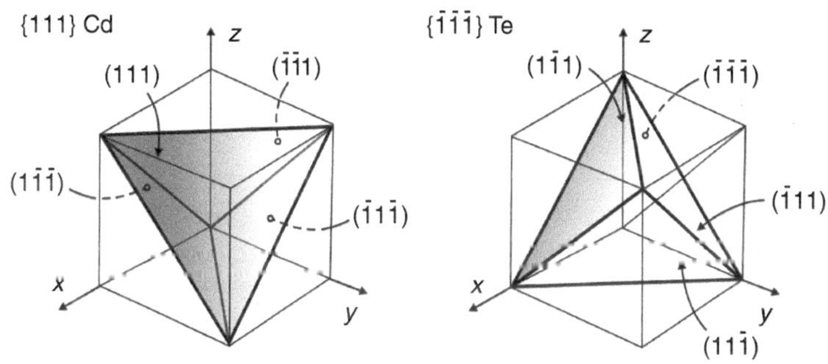

Figure 5 Tetrahedra comprising the sets of Cd- and Te-terminated {111} planes. It is the external surfaces of the tetrahedral that are indexed – this is most easily appreciated by considering a tetrahedral volume with the origin at its centre. The set of {111}Cd surfaces is (111), $(\bar{1}1\bar{1})$, $(1\bar{1}\bar{1})$ and $(\bar{1}\bar{1}1)$ and that of $\{\bar{1}\bar{1}\bar{1}\}$Te surfaces is $(\bar{1}\bar{1}\bar{1})$, $(1\bar{1}1)$, (111) and $(\bar{1}\bar{1}1)$.

the crystal structure it is recommended that the colour convention A (alabaster) and B (black) be used, as suggested by Holt [44].

Whereas polar differences are best known and strongest on {111} surfaces, others may be polar, for example {211} and {311}. Both the Cd- and Te-terminated variants of these surfaces of CdTe have been explored for use in epitaxy. The degree of polarity may be expressed by the 'polarity index' [44] which is defined as:

$$P = \frac{n_A - n_B}{n_T}, \tag{3}$$

where n_A and n_B are the numbers of A and B atoms, and n_T the total.

By this definition, the (110) and (100) surfaces are non-polar ($P = 0$). Polarity is nevertheless evident in the orthogonal [110] and [$1\bar{1}0$] *directions* in the (100) plane, and these reflect the underlying tetrahedral symmetry of the bonding. This is manifest in the distribution of stacking faults and micro-twins in (001) epilayers grown on CdTe – these tend to align with the [$1\bar{1}0$] direction rather than [110] (see the review in Ref. [68]).

Holt [44] gives a general review of the macroscopic aspects of polarity in semiconductors. Crystallographic polarity is known to determine significant differences in the chemical behaviour of opposing (*hkl*) and ($\bar{h}\bar{k}\bar{l}$) facets of many semiconductors. For CdTe epitaxial growth is well known to give better quality crystals on the {$\bar{1}\bar{1}\bar{1}$}Te surfaces of CdTe, whereas the reliable counting of dislocations from etch pitting experiments is best done on {111}Cd.

A further consequence of polarity is that many for the crystal defects in the sphalerite structure have polar variants, for example dislocations may be terminated by either Cd or Te dangling bonds, whereas there are no such variants in silicon (see Section 2.2.1).

Given its importance, it is essential to have easily accessible methods of determining crystallographic polarity by experiment. In practice this is done by taking the result an absolute determination – from anomalous X-ray diffraction or convergent beam electron diffraction – and referencing it to an easily reproduced etching procedure that shows a distinct difference between the polar faces. A brief review of polarity determination now follows. Particular attention is drawn to the X-ray/etching method and the fact that the result in general use is due to Fewster and Whiffin [69], rather than that due the earlier work of Warekois et al. [70] which contains an error.

Generally in X-ray diffraction experiments the intensities of polar surfaces I_{hkl} and $I_{\bar{h}\bar{k}\bar{l}}$ are equal; this is Friedel's law. However, when an X-ray wavelength is used that is absorbed more strongly by one component atom than another, then Friedel's law breaks down – hence measurement of the intensities I_{hkl} and $I_{\bar{h}\bar{k}\bar{l}}$ and comparison to structure factor calculations may discriminate between the polar faces. This was first done for CdTe by Warekois et al. [70], who also included ZnS, ZnTe, CdS, CdSe, HgSe, HgTe in their study. For CdTe they used a CrKα source ($\lambda = 0.2291$ nm) which is absorbed by Te, the TeL$_1$ edge being at 0.2510 nm. {111}, {222} and {333} reflections were investigated and the result correlated with etchants. Both (48%HF: 30% H_2O_2; H_2O, 3:2:1 v/v for 2 min at 25°C) and (HCl: conc. HNO_3, 1:1 v/v) were found to pit one face only, but it was a different face for each. This result was widely used in the literature, but in 1983, Fewster and Whiffin [69] noticed that the Warekois result for CdTe, and that for $Cd_xHg_{1-x}Te$ were inconsistent: epilayers of the latter were indicated to have changed polarity from their CdTe substrates. Since this was energetically unlikely, Fewster undertook a new study using the X-ray/etching method. Whereas the X-ray results of both Warekois and Fewster are self-consistent, the etching behaviour reported in Ref. [70] was shown to be in error. Moreover, the use of Nakagawa's etchant [7] (see Table 7) was recommended as giving more distinctive pits than the earlier etchants, these being on the (111)Cd face. Fewster also presented data that indicated that the 111 and 333 reflections gave the most reliable results, that the sample quality could affect the result, and that it was important to make a statistically relevant number of intensity measurements.

The subject was revisited in 1989 by Brown et al. [71] since both of the above conventions remained in common use. Brown et al. used electron micro-diffraction to assign polarity and concurred with the results of Fewster. Brown also reports upon the use of alternative etchants that discriminate between polar faces with more distinctive results. It was found [71] that both $HF:HNO_3$:acetic acid [72, 43] and $HF:HNO_3$:lactic acid [73] (1:1:1 v/v for both) yield a matt black surface on (111)Cd and a bright shiny surface on $\{\bar{1}\bar{1}\bar{1}\}$Te. (These etchants work best when washing with water is immediate.) Both have the advantage that the result is very clear to the unaided eye, and that they may be applied to the edge of a wafer as a liquid drop. Brown et al. conclude with a series of recommendations as given below.

2.1.3.1. Recommendations regarding polarity determination for CdTe using etchants The following are recommended [71]:

(i) The use of the Gatos indexing convention, i.e. the Cd-face is the {111} or A-face; the Te-face is the $\{\bar{1}\bar{1}\bar{1}\}$ or B-face. This is almost universal.
(ii) The identification of {111} faces of CdTe in accordance with the findings of Fewster and Whiffin [69].
(iii) The use of reliable discriminatory etchants for CdTe, the best being 'black-white' etchants consisting of 1:1:1 mixtures of HF, HNO_3 and either lactic or acetic acid. These both leave the {111} Cd face matt black and the $\{\bar{1}\bar{1}\bar{1}\}$Te face bright and reflecting. Inoue's etchants and Nakagawa's reagent (see Table 7) are less reliable.

Where polarity is referred to in this chapter, the author's original designations are used.

2.1.4. Cleavage planes

CdTe generally cleaves on {110}, and to a lesser extent on {111}.

Wolff and Broder [74] made a wide study of cleavage in semiconductors to investigate bonding phenomena in diamond-like materials: covalent materials including diamond itself cleave on {111} whereas sphalerite materials with ionic character cleave on {110}. The study was done by optical evaluation of rough-ground spheres. The preference of CdTe to cleave on {110} but still to cleave on {111} leads to an estimate of its ionicity to be ~60% (Philips estimates 50% [75]). Wolff and Broder remarked that {111} cleavage was more favoured in impure and high resistivity CdTe. They also found evidence of micro-cleavage on {*hhl*} planes for which $h > l$.

2.2. Dislocations

2.2.1. Dislocations in the sphalerite structure

For a general review of dislocations the reader is referred to the text by Hirth and Lothe [76], while for a specific description of dislocation phenomena in semiconductors see Holt and Yacobi's comprehensive book [77].

Dislocations in the sphalerite structure are best understood in terms of those in fcc metals, and in diamond, to which they are related. These comprise (non-exhaustively) perfect dislocations of the edge, screw and 60° types (Burgers vector $b = ½<110>$), the latter being the most common. The slip system in CdTe is [110]{111}.

Of the partials, the Shockley is perhaps the most important and arises from the dissociation of the perfect 60° dislocation as follows:

$$\frac{1}{2}[1\bar{1}0] \rightarrow \frac{1}{6}[1\bar{2}1] + \frac{1}{6}[2\bar{1}\bar{1}] \tag{4}$$

These two have 30° and 90° character, and each may be associated with a slip vector which translates atoms in a close-packed plane between site types, i.e. from A → B etc. (in the notation of Section 2.1.2)). Hence Shockley partial dislocation is associated with stacking faults that are generated by slip. Where a perfect dislocation dissociates into a pair of partials as above, then they repel – and move apart until the force is balanced by the surface tension of the 'ribbon' of stacking fault between them. A description of stacking faults is given in Section 2.2.2.

An alternative case is for which stacking faults are formed by either the condensation of vacancies (intrinsic fault – corresponding to a missing plane) or of interstitials (extrinsic fault – corresponding to an extra plane). Such stacking faults are bounded by a dislocation loop (Frank loop) having $b = 1/3\langle 111\rangle$. Such dislocations are 'sessile', i.e. they may climb but not glide. Partial dislocations are discussed in more detail by Read [78].

The most significant difference between dislocations in the elemental semiconductors such as silicon, and the compound adamantine (diamond-like) semiconductors, is the influence of crystallographic polarity. The cores of dislocations may comprise dangling bonds of either Cd or Te atoms.

Moreover, when ball and stick models of dislocations in sphalerite are considered, then two kinds of dislocation core may be drawn on the structure, these being known as the 'glide' and 'shuffle' types. The

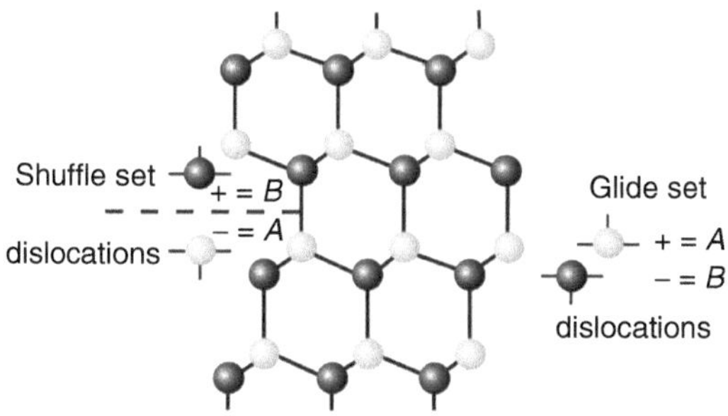

Figure 6 Convention for labelling dislocations according to (A) the position of the 'extra half plane' of atoms – in the 'shuffle' or 'glide' position and (B) the identity of the terminating atom – A (Cd) or B (Te). Dislocations are identified as Cd(g), Te(s) etc. The assignation is sensitive to the orientation of the extra half plane, i.e. whether it is in the upper (+ve dislocation) or lower (−ve dislocation) part of the diagram. Diagram redrawn from Holt and Yacobi [77].

designation describes the position at which the 'extra half plane' of atoms (that constitutes the dislocation core) is terminated. This is shown in Fig. 6. From the diagram it may be seen that for positive dislocations, i.e. those for which the extra half plane is in the upper part of the diagram the core type may be B or A depending on whether the dislocations are of the shuffle or glide sets, respectively. Similar considerations apply to negative dislocations.

In practical experiments, the type of dislocation set is not always known. In identifying dislocations, workers must therefore be sure to state their assumptions. The convention was formalised in 1979 at the International Symposium on Dislocations in Tetrahedrally Coordinated Semiconductors [79], as discussed by Holt and Yacobi [77]. In particular, they advocated clear labelling and a clear distinction as to the meaning of the earlier notation using Greek letters. The recommendation was:

> 'To avoid confusion in nomenclature in polar AB compounds the participants in the symposium recommended the use of the term "A- or B dislocation" for a dislocation with A or B atoms in the most distorted core positions... in the so-called shuffle set (s)' or 'the glide set (g) of {111} planes.... It is hoped that future publications will state clearly whether they base their discussion on A(g) and B(g) dislocations or A(s) and B(s) = α and β-dislocations.'

Accordingly the use of the terms 'α' and 'β' dislocation (Fig. 7) is falling into disuse. A third system is sometimes encountered in the CdTe literature: The close-packed stacking sequence AαBβCγ.... (see Section 2.1.2) is sometimes written AaBbCc. Identification of dislocations by the bond type that they break leads to the designations aB (for glide) and Bb (for shuffle) as used in Ref. [80], for example.

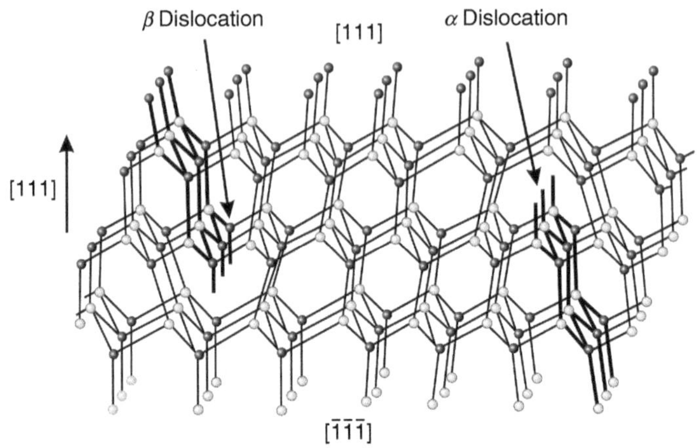

Figure 7 Polar variants of 60° dislocations in the sphalerite structure. Cd (A) atoms are shown in white and Te (B) in black. The two polar variants shown are Cd-terminated (α) and Te-terminated (β). Using modern conventional labelling, these are 'shuffle' set dislocations of the types Cd(s) and Te(s) (see text in this section). Note that most experimental evidence is for 'glide' set dislocations in CdTe. Redrawn from Inoue et al. [39].

Where the dislocation type is stated in this book, the author's original designations are used. Most experimental evidence indicates that dislocations in CdTe are of the glide type.

2.2.2. Experimental evaluation of dislocation types in CdTe by bending, indentation and high-resolution electron microscope studies

2.2.2.1. Bending tests and dislocation indentation rosettes To establish the fundamental properties of dislocation crystallography, both the bending and indentation methods have been used. The philosophy is that the deformation shall introduce dislocations, and that by coupling detailed knowledge of the crystallography with that of the observed dislocations, the character and type of the dislocations can be determined.

The earliest experiments used 4-point bending with etching. Inoue et al. [42] reported that bending introduced slip bands, and that two variants of their etchant, EAg-1 and EAg-2, revealed different types of dislocations, i.e. α and β dislocations. In regions where there were unexpected types of dislocations (i.e. not those expected from the sign of the bending), then they were seen to be annihilated upon annealing: this gave weight to the designation of α and β. Room temperature bending tests were also undertaken by Maeda et al. [81], *in situ* in a scanning electron microscope (SEM) – with CL microscopy being used to observe the dislocations. Fresh dislocations appeared as dark spots, and slip bands were seen to develop.

Formation of indentation dislocation rosettes – by use of, for example a Vickers hardness tester – gives a more complex strain field, but use of etching or CL microscopy nevertheless allows for the dislocations to be observed and analysed. For CdTe a number of such studies have been reported, including the following: [82–86]. Here we highlight the work of Schreiber and colleagues using indentation and CL, with crystallographic polarity being identified by X-ray means. Figure 8 shows the important slip directions and planes for deformation of a $\{\bar{1}\bar{1}\bar{1}\}$ surface. For each slip direction, the glide planes form a 'glide prism', the walls of which are polar faces that may be used to infer the dislocation core type, i.e. A(g) or B(g), with glide dislocations being assumed throughout. Figure 9 shows

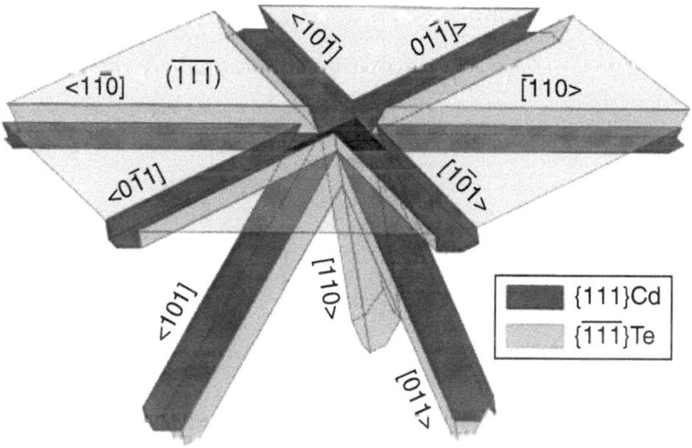

Figure 8 Three-dimensional glide prism system showing the important slip directions and planes for indentation of a $\{\bar{1}\bar{1}\bar{1}\}$ Te surface. Figure from Schreiber et al. [86]. *Copyright Wiley-VCH Verlag GmbH & Co. KGaA. Reproduced with permission.*

Figure 9 Cathodoluminescence micrograph of an indentation in a $\{\bar{1}\bar{1}\bar{1}\}$ Te surface. The light and dark branches correspond to radiative recombination at Te(g) and non-radiative recombination at Cd(g) dislocations, respectively. Figure from Schreiber et al. [86]. *Copyright Wiley-VCH Verlag GmbH & Co. KGaA. Reproduced with permission.*

an experimental result of CL microscopy of an indentation rosette on the same surface. The pairs of arms clearly contain bright and dark contrast, the former being for Te(g) and the latter for Cd(g) dislocations. Hence it is inferred that the Te(g) dislocations give radiative recombination, whereas the Cd(g) ones promote non-radiative recombination.

2.2.2.2. 'Weak beam' and high-resolution transmission electron microscopy of dislocations Hall and Vander Sande [80] made a 'weak beam' study of the structures of dislocations in CdTe that had been deformed at 200 and 300°C. They found that almost all such dislocations are dissociated into Shockley partials (hence allowing the stacking fault energy to be measured see Section 2.3.1). Dissociation of Bb (shuffle) dislocations is considered energetically unfavourable (it creates Cd–Cd or Te–Te bonds) whereas aB (glide) types are expected to dissociate easily. The widespread dissociation observed was taken as evidence that the dislocations were of the aB (glide) type.

From the early 1980s onwards, multi-beam high-resolution transmission electron microscopy (HREM) began to allow for lattice imaging in CdTe – and hence the direct observation of the atomic configurations of its defects. Many of the dislocation phenomena expected for the sphalerite structure have been observed in this way in CdTe. However, the resolving power of the microscopes and the use of simulation to interpret the image details are key factors in the success of the HREM approach.

Ponce et al. [87] made an early HREM study of CdTe and observed intrinsic stacking faults terminated by Shockley partials. However, the resolution available did not permit identification of the core type. Lu and Smith made a later and correspondingly more detailed investigation, and reported both un-dissociated and dissociated dislocations. Un-dissociated 60° dislocations were seen to be of the glide type. Also, 60° dislocations were observed that had dissociated into 30° and 90° partials bounding an intrinsic stacking fault. The 30° ones were assigned to the glide set, but the 90° ones were kinked and were not matched definitively to the simulations. Screw dislocations were shown to have dissociated into the two polar variants of the 30° partials, and although polarity could not be assigned, both were shown to be glide set types. Nevertheless, Hutchison et al. [88] have demonstrated that polarity of surface atoms could be assigned by HREM. Shockley partials (30°) associated with a twin boundary were seen to be of the glide type and were tentatively assigned as Cd(g). However, un-dissociated dislocations of the $b = 1/2\langle 110\rangle 60°$ type were not identified.

At the time of writing rapid advances have been made in aberration corrected HREM, and it would seem likely that further details of the atomic configurations of defects in CdTe shall be made in the relatively near future should they be sought.

2.2.3. Dislocation phenomena: Slip bands, polygonisation and mosaicity

Under conditions of general plastic deformation, dislocations form in semiconductors and propagate over the slip planes – providing that the CRSS for slip is exceeded. The intersections of these heavily dislocated slip planes with the crystal surface are termed 'slip bands', and they are readily observed by etching and cathodoluminescence microscopy – for example on the surfaces of bars of CdTe after bending [42, 81]. Slip bands may occur in as-grown CdTe crystals, but are rarely reported. This is possibly because they are not often recognised, but also because the dislocations reorganise themselves at moderate temperatures and hence slip bands may disappear during crystal growth. An example from vapour-grown bulk CdTe is shown in Fig. 10. The process of reorganisation of stress induced dislocations by slip and climb is known as 'polygonisation'. It is known to occur at, for example 500°C in CdTe and can take place dynamically during straining events [18]. During polygonisation, dislocations rearrange so as to minimise the overall strain field of the crystal. In samples that contained dislocations having only a single orientation and burgers vector, then the arrays that form at equilibrium are highly ordered as illustrated in Fig. 11. The boundaries are known as 'polygonisation walls', 'low-angle grain boundaries' or 'sub-grain boundaries' – the latter being most common in the CdTe literature. Sub-grain boundaries support mis-orientations as follows:-

$$\varphi = \frac{b}{s}, \tag{5}$$

Figure 10 Slip bands in vapour-grown CdTe oriented close to {111} [43]. Since the crystal was grown at ~780 °C – at which temperature the dislocations would polygonise – the slip bands are likely to have formed as a result of thermal stress on cooling in contact with a quartz capsule. They were revealed by etching in 0.5% bromine in methanol (v/v) under tungsten illumination for 5 min (The dark area is a large depression).

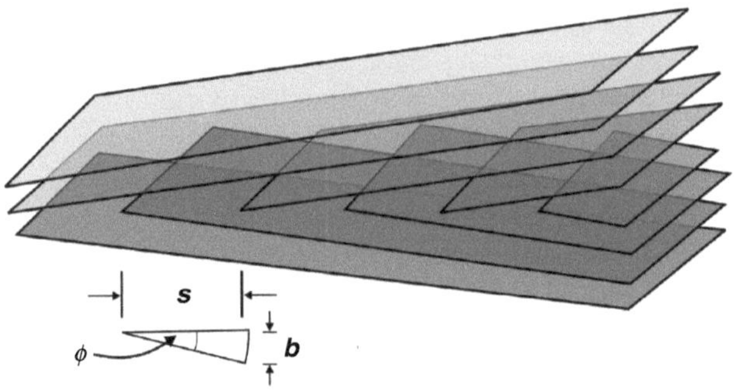

Figure 11 Low-angle boundary supported by a dislocation array.

where b is the burgers vector magnitude, and s is the spacing between dislocations. In general, such boundaries adopt many forms and may comprise multiple types of dislocations, widely spaced dislocations or dense tangles.

Sub-grains are endemic in CdTe, and much effort is expended in growing crystals with sufficiently low dislocation density as to reduce their incidence or minimise their impact. They are found to range in size from the micron to the mm scale and support mis-orientations (mosaicity) from zero up to $\sim 1°$ – see Section 3 for a survey of their incidence in bulk CdTe.

2.2.4. Critical-resolved shear stress for slip

Measurement of CRSS as a function of temperature is reviewed by Balasubramanian and Wilcox [89]. CRSS is generally evaluated from engineering stress–strain curves – from the point at which the linear elastic region gives way to plastic deformation. Their measurements show a rapid fall-off in CRSS with temperature, followed by a regime of slower decline (300 K – 4.8 MPa; 373 K – 1.8 MPa; 473 K – 1.8 MPa; 873 K – 0.9 MPa). Their values are comparable to those of Imhoff (see data in Ref. [89]), but are consistently greater than those of Rai et al. [18]. The differences might be attributed to the samples or to experimental factors. Balasubramanian also draws attention to the fact that X-ray topographic imaging under strain (as a function of temperature) indicates that dislocation motion occurs *before* the onset of macroscopic plastic deformation as revealed from engineering stress–strain curves. Balasubramanian and Wilcox [89] give data for Si and GaAs in this regard, and Table 1 compares CRSS values for some important semiconductors.

2.2.5. Micromechanics under controlled stress testing

Correlations have been made between the various stages of macroscopic stress–strain behaviour and the microstructure of CdTe. In the earliest of these, dislocations were monitored on the surfaces of bent bars using etching [42] and CL microscopy [81]. For room temperature deformation, slip bands on {111} were revealed, with annihilation of opposing types of dislocation occurring after annealing at 650 °C for 12 h.

Hall and Vander Sande [90] report a combined TEM and strain curve study, with the dislocation phenomena at each of stages I, II and III of deformation being reported. The bars were oriented for single slip, ⟨123⟩, and the primary slip plane was imaged in plan view by TEM. The following description is for 2–300 °C: 'Stage I' is characterised by single slip, and 98% of the dislocations were primaries, with long dislocations and loops being elongated along ⟨112⟩. There were dislocation multipoles. Hardening was considered to be consistent with Argon's theory (see Ref. [90]) of dislocation motion having to overcome obstacles at the multipoles. The predominance of 60° and edge dislocations was taken to indicate a screw annihilation mechanism. 'Stage II' is characterised by the onset of multi-slip and the fraction of primaries was reduced to 60%. Dislocation features now aligned on both ⟨112⟩ and ⟨110⟩, the latter being the intersection of slip planes. Dislocation tangles formed, as the increased stress released dislocations trapped at multipoles. In 'stage III', glide dislocations overcome the earlier limiting obstacles by cross-slip. All of these stages had strong temperature dependence indicating that the processes of overcoming the barriers are thermally activated.

At 500 °C, the stress–strain curve was much flattened and screw dislocations predominated, these being formed into very well developed sub-grain boundaries with clear zones in-between. A powerful recovery mechanism operates at 500 °C, and there is much reorganisation of dislocations and annihilation of dislocations with opposing signs. Cross-slip and climb operate and recovery processes take place dynamically with deformation.

TEM observations made of deformed CdTe by Rai et al. [18] generally fall into the above scheme, and the same paper gives results for (Cd,Zn)Te.

2.2.6. Mechanical polishing damage

It is well known that to prepare damage-free surfaces, CdTe wafers that have been sawn or mechanically polished should be chemically polished. There have been a few studies of the nature of the damaged layer arising from mechanical polishing, and its depth. The work of Hähnert and Wienecke and that of Weirauch indicate models that differ in their description of the uppermost damaged layer [91, 92].

Weirauch [92] demonstrated that abrasive particles create surface scratches and micro-cracks that can be covered up after formation by transfer of material on the surface. This was presumed to take place by the formation (and redistribution) of an amorphous 'Beilby' layer. Chemical etching was used to show that beneath these features dislocation arrays resulting from mechanical polishing persisted to depths of between 1 and 10 times the grit diameter used.

From RHEED examination of mechanically polished surfaces, Hähnert and Wienecke [91] showed that the surfaces were polycrystalline, with the uppermost surface contaminated with grains of the abrasive (see also Refs. [43] and [92]). This region persisted to a depth of ~10 μm. Bevel-etching and CL microscopy demonstrated that beneath that, excess dislocations were present to a depth of ~60μm, after which the material was undisturbed.

2.3. Planar defects: Stacking faults, grain boundaries and twin boundaries

2.3.1. Stacking faults

There are two kinds of stacking faults: intrinsic, in which a plane is missing, and extrinsic, in which there is an extra plane. The stacking sequences are

Wurtzite	AαBβAαBβAαBβ...
Sphalerite	AαBβCγAαBβCγAαBβCγ...
Sphalerite	with intrinsic fault AαBβCγBβCγAαBβCγ...
Sphalerite	with extrinsic fault AαBβCγAαCγBβCγAαBβCγ...

The termination of these faults with either Shockley partial dislocations (in the case of slip) or Frank partials (for vacancy or interstitial condensation) has been outlined in Section 2.2.1. Both types of stacking fault are observed in CdTe.

Stacking fault energies for semiconductors, and their trends are reviewed by Takeuchi et al. [93] (see also Ref. [77]). As mentioned earlier, dissociated dislocations in the sphalerite structure separate until their repulsive force are balanced by the surface tension of the stacking fault ribbon between them. Hence measurement of the separation of Shockley partials from dissociated dislocations (using weak beam TEM) allows for measurement of the stacking fault energy. Hall and Vander Sande [80] measured $10.1 \pm 1.4\ mJ/m^2$ and Lu and Cockayne $9.7 \pm 1.7\ mJ/m^2$ for CdTe. Takeuchi et al. [93] explore the relationship between stacking fault energy and the bond ionicity – and the use of this to rationalise the sphalerite – wurtzite transition in semiconductors. A charge redistribution index was invoked in the study. While CdTe is stabilised in the sphalerite form, it is less stable than many other semiconductors: its low stacking fault energy contributes to the high incidence of twins and stacking disorder in CdTe.

2.3.2. Grain and twin boundaries

2.3.2.1. Grain boundaries – and twinning issues for CdTe For an arbitrary or 'random' grain boundary, there may be any relationship between the two grains, and the plane that separates them may assume any orientation. Grain boundaries have five degrees of freedom.

Grain boundaries in bulk CdTe are a serious problem, with most large scale bulk growth methods yielding polycrystalline boules. It is common industrial practice to select or 'mine' single grained wafers from the boules – this is wasteful and adds to costs. Whereas random grain boundaries are common in CdTe, twins – that is grain boundaries with special orientations – are very common indeed. For this reason, they have received particular attention.

2.3.2.2. Twinning in metals and semiconductors Twins may be defined as grains having a highly symmetric crystallographic relationship to their parent grain. The phenomenon is well known in mineralogy, and there is an extensive literature on twinning in fcc metals, group IV crystals (diamond [94], Si, Ge [95, 96]), group III–V semiconductors (InSb [97], GaAs [98]) and the sphalerite group II-VI semiconductors [54]. For these cubic materials, the symmetry of the most common form of twin is high, the boundary lies on {111} in both the twin and host lattice orientations, and hence the boundary energy is low. Accordingly the twin boundaries usually observed are strikingly long and straight as may be seen in Fig. 12.

Figure 12 Twin bands visible on the growth surface of a 29 mm diameter vapour-grown boule of CdTe [43].

2.3.2.3. Terminology in twinning The terminology used in the literature is now briefly outlined. The orientation of the largest grain in a twinned volume is referred to as the 'host', 'matrix' or 'parent' lattice. 'Twin band' and 'twin lamellae' refer to slabs of material having the twin orientation and are often used to describe the visible intersection of such a slab with a surface. Twinning may occur on any scale, with twin lamellae ranging from a few atomic layers to many cm in size. 'Micro-twin' refers to small twins. 'Double positioning twinning' is an epitaxial phenomenon in which two orientations of islands (one twin and one matrix) nucleate on the substrate. The 'twin boundary', the interface between twin and matrix, may in principle take any orientation. 'Coherent' twin boundaries have perfect bonding and have low energy – in CdTe they lie on $\{111\}$ planes. 'Incoherent' or 'lateral' twin boundaries lie on other planes and have wrong bonds.

2.3.2.4. Twins as special tilt boundaries Considering the dislocation array in Fig. 11 it may be seen that increasing dislocation densities may support increasing tilt angles θ. Hence the energy of the boundary increases with tilt angle – this was shown by Read and Shockley to vary as

$$E = E_0\theta(A_0 - \ln\theta), \tag{6}$$

where A_0 is related to the core energy of the dislocations, and E_0 is related to the elastic constants of the material. Comparison with experimental values (e.g. for Pb [99]) indicates that the relation holds, even at high values of tilt for which the assumption that the dislocations are discreet breaks down. However, cusps or minima are expected when values of θ corresponding to 'special' grain boundaries with high symmetry are encountered. These are the 'coincidence' boundaries for which the superimposition of the atoms from the two grains generates a set of lattice points which are common to both grains, the coincidence site lattice (CSL). The ratio of the density of lattice points to that of CSL points is the Friedel index, Σ [77]; this defines the relationship between the two grains. Brandon et al. [100] list the angles corresponding to coincidence boundaries for a variety of low index tilt axes. For the $\langle 110\rangle$ tilt axis in fcc metals, the following angles are associated with CSLs: $\Sigma = 3$, 70.5° (70°32′); $\Sigma = 9$, 38.9°; $\Sigma = 11$, 50.5°; $\Sigma = 17$, 86.6°, $\Sigma = 19$, 26.5°. Twins may be defined as the $\langle 110\rangle$ tilt boundaries having Friedel indices $\Sigma = 3^n$, where n is the order of twinning, first, second, third and so on. For some metals there is experimental evidence that the energies of such boundaries are between 0.01 and 0.04 of those of random grain boundaries [101]. This is shown schematically in Fig. 13.

2.3.2.5. Crystallography of twins in the sphalerite structure The CSL model was applied to diamond to evaluate the structures of $\langle 110\rangle$ tilt boundaries [102], twins [103] and also the $\langle 100\rangle$ tilt boundaries [104].

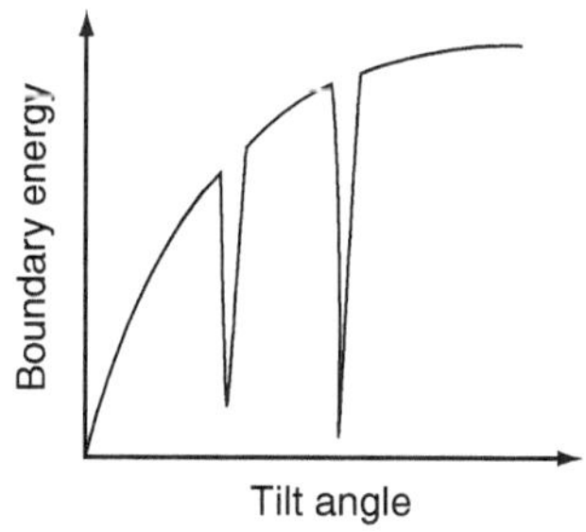

Figure 13 Energy versus tilt angle for grain boundaries. 'Coincidence boundaries' occur at special angles corresponding to a highly symmetric relationship between the grains – these include the twin boundaries. They have lower energies than randomly oriented boundaries.

For $\Sigma = 3$ twin boundaries lying on {111} (i.e. ⟨110⟩ tilt boundaries with a tilt angle of 70 °32′) the bonding is perfect, and the stacking disorder is only apparent in second nearest neighbours. However, as with the case of dislocations, the polarity of the sphalerite structure generates two possible twin configurations [105]. In the first, the crystallographic polarity is preserved upon crossing the {111} boundary plane – this is the 'ortho-twin' having tilt angle $\Theta = 180 + 70\ ^\circ 32' = 250\ ^\circ 32'$. In the second the polarity is reversed – this is the 'para-twin' having tilt angle $\theta = 70\ ^\circ 32'$.

Ball and stick models of the two indicate that while for the ortho twin, the bonding may be perfect, for the para-twin the bonding must contain a plane of either Cd–Cd or Te–Te wrong bonds. The latter is therefore not expected and has not been observed, in experiments: By cutting a crystal so as to intersect a {111} oriented $\Sigma = 3$ twin boundary at a grazing angle, it was possible to use chemical etching to determine the polarity on both sides [43, 106]. Figure 14 shows how (i) the polarity is preserved, and (ii) the twin transformation corresponds to a rotation of 180° about ⟨111⟩. This is geometrically equivalent to a tilt of 250 °32′ about ⟨110⟩ – indeed etching of twin-containing {110} surfaces with Inoue's EAg-1 shows that the pits are rotated by this angle and *not* 70 °32′. It is the 'ortho-twin' that occurs in CdTe, and by extension in the sphalerite structure. (Similar results were found for InSb [97], InP [107], GaAs [98] and ZnSe [108].)

Hence the symmetry relation for first-order twinning in sphalerite may be described equivalently as (i) a rotation of 250 °32′ about ⟨110⟩, (ii) a rotation of 70 °32′ about ⟨111⟩ or (iii) a shear of 1/6 ⟨211⟩ per atomic plane with {111} being invariant. The latter corresponds to the passage of a Shockley partial over successive close-packed planes, as in deformation twinning of metals. The twinning relation may be described by matrices [43, 97], and for the four ⟨111⟩ rotation axes they are:

$$T_{111} = \begin{bmatrix} -1 & 2 & 2 \\ 2 & -1 & 2 \\ 2 & 2 & -1 \end{bmatrix}$$

$$T_{\bar{1}11} = \begin{bmatrix} -1 & -2 & -2 \\ -2 & -1 & 2 \\ -2 & 2 & -1 \end{bmatrix}$$

$$T_{1\bar{1}1} = \begin{bmatrix} -1 & -2 & 2 \\ -2 & -1 & -2 \\ 2 & -2 & -1 \end{bmatrix}$$

$$T_{11\bar{1}} = \begin{bmatrix} -1 & 2 & -2 \\ 2 & -1 & -2 \\ -2 & -2 & -1 \end{bmatrix}$$

For example,

$$T_{\bar{1}11}\begin{bmatrix} 1 \\ 1 \\ 1 \end{bmatrix} = \begin{bmatrix} -5 \\ -1 \\ -1 \end{bmatrix}$$

that is twinning of the (111)Cd plane on the $[\bar{1}11]$ axis generates a twin with a surface orientation of $(\bar{5}\bar{1}\bar{1})$. The orientations of multiply and

Figure 14 An etched surface (Inoue EAg-1 – see Table 2) of CdTe that had been cut so as to intersect the $\{\bar{1}\bar{1}\bar{1}\}$Te plane at grazing incidence (~3 °). The crystal contains a twin, and the left- and right-hand sides of the image are related by the twinning transformation. On each side, the flat bottomed triangular pits indicate that the ($\{\bar{1}\bar{1}\bar{1}\}$Te) polarity is preserved upon crossing the twin boundary. The pits also show that the twin and matrix are related by a rotation of 180° about ⟨111⟩. (The wide diagonal band of disrupted hexagonal pits represents the twin boundary where it intersects the surface). Figure from Refs. [43, 106].

higher order twinned grains may be determined by successive application of the matrices. Slawson [94] provides tables of the orientations expected for diamond. {100} twins to {122}, {110} twins to {110} and {114}, {111} twins to {111} and {115}, and subsequently second-order twins to {111}, {115}, {1 1 11} and {5 7 13}. These are the combinations of orientations that might be expected on the surfaces of twinned wafers.

2.3.2.6. CSL models for twin boundaries in the sphalerite structure; experimental observations and predictions for the boundary properties of CdTe Having defined the symmetry relation it was possible to use the CSL model to postulate models of the boundary structure variants for first order, first-order lateral and second-order twins [43, 109, 110].

For $\Sigma = 3$ (first order) twins, the most common and lowest energy boundary lies on {111} in both the twin and matrix orientations. These are the long, straight 'coherent' boundaries routinely seen in CdTe crystals. Other boundary orientations, i.e. short lateral boundaries, have been identified and drawn [43, 109] by analogy with those in diamond by Ellis and Treuting [103]. Using the rules in Ref. [103], only four kinds of lateral twin boundary were identified for sphalerite, these being defined by their planes in the twin and matrix orientations as shown in Table 4.

In an experimental study on CdTe, these – but no others, were identified in etching experiments on oriented surfaces [43, 109]. In all cases the boundaries had disrupted bonding, and the models could not be constructed without there being wrong-bonds, for example Cd–Cd or Te–Te in each boundary period.

Similarly, and also by analogy with diamond, the boundary planes and structures of $\Sigma = 9$ (second order) twins were investigated [43, 110]. Of the two variants of the CSL, that with the highest density of CSL points was used to gain the correct polarity for the boundary. The boundary

Table 4 Tilt angles and boundary planes for the first- and second-order twins in the sphalerite structure as determined from coincidence site lattice models [43, 109, 110]

$\Sigma = 3$	$\Theta = 250\,°32'$	First order
	Coherent	{111}–{111}
	Lateral	{111}–{115};
		{112}–{112}
		{001}–{221}
		{110}–{114}
$\Sigma = 9$	$\Theta = 218\,°57'$	Second order
		{111}–{115}
		{114}–{114}
		{221}–{221}

planes are shown in Table 4. All of the boundaries contained disrupted and wrong bonding.

The models drawn using the CSL approach may be used to infer facts about the crystallography, energies and electrical activities of twin boundaries in CdTe. Firstly, since only those lateral boundaries predicted by the CSL model were seen in experiments, this gives some confidence in the approach. Secondly, the qualitative forecast that the coherent first-order boundaries have low energy, while the laterals and second-order twins are of higher energy, has some experimental backing: only the first-order coherent boundary appears in such long lengths. Also, the trend of Te segregation to these boundaries (Section 3.4) adds further evidence to the postulate. Thirdly, it may be expected that those boundaries with more disrupted bonding are intrinsically electrically active. Therefore, while first-order coherent boundaries (which may have perfect bonding) should not always be electrically active, the laterals and second-order boundaries and random grain boundaries are all expected to be electrically active and to act as recombination centres.

This has been investigated by electron beam methods. Figure 15 shows an EBIC micrograph of a twin band terminated by a lateral twin boundary in vapour-grown CdTe. The lateral segment shows stronger contrast than the coherent segment, indicating enhanced recombination. In another study, the 'R-EBIC' (remote EBIC) method was used to show that random grain boundaries in a bulk CdTe sample had charge separation and recombination behaviour consistent with downwards band bending [11].

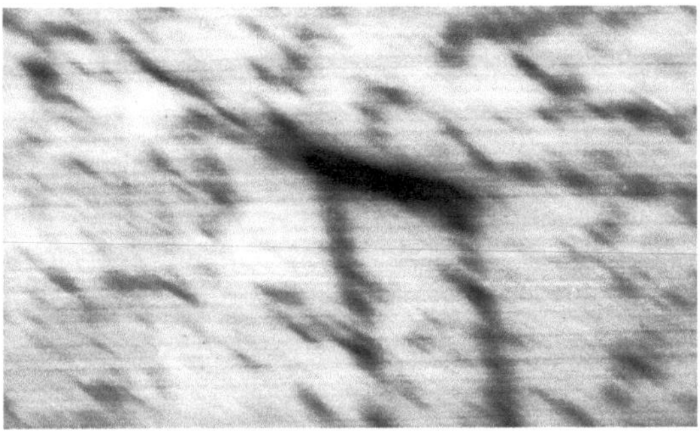

Figure 15 EBIC micrograph showing enhanced recombination (dark) contrast at a $\Sigma = 3$ lateral twin boundary segment compared to that at a coherent segment. This confirms the findings from CSL boundary models that the lateral boundaries may be expected to have disrupted bonding. The scattered contrast dots are from dislocations and dislocation arrays. The weak contrast at the coherent boundary (parallel lines) is probably from boundary dislocations that were shown to be present by TEM. The twin band itself is 50 μm wide. Figure from Ref. [43].

Despite its successes the CSL model cannot be guaranteed to give an accurate description of the atomic arrangements at twin boundaries. Some of the models contain saw-tooth steps which look unlikely when compared to the ring structures at boundaries in silicon observed by HRTEM. Additionally, rigid body translations between twin and matrix are not accounted for by the CSL model – such boundaries differ in points of detail [111]. The full details of boundary structure in CdTe will not be established without resort to high-resolution microscopy and perhaps relaxation calculations.

2.3.4. Origins of twinning

It is well known from crystal growth results that twinning is very common in CdTe, and it is generally accepted that this is a result of its low twin energy. It is however more difficult to demonstrate the origin of twinning, which might, in principle, be caused by either growth or strain mechanisms.

2.3.4.1. Stress twin model Stress-induced twinning is well known in metals. It is considered to arise when, for energetic reasons, a material deforms by twin nucleation and propagation in preference to dislocation slip. For twins to form in this way, a dislocation source must nucleate and then systematically issue Shockley partial dislocations over successive {111} planes. In this way the shear transformation that represents a twin is then built up in a volume of material. In considering twinning in fcc metals and alloys, Venables [112] describes one such source, and there are other models. Similarly the *obstacles* to twin propagation in fcc metals were considered, and dislocation reactions of the inverse type to those invoked in dislocation formation [112] were considered significant. To establish the relative probability of slip and twin nucleation/propagation a full evaluation would be required, and its relation to, for example stacking fault energy, understood for CdTe.

In an experimental study, Vere et al. [113] attempted to induce twinning in ⟨123⟩ oriented bars of CdTe at temperatures up to 500 °C. No twins were observed by etching and optical/scanning electron microscopy. Similarly, work on bending, indentation and compression testing (as described in Section 2.2.2) does not report deformation twinning in optical SEM and TEM measurements. However, it is recognised that the failure of experiments to induce deformation twins in CdTe is not in itself evidence that it is not possible. Indeed, in the case of (Cd,Zn)Te, compression stress testing did induce micro-twinning that could be observed in TEM [18]. This may indicate that different stress-relief mechanisms may operate in the solid solution than in pure CdTe.

A final piece of evidence concerning the stress-twinning mechanism comes from consideration of the shear transformation itself. It is well known that twins up to some cm in size may occur in crystal boules that grow from either the vapour or the melt while contained in close fitting quartz capsules. If such boules were to grow perfectly, and then become twinned by deformation (thermal stress or constraint from the container), then there would be a large volume of displaced material. The constraints of the container would make such a straightforward shear impossible.

*2.3.4.2. **Growth twinning model*** The growth twinning model demands that the usual sequence of stacking be disrupted in favour of the stacking sequence in a twin orientation. This would be presumed to occur in response to instability during growth, or a new orientation being adopted for energetic reasons.

Hurle [114] presents a thermodynamic analysis of twin formation at facets during Czochralski growth of III–V semiconductors. It was considered that twin orientations (having low-energy faces) may arise as a thermodynamically favourable response to the formation of the facets, i.e. by a growth mechanism.

CdTe does not display the facet formation phenomenon in Czochralski growth [113]. Nevertheless, it is know that the conditions of growth can influence the density of planar faults. For example in the sublimation growth of CdTe films, the density of planar faults and twins within the grains decreases with increasing substrate temperature, as shown by Al-Jassim et al. [115] in a TEM of a sequence of layers grown at 425, 525 and 625°C. Presuming temperature to be the only variable, this indicates a growth-related mechanism for faulting and twinning.

Overall the present author considers that the evidence presented above points to growth mechanisms as being more likely to be responsible for twinning than deformation mechanisms.

2.4. Second phases: Precipitates and inclusions

CdTe has a strong tendency to become Te-rich at high temperatures since the Cd overpressure exceeds that of Te. When a stoichiometric excess of Te is present, its emergence as a second phase in solidified crystals is inevitable. In the unusual case of there being excess Cd, second phase Cd may be present – a description is deferred to the end of this section. For an alternative review of both Te and Cd second phases the reader is referred to that of Williams [116].

Second phase Te (we have been careful not to name it so far) is considered to present itself into crystalline CdTe by two distinct mechanisms – at least in the case of bulk growth from the melt [117], and probably generally (see Fig. 16). Firstly, large *inclusions* 1–2 μm in size (occasionally 10–20 μm) are considered to arise from instability at the liquid–solid interface. For example if the interface has a dip resulting from a grain

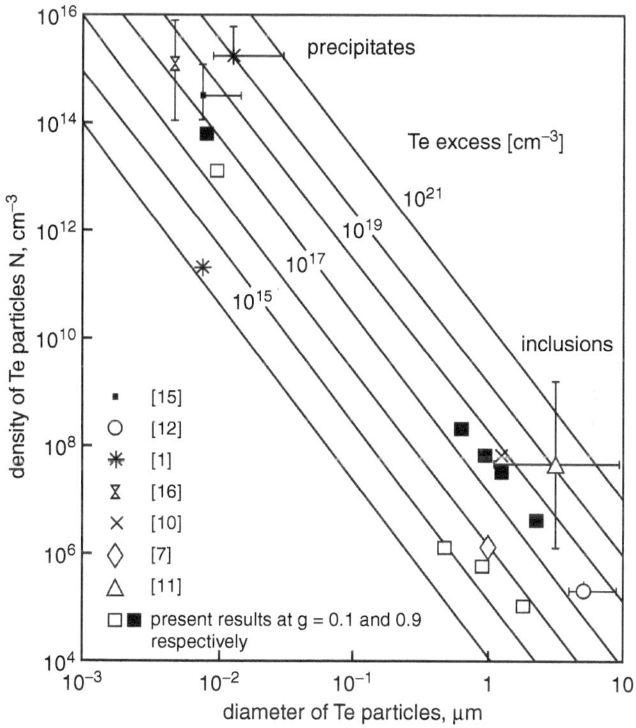

Figure 16 An experimental plot of the density versus diameter of second phase Te in CdTe shows two size clusters. Rudolph et al. classify the second phase particles as *inclusions* and *precipitates*, and this is generally accepted. Redrawn from Ref. [117]. The references refer to the original source.

boundary, then excess Te may become trapped there and incorporated in the bulk of the crystal. Such inclusions are easily detected by infrared microscopy. Secondly, formal *precipitation* of Te is prone to occur in Te-rich CdTe upon cooling; the *T–x* phase diagram indicates that the solubility of Te decreases with decreasing temperature (this is the so-called retrograde solid solubility). Experiments show such precipitates to be 10–50 nm in size [117, 118], while calculations suggest that they form in about 100 s and are separated by tens of nm. Since the amount of Te that can be precipitated in this way is dependent only on the shape of the solubility limit line on the phase diagram, then the amount of Te precipitated is likely to be independent of the amount of Te-excess, providing that it is greater than $\sim 10^{18}$ cm^{-2}. Nevertheless, the cooling rate at high temperatures is known to influence their size and density [118]. Such precipitates cause optical extinction [119] that may be eliminated by annealing in Cd vapour. Both Rudolph [117] and Yadava report optical methods for the detection of Te precipitates. Yadava et al. [118] give some detail of the simulation of Mie scattering in this respect. Results for vertical Bridgman (VB) samples are given in Section 3.2.

Regardless of the mechanism of their formation, the amount of excess Te may be expressed in terms of the numbers and dimensions of the precipitates/inclusions. Rudolph et al. [117] give this as a summation over the number of precipitates/inclusions, taking into account their radii, the relative atomic mass and density of Te (A_{Te} and ρ_{Te}), and Avogadro's number:

$$N_{Te} = \frac{4\pi\rho N_A}{3A_{Te}} \sum_{i=1}^{n} r_i^3 \rho_i \tag{7}$$

Inclusions (and precipitates) may be expected to be associated with other crystallographic defects in the lattice, i.e. they may decorate them. The driving force for this is the reduction of the energy of such defects – similar phenomena are reported for inclusions in metallic systems [4]. Where crystal growth has taken place in conditions that encourage equilibrium, then the density of Te inclusions at boundaries is seen to follow the trend of the boundary energies themselves. For example, in the case of slow vapour growth of CdTe at high temperatures [43, 120] (Fig. 17) the density decreases in the sequence:

Random boundaries > second-order twins > first-order lateral twins > first-order coherent twins.

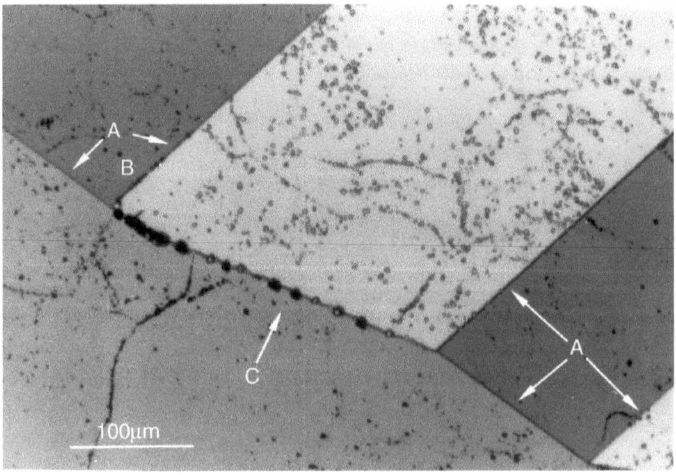

Figure 17 Te inclusions associated with boundaries in vapour-grown CdTe. Positions A are first-order twin boundaries, which are largely undecorated, except at B. A second-order twin boundary (C) is highly decorated. The main grain was $\{\bar{1}\bar{1}\bar{1}\}$Te oriented and etched with Nakagawa's reagent. From Ref. [43].

The inclusions may perhaps nucleate at planar boundaries, but migration of inclusions under the influence of temperature gradients so as to cause decoration is also plausible.

A discussion of the shapes of Te inclusions in terms of the energies of low index planes is given in Ref. [121]. Smaller Te volumes are rounded in shape; larger ones have facets. The shape presented depends on the crystallographic projection; distorted hexagons are commonly seen (Cd inclusions differ – see below).

There is evidence that both inclusions and precipitates are associated with stress, both in the Te and in the surrounding lattice. Where the critical stress for dislocation nucleation is exceeded, dislocations form and are associated with the inclusion or precipitate. There are two possible mechanisms:

(a) The thermodynamic driving force for Te precipitation is considered sufficient to generate stresses that exceed those required to form dislocations [121], i.e. as precipitates grow, *compressive* stress within them increase, and strain relief takes place in the surrounding CdTe by plastic deformation. *Growth* is opposed by the mounting compressive stress in the precipitates, and the resulting overpressurisation causes dislocation 'loop punching' – both operate in tandem during precipitation.

(b) Differential thermal expansion between the Te inclusion and the surrounding CdTe matrix generates stress, and dislocations result from the so-called 'punch-out' or 'loop punching' mechanism. The thermal expansion coefficients for CdTe, Te and Cd are given in Table 5. As the solid cools the inclusions freeze, and both the inclusions and the surrounding matrix are put under *tensile* stress [8, 122]. Dislocation rosettes similar to those resulting from indentation studies result.

Some authors suggest that precipitation may account for the overall dislocation density in CdTe (see Ref. [121]). For example Wada and Suzuki [123] presents a combined Nakagawa etching and infrared microscopy study in which the areal density of inclusions correlates linearly with that of dislocations in the range $\sim 10^5$–10^6 cm^{-2}. However, the

Table 5 Thermal expansion coefficients and melting points of CdTe and its elements

	Thermal expansion coefficient (10^{-6} K^{-1})	Melting point (°C)
CdTe	4.5	1091
Te	14.7	450
Cd	31	321

Differential thermal contraction may be expected to be greater in the case of Cd rather than of Te inclusions.

imaging of a single area using both methods did not demonstrate a good spatial correlation. While it may be expected that precipitates make a contribution to dislocation density, it is unlikely that all dislocations are precipitate-related: other mechanisms (e.g. direct growth stresses) are known to be important.

Shin et al. [119] present X-ray and Raman evidence showing that at room temperature, Te inclusions are present in the high-pressure rhombohedral phase. This was confirmed by Schaake [127]. Yadava's calculations [121] indicate that the 70 kbar (7000 MPa) required is unlikely to be achieved, and that the more likely phase is the intermediate monoclinic phase known for pressures >40 kbar (4000 MPa). In a high-pressure XRD study, Aoki [125] presents evidence that the monoclinic phase persists to pressures up to 100 kbar (10,000 MPa) (Table 6), and that there is no intermediate phase. In any event, observation of any high-pressure phase indicates that Te inclusions remain under a *compressive* stress at room temperature. This is apparently in contradiction of the thermal contraction model, for which tensile stress is predicted. Overall it seems likely that the stress around second phase Te will depend on a combination of factors: its formation mechanism (inclusion or precipitate); for inclusions, the possible subsequent contribution from precipitation adding to their volume; and the effects of differential thermal contraction.

Inclusions and precipitates may influence optical transmission, electrical recombination, mobility-lifetime product, introduce dislocations

Table 6 The phases of tellurium as a function of pressure

Phase	Pressure	Unit cell information, lattice parameters (nm) (intensity), Refs.
Rhombohedral	$p < 40$ kbar, $p < 4000$ MPa	$a = 0.4457$ nm, $b = 0.4457$ nm, $c = 0.5929$ nm, $\alpha = 90.0°$, $\beta = 90.0°$, $\gamma = 120.0°$ [124]
Monoclinic	$(40–45) < p$ (kbar) < 70, $(4000–4500) < p$ (MPa) < 7000	$a = 0.3104$ nm, $b = 0.7513$ nm, $c = 0.4760$ nm, $\beta = 92.71°$ 0.2948 nm (100%), 0.2873 nm (53%), 0.2418 nm (14%), 0.1835 nm (16.9%), 0.1793 nm (19.4%) [125]
Rhombohedral	$p > 70$ kbar, $p > 7000$ MPa	$a = 0.3002$ nm, $\alpha = 103.3°$ 0.278 nm (100%), 0.236 nm (61%), 0.178 nm (26%) [126]

Aoki's results [125] indicate that the monoclinic phase persists up to 100 kbar (10,000 MPa), and that the transition at ~70 kbar (7000 MPa) is electronic but not structural. Some authors report high-pressure phases of Te in inclusions in CdTe.

and interference with over-layer growth. There has therefore been much research into eliminating – or avoiding the formation of – inclusions. The main themes of the work are:

(a) Reduction in the density of inclusions arising in melt growth by stirring, as in the accelerated crucible rotation technique (ACRT) [34].
(b) *Post-growth annealing under Cd vapour*. Using a defect equilibrium model, Vydyanath [1] was able to calculate the minimum deviation from stoichiometry of CdTe as a function of temperature and as a function of an applied overpressure of Cd vapour. In this way it was possible to calculate the temperature of a Cd reservoir required to achieve optimal correction of the stoichiometry as a function of a crystal's temperature. Annealing under conditions determined in this way was found to be effective in reducing the concentration of Te inclusions >1 μm in size, but was less effective in removing smaller ones. While in-diffusion of Cd might be expected to be significant, thermal migration of the larger inclusions towards regions of higher temperature was also considered to be an important mechanism for removing the larger, but not smaller, inclusions.
(c) *Growth under excess Cd to correct the stoichiometry*. Szeles et al. [128] have extended the method in (b) using apparatus capable of controlling the Cd vapour pressure over the crystal at all temperatures during both the growth of the crystal and its cooling from the growth temperature, i.e. programmed annealing during cool-down. Significant improvements in the density of inclusions are reported for (Cd,Zn)Te, and a concomitant decrease in the density of associated dislocations expected.
(d) *Low-temperature growth from the vapour*. This allows for (i) control of the vapour stoichiometry and (ii) reduction of the extent of Te precipitation – cooling commences from a lower temperature point on the solubility limit curve of the *T*–*x* phase diagram. Vapour growth experiments are reported extensively, and significant progress is being made in the engineering of control schemes for the elimination of second phases.

Under conditions of excess Cd – either from intentional excess in the starting material or from over annealing in Cd vapour, Cd inclusions result. While they may result from the same mechanisms as invoked above for Te, Williams considers that constitutional supercooling may play a role [116]. Cd inclusions are identified by EDX [129] and revealed easily by selective etching and infrared microscopy as demonstrated by Watson [8, 130]. Etching reveals extensive dislocation rosette arms with crystallographic shape – these are visible in infrared microscope images, and it is inferred that they are decorated with Cd. This was attributed to differential thermal contraction upon cooling of the inclusions below the melting point of Cd (321°C). Since Watson's paper contains a figure labelling error (Te for Cd), the more complete study of Brion [122] is more regularly cited. Both Cd and Te precipitates are compared – those of Cd have a more fully developed dislocation rosette structure, this being explored in the infrared microscope and confirmed as being crystallographically related. The higher level of dislocation at Cd-compared to Te precipitates is attributed by Brion et al. [122] to the higher level of tensile stress expected from cooling (see thermal expansion coefficients in Table 5). Certainly Williams demonstrated a clear strain field at Cd inclusions using birefringence [116].

3. DEFECTS IN BULK CRYSTALS OF CdTe

3.1. General observations

Each of the principal methods of bulk growth for CdTe – melt, solution, and vapour growth – has its individual characteristics. These include temperature, temperature field, temperature–time profile of the process, relationship to the *T–x* phase diagram, overpressure of inert or constituent gas and contact with seeds or the container. Accordingly the type density and distribution of extended defects may be expected to be dependent on the growth method. Additionally it is expected that different embodiments of each principal growth method (e.g. vertical and horizontal Bridgman) may introduce characteristic variations. However, a review can only represent the present state of a particular technology at the present time, rather than its fundamental limit. With that in mind, the reader should temper the following general observations about the relative merits of the main technologies.

In melt growth the latent heat of fusion is generally conducted away via the solidified crystal itself, this being a particular limit for CdTe, which has very low thermal conductivity. The thermal gradients encountered in Czocharalski growth may be especially severe and polycrystallinity results, making the method unsuitable for CdTe. Bridgman (directional freezing) methods much more successful, and the vertical Bridgman method is the industry standard for the production of substrates. Nevertheless there are issues with single crystal yield, with wafers being extracted from the grains in (large) multi-crystalline boules. The dislocation density may be low, but is often variable over a wafer and is commonly at the 10^4–10^5 cm^{-2} level. Distribution of dislocations into subgrains (polygonisation walls) is influenced by strain and thermal history and is also often variable over a wafer. Large inclusions of Te may be present. Smaller precipitates are endemic.

CdTe grown from THM using a Te molten zone has generally (but not universally) poorer crystallinity than Bridgman material and is prone to Te inclusion from the molten zone. Solvent evaporation (SE) growth may be expected to give similar issues, but has nevertheless demonstrated an ability to generate some spectacularly large polycrystal plates.

Vapour growth offers the advantages of lower temperature growth and stoichiometric control, with some schemes allowing for growth free from the walls of the container. Control of seeding is approached variously with various degrees of success. Low dislocation densities ($\sim 10^3$ cm^{-2}) have been reported more frequently than for the melt methods, but depend on the technology (as for other methods). There is also some evidence that the tendency for dislocations to cluster in CdTe is less marked in vapour – than in Bridgman-grown – CdTe [48]. Although the lower temperatures used might be expected to give a reduced incidence of Te precipitation, freedom from second phase Te is by no means guaranteed. For vapour growth there is a multiplicity of embodiments of the technology – reflecting the fact that while vapour growth is attractive for fundamental reasons, it remains complex to engineer optimally.

3.2. Melt-grown CdTe

3.2.1. Czochralski-grown CdTe

Liquid encapsulated Czochralski (LEC) growth is not a suitable method for CdTe [25, 28]. Its low thermal conductivity (lower than the encapsulant), and the high thermal gradient encountered in the method generates small-grained polycrystalline boules. Thomas et al. [28] show a photograph of an axial cross section of a 1.5 kg boule 50 mm in diameter which is entirely polycrystalline. The grains are in the range 5–10 mm in size and are mostly twinned.

3.2.2. Vertical Bridgman (VB) and vertical gradient freeze (VGF) – grown CdTe

Vertical Bridgman growth is the principal industrial route used for bulk CdTe growth [22] and accordingly there is much published on not only the trends in the types and densities of defects found in it, but also on the mechanisms by which they are introduced.

The brief review that follows is presented in the order: grains/twins, dislocations and finally inclusions/precipitates.

3.2.2.1. VB: Grains and twins It is generally accepted that the success of the vertical Bridgman over the Czochralski method in producing large grains stems from the ability to adjust the temperature gradient in the former. Several authors (see, e.g. Refs. [131, 132]) report that use of low-temperature gradients, of the order of 10 K cm^{-1} [132], are needed for the growth of single crystals – or at least of large grained boules with a reduced incidence of twinning.

During the growth of long rods (e.g. 20 cm) of CdTe by VB, the crystallinity varies along their length [132, 133]. Generally the first 15% is small grained, the middle comprises two or three grains (often with twins), while the last 15% to grow has grains, although larger than at the start. Pfeiffer and Mühlberg [132] explored the melt–solid interface shape both experimentally and theoretically. They concluded that the interface shape changes as growth progresses. Moreover, the regions of polycrystallinity correlated with the regions in which the interface shape was changing most rapidly. Supercooling is also implicated in determining the crystallinity at the first-to-freeze end of the boule [134]: low superheating of the melt encourages monocrystalline growth at the tip, but the crystallinity degrades thereafter, whereas higher superheating encourages polycrystalline nucleation – but thereafter the crystallinity improves with further growth. Attempts to seed the growth met with mixed success. Azoulay et al. [135] report some improvement in grain size with seeding, while Pfeiffer and Mühlberg [132] point out that the result is not reproducible on account of supercooling phenomena.

3.2.2.2. VB: Dislocations and polygonisation walls Dislocation phenomena in VB material have also been studied in some detail, and the results are shown in the survey presented in Table 7. The dislocation density – and its distribution into sub-grain boundaries (in a given sample) – is influenced by axial and radial stress, temperature gradients and stirring. Generally,

Table 7 Summary of some dislocation phenomena reported for vertical Bridgman (VB)- and vertical gradient freeze (VGF)-grown CdTe bulk crystals

Authors	Observation of dislocation phenomena
Lu, 1986 [137]	epd 10^5–10^6 cm^{-2} (Inoue) Mosaicity 200 – 500″ Sub-grain size 500–1000μm Slip bands
McDevitt, 1986 [138]	epd 2 × 10^5 cm^{-2} (Nakagawa) sub-grain size 150–400 μm
Tanaka, 1987 [45], VGF	epd min 10^4, average 10^4 cm^{-2} (Nakagawa) Mosaicity ~300″ Sub-grain size 200–300 μm Slip bands epd and sub-grain structure depend on T gradient

(*continued*)

Table 7 *(continued)*

Authors	Observation of dislocation phenomena
Sen, 1988 [139]	epd 2×10^4 in bulk, 10^5–10^6 cm^{-2} near boule surfaces (Nakagawa)
	Sub-grains prevalent near to boule surfaces
Song, 1986 [140]	epd 10^3–10^5 cm^{-2} (Inoue)
Sabinina, 1991 [141]	TEM of dislocation arrays in sub-grain boundaries
Azoulay, 1990, VGF [135]	For seeded {111} – epd ~10^4 cm^{-2} (Nakagawa), FWHM 20–40″
	For unseeded – epd ~10^5–10^6 cm^{-2} (Nakagawa), FWHM 30–50″
Mühlberg, 1990 [133]	Polygonisation dependent on axial temperature gradient G
	$G \geq 10$ K cm^{-1}, sub-grain size 500–1000 μm, mosaicity 30–120″
	$G \sim 4$–5 K cm^{-1}, no polygonisation
Becla [131]	Dislocation density increase with growth rate and temperature gradient
Capper, 1993 [34]	Use of ACRT reduced dislocation density from 9.6×10^5 to 3.3×10^4 cm^{-2}, HRXRD shows that tilt FWHM changed little (17–22″); that for strain is reduced from 40″ to 26″
Casagrande, 1993 [46]	epd 10^4 min, 1.4 (±0.7) $\times 10^5$ cm^{-2} typical (Nakagawa)
	Sub-grain size 400–600 μm
	FWHM 9.4″best, 12–16″ typical, 5.1″ theoretical
Guegori, 1994	epd 5 (±1) 10^5 cm^{-2} (Inoue)
	sub-grain size 80 μm
	FWHM 100″
Hähnert, 1994 [47]	epd 10^5 cm^{-2} (Hähnert)
	Dislocation density 10^5 cm^{-2} (X-ray topography)
Everson, 1995 [41]	epd 10^5–10^6 cm^{-2} (Everson)
Shetty, 1995 [136]	epd 1.5–8.6 $\times 10^5$ cm^{-2} (Nakagawa)
	epd depends on ampoule coating (correlates to work of adhesion)
Kumaresan, 2000 [142]	FWHM 20″ on boule axis, 72″ near wall

All material is VB-grown unless stated. Where etch pit densities (epd) are reported, the etchant used is indicated in brackets – further details are given in Table 2. FWHMs refer to double crystal X-ray rocking curves unless stated. Where tilts between sub-grains (mosaic blocks) are reported, then they are listed as 'mosaicity' followed by a tilt value.

the dislocation densities in VB CdTe are of the order of 10^5 cm^{-2}, with some reports in the 10^4 cm^{-2} range, and exceptionally as low as 5×10^3 cm^{-2} [45]. There is a similarly large range of sub-grain (mosaicity) behaviour reported, ranging from the absence of sub-grains to those in the range 80–1000 μm in size and with tilt distributions ranging from a DCXRD FWHM of 9.4″ up to tilts between individual sub-grains of 120″.

Tanaka et al. [45] demonstrated a clear dependence of the dislocation density and distribution on the temperature gradient, with values of 7.5, 15 and 25 K cm^{-1} being studied. The higher gradient gave dislocation densities of $\sim 10^6$ cm^{-2}, while the lowest reduced this to $\sim 10^5$ cm^{-2}. Moreover, the tendency to form well-defined polygonisation walls became reduced as the temperature gradient was lowered, as shown in Fig. 18.

Other workers reported similar results: Mühlberg et al. [133] found that for temperature gradients ≥ 10 K cm^{-1}, sub-grains having sizes in the range 500–1000 μm – and supporting tilts from 30 to 120″ – were generated. Reduction of the gradient to 4–5 K cm^{-1} eliminated the polygonisation. Becla et al. [131] reported an increase in the dislocation density with increased growth rate.

A correlation between high dislocation densities, increased polygonisation, and the work of adhesion between CdTe and a variety of ampoule coatings has been demonstrated by Shetty et al. [136]. It is a clear demonstration that high stresses that arise during growth give rise to high dislocation densities and also encourage polygonisation.

Stirring of the melt during growth – using the ACRT [34] – has some benefits over Bridgman growth in otherwise comparable conditions: it has been shown to reduce the dislocation density from 9.6×10^5 to

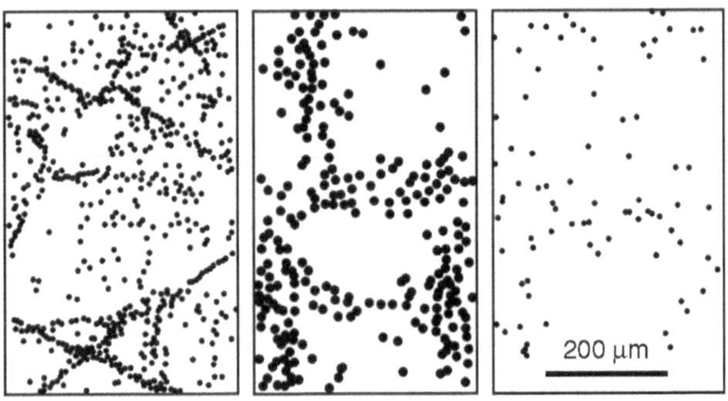

Figure 18 Etch pit distributions for CdTe grown using three different temperature gradients in the vertical gradient freeze configuration. Left 25 K cm^{-1}, middle 15 K cm^{-1}, right 7.5 K cm^{-1}. Reduction of the temperature gradient encouraged fewer dislocations and less pronounced polygonisation. Etch pit positions redrawn from the micrographs in Tanaka et al. [45].

3.3×10^4 cm^{-2}. While high-resolution XRD shows that the tilt FWHM is relatively unaltered (17–22″), that for strain is reduced from 40 to 26″. In other experiments [135], seeding is reported to reduce the dislocation density and double crystal FWHM from 10^5–10^6 cm^{-2}/30–50″ to ~10^4 cm^{-2}/20– 40″, respectively.

3.2.2.3. VB: Slip bands A limited number of authors [45, 137] report the incidence of slip bands (see Section 2.2.3 and Fig. 10). It is unclear from the literature whether this is because they have low incidence – or whether they are under-recognised resulting in their being under-reported. Where they are observed they are an indication that stress has been encountered during cool-down, without there having been an opportunity for polygonisation.

3.2.2.4. VB: Second phase Te Since at the melt temperature of CdTe there is a significant overpressure of Cd, VB growth is susceptible to the presence of both included and precipitated Te. If the stoichiometry is not corrected, then aggregation of the excess Te is inevitable, either in the form of inclusions or precipitates (see Section 2.4).

For VB material inclusions are generally 1–2 μm in size, but may be as large as 10–20 μm, while precipitates are typically 10 nm in diameter and present at densities of mid-10^{17} cm^{-3} [117]. Since inclusions are considered to arise from instability in the solid–liquid growth interface, agitation of the melt by the ACRT method may be considered a means of reducing them. Indeed it is successful in eliminating Te inclusions having diameters ≥10 μm, although smaller inclusions are reported as having the same density with and without ACRT [34]. In normal VB growth, the density of inclusions may decrease as freezing progresses along the ingot – as shown by Rudolph et al. [117].

Precipitates are considered to form as a result of the solidified melt being Te-rich, and of the solid solubility of Te decreasing with temperature. Hence Rudolph et al. [117] point out that for melt growth the amount of precipitation is a constant, providing that the excess of Te is greater than 10^{18} cm^{-3}. Yadava et al. [118] describe the basis of optical extinction for measuring Te precipitation and give experimental data for VB samples. Cooling in the range 1100–850 °C was considered important, and precipitates were recorded in various samples as having a minimum size of 5.7 ± 1.6 nm (density 2.1×10^{15} cm^{-3}) and a maximum of 51.5 ± 17.9 nm (density 1.4×10^{12} cm^{-3}) – the errors are the standard deviations. Presumably the reduction in density of the large precipitates reflects the constant amount of Te available for precipitation referred to above.

3.2.3. Horizontal Bridgman

Oda [22] points out that the horizontal Bridgman (HB) configuration – in which the melt lies in a boat – is subject to more complex thermal and stress fields than the vertical Bridgman configuration: radiation may take place from the free surface of the melt, while heat loss through the (quartz) boat may be by conduction and radiation. Nevertheless, HB growth has been developed into a successful commercial technology with the variants of a high overpressure of Ar and an overpressure of Cd vapour having been explored, as has seeded growth.

Early experiments showed the effect of the boat on the crystallinity: while the free surfaces of ⟨110⟩ seeded boules were monocrystalline, the bottom 20% of the volume had grains [143]. For unseeded growth the single-grain fraction was typically 40% [36]. Others report polycrystalline growth with grains being several cm^3 in volume [144]; multiple twinning may occur.

The best parts of HB – grown CdTe display Nakagawa etch pit densities and DCRC rocking curves as low as 5×10^4 cm^{-2} and 9.5″ [143]. Everson [41] reports 3–5 $\times$ 10^4 cm^{-2}. There is generally considerable variation across a boule: the Nakagawa etch pit density has been reported as being low ($\sim 10^4$ cm^{-2}) in the centre of an ingot and higher ($\sim 10^5$ cm^{-2}) at the ends. Systematic variations in the etch pit *distribution* have also been demonstrated, with the bottom and middle points of a boule being more prone to sub-grain boundary formation than others [36]. Variations in etch pit density, distribution type and FWHM over a single wafer may be significant. Lay [36] reports dislocation density in the range 3–8 $\times$ 10^4 cm^{-2}, random to highly polygonised distributions and 11–37″ FWHMs with single and multiple peaks. Johnson [35] reports similar results, with dislocation density extending up to 6.6×10^5 cm^{-2}. HB CdTe has been used to study the relationship between DCRC FWHM and dislocation density [35, 145], and a meta-analysis of the data is presented in Section 1.3.1.

The incidence of Te inclusions is also related to the position in the boule. The density of large inclusions (<10 μm) may be greater near to the part of the boule in contact with the boat [145] and may increase along the direction of growth [36]. Smaller inclusions (~1 μm) have been observed to be associated with extended defects [36]. Voids have also been reported. They have high incidence at the bottom of boules [143]; this can be reduced by using vitreous carbon rather than quartz boats [144]. The voids increase in density along the length of the boule as freezing progresses [36, 144].

3.3. Travelling heater method- and solvent evaporation-grown bulk CdTe

3.3.1. Travelling heater method using Te

The travelling heater method (THM) is used predominantly to grow high resistivity CdTe for detector applications. Its principal issues for crystal quality are crystallinity and the incorporation of Te inclusions from the mother liquid. Generally there is less reported about microstructural characterisation than for Bridgman-grown CdTe, and the FWHM of radiation detector response is commonly used as a measure of the material quality. This is a reflection of the diminished importance of extended defects compared to the highly demanding application of CdTe as a substrate. Nevertheless, the crystallinity is important, not least in processing, and Te inclusions/precipitates are known to degrade uniformity and detector performance. A brief outline of the main issues for crystallinity and inclusions follows.

It has been pointed out [35] that the crystallinity of THM-grown CdTe can be poor and is often inferior to Bridgman material, and accordingly a number of reports of attempts to improve the grain structure have been made. Seeded growth is often used, and twins in seeds are seen to propagate into the growing crystals [146, 147] (see Fig. 19). Choice of the dimensions of the Te zone and control of its interface shape are considered essential in determining the crystallinity. For example, for the growth of rods 10 mm in diameter and ~60 mm long, Schoenholtz et al. [146] demonstrates that a 3 mm wide Te zone generates polycrystalline boules, the grain size being ~0.5 of the diameter of the rod. Use of a 4 or 6 mm wide zone generated single crystal rods, albeit with twins (see Fig. 19). Multiple passes of Te zones were not found to eliminate grains and twins [148].

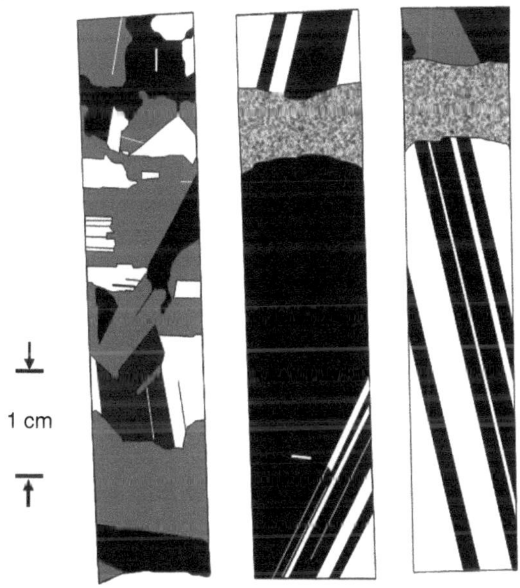

Figure 19 Sections of seeded THM-grown CdTe showing the influence of the width of the Te zone. Left 3 mm, centre, 4 mm and right 6 mm. While the narrowest zone caused polycrystallinity, when its width was 4 mm or greater, the boules had single grains (with twins). Drawn from the photographs of etched boules in Schoenholtz et al. [146].

Funaki et al. [149] extended the diameter of THM CdTe to 50 mm and found that a Te zone in the range 30–50 mm was most effective, with the very best results being obtained for a 40 mm zone. It was noted that for zones of greater or smaller width, the convex interface shape included inflection points which were likely to introduce grain boundaries.

Funaki et al. [149] also report a study of the effect of seed orientation on the crystallinity. Use of the (111)Cd orientation encouraged the generation of sets of parallel twin lamellae on more than one set of {111} planes. The boules were also reported to have sub-grain boundaries. For the $(\bar{1}\bar{1}\bar{1})$ Te seed orientation, the character of the twins changed; twinned grains were present, these being equally distributed among the three inclined and equivalent {111} planes. To control this, $(\bar{1}\bar{1}\bar{1})$Te seeds off-cut by 5° towards the nearest ⟨110⟩ were used to incline one of the three {111} planes at a more shallow angle to the growth interface than the others. This had the effect of confining twinning to a single set of {111} planes. Boules grown in this way could therefore be sliced parallel to the twin planes yielding wafers with no intersecting twins.

Inclusions are known to decrease the resistivity of THM CdTe [150]. The inclusion problem in CdTe is compounded in THM growth by direct inclusion of Te from the melt zone, and accordingly inclusions ranging from 1 to several hundred μm are reported [151]. The presence of very large inclusions may be more severe than in the case of Bridgman growth. Schwartz [147] attributes their presence to temperature instabilities at the growth interface. This was confirmed by switching off the heater for brief periods, disturbances which generated planes of inclusions. Unintended fluctuations were considered to be responsible for the lower density of inclusions generated during normal growth.

In THM growth, Te inclusions may be considered as droplets which are trapped or left behind in the solid due to some irregularity of the growth interface [151]. Their shape is determined by the freezing of the matrix around the inclusion, i.e. the shapes of inclusions may be more properly considered to reflect the habit of a void in the frozen matrix, within which the Te inclusion is the last part to freeze [151]. This implies that the shape is a property of the matrix rather than that of the inclusion, and where there is a distinct shape to the inclusions they are oriented to the host lattice [147]. The shape (triangular or hexagonal) and size of Te inclusions has been observed to be dependent on the temperature gradient used, i.e. on the cooling rate of the region where the inclusions form [147].

3.3.2. Travelling heater method using Cd

Since Te *inclusion* from the melt is a regularly encountered using a Te molten zone in THM growth, the use of a Cd molten zone has been investigated. This has the added advantage that growth proceeds from the Cd-rich side of the T–x phase diagram therefore avoiding the possibility of Te *precipitation*.

In a study of THM growth using a Cd molten zone, Triboulet et al. [152] reported that the degree of crystallinity was related to the growth rate. For 15 mm diameter boules, 2 mm/day yielded grains ~2 mm in diameter, whereas at 1 mm/day this increased to ~5 mm. Twinning was prevalent in the grains. Cathodoluminescence revealed isolated dislocations and highly developed sub-grain structures, the latter also being revealed in X-ray topographs. Triboulet et al. [152] de not report further on inclusions, but comment that the crystallinity of Cd (and Te) zone THM CdTe is poor.

3.3.3. Solvent evaporation from Cd

Mullin et al. [153] grew short CdTe rods from solution in Cd in a vertical geometry. About 10% of the crystals grown had a single twin-free grain dominating 90% of the crystal. The etch pit density was in the range 2.0×10^2 cm^{-2} to 5×10^3 cm^{-2}, but the etchant was not named. Preliminary DCRC gave a FWHM of ~82″ (this suggesting a higher dislocation density than was revealed by etching).

3.3.4. Solvent evaporation from Te

Pelliciari et al. [154] demonstrate the growth of plates of CdTe up to 300 mm in diameter using a SE method. The as-grown surfaces of these plates reveal a spectacular large grain and twin structure directly. For both 65 and 300 mm diameter growth, the largest grain has a width of about half of the diameter of the full plate. Te inclusions might be expected to be problematic for this method.

3.4. Vapour-grown bulk CdTe

3.4.1. Piper–Polich (transport in a simple capsule) and its variants

In Piper and Polich's method [155], II–VI crystal growth is achieved by transporting material from one end of an evacuated sealed tube to the other in a temperature gradient. In addition to the simple method, variants have been used for CdTe growth, for example growth with seeds, and with the use of unclosed tubes with holes or capillaries which may self-seal when transport starts.

Growth in uncontrolled conditions can lead to polycrystals with fine grains, and with voids [43]. When large grains are achieved, the method favours the [100] and $[\bar{1}\bar{1}\bar{1}]$Te directions [156]. Success in growing large-grained boules (with seeded growth) has been linked to the suppression of constitutional supercooling at the growth front [156, 157]. Wiedemeier et al. [157] suggest that this may be achieved by use of a sufficiently high temperature gradient: this stabilises the growth interface and ensures that it is convex [156]. For each temperature gradient investigated, the rate of pulling through the furnace influenced the grain boundary and twin content, this being shown as a process 'map' in Ref. [157]. Temperature gradients of $<10\ \mathrm{K\ cm^{-1}}$ always gave small-grained boules. Intermediate gradients yielded single-grained boules with coherent and lateral twin boundaries, while the highest ($\sim 40\ \mathrm{K\ cm^{-1}}$) yielded single grains with only the lowest energy (coherent) twin boundaries. It was concluded that the increasing degree of interface stabilisation at higher temperature gradients successively eliminated the more energetic boundaries, until only coherent twin boundaries remained. Mostly the twins were present in the form of closely spaced multiple thin lamellae extending up to the full diameter of the boules. However, for the highest pulling rates used (at each temperature gradient), etching revealed additional dense localised arrays of planar faults identified as microtwins. The authors speculated that both growth and deformation play a role in twinning.

Boone et al. [158] report seeded growth to give squat single crystal boules of up to 45–50 mm diameter. Nakagawa etch pit densities are reported by Boone in the range $3\text{–}7 \times 10^4\ \mathrm{cm^{-2}}$ [158] with rocking curves in the range 8.6–19.5″, (8.5″ being the theoretical value). The dislocation density was related to growth rate, being $\sim 10^4\ \mathrm{cm^{-2}}$ for growth at 15 mm/day and $\sim 10^5\ \mathrm{cm^{-2}}$ at 35 mm/day [156]. TEM has shown that there may be sub-grain boundaries with loose arrays of dislocations [43, 120]. The same references also show TEM of Te precipitates $>1\ \mu\mathrm{m}$ in size and their associated strain fields. This is for material grown in self-sealing tubes over several days.

3.4.2. Growth in a closed capsule with a separate Te cold zone ('Durham method')

Developed for CdS in the early 1960s, the so-called 'Durham' method is a variant of the Piper–Polich method in which the capsule is connected by a capillary to a small reservoir containing one of the elements and held at a different temperature. This is to afford some stoichiometric control. Growth took place at $\sim 780°\mathrm{C}$ by drawing the capsule through the furnace over 10–14 days with cooling taking 3 days. Nucleation was into a conical silica tip and proceeds in contact with the walls.

Boules up to 29 mm in diameter (Fig. 12) and weighing several hundred grams were grown, but grains rarely exceeded 80% of the boule volume and polycrystallinity was rife. Often, the predominant grains were twin related. Of hundred or so growth runs reported, all contained twins on one or more {111} planes [110].

The dislocation structure was of well-developed tight arrays of dislocations, with densities of 8–16 dislocations per μm being observed in TEM, these supporting measured tilts of up to 0.5° [120, 159]. Sub-grains were in the range 5–500 μm in width, with an average of 150 μm [43, 159].

Overall it was concluded that contact with the walls and soaking for long periods of time at high temperature with slow cooling acted to promote slip, and very advanced polygonisation of dislocations to form well developed sub-grain boundaries and mosaicity. Some samples showed clear evidence of slip bands (Fig. 10) [43], presumably from the later stages of cooling. Durose and Russell [120] make comparison with the more rapidly grown Piper–Polich CdTe.

Inclusions ~10 μm in size were routinely present and, in accordance with the thermal history of the boules (above), were segregated to planar defects. The density of decoration declined with the boundary energy [120] in the sequence described in Section 2.4 (see Fig. 17).

3.4.3. Markov–Davydov and related methods

In this method [160, 161] vapour transport takes place onto a seed plate (silica or a wafer) which is itself removed from the walls of the tube by an annular gap. Growth is therefore free from the walls. Beyond the gap, on the other side of the crystal from the source, there is a volume (sometimes pumped) where condensation of excess components may take place.

Laasch et al. [162] have achieved excellent results: Unintentional contact with the walls causes parasitic nucleation at the edges of the crystals, and the result is a central grain surrounded by smaller grains. Use of a sink at ~700 °C gives single crystals, but they are twinned. A sink temperature of ~100 °C stabilises the interface both thermally and morphologically: as a result the growth is twin-free.

Palosz et al. [50] report the effect of post-growth cooling rate on the dislocation density in CdTe boules grown in Markov-like capsules (see Table 8). Generally, the faster the cooling, the higher the dislocation density resulted from it. However, the lowest dislocation density was found to be for free-cooling in the furnace rather than a somewhat slower programmed cool-down; perhaps the smoothness of the cooling is better when there is no active temperature control. In the same work, the spatial pattern of the dislocations was analysed quantitatively. None of the distributions were random. Where the cooling was slowest, the spatial

Table 8 Relationship between the dislocation density and cooling rate in Markov-like growth on a quartz crystal holder [50]

Cooling method	Dislocation density (10^4 cm^{-2})
Quenched in water	40.9 ± 3.5
Quenched in air	18.4 ± 0.8
Switch off furnace ~260 °C/h	2.8 ± 0.4
Controlled cooling 10 °C/h	5.0 ± 1

Faster cooling causes higher dislocation densities.

association of the dislocations was greatest: this probably reflects the greater extent of annealing processes and polygonisation taking place.

It was also demonstrated [50] that unintended contact of the boules with the walls caused the dislocation density to be increased by approximately an order of magnitude.

3.4.4. Grasza's variation of Markov-like growth

In Grasza's variation [163], sublimation takes place on a seed that is created *in situ* on a quartz growth surface within the (vertical) growth tube itself. The temperature profile of the charge is manipulated to create a stalagmite that makes contact with, and wets, the quartz growth surface above. Then the temperature profile is further manipulated so as to break the stalagmite, leaving a globule of CdTe that acts as a seed. During the growth phase a third temperature profile is arranged to promote sublimation from the charge and condensation on the seed. Growth is free from the walls.

When a good nucleus was obtained, single crystals of ~55 mm diameter and several tens of mm tall were demonstrated; these had bright convex surfaces and occasionally small facets [164]. The twin fraction has been reported in the range 30–80% of the volume of the boule [165]. In a comparisons of the dislocation density and distribution achieved with Bridgman and other vapour methods,Walker et al. [48] report an etch pit density of 5×10^4 cm^{-2}, this being an order of magnitude better than the comparison samples. Moreover, quantitative measures of distribution indicated that dislocations in this material are not significantly clustered.

3.4.5. Multi-tube vapour transport

Multi-tube vapour transport [166] is related to the Markov method, but differs significantly in that the seed wafer is separated from the charge by a flow restrictor, making it possible to decouple the transport rate from the temperature difference between source and crystal. Growth is free from the walls.

Flat plates 50 mm in diameter have been demonstrated, and single crystal growth has been achieved. The density of dislocations and X-ray FWHM values improve as the crystal thickness increases. An example from [49] is that over a 6 mm thickness, the $\theta - 2\theta$ X-ray peak changed from being multi-peaked with an FWHM of 107″–43″. The $FeCl_3$ etch pit density was $6 \pm 3 \times 10^4$ cm^{-2}, the best values being furthest from the substrate. With improvements to the technology, FWHMs in the low 20s of arc seconds are now routinely achievable [167].

3.4.6. Self-selecting vapour growth method (Szczerbakow)

The method is different in concept from all of the above: sublimation and condensation takes place between the bottom (hot) and top (cooler) surfaces *of the charge material itself*. In this way, the growth is completely free from the walls, and the sublimation/condensation of the same mass may be extended indefinitely, leading to continual refinement of the crystallised product. The temperature gradient may be held to be very low and may be dominated by thermal radiation. A comprehensive review may be found in Ref. [168].

Self-selecting vapour growth (SSVG) has been used to date to grow small (~20 mm) crystals of CdTe, and these display facets indicating single crystal growth [169]. The dislocation density is reported in the range $1.2–3 \times 10^5\ cm^{-2}$ [48, 51, 170]. However, more generally the method has achieved dislocation densities for other binary compounds as low as $10^3\ cm^{-2}$ and FWHMs of ~30″. SSVG deserves to be scaled up, particularly for the growth of ternary compounds for which high compositional uniformity has been demonstrated.

ACKNOWLEDGEMENTS

The author would like to thank Andy Brinkman for pointers towards current literature on crystal growth. The author would also like to apologise to those authors whose work is not mentioned here or which has been overlooked.

LITERATURE ON EXTENDED DEFECTS IN CdTe

A number of book reviews, data books and ongoing conference series have general relevance to the subject of extended defects. While the list that follows is not intended to replace the references, the reader's attention is drawn to it a source list, some of it ongoing in the form of conference series.

M. Aven, J.C. Prener (Eds.), The Physics and Chemistry of II–VI Compounds, North-Holland, Amsterdam, 1967.

K. Zanio, in: R.C. Willardson, A.C. Beer (Eds.), 'Cadmium Telluride' in Semiconductors and Semimetals, vol. 13, Wiley, New York, 1978.

P. Capper (Ed.), Properties of Narrow Gap Cadmium-Based Compounds, EMIS Datareview Series No 10, IEE/Inspec, 1994. (Comprehensive multi-author review including chapters on Cathodoluminescence – B3.5, Annealing – B5.1, Dislocations – B5.4, Preicipitation – B5.5, Defect etching – B5.6, X-ray rocking curve widths – B5.7 and X-ray topography – B5.8).

M. Hage-Ali, P. Siffert, 'Growth Methods of CdTe Nuclear Detector Materials', 'Characterization of CdTe Nuclear Detector Materials' and R.B. James, T.E. Schlesinger, J. Lund, M. Schieber, in: A.C. Beer, R.K. Willardson, E.R. Weber, (Eds.), Semiconductors for Room Temperature

Nuclear Detector Applications: Semiconductors and Semimetals, vol. 43 Eds Academic Press, New York, 1995.

K. Durose, in: P. Capper (Ed.), Narrow-gap II–VI Compounds for optoelectronic and Electromagnetic Applications (Electronic Materials Series vol. 3), Chapman and Hall, London, 1997.

(Review of the systematic behaviour of extended defects in narrow gap II–VI semiconductors).

O. Oda, Compound Semiconductor Bulk Materials and Characterisations, World Scientific Ltd, New Jersey, 2007. (A wide ranging and well-referenced review of the growth and characterisation of II–VI and III–V compound semiconductors).

D.B. Holt, B.G. Yacobi, Extended Defects in Semiconductors: Electronic Properties, Device Effects and Structures, Cambridge University Press, New York, 2007 (Comprehensive treatise on extended defects in diamond-like semiconductors).

International Conference on II–VI Compounds. (Held bi-annually. Published in special editions of Journal of Crystal Growth and Phys. Status. Solidi A).

Proceedings of the MCT Workshop, latterly the 'US Workshop on the Physics and Chemistry of II–VI Materials', held annually.

E-MRS Spring Meeting, held annually, most often in Strasbourg.

Microscopy of Semiconducting Materials Conf., Institute of Physics/ Royal Microscopical Society (Held bi-annually in Oxford or Cambridge. Specialist conference on semiconductor microscopy).

REFERENCES

[1] H.R. Vydyanath, J. Ellsworth, J.J. Kennedy, B. Dean, C.J. Johnson, G.T. Neugebauer, J. Sepich, P.K. Liao, J.Vac. Sci. Technol. B 10(4) (1992) 1476–1484.

[2] A.F.W. Willoughby, In: P. Capper (Ed.), Narrow-Gap II–VI Compounds for Optoelectronic and Electromagnetic Applications, vol. 3, Chapman and Hall, London, 1997, pp. 268–290.

[3] D.W. Lane, J.D. Painter, M.A. Cousins, G.J. Conibeer, K.D. Rogers, Thin Solid Films 431 (2003) 73–77.

[4] H. Gleiter, Acta Metallurgica 18 (1)(1970) 117–121.

[5] S.A. Galloway, P.R. Edwards, K. Durose, Inst. Phys. Conf. Ser. No 157 (1997) 579–582.

[6] L.O. Bubulac, J. Bajaj, W.E. Tennant, P.R. Newman, D.S. Lo, J. Cryst. Growth 86 (1–4) (1988) 536–543.

[7] K. Nakagawa, K. Maeda, S. Takeuchi, Appl. Phys. Lett. 34 (1979) 574–575.

[8] C.C.R. Watson, K. Durose, A.J. Banister, E. O'Keefe, S.K. Bains, Mater. Sci. Eng. B16 (1993) 113–117.

[9] S.A. Galloway, J. Holland, K. Durose, J. Cryst. Growth 159 (1996) 925–929.

[10] T.P. Thorpe, A.L. Fahrenbruch, R.H. Bube, J. Appl. Phys. 60 (10) (1986) 3622–3630.

[11] K. Durose, J.R.E. Sadler, A.J.W. Yates, A. Szczerbakow, 28th IEEE Photovoltaic Specialists Conference, IEEE, 2000, pp. 487–490.

[12] J.D. Major, PhD Thesis, CdTe solar cells: growth phenomena and device performance, Durham University, UK, 2008.

[13] S.M. Johnson, D.R. Rhiger, J.P. Rosbeck, J.M. Peterson, R.E. Taylor, M.E. Bond, J. Vac. Sci. Technol. B10 (1992) 1499–1506.
[14] I. Visoly-Fisher, S.R. Cohen, A. Ruzin, D. Cahen, Adv. Mater. 16 (11) (2004) 879–883.
[15] D.W. Niles, X. Li, P. Sheldon, J. App. Phys. 77 (9) (1995) 4489–4493.
[16] S.A. Galloway, P.R. Edwards, K. Durose, Solar Eng. Mater. Solar Cells 57 (1999) 61–74.
[17] S.L. Bell, S. Sen, J. Vac. Sci. Technol. 3 (1) (1985) 112–115.
[18] R.S. Rai, S. Mahajan, D.J. Michel, H.H. Smith, S. McDevitt, C.J. Johnson, Mater. Sci. Eng. B10 (1991) 219–225.
[19] C.Y. Yeh, Z.W. Lu, S. Froyen, A. Zunger, Phys. Rev. B 46 (16) (1992) 10086–10097.
[20] Y. Okada, Y. Tokumaru, J. Appl. Phys. 56 (2) (1984) 314–320.
[21] R. Triboulet, A. Tromson-Carli, D. Lorans, T.N. Duy, J. Electron. Mater. 22 (8) (1993) 827–834.
[22] O. Oda, Compound Semiconductor Bulk Materials and Characterisation, World Scientific Publishing Co. Pte. Ltd, Singapore, 2007.
[23] T.-C. Yu, R.F. Brebrick, in: P. Capper (Ed.), Properties of Narrow Gap Cadmium-Based Compounds, vol. 10, IEE, London, 1994, pp. 413–419.
[24] T. Jasinsky, A.F. Witt, J. Cryst. Growth 71 (1985) 295–304.
[25] H.M. Hobgood, B.W. Swanson, R.N. Thomas, J. Cryst. Growth 85 (1987) 510–520.
[26] S. Rolland, in: P. Capper (Ed.), Properties of Narrow Gap Cadmium-Based Compounds, vol. 10, IEE, London, 1994, pp. 44–47.
[27] R. Triboulet, A. Lashley, B. Toulouse, R. Granger, J. Cryst. Growth 79 (1986) 695–700.
[28] R.N. Thomas, H.M. Hobgood, P.S. Ravishankar, T.T. Braggins, J. Cryst. Growth 99 (1990) 643–653.
[29] H. Maleki, L.R. Holland, in: P. Capper (Ed.), Properties of Narrow Gap Cadmium-Based Compounds, vol. 10, IEE, London, 1994, pp. 408–412.
[30] P.F. Fewster, Progr. Cryst. Growth Charact. Mater. 48–49 (2004) 245–273.
[31] M.J. Hordon, B.L. Averbach, Acta Metallurgica 9 (1961) 237–246.
[32] P. Gay, P.B. Hirsch, A. Kelly, Acta Metallurgica 1 (3) (1953) 315–319.
[33] P. Gay, P.B. Hirsch, A. Kelly, Acta Cryst. 7 (1) (1954) 41–49.
[34] P. Capper, J.E. Harris, E. O'Keefe, C.L. Jones, C.K. Ard, P. Mackett, D. Dutton, Mater. Sci. Eng. B16 (1993) 29–39.
[35] C.J. Johnson, Future Infrared Detector Mater. SPIE 1106 (1989) 56–68.
[36] K.Y. Lay, D. Nichols, S. McDevitt, B.E. Dean, C.J. Johnson, J. Cryst. Growth 86 (1988) 118–126.
[37] P.T. Bowman, E.I. Ko, P.J. Sides, J. Electrochem. Soc. 137 (4) (1990) 1309–1311.
[38] N.G. Chew, A.G. Cullis, Ultramicroscopy 23 (2) (1987) 175–198.
[39] M. Inoue, I. Teramoto, S. Takayanagi, J. Appl. Phys. 33 (8) (1962) 2578–2581.
[40] I. Hähnert, M. Schenk, J. Cryst. Growth 101 (1990) 251–255.
[41] W.J. Everson, C.K. Ard, J.L. Sepich, B.E. Dean, G.T. Neugebauer, H.F. Schaake, J. Electron. Mater. 24 (5) (1995) 505–510.
[42] M. Inoue, I. Teramoto, S. Takayanagi, J. Appl. Phys. 34 (2) (1963) 404–405.
[43] K. Durose, PhD Thesis, Structural defects in CdTe, Durham University, UK, 1986.
[44] D.B. Holt, J. Mater. Sci. 23 (1988) 1131–1136.
[45] A. Tanaka, Y. Masa, S. Seto, T. Kawasaki, Mater. Res. Soc. Symp. Proc. 90 (1987) 111–118.
[46] L.G. Casagrande, D. Di Marzio, M.B. Lee, D.J. Larson Jr., M. Dudley, T. Fanning, J. Cryst. Growth 128 (1993) 576–581.
[47] I. Hähnert, M. Mühlberg, H. Berger, C. Genzel, J. Cryst. Growth 142 (3–4) (1994) 310–314.
[48] E.M. Walker, D.J. Buckley, N.M. Boyall, K. Durose, K. Grasza, A. Szczerbakow, W. Palosz, Inst. Phys. Conf. Ser. No. 169 (2001) 215–218.
[49] N.M. Aitken, M.D.G. Potter, D.J. Buckley, J.T. Mullins, J. Carles, D.P. Halliday, K. Durose, B.K. Tanner, A.W. Brinkman, J. Cryst. Growth 198–199 (1999) 984–987.

[50] W. Palosz, K. Grasza, K. Durose, D.P. Halliday, N.M. Boyall, M. Dudley, B. Raghotamachar, C. Lai, J. Cryst. Growth 254 (3–4) (2003) 316–328.
[51] D. Rose, K. Durose, W. Palosz, A. Szczerbakow, K. Grasza, J. Phys. D Appl. Phys. 31 (1998) 1009–1016.
[52] A.J.W. Yates, K. Durose, A. Szczerbakow, J. Domagala, Inst. Phys. Conf. Ser. No 168 Sect. 10 (2001) 445–448.
[53] N.M. Boyall, K. Durose, T.Y. Liu, A. Trampert, C. Liu, L.M. Watson, Inst. Phys. Conf. Ser. No 179 Sect. 2 (2003) 61–66.
[54] M. Aven, J.S. Prener, The Physics and Chemistry of II–VI Compounds, North-Holland Publishing Company, Amsterdam, 1967.
[55] K. Zanio (Ed.), Cadmium Telluride in Semiconductors and Semimetals, R.C. Willardson, A.C. Beer (Eds.), Wiley, New York, 1978.
[56] P.L. Smith, J.E. Martin, Phys. Lett. 6 (1963) 42.
[57] N.B. Owen, P.L. Smith, J.E. Martin, A.J. Wright, J. Phys. Chem. Solids 24 (1963) 1519–1520.
[58] A.N. Mariano, E.P. Warekois, Science 142 (1963) 672–673.
[59] K. Durose, A.T. Fellows, A.W. Brinkman, G.J. Russell, J. Woods, J. Mater. Sci. 20 (1985) 3783–3789.
[60] P.N. Gibson, M.E. Ozsan, J.M. Sherbourne, D. Lincot, P. Cowache, 2nd World Conference on Photovoltaic Solar Energy Conversion, WIP, 1998, , pp. 1089–1095.
[61] M. Weinstein, G. Wolff, B.N. Das, Appl. Phys. Lett. 6 (1965) 73–75.
[62] A.W. Brinkman, H.M. Al-Allak, G.R. Awan, P.D. Brown, K. Durose, C. Ercelebi, M.Y. Simmons, J. Woods, Int. J. Solar Energy 12 (1–4) (1992) 233–245.
[63] I. Spinulescu-Carnaru, Phys. State Solidi 15 (1966) 761–765.
[64] K. Fujita, Y. Furukawa, C. Kaito, J. Cryst. Growth 71 (1985) 821–823.
[65] V. Barrioz, Univ. Wales, Bangor, UK, Personal Communication, 2008.
[66] S. Chandramohan, R. Sathyamoorthy, S. Lalitha, S. Senthilarasu, Solar Eng. Mater. Solar Cells 90 (2006) 686–693.
[67] H.C. Gatos, In: M.C. Lavine, J. Electrochem. Soc. 107 (1960) 427–433.
[68] K. Durose, In: P. Capper (Ed.), Narrow-Gap II–VI Compounds for Opto-electronic and Electro-magnetic Applications, Chapman and Hall, London, 1996, pp. 327–359.
[69] P.F. Fewster, P.A.C. Whiffin, J. Appl. Phys. 54 (8) (1983) 4468–4670.
[70] E.P. Warekois, M.C. Lavine, A.N. Marinano, H.C. Gatos, J. Appl. Phys. 33 (2) (1962) 690–696.
[71] P.D. Brown, K. Durose, G.J. Russell, J. Woods, J. Cryst. Growth 101 (1989) 211–215.
[72] M. Astles, Personal Communication, 1982.
[73] T.H. Myers, J.F. Schetzina, T.J. Magee, R.J. Ormond, J. Vac. Sci. Technol. A 3 (1983) 1598–1603.
[74] G.A. Wolff, J.D. Broder, Acta Cryst. 12 (1959) 313–323.
[75] J.C. Phillips, Bonds and Bands in Semiconductors, Academic Press, New York, 1973.
[76] J.P. Hirth, J. Lothe, Theory of dislocations, McGraw-Hill, New York, 1968.
[77] D.B. Holt, B.G. Yacobi, Extended Defects in Semiconductors: Electronic Properties, Device Effects and Structures, Cambridge University Press, Cambridge, 2007.
[78] W.T. Read, Dislocations in Crystals, McGraw-Hill, New York, 1953.
[79] P.B. Hirsch, J. Phys. Colloq. 40 (1979) C6–27–C6-32.
[80] E.L. Hall, J.B. Vander Sande, Phil. Mag. 37 (1) (1978) 137–145.
[81] K. Maeda, K. Nakagawa, S. Takeuchi, K. Sakamoto, J. Mater. Sci. 16 (1981) 927–934.
[82] C. Braun, H.W. Helberg, A. George, Phil. Mag. A 53 (2) (1986) 277–284.
[83] K. Guergouri, N. Brihi, R. Triboulet, J. Cryst. Growth 209 (4) (2000) 709–715.
[84] S. Hildebrandt, H. Uniewski, J. Schreiber, H.S. Leipner, J Phys. III France 7 (1997) 1505–1515.

[85] J. Schreiber, H. Uniewski, S. Hildebrandt, L. Höring, H.S. Leipner, Inst. Phys. Conf. Ser. No 157 (1997) 651–654.
[86] J. Schrieber, L. Höring, H. Uniewski, S. Hildebrandt, H.S. Leipner, Phys. State Solidi (a) 171 (1999) 89–96.
[87] F.A. Ponce, T. Yamashita, R.H. Bube, S.R., Defects in Semiconductors, vol. 2, North-Holland, Amsterdam, 1981, pp.503–508.
[88] J.L. Hutchison, M. Lyster, G.R. Booker, Inst. Phys. Conf. Ser. No 100 (1989) 29–32.
[89] R. Balasubramanian, W.R. Wilcox, Mater. Sci. Eng. B16 (1993) 1–7.
[90] E.L. Hall, J.B. Vander Sande, J. Amer.Ceram. Soc. 61 (9–10) (1978) 417–425.
[91] I. Hähnert, M. Wienecke, Mater. Sci. Eng. B16 (1993) 168–171.
[92] D.F. Weirauch, J. Electrochem. Soc. 132 (1) (1985) 250–254.
[93] S. Takeuchi, K. Suzuki, K. Maeda, H. Iwanaga, Phil. Mag. A 50 (1984) 171–178.
[94] C.B. Slawson, Am. Mineral. 35 (1950) 193–206.
[95] W.C. Ellis, J. Metals 188 (1950) 886.
[96] W.C. Ellis, J. Fageant, J. Metals 200 (1954) 291–294.
[97] K.F. Hulme, J.B. Mullin, Solid State Electron. 5 (1962) 211–247.
[98] A. Steinemann, U. Zimmerli, Solid State Electron. 6 (1963) 597–604.
[99] K.T. Aust, B. Chalmers, Proc. Roy. Soc. A204 (1950) 359–366.
[100] D.G. Brandon, B. Ralph, S. Ranganthan, M.S. Wald, Acta Metallurgica 12 (1964) 813–821.
[101] L.E. Murr, Interfacial Phenomena in Metals and Alloys, Addision-Wesley, Redwood City, 1975.
[102] J. Hornstra, Physica 25 (1959) 409–422.
[103] W.C. Ellis, R.G. Treuting, Trans. Am. Inst. Min. (Metall) Eng. 191 (1951) 53–55.
[104] J. Hornstra, Physica 26 (1960) 198–208.
[105] D. Holt, J. Phys. Chem. Solids 25 (1964) 1385–1395.
[106] K. Durose, G.J. Russell, J. Woods, Inst. Phys. Conf. Ser. No 76 (Sect.) 6 (1985) 233–238.
[107] S. Tohno, A. Katsui, J. Cryst. Growth 74 (1986) 362–374.
[108] J.R. Cutter, G.J. Russell, J. Woods, J. Cryst. Growth 32 (2) (1976) 179–188.
[109] K. Durose, G.J. Russell, Inst. Phys. Conf. Ser. No 84(Sect.) 4 (1987) 327–332.
[110] K. Durose, G.J. Russell, J. Cryst. Growth 101 (1989) 246–250.
[111] R.C. Pond, J. Phys. C1 (43) (1982) 51–56.
[112] J.A. Venables, J. Phys. Chem. Solids 25 (1964) 685–692.
[113] A.W. Vere, S. Cole, D.J. Williams, J. Electron. Mater. 12 (3) (1983) 551–561.
[114] D.T.J. Hurle, J. Cryst. Growth 147 (3/4) (1995) 239–250.
[115] M.M. Al-Jassim, R.G. Dhere, K.M. Jones, F.S. Hasoon, P. Sheldon, 2nd World Conf and Exhibition on Photovoltaic Solar Energy Conversion, WIP, vol 1, 1998, pp. 1063–1066.
[116] D.J. Williams, In: P. Capper (Ed.), Properties of Narrow Gap Cadmium-Based Compounds, vol. 10, IEE, London, 1994, pp. 510–515.
[117] P. Rudolph, M. Neubert, M. Mühlberg, J. Cryst. Growth 128 (1993) 582–587.
[118] R.D.S. Yadava, B.S. Sundersheshu, M. Anandan, R.K. Bagal, W.N. Borle, J. Electron. Mater. 23 (12) (1994) 1349–1357.
[119] S.H. Shin, J. Bajaj, L.A. Moudy, D.T. Cheung, Appl. Phys. Lett. 43 (1) (1983) 68–70.
[120] K. Durose, G.J. Russell, J. Cryst. Growth 86 (1988) 471–476.
[121] R.D.S. Yadava, R.K. Bagai, W.N. Borle, J. Electron. Mater. 21 (10) (1992) 1001–1016.
[122] H.G. Brion, C. Mewes, I. Hahn, U. Schaufele, J. Cryst. Growth 134 (3–4) (1993) 281–286.
[123] M. Wada, J. Suzuki, Jpn. J. Appl. Phys. 27 (1988) L972–L975.
[124] P. Cherin, P. Unger, Acta Cryst. 23 (1967) 670–671.
[125] K. Aoki, O. Shimomura, S. Mimomura, J. Phys. Soc. Jpn. 48 (1980) 551–556.
[126] J.C. Jamieson, D.B. McWhan, J. Chem. Phys. 43 (4) (1965) 1149–1152.
[127] H.F. Schaake, J.H. Tregilgas, J. Electron. Mater. 12 (6) (1983) 931–945.

[128] C. Szeles, S.E. Cameron, J.O. Ndap, W.C. Chalmers, IEEE Trans. Nucl. Sci. 49 (5) (2002) 2535–2540.
[129] J. Shen, D.K. Aidun, L. Regel, W.R. Wilcox, Mater. Sci. Eng. B16 (1993) 182–185.
[130] C.C.R. Watson, PhD Thesis, Structural defects in CdTe and related materials, Durham University, UK, 1993.
[131] P. Becla, A.F. Witt, C.F. Dewey, J. Vac. Sci. Technol. B 10 (4) (1992) 1594–1598.
[132] M. Pfeiffer, M. Mühlberg, J. Cryst. Growth 118 (1992) 269–276.
[133] M. Mühlberg, P. Rudolph, C. Genzel, B. Wermke, U. Becker, J. Cryst. Growth 101(1990) 275–280.
[134] M. Mühlberg, P. Rudolph, M. Laasch, E. Treser, J. Cryst. Growth 128 (1993) 571–575.
[135] M. Azoulay, A. Raizman, G. Gafni, M. Roth, J. Cryst. Growth 101 (1989) 256–260.
[136] R. Shetty, W.R. Wilcox, L. Regel, J. Cryst. Growth 153 (3–4) (1995) 103–109.
[137] Y.-C. Lu, R.S. Feigelson, R.K. Route, Z.U. Rek, J. Vac. Sci. Technol. 4 (4) (1986) 2190–2194.
[138] S. McDevitt, B.E. Dean, D.G. Ryding, F.J. Scheltens, S. Mahajan, Mater. Lett. 4 (11–12) (1986) 451–454.
[139] S. Sen, W.H. Konkel, S.J. Tighe, L.G. Bland, S.R. Sharma, R.E. Taylor, J. Cryst. Growth 86 (1988) 111–117.
[140] W.-B. Song, M.-Y. Yu, W.-H. Wu, J. Cryst. Growth 86 (1988) 127–131.
[141] I.V. Sabinina, A.K. Gutakovski, T.I. Milenov, N.N. Lyakh, Y.G. Sidorov, M.M. Gospodinov, Cryst. Res. Technol. 26 (8) (1991) 967–972.
[142] R. Kumaresan, R. Gopalakrishnan, S.M. Babu, P. Ramasamy, P. Zaumseil, M. Ichimura, J. Cryst. Growth 210 (1–3) (2000) 193–197.
[143] A.A. Khan, W.P. WAllred, B. Dean, S. Hooper, J.E. Hawkey, C.J. Johnson, J. Electron. Mater. 15 (1986) 181–184.
[144] P. Cheuvart, U. El-Hanani, D. Schneider, R. Triboulet, J. Cryst. Growth 101 (1990) 270–274.
[145] W.P. Allred, A.A. Kahn, C.J. Johnson, N.C. Giles, J.F. Schetzina, Mater. Res. Soc. Symp. Proc. 90 (1987) 103–109.
[146] R. Schoenholtz, R. Dian, R. Nitsche, Nucl. Inst. Methods Sect. A 72 (1985) 72–79.
[147] R. Schwarz, K.W. Benz, J. Cryst. Growth 144 (1994) 150–156.
[148] R. Triboulet, Y. Marfaing, J. Cryst. Growth 51 (1981) 89–96.
[149] M. Funaki, T. Ozaki, K. Satoh, R. Ohno, Nucl. Inst. Methods Sect. A 436 (1–2) (1999) 120–126.
[150] M. Ayoub, M. Hage-Ali, A. Zumbiehl, R. Regal, J.M. Koebel, C. Rit, P. Fougeres, P. Siffert, IEEE Trans. Nucl. Sci. 49 (4) (2002) 1954–1959.
[151] R.U. Barz, P. Gille, J. Cryst. Growth 149 (3–4) (1995) 196–200.
[152] R. Triboulet, R. Legros, A. Heurtel, B. Sieber, G. Didier, D. Imhoff, J. Cryst. Growth 72 (1985) 90–96.
[153] J.B. Mullin, C.A. Jones, B.W. Straughan, A. Royle, J. Cryst. Growth 59 (1982) 135–142.
[154] B. Pelliciari, F. Dierre, D. Brellier, L. Verger, F. Glasser, B. Schaub, J. Electron. Mater. 34(6) (2005) 693–698.
[155] W.W. Piper, S.J. Polich, J. Appl. Phys. 32 (7) (1961) 1278–1279.
[156] H. Wiedemeier, G.H. Wu, J. Electron. Mater. 22 (9) (1993) 1121–1127.
[157] H. Wiedemeier, G.H. Wu, J. Electron. Mater. 24 (8) (1995) 1007–1015.
[158] J.L. Boone, J.L. Cantwell, W.C. Harsch, J.E. Thomas, B.A. Forman, J. Cryst. Growth 139(1994) 27–36.
[159] K. Durose, G.J. Russell, J. Woods, J. Cryst. Growth 72 (1985) 85–89.
[160] E.B. Markov, A.A. Davidov, Inorg. Mater. 7 (4) (1971) 575–579.
[161] E.V. Markov, A.A. Davidov, Inorg. Mater. 10 (1975) 1755–1758.
[162] M. Laasch, T. Kunz, C. Eiche, M. Fiederle, W. Joerger, G. Kloess, K.W. Benz, J. Cryst. Growth 174 (1997) 696–707.

[163] K. Grasza, U. Zuzuga-Grasza, A. Jedrzejczak, R.R. Galazka, J. Majewski, A. Szadkowski, E. Grodzika, J. Cryst. Growth 123 (1992) 519–528.
[164] K. Grasza, R. Schwarz, M. Laarsch, K.W. Benz, M. Pawlowska, J. Cryst. Growth 151 (1995) 261–266.
[165] K. Grasza, E. Janik, A. Mycielski, J. Bak-Misiuk, J. Cryst. Growth 146 (1995) 75–79.
[166] J.T. Mullins, J. Carles, N.M. Aitken, A.W. Brinkman, J. Cryst. Growth 208 (1–4) (2000) 211–218.
[167] J.T. Mullins, Personal Communication, 2008.
[168] A. Szczerbakow, K. Durose, Progr. Cryst. Growth Charact. Mater. 51 (2005) 81–108.
[169] A. Szczerbakow, Z. Golacki, Mater. Sci. Eng. B 16 (1993) 68–70.
[170] A. Szczerbakow, J. Domagala, D. Rose, K. Durose, V.Y. Ivanov, A. Omeltchouk, J. Cryst. Growth 191 (1998) 673–678.

CHAPTER **VB**

Inclusions and Precipitates in CdZnTe Substrates

Jean-Olivier Ndap

1. INTRODUCTION

CdZnTe substrates can be obtained as slices cut from bulk ingots grown from the liquid, the solid, or from the vapor phase. Relatively higher single crystal yields make growth from the liquid phase the most attractive approach to produce large CdZnTe substrates. Liquid growth techniques involve solution/flux growth methods, such as the traveling heater method (THM), and melt growth methods, such as the conventional Bridgman technique and all its modified versions. Usually, the crystals produced via these techniques hardly yield epi-ready substrates due to the presence of relatively large densities of structural defects such as dislocations, twins, and second phase microparticles. They form during crystallization of the liquid and cooling down of the crystallized bulk ingots [1, 2]. It is well documented that these structural defects adversely affect the structural quality and electrical performances of the HgCdTe (or HgZnTe) epi-layers. [3–8]

The second phase microparticles have extensively been studied using characterization techniques such as infrared optical transmission microscopic imaging [9–16], secondary electron microscopy (SEM) [10, 11, 15, 17] and transmission electron microscopy (TEM) [18–20]—for assessment of their structure—, energy dispersive X-ray (EDX) analysis [10, 11, 15], differential scanning calorimetry (DSC) [21, 22], and Auger spectroscopy [17–23]—for assessment of their chemical composition. Their shapes, size and densities may also be related to the growth history.

eV Products (a division of II-VI Incorporated), Saxonburg, PA 16056, USA

It appears from the studies mentioned above that these particles are in fact precipitates and inclusions of tellurium or cadmium. The difference between an inclusion and a precipitate depends upon their size and their mode of formation; inclusions are considered to be much larger than precipitates. It has been shown [10, 17, 22, 24, 25] that Te particles getter impurities originating from various sources such as the Cd, Zn, and Te starting elements (by-products of copper mining), the containers used during crystal growth and eventual contamination during material preparation prior to growth. During epitaxial growth, these impurities can easily be freed from the heated inclusions, expand all over the substrate, diffuse out into the purer epi-layer [17, 19, 25–27] and seriously affect their electronic properties [3–8, 28]. Dislocations associated with these particles [11] have also been reported not only to propagate into the epi-layers but also to serve as channels for fast out-diffusion of Hg from the epi-layer to the substrate [7].

A relatively large amount of work has been carried out in the purpose of producing CdZnTe substrates free of these second phase particles. The techniques used involve *in situ* control of their formation during crystal growth and postgrowth processing of the wafers.

In *in situ* techniques, the most efficient control of formation of second phase Te particles is achieved through saturation of the vapor phase above the melt with the most volatile component (Cd). This can be done through dynamic control of the vapor pressure, using a Cd source contained in a separated reservoir communicating with the reaction chamber [12, 18, 29–34] or through passive control of Cd evaporation by addition of a surplus of cadmium in the source charge; the excess Cd can be calculated as function of the free volume above the melt, the melt temperature, and the corresponding equilibrium vapor pressure [12, 35–37]. The liquid B_2O_3 encapsulating approach was used to prevent evaporation of the melt components [38–41]. The drawback of this technique is contamination of the crystals with boron. The utilization of highly pressurized inert gases in autoclaves has not shown any success in preventing components evaporation [30, 42, 43]. Relatively large amounts of carbon are usually found in these crystals as contamination from the graphite crucible and heater elements.

Postgrowth processes involve annealing the wafers of the compound in a controlled atmosphere, usually saturated with the deficient component(s), to get rid of the excess component second phase microparticles; in fact, a reaction between in-diffused atoms from the gas phase and a microparticle produces on the site of the latter a crystalline compound. This may also be done on the whole ingot during its cooling down after growth [1, 2, 44], however, with low efficiently, because of short diffusion lengths within the large solid boule.

The removal of the tellurium second phase particles during the postgrowth annealing process is usually accompanied with the release of impurities they contained. The latter will subsequently out-diffuse into the epi-layers. Purification of postgrowth annealed wafers is therefore a logical step within this process.

In the following sections, each type of second phase microparticles will clearly be identified and descriptions of their origin will be given. A few approaches used to avoid their formation as well as their post-growth elimination by annealing will also be presented. And last, some interesting purification processes for cleaning up impurities released during postgrowth annealing will be presented.

2. SECOND PHASE PARTICLES: FORMATION AND IDENTIFICATION

The second phase particles expected to form in undoped or not heavily doped CdZnTe are known to be precipitates and inclusions of Te or Cd. Their formation is well explained by the retrograde shape of the solidus and can be strongly influenced by such factors as the melt stoichiometry and the growth process's history.

Figure 1 shows the solidus lines of CdZnTe alloys of four different zinc compositions (0, 4, 10, and 15%), determined from partial pressure

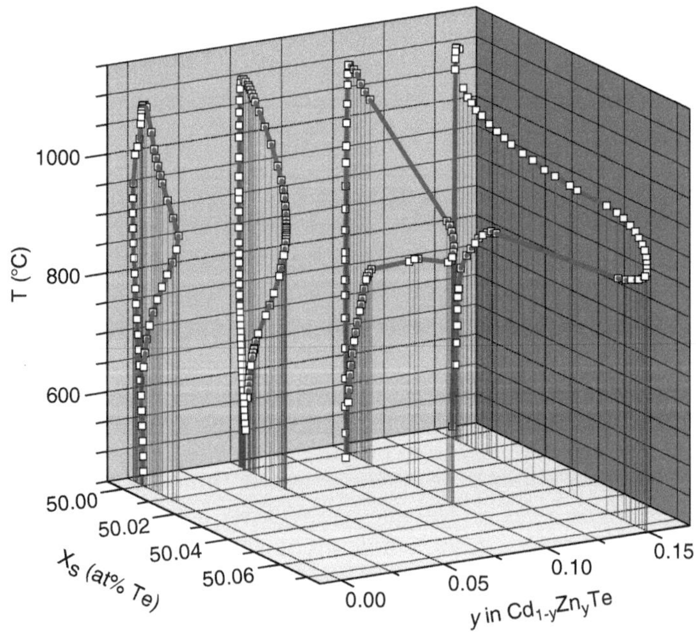

Figure 1 Solidus line of CdZnTe with four different zinc compositions (0, 0.04, 0.1, and 0.15)—Greenberg et al. [45].

scanning measurements [45, 46]. As the zinc composition increases, so does the solubility limit of tellurium and the retrograde shape of the solidus on the tellurium side becomes more accentuated. On the other hand, on the Cd side, the solidus line progressively flattens out and then shifts toward the Te-rich region. Greenberg et al. [45] show that a complete shift of the whole solidus boundary line well into the Te side occurs for zinc compositions greater or equal to 0.15.

Nonstoichiometry of the melt can result from two factors: (i) errors introduced during weighing the starting elements prior to synthesis of the alloy. (ii) Incongruent evaporation of the components in the evacuated free volume during high temperature processing.

Weighing errors are usually due to the experimentator not weighing the elements to the absolute exact stoichiometry needed and also to built-in errors in the scale used. Even the most state-of-the-art electronic scale has accuracy limited to up to just few thousandths of a decimal. Rudolph et al. [12, 13] estimated the stoichiometric deviation resulting from weighing errors in the case of CdTe, using the following expression:

$$N_w \approx 2\delta y N_o \tag{1}$$

N_w is the density of excess tellurium. $N_o = 1.486 \times 10^{22}$ atom/cm^3 is the atomic density of the stoichiometric CdTe alloy. δy is the stoichiometry deviation of the $Cd_{0.5-\delta y}Te_{0.5+\delta y}$ alloy, due to weighing errors.

The application of this equation could be extended to $Cd_{1-x}Zn_xTe$ where the nonstoichiometric alloy would be $(Cd_{1-x}Zn_x)_{0.5-\delta y}Te_{0.5+\delta y}$, assuming that the atomic density of stoichiometric (face-centered cubic) $Cd_{1-x}Zn_xTe$ may be expressed as a function of the zinc composition as:

$$N_o(x) = \frac{4}{(a(x))^3} \tag{2}$$

where $a(x) = (6.4833 - 0.3501x - 0.0283x^2) \times 10^{-8}$ (in cm) is the room temperature lattice parameter of stoichiometric $Cd_{1-x}Zn_xTe$ ($0 \leq x \leq 1$), determined from the lattice parameters of CdTe and ZnTe, and assuming the validity of Vegard's low.

A stoichiometric CdZnTe melt evaporates incongruently. The gas phase in the free volume above the melt comprises Cd, Zn, and Te_n ($n = 1, \ldots, 7$) [47], Te_2 monomer being the most abundant among the Te species. Peters et al. [48] show through partial pressure measurements that about 96% of the vapor phase in equilibrium with molten CdTe at 1473 K is cadmium. The number of atom-grams (or moles) of Cd evaporated in a free volume V_v at temperature T can be expressed, assuming an ideal gas, as:

$$\nu_{Cd} = \frac{p_{Cd} V_v}{RT} \tag{3}$$

where P_{Cd} is the partial pressure of Cd and R is the universal gas constant. The number of atom-grams of Zn and Te may be evaluated similarly, using P_{Zn} and $\sum_n (nP_{Te_n})$ as partial pressures of Zn and Te, respectively. The total pressure of the gas phase in the free volume above the melt is the sum of the partial pressures of the constituents. The mole fraction of Te in the resulting crystal may be expressed as:

$$X_S^{Te} = \frac{N_{Te} - v_{Te}}{\sum_j (N_j - v_j)} \tag{4}$$

N_j and v_j (j = Cd, Zn, Te) are the number of moles of the elements, respectively, in the stoichiometric charge and evaporated in the free volume. To evaluate this expression, it is imperative to know the partial pressures of each of the components of the vapor phase above the melt as well as above the solid (during cool down). Optical density measurements [49] and vapor pressure scanning [45–47] are the few techniques utilized to determine the partial pressures of Cd, Te, and Zn components over solid or liquid CdTe and CdZnTe alloys. It is in this framework that Greenberg [47, p. 121] evaluated the Cd partial pressure over a nearly stoichiometric CdTe melt (X_L = 50.0003 at.%Te) near its melting point (1364.7 K) to be ~1007 mmHg (1.33 at.); the vapor phase was constituted of 99.6% of Cd.

A compilation of data from Refs. [45–47] shows in Fig. 2 the estimated variation of the maximum weight percent of excess Te soluble in a CdZnTe alloy at the maximum stoichiometry, for zinc composition varying between 0 and 0.15. One can see, for example, that a weight of excess Te higher than only ~0.035 g will not be completely soluble in 100 g of a solid $Cd_{0.95}Zn_{0.05}Te$ alloy. This amount of excess Te could well be within experimental weighing errors.

The same consideration may not hold in the case of excess Cd because Cd being the most volatile, it will evaporate into the free volume above the melt until a thermodynamic equilibrium between the two phases is established.

The equilibrium segregation coefficients of both Cd and Te, evaluated from the solidus and liquidus of the pseudo-binary phase diagram, have values less than 1 and are temperature dependent. The melt will therefore continuously be enriched with the excess element (Cd or Te) at the expense of the solid. This has the effect of lowering the crystallization temperature, which results in constitutional supercooling and entrapment of Cd or Te-rich CdZnTe droplets, if the axial temperature gradient at the crystallization interface is below a critical threshold for a given crystallization rate.

The second phase particles resulting from stoichiometry deviation of the CdZnTe melt are classified into two genres, based on their mode of

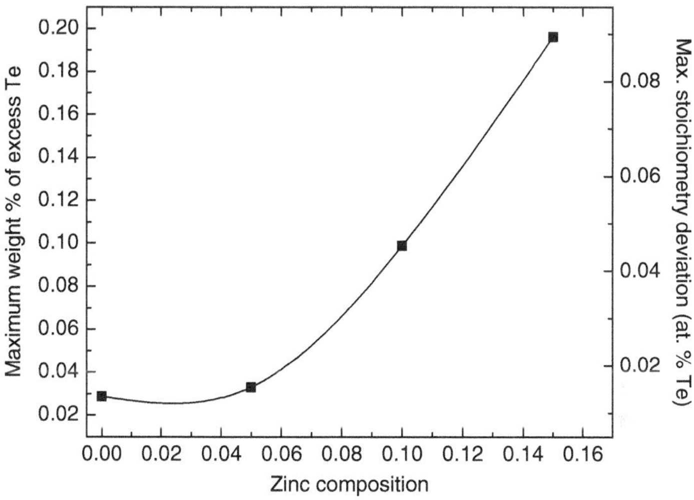

Figure 2 Percentage weight of maximum soluble excess Te as function of zinc composition in CdZnTe. The data points were compiled from data in Refs. [45–47]. The line is a guide to the eye. Weight % excess Te is the ratio of the weight of excess Te over the weight of stoichiometric CdZnTe.

generation and size. These two classes of particles are precipitates and inclusions. It is argued that precipitates are smaller in size than inclusions. Rudolph et al. [14] even recommend in the case of Te precipitation in CdZnTe that particles of size exceeding 1 μm, lateral optical resolution in infrared microscopy, should be called Te inclusions.

How do precipitates and inclusions form? How do they morphologically distinguish from each others? Little is known on the generation mechanisms of these respective defects. Few models have however been proposed in the purpose of understanding their formation mechanisms.

2.1. Precipitates

The shape of the T-X projection of the solidus surface (Fig. 1) delimits the region of existence (or homogeneity) of the CdZnTe solid alloy. The shape of this region and particularly the maximum stoichiometry deviation (or solubility limit) dictates the formation of precipitates. In fact, precipitation of the excess component occurs when the solidified non-stoichiometric alloy crosses the solidus during cool down. The amount of precipitation cannot exceed the solubility limit.

Precipitation (meaning here formation of precipitates) will be as easy as the solidus is pronouncedly retrograde; the density of precipitates increases with the maximum stoichiometry deviation of the alloy.

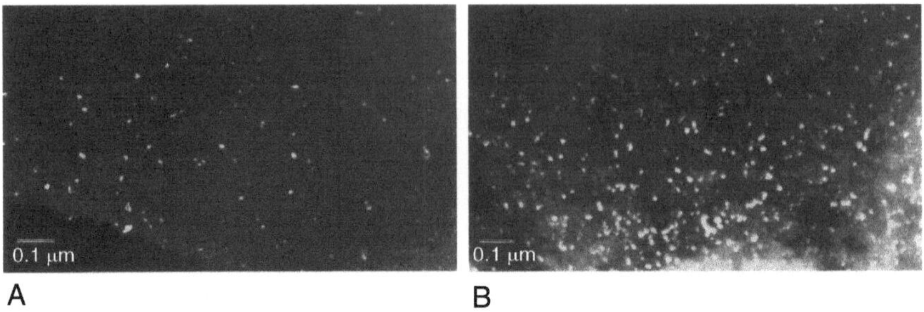

Figure 3 Dark field transmission electron micrographs showing bright spots of Te precipitates in (A) CdTe and (B) CdZnTe of unknown Zn composition [18]. They are randomly distributed and have irregular shapes. The density of precipitates is larger in CdZnTe as compared to CdTe. The average size and density of Te precipitates measured by the authors are 10 nm and 6×10^{15} cm^{-3} for CdTe and 20 nm and 1.3×10^{16} cm^{-3} for CdZnTe.

Cd precipitates resulting from retrograde solubility are therefore more unlikely than Te precipitates [13, 17, 24, 50, 51]. Furthermore, Rudolph et al. [12] suggest that Te precipitates are also generated in the course of condensation of Cd vacancies.

Te precipitates were distinctly detected from observations of thinned wafers under high magnification transmission electron microscopes [18], as shown in Fig. 3, were they appear as randomly distributed bright spots with irregular shapes. Becker et al. [52] earlier utilized infrared extinction spectroscopy to detect their presence from quantification of the energy dependence of their scattering cross sections.

Shen et al. [11], as they examine Te precipitates with an IR transmission microscope found that they appear as below 1 μm-size particles decorating dislocations, stacking faults, and subgrain boundaries. This is illustrated in Fig. 4 where IR transmission micrographs show needle clusters and clouds of Te precipitates along the $\langle 211 \rangle$ directions within the $\{111\}$ plane. These directions are identical to the locations of dislocations and stacking faults within $\{111\}$ planes [25, 53]. It was concluded from detailed observations under TEM that the size of Te precipitates varies from 10 to 50 nm [12, 13, 23].

Rudolph reports [13] that Te-precipitation does not depend on the initial excess of Te in the melt, but rather on the shape of the solidus. He furthermore showed, as depicted in Fig. 5, that the axial distribution of Te precipitates in crystals grown from the melt followed a profile that agreed with Pfann's model [54], given by the distribution function (5):

$$C_{\mathrm{Te,S}}(g) = C_0[1 - (1 - k)\exp(-kg)] \tag{5}$$

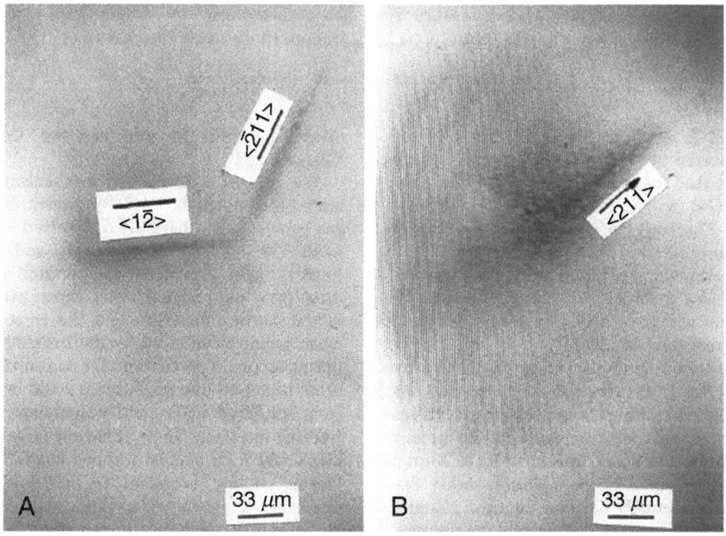

Figure 4 IR transmission micrographs observed from the direction ⟨111⟩ show needle cluster (A) and cloud-shape (B) Te precipitates along the ⟨211⟩ directions in CdZnTe (4% Zn); Ref. [11].

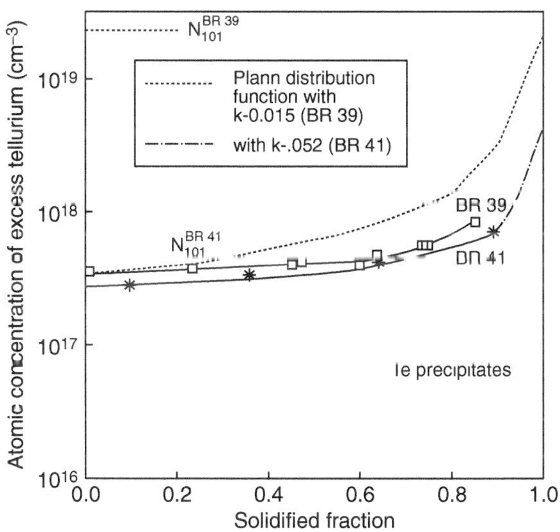

Figure 5 Axial distribution of excess Te as precipitates in CdTe crystals grown from the melt [13].

where C_0 is the initial concentration of Te inclusions in the melt, k is the equilibrium segregation coefficient of Te, and g is the solidified fraction.

The atomic concentration $C_{Te,S}$ (per unit of volume) of excess Te formed as second phase particles in the crystals was estimated using the following equation, with the assumption that the second phase particles are spheres:

$$C_{Te,S} = \frac{4\pi d_{Te} N_a}{3M_{Te}} \sum_{1=1}^{n} r_i^3 \rho_i \tag{6}$$

ρ_i and r_i are, respectively, the density and radius of particles of class of diameter i in the crystal. M_{Te} and d_{Te} are, respectively, the atomic mass and mass density of Te. N_a is the Avogadro's number.

The author observed that for melts with small deviation from stoichiometry, the equilibrium segregation coefficient of Te was nearly constant. However, it could vary during the process (as expected from the binary phase diagram) for melts with larger deviation from stoichiometry.

2.2. Inclusions

Rudolph et al. [12] differentiated Te inclusions from Te precipitates by their mode of generation and by their size. Te inclusions result from Te-rich melts when, due to morphological instabilities at the crystallization interface, large droplets of Te-rich CdZnTe are captured from the Te-rich diffusion layer at the interface's front. Their size (≥ 1 μm) and morphology depend on the experimental conditions. Large-elongated Te inclusions with sizes up to 3 mm long and 0.25 mm large were reported in $\langle 111 \rangle$ oriented THM μg-grown CdZnTe crystals [14]. Cd inclusions may form in a similar way by entrapment of Cd droplets from the Cd-rich diffusion layer.

Interface instabilities result from factors such as: (i) constitutional supercooling that will hardly be avoidable during crystallization of Te-rich melts, if the ratio of the axial temperature gradient at the interface and the growth velocity does not fulfill the conditions for interface stability [55–58], (ii) grain boundaries and twins intersecting the interface, and (iii) poor control of the process's temperature. An evident correlation between the density of Te inclusions and the growth velocity is given in Fig. 6, for CdTe crystals grown from Te-rich melts in a vertical Bridgman furnace with low axial temperature gradients [13–59]. It is clear in this example that as the growth rate increases, so does the density of trapped droplets. One can also see in the inserted graph, as expected, a decrease in Te inclusions density with the axial temperature gradient.

A considerable amount of work has been done in the purpose of studying Te inclusions; mainly their mode of generation, their influence

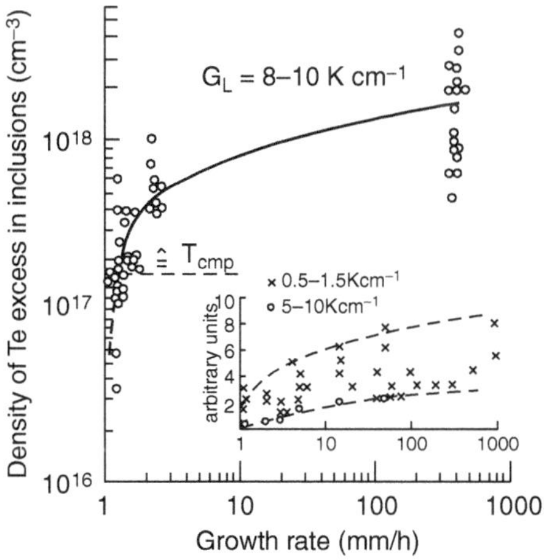

Figure 6 Density of Te excess in inclusions (1-10 μm) in CdTe crystals grown from Te-rich melts at various growth rates. Data are from Ref. [13], where the samples characterized were taken from the tip (first-to-freeze) region of the ingot. The data in the insert are from Ref. [59].

on the physical properties CZT crystals as well as their own physical properties. Infrared optical transmission microscopy has been the most widely used technique for observation of these particles. This technique, along with the seldom utilized cathodoluminescence [17, 60], is limited by its restricted resolution, which does not give a detailed morphology of inclusions.

SEM on the other hand is a more suitable technique for observation of fine features of inclusions. Actually, the incident electron beam of a SEM can probe at best, up to 5 μm below the surface of the substrate. It is therefore necessary to expose the inclusion, without considerably altering its morphology, by appropriate surface preparation. A variety of CdZnTe surface processing techniques have thus been developed. They involve, for example: (i) fine mechanical polishing of the substrate on soft polishing pads using micron to submicron grain-size alumina (Al_2O_3) pastes, slurries, or diamond suspensions, (ii) chemo-mechanical polishing in diluted solutions of bromine in methanol, and (iii) E-solution (HNO_3 + $K_2Cr_2O_7$ + H_2O), to name only these few. Shen et al. [11] give a list of solutions frequently used for exposure of Te inclusions and furthermore introduce their own process; they suggest using the same ingredients to expose Cd inclusions. Identification of the inclusions was carried out by measurement of their composition using EDX spectroscopy [10, 11], DSC [21, 22], or Auger spectroscopy [17, 23].

In 3D, Te inclusions would be viewed in CdZnTe crystals grown by the vertical Bridgman or the vertical gradient freeze technique as polyhedrons, spheres, or irregular geometrical particles. Barz et al. [61] suggest that the entrapped Te-rich droplet freezes radially and the ultimate inclusion owns its shape only to the surrounding matrix. 2D imaging generally shows (Fig. 7A) Te inclusions shaped as triangles, circles, hexagons, and simply irregular figures [10, 11, 14–16]. In Fig. 7B is displayed an IR image of randomly distributed multishaped Te inclusions in a CdZnTe crystal grown from a Te-rich melt. In polycrystalline material, Te inclusions mainly decorate grain boundaries (Fig. 7C), as well as twin boundaries. Six-branch star-shape features can also be found in crystals grown by vertical Bridgman or by vertical gradient freeze in very low axial temperature gradients.

These star-shape features were first reported by Rudolph et al. [14] in {111} oriented and disoriented slices of CdTe. A closer look at the image in Fig. 7D reveals that they consist in fact of star-like arrangements of small Te droplets surrounding a hexagonally shaped larger one. Rudolph et al. [14] give a quite convincing explanation about their genesis: if the low temperature gradient (temperature plateau) is maintained during cool

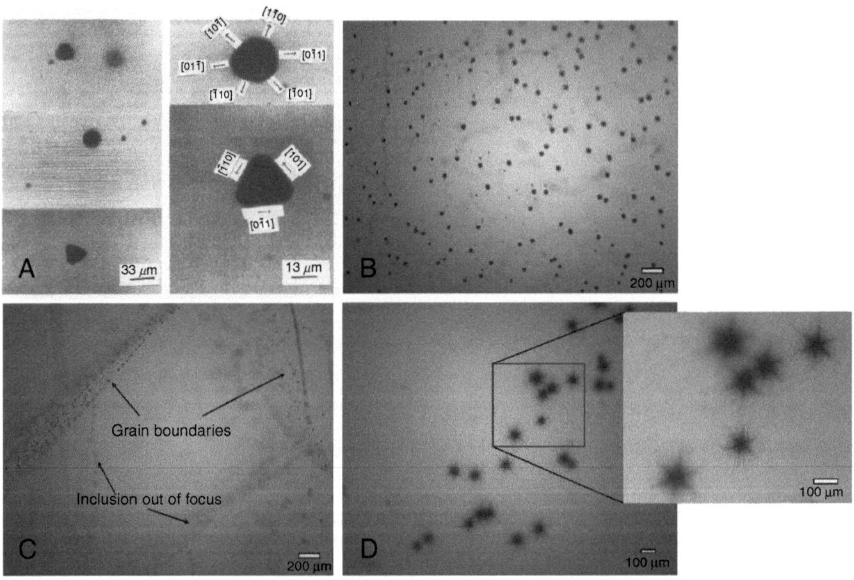

Figure 7 IR images of CdZnTe crystals grown from Te-rich melts by vertical Bridgman stockbarger (A) and Vertical gradient freeze (B)-(D). (A) Shows on a {111} the morphology of Te inclusions [11]. (B) Random distribution of Te inclusions, (C) segregation of Te inclusions along grain boundaries—the crystals grown in a high-pressure environment. (D) Star-shape arrangement of Te inclusions—crystal grown in a low-pressure chamber and low axial temperature gradient (this work). The insert is a magnified region of image (D); one can see 2-4 µm size inclusions smeared within the crystal along the star features.

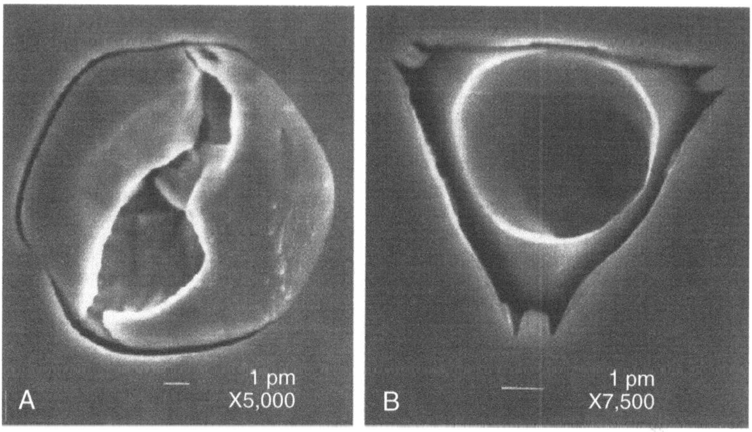

Figure 8 SEM images showing voids within a hexagonal and a triangular Te inclusion [11].

down of the as-grown crystal, star-shaped fields of Te inclusions will form by a symmetrical inward crystallization of supersaturated droplets.

Upon observations at room temperature under high magnification, Shiozawa et al. [62] reported that Te inclusions are actually voids partially filled up with Te. This was later confirmed by Sen et al. [11] and is shown in Fig. 8. Owing to the difference in thermal expansion coefficients between CdZnTe and Te, this is somewhat expected at room temperature, after the entrapped liquid droplet has completely solidified.

Barz et al [61] in their proposed model of inclusion formation come to the conclusion that inclusions in CdZnTe crystals grown from Te-rich liquids content more than 99% pure Te and less than 1% each, Cd and Zn. Fig. 9 shows a linear distribution of Zn in a region surrounding a Te inclusion in a THM-grown CdZnTe crystal. The scattered points are experimental data measured by electron probe microanalysis (EPMA) and the solid line represents calculated values from the theoretical model.[1]

Figure 10 shows a comparison of size and density of Te precipitates and inclusions in CdTe crystals grown from Te-rich melts [12]. There is a clear difference between these two classes of particles.

Cd inclusions have not been studied as much as Te inclusions. The reason is their uncommon presence in most bulk CdZnTe crystals usually grown from Te-rich melts. They have however been observed in crystals grown by vertical Bridgman and gradient freeze from Cd-rich melts.

[1] Hypothesizing a spherical geometry of the inwards solidification of the droplet, the authors obtain the Gulliver-Scheil equation corresponding to this geometry. The calculated data were obtained assuming $k_{Zn} = 2$ (segregation coefficient of Zn in the droplet) $x_{Zn,S} = 0.02$ (minimum mole fraction of Zn in the crystallized droplet). For more details, read the article (Ref. [61]).

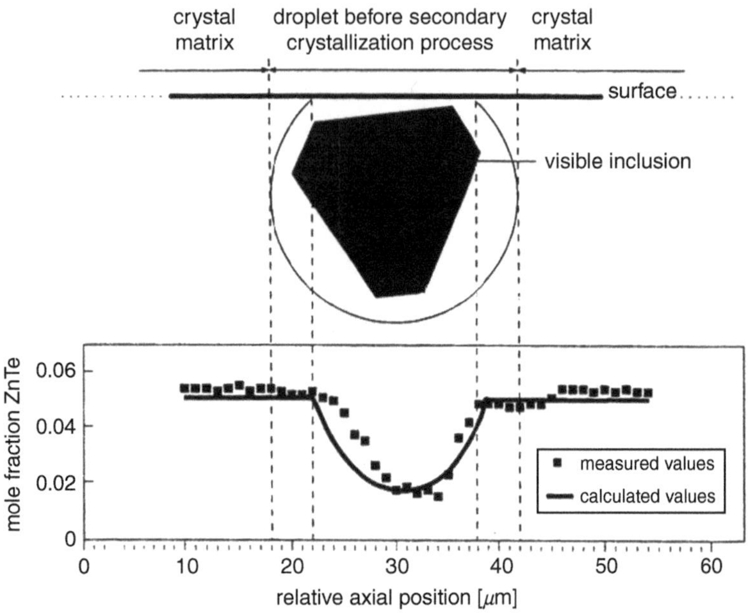

Figure 9 Mole fraction distribution of Zn in the region surrounding a Te inclusion in a THM grown CdZnTe crystal. The scattered data were obtained from EPMA and the solid line represents calculated data [61].

It can be implicitly admitted that similarly to Te inclusions, Cd inclusions occur by entrapment of Cd-rich CdZnTe droplets from the diffusion layer ahead of the crystallization interface. A Cd-rich melt can be maintained during the growth process through passive or dynamic control of Cd evaporation. During cool down of a vertically grown ingot, convectional exchanges between the gas phase and the solid phase occur at the free (top) surface of the boule. The probability of entrapped Cd droplets escaping the ingot is rather small, mainly because of very short diffusion lengths of Cd atoms in the solid alloy.

Cd inclusions usually appear on IR microscopy imaging of CdZnTe crystals as six-point stars (Fig. 11) with sizes as large as 10-200 μm [10, 15, 16]. However, circular features were also observed [10, 15, 16] and were proved to be simply voids [10].

Some visible features, nevertheless subjective, may differentiate Cd stars from Te stars: the points of Cd stars are rounded, whereas they appear sharp for Te stars. The branches of Te stars are contrasted along their centerline, but are more homogeneous for Cd stars. However, decisive identification of Cd and Te inclusions is achieved through nonsubjective analysis of their chemical composition or by comparing their physical properties to those of the pure elements, using analytical tools such as EDX, DSC, Auger spectroscopy, X-ray diffraction or micro-Raman spectroscopy [23].

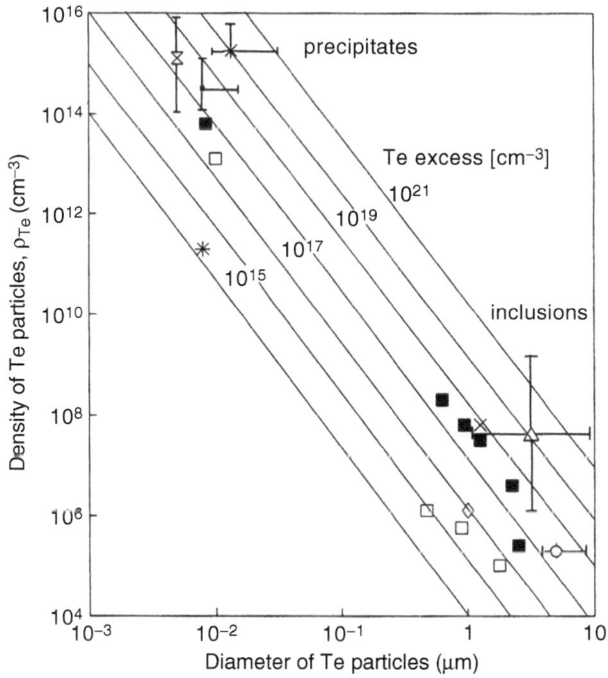

Figure 10 Comparison [12] of size and density of Te precipitates and inclusions in CdTe crystals grown from Te-rich melts. The scatted data were obtained from measurements by different authors. The solid lines are calculated data using Eq. (6) in the previous paragraph with the assumption of spherical particles.

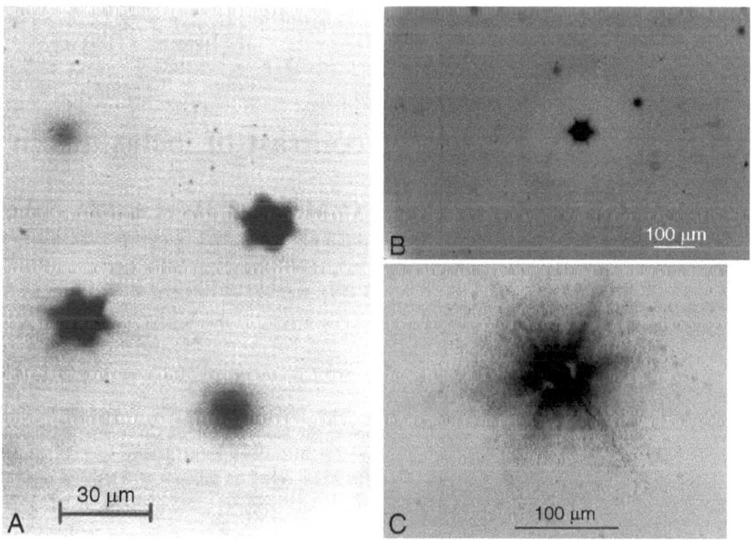

Figure 11 IR microscope image showing six-point star Cd inclusions in CdZnTe crystals grown from a Cd-rich melts by vertical Bridgman (A) Brion et al. [15] and vertical gradient freeze (B), (C) Belas et al. [16]. The contrast of Cd stars is uniform. The rounded features seen in the pictures are voids as pointed out by Brion et al.

3. HOW TO PRODUCE PRECIPITATE AND INCLUSION FREE CdZnTe SUBSTRATES

It was shown in the previous section that the formation of precipitates and inclusions in CdZnTe substrates grown from the melt is mainly governed by the melt's stoichiometry. Control of melt stoichiometry is the prime criterion for managing the development of these defects during crystallization. As condensation of vacancies results in precipitation of excess components, and because CdZnTe sublimes incongruently, a careful design of the crystal's cool down is therefore an equally important step.

In the following subsections *in situ* methods for control of the formation of second phase particles in melt-grown CdZnTe crystals and post-growth annealing processes of the wafers for suppression of these second phase microparticles will be introduced.

3.1. *In situ* control of formation of the second phase inclusions in melt-grown CdZnTe crystals

Control of melt stoichiometry is the key factor that allows managing the formation of second phase microparticles during growth of CdZnTe crystals. However, let us note that Jayatirtha et al. [21] observed no inclusions in their Cl:CdTe (1000 ppm of Cl) crystals grown by THM using a Te solvent, but abundance of them in undoped crystals grown under the same conditions. The raison for this remains unclear.

In vacuum, a stoichiometric CdZnTe melt will incongruently evaporate, losing mostly Cd in the free volume above the melt, until equilibrium. The melt as a consequence gets enriched with Te, the excess of which will eventually convert into Te precipitates and inclusions. It is worth noting that a stoichiometric CdZnTe crystal will also incongruently sublime in vacuum at relatively elevated temperatures if the total pressure of the subliming elements in the chamber stays below the minimum pressure (at congruent sublimation). Mostly Cd vapor will fill up the free volume, leaving in the crystal an excess of Cd vacancies that can condense and generate Te precipitates.

Earlier in this chapter, were briefly introduced various techniques so far used for control or prevention of Cd evaporation from CdZnTe melts. The most effective among them is the cadmium overpressure method, which consists in passively or dynamically filling-up the free volume above the melt with Cd vapor. The Cd vapor arises from an excess of Cd in the melt (passive control) or a separated Cd source, the temperature of which is independently controlled (dynamic control).

Therefore, this section is exclusively consecrated to the cadmium overpressure methods for control of melt stoichiometry.

Production of CdZnTe crystals from the melt is generally performed in two steps: (i) the first is the compounding step, where the material is synthesized by reaction of the Cd, Zn, and Te elements at relatively low temperature and (ii) the second is the growth step, where the synthesized charge is then liquefied at a temperature above the compound's melting point and subsequently solidified to form the crystal. These two steps can be performed in the same or in two different chambers. In both cases, control of Cd evaporation and sublimation is necessary to avoid, latter in the process, formation of Te second-phase particles in the crystal.

3.1.1. Passive control of Cd pressure

In this method, the excess of Cd added to the charge is usually estimated from Eq. (3), with the assumption that it is equivalent to the amount of Cd that will evaporate from the melt in the free volume of the evacuated chamber, at the homogenization temperature. Convective exchanges between the vapor and the melt take place at the melt's free surface, the temperature of which determines the amount of Cd that evaporates in the free volume. During cool down, control of Cd sublimation will depend only on the difference between the pressure of Cd already evaporated in the free volume and the partial pressure of Cd subliming at the solid/vapor interface.

3.1.2. Dynamic control of Cd pressure

Dynamic control of Cd evaporation is relatively more evolved than the former technique. It possesses an additional degree of freedom, which is an independent control of Cd evaporation/sublimation from a separated Cd source. To achieve a solid CdZnTe with no deviation from stoichiometry due to Cd evaporation, a pressure of Cd vapor has to be applied over the melt's free surface from the source of pure Cd. This pressure should be equivalent to the partial pressure of Cd evaporating from the melt, in the free volume for a temperature T_h at the vapor-liquid interface (see illustration in Fig. 12). The amplitude of the overpressure is controlled by the temperature of the Cd source. Rudolph [13] suggests an approximated calculation of that temperature in the case of CdTe. The same approximated calculation can be applied in the case of CdZnTe, with the same assumptions of ideality of the molten alloy, thus validity of Raoult's law. Considering the respective relative volatility of Cd and Zn to be expressed as:

$$\alpha_{Cd}^{Te_2} = \frac{P_{Cd}^{o}}{P_{Te_2}^{o}} \text{ and } \alpha_{Zn}^{Te_2} = \frac{P_{Zn}^{o}}{P_{Te_2}^{o}} \tag{7}$$

where $P_i^o(i = Cd, Zn, \text{and } Te_2)$ is the pressure of each of the pure elements. The concentration in mole fraction of Cd evaporated in the free volume can then be expressed as:

$$X_{Cd}^{V} = \frac{X_{Cd}^{L}\alpha_{Cd}^{Te_2}}{X_{Cd}^{L}\alpha_{Cd}^{Te_2} + \left(X_{Te}^{L} + X_{Zn}^{L}\alpha_{Zn}^{Te_2}\right)} \tag{8}$$

Furthermore, $\sum_i X_i^L = \sum_i X_i^v = 1$, where X_i^L and X_i^v are the mole fractions of component i, respectively, in the liquid and the vapor phase. The partial pressure of Cd evaporating from the melt would then be:

$$P_{Cd} = X_{Cd}^{v} P_{Tot} \tag{9}$$

The pressure (in atmosphere) of the pure elements as function of temperature can be evaluated from the following expressions [13, 63, 64]:

$$\left.\begin{aligned} P_{Cd}^{o} &= 10^{-(5319/T)+5.1368} \\ P_{Zn}^{o} &= 10^{-(6242/T)+5.2953} \\ P_{Te_2}^{o} &= 10^{-(5960/T)+4.7191} \end{aligned}\right\} \tag{10}$$

The volatilities $\alpha_i^{Te_2}$ are quasiconstant over the range of temperature considered (1320–1450 K). Therefore, the temperature (in Kelvin) of the Cd in the reservoir can be calculated from Eqs. (8) to (10) as:

$$T_{Cd} = \frac{5319}{5.1368 - \log\left[\dfrac{X_{Cd}^{L}\alpha_{Cd}^{Te_2}}{1 + X_{Cd}^{L}\left(\alpha_{Cd}^{Te_2} - 1\right) + X_{Zn}^{L}\left(\alpha_{Zn}^{Te_2} - 1\right)} P_{Tot}\right]} \tag{11}$$

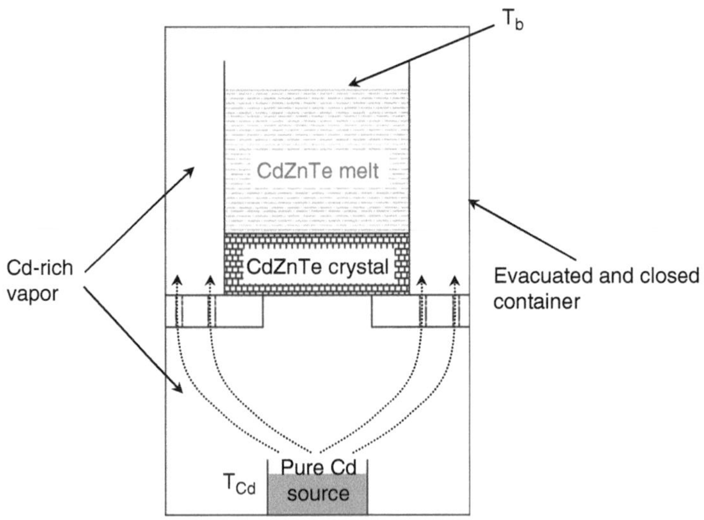

Figure 12 Illustration of a setup for a dynamic Cd-overpressure control applied in a CdZnTe crystal growth process.

P_{Tot} is the total pressure of evaporated elements over the CdZnTe melt. It depends on the temperature T_b at the vapor-liquid interface and can be approximately evaluated from the following Clausius-Clapeyron equation:

$$P_{Tot} = P_m \exp\left[-\frac{\overline{\Delta H}_{mv}}{R}\left(\frac{1}{T_b}-\frac{1}{T_m}\right)\right] \tag{12}$$

Where P_m and $\overline{\Delta H}_{mv}$ are, respectively, the total pressure and heat of vaporization per mole at the melting temperature T_m. $\overline{\Delta H}_{mv}$ can be roughly approximated as being identical to the enthalpy of vaporization, which can be estimated from the slope of the congruent sublimation ($S = L$) lines [45] in the *P-T* coordinates. The values obtained vary between 185 and 212 kJ · mol^{-1} as the zinc composition varies between 0 and 1. P_m varies between 1 and 2.6 atmospheres for quasi-stoichiometric material. T_m can be obtained from the available phase diagrams.

The density of Cd atoms evaporated from the melt into the free volume or incorporated into the melt may then be evaluated as a function of the temperature of the Cd source, assuming that the gas is ideal. In the example given in Fig. 13 where the starting material is stoichiometric $Cd_{0.96}Zn_{0.04}Te$, the temperature $T_{Cd\_stoi} = 836°C$ of the separated Cd

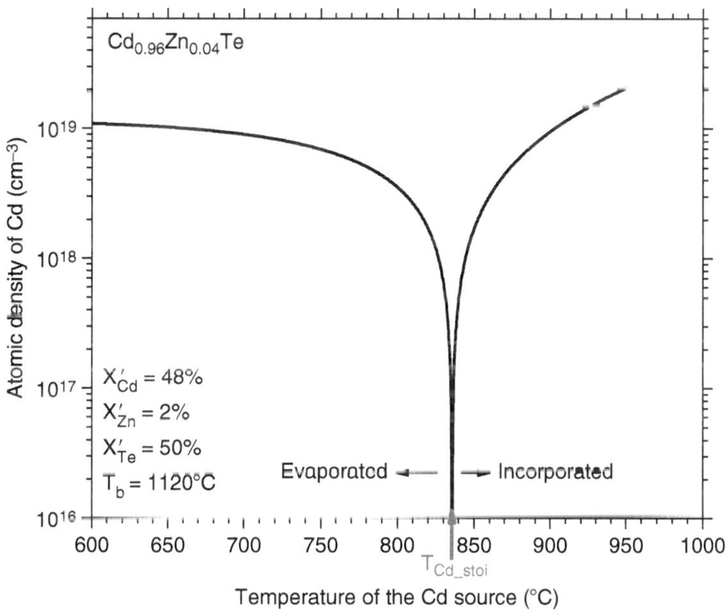

Figure 13 Density of Cd atoms evaporated or incorporated versus the temperature of the Cd source controlling the vapor pressure over a stoichiometric CdZnTe (4% Zn) melt. It was assumed that the total pressure (P_m) at the melting point was ~2 at.

source will keep the melt stoichiometric. Evaporation of Cd from the melt will occur for temperatures below that value, while for higher temperatures, Cd will be incorporated into the melt.

Ideally, to avoid entrapment of Cd or Te droplets at the crystallization interface, the Cd source temperature should be kept as close as possible to T_{Cd_stoi}. CdZnTe melt stoichiometry may be adjusted and controlled as follows: choose a desired stoichiometry and set the temperature of the Cd source using Eqs. (11) and (12). However, for too large stoichiometry deviation on the Te side, readjustment of melt stoichiometry may not be completely achievable [34], probably because of slower kinetic processes in Te-rich melts.

Dynamic control of Cd evaporation can already be done during synthesis of the feed material from pure Cd, Zn, and Te elements, as demonstrated by Ndap et al. [34]. Using the setup describe in Fig. 14, the authors were able to synthesize as large as 4 kg ingots. Stoichiometric or Cd rich feed material is practically preferable, for the reason that it is easy in this case to adjust and control the melt stoichiometry during the subsequent growth process.

The cool down is also an important step of the process, since the crystal's stoichiometry may change due to incongruent sublimation. Equation (11) could well be used to design a cool down scheme that would prevent preferential sublimation of Cd. However, because the assumption of validity of Raoult's law used here in the development of the above equation holds only for solutions, the temperature of the Cd source may therefore be overestimated, which would result in an alteration of the crystal's stoichiometry over a diffusion layer extending some distance beneath the free surface.

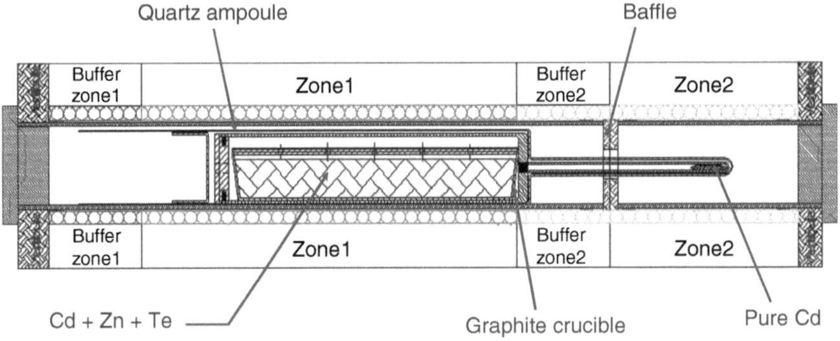

Figure 14 Schematic of an experimental setup for synthesis of CdZnTe material under a dynamic control of Cd evaporation. The pure Cd source is contained in the extension reservoir. At least 4 kg of material could be synthesized in this system. Heat up must be slowed as the amount of material to synthesize is increased.

Greenberg et al. [45–47, 65, 66] utilizing the vapor pressure scanning technique, determined isopleths of partial pressures over CdZnTe crystals of 5%, 10%, and 15% zinc as function of the crystal's stoichiometry within its domain of existence. Such isopleths, shown in Fig. 15 for the case of crystals of 5% and 10% Zn give better representations of SV equilibriums and should be used to design stoichiometry-preserved cool downs of the crystal in a Cd-rich atmosphere.

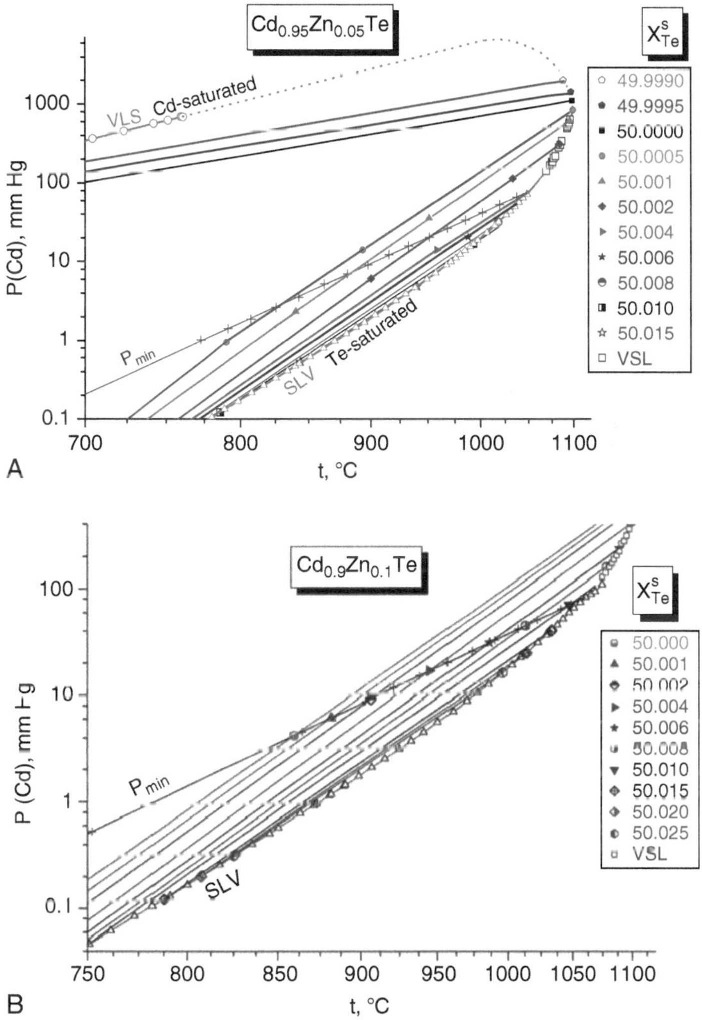

Figure 15 P(Cd)-T projection of $Cd_{1-x}Zn_xTe_{1-\delta}$ solidus for: (A) $x = 0.05$ and (B) $x = 0.1$ [45, 46]. The solid's stoichiometry given in the legend is in atom% Te.

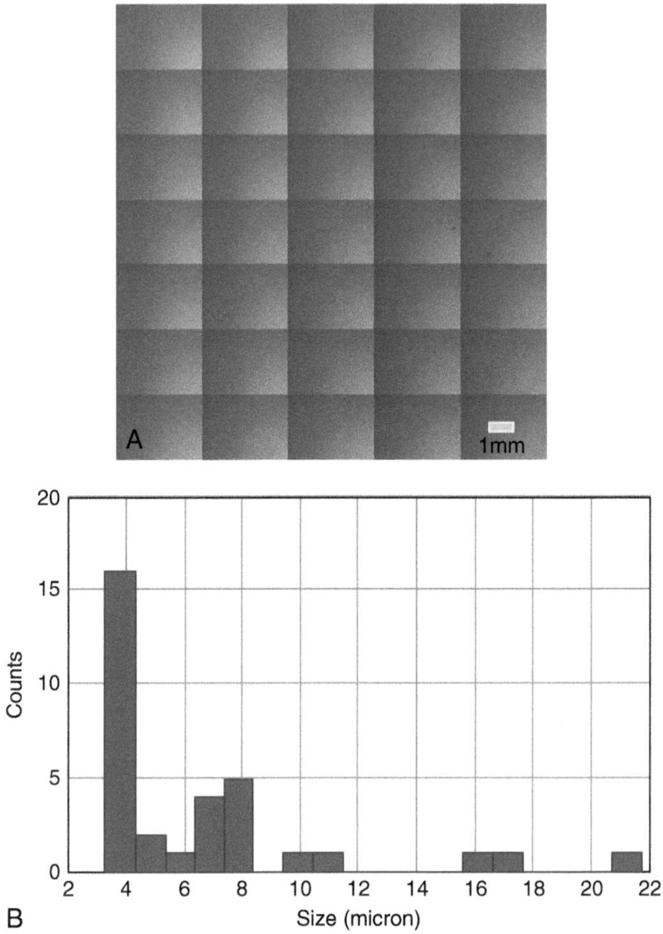

Figure 16 (A) IR micrograph of a portion of slice cut from a $Cd_{0.9}Zn_{0.1}Te$ ingot grown by direct solidification of the melt using the electro-dynamic gradient freeze technique in a vertical configuration [34]. Control of Cd evaporation and sublimation was achieved according to the protocol described in the text. In (B) is presented a histogram of inclusions' size within the mapped region. There are only few residual second phase microparticles believed to be Te inclusions.

The effectiveness of this approach is demonstrated in Fig. 16 where an IR micrograph (a) of a slice cut from a doped[2] CdZnTe (10%Zn) ingot shows only few, mostly, small second phase particles, as evidenced in the histogram (b). The ingot was grown and cooled down in a system where dynamic control of Cd evaporation and sublimation was achieved through a protocol designed according to the description above [34]. One should however note that for doping a crystal, Cd-substitutional

[2] The ingot was doped in order to achieve semi-insulating compensated crystals.

dopants will effectively be soluble into the crystal lattice only if the temperature of the Cd source is appropriately set to a value below T_{Cd_stoi}, which was the case for this ingot.

3.2. Postgrowth wafer annealing

Postgrowth wafer annealing has proved to help dramatically reduce in size and to some extent, completely annihilate existing second phase particles in CdZnTe crystals grown from nonstoichiometric melts. However, inclusions located along grain boundaries and within twin fields are the most difficult to completely eliminate without practically liquefying the CdZnTe wafer. The commonly applied approach consists in annealing the crystals in an atmosphere saturated with vapors of the most deficient elements of the crystalline compound.

Since melt-growths are usually performed under uncontrolled Cd evaporation, postgrowth wafer annealing in Cd-saturated atmosphere is a subsequent logical step done in the purpose of removing Te inclusions. Such a technique has been applied extensively under a relatively wide range of temperatures (400-950°C) and annealing times (from a few of hours to a few weeks). A typical experimental setup is sketched in Fig. 17. It is however required during annealing and cool down to place the CdZnTe wafer under conditions of Cd saturation. There is no standard recipe of annealing conditions (temperature, pressure, time) for complete elimination of Te inclusions. However, any temperature T_{Cd} ($\leq T_w$) generating a Cd pressure (P^o_{Cd}) higher than a given isopleth (Fig. 15) of Cd partial pressure (P(Cd)) over the solid should promote incorporation of Cd into the crystal, size reduction and eventually, annihilation of Te inclusions and precipitates. The annealing time may be estimated from a simple Fick's diffusion profile or more complex diffusion models. The experimentator will find the annealing conditions best suited to his/her wafers' dimensions and history. Vydyanath et al. [9] recommend step annealing as the most effective way to eliminate Te inclusions.

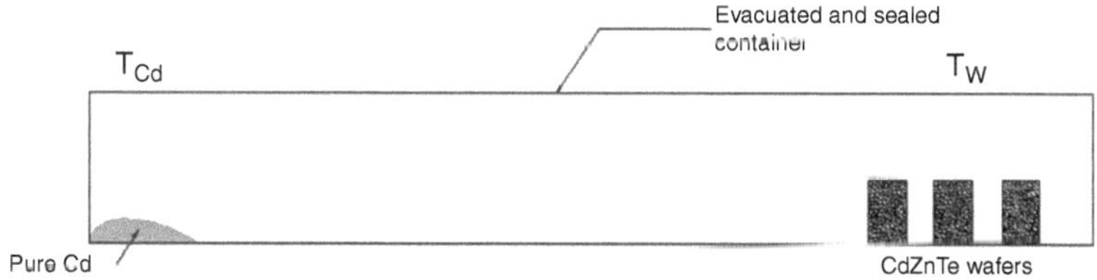

Figure 17 Sketch of an experimental setup for postgrowth wafer annealing. For annealing in a Cd-saturated atmosphere, the following conditions may applied: $T_{Cd} \leq T_w$ and $P^o_{Cd} > P(\mathrm{Cd})$.

Belas et al. [16] instead, argue that single temperature annealing and step annealing achieve similar results.

The cyclic-annealing may also be applied on CdZnTe. This technique was previously utilized in Zn-saturated annealing of ZnTe bulk crystals by Yoshino et al. [67]. The authors report that cyclic changes of the annealing temperature (cyclic-annealing) of their ZnTe wafers in a Zn-saturated atmosphere promoted the formation of single crystalline ZnTe structures at the place of Te inclusions; the residual IR-absorbing polycrystalline structures they observed after annealing at a constant temperature in a similar Zn atmosphere had completely vanished. The authors argue that the fast ramp-down and ramp-up of the temperature (cf. Fig. 18) create a temperature gradient across each Te inclusion, which induces localized solution growths of monocrystalline ZnTe (in Te solvent). Ndap observed unnoticeable differences in CdZnTe wafers processed using this technique or the constant temperature annealing approach in Cd atmosphere [68].

Furthermore, there are reports on the usage of $Cd_{1-x}Zn_x$ alloys instead of pure Cd in order to prevent evaporation of Zn from the wafers [9, 10, 16, 69, 70]. In their work, Sen et al. [10] show that this eliminates only the contribution of Zn loss into surface damage, which is therefore only reduced, suggesting that Te is also escaping the wafer; this was later confirmed by Greenberg et al. [46, 47] from vapor pressure scanning measurements. The authors [10] evaluated the damaged layer to be as thick as few hundred micron below the surface, for annealing conditions of 700°C/72-h. One expects the depth of this damage layer to increase with the annealing temperature and duration, identically to a diffusion front.

Cooling down the crystals in an under-saturated Cd atmosphere will lead to Cd-evaporation and eventually to formation of Te precipitates. It may therefore be preferable to perform this step under Cd (or CdZn)

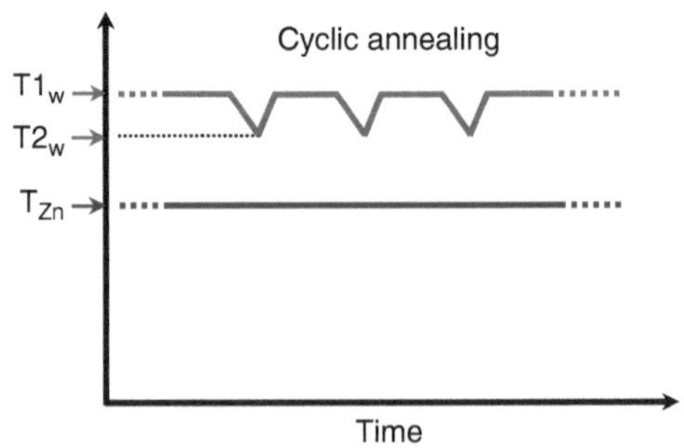

Figure 18 A cyclic-annealing process. The temperature of the wafer cycles between $T1_w$ and $T2_w$ during a regular time interval.

saturated conditions. Quenching the wafers from high temperature will obviously prevent Te precipitation but may not be suitable for large wafers due to formation of additional dislocations or/and wafer-cracking as a result of thermal shock.

Annealing CdZnTe wafers in gallium may promote elimination of Te inclusions. In fact, Sochinskii et al. [60] report that annealing CdTe wafers in liquid gallium at 600 °C for 2-24 h had a very high rate of removal of Te inclusions. The authors report obtaining Te-inclusion free wafers after only 24 h annealing. The drawback for this technique is the resulting gallium doping of the wafer.

It was mentioned earlier that Cd inclusions could form in crystals grown from Cd-rich melts. Their elimination can also be achieved by postgrowth annealing in a Te-saturated vapor. Belas et al. [16] report that 700 °C/7 $\times$ 10^{-3} at.-Te gave the best result for elimination of Cd inclusions. Vacuum annealing will also promote a similar result; however, one should be aware of wafer decomposition due to fast loss of Cd by sublimation. Overall, the wafers' dimensions and history will dictate the optimal annealing conditions. Te precipitates may also form during Cd inclusion elimination. Indeed, Belas et al. [16] observed their appearance and the resulting reduction of their wafers' IR transmittance. A subsequent annealing in Cd-saturated atmosphere usually improves the material's properties.

A straight annealing in Cd atmosphere does not eliminate Cd inclusions. Their size may only be reduced to a limit that will not change even after longer annealing times or higher annealing temperatures [10].

Te and Cd inclusions in CdZnTe crystals are impurity gettering particles. Sen et al. [10] measured concentrations of Cu 40+ and 100+ times more in Cd and Te inclusions than in the CdZnTe matrix. While postgrowth annealing annihilates the second phase microparticles, all the impurities trapped therein are released into the CdZnTe matrix during the process. These impurities can subsequently easily diffuse out into an epi-layer. Purification is a necessary step that may help extract most of these defects from CdZnTe substrates. In the following section, some useful substrates purification techniques will be presented.

4. CdZnTe WAFER PURIFICATION

4.1. By liquid phase diffusion

In the early sixties Aven et al. [71] pioneered the first technique for purification of II-VI compounds (CdS, CdSe, CdTe, ZnTe, ZnSe, and ZnS) they were working on. The authors were unable to control doping of these crystals because of the high concentration of residual impurities therein, stemming from leftovers in the starting material and exo-diffusion from the container material during crystal growth at high temperature. As a result, they developed the solvent extraction technique (or liquid diffusion technique) that consists in annealing the crystal immersed in a liquid made of one of the constituents of the alloy, preferably a group II constituent: for example, molten Cd for CdS or molten Zn for ZnTe and ZnSe. The evacuated annealing chamber may be back filled with an inert gas and sealed off. During annealing at reasonably high temperature (avoid melting or seriously dissolving the crystal), numerous residual impurities diffuse out of the crystal into the molten material that must be, in a further step, separated from the crystal. Figure 19 shows a schematic of an annealing chamber similar to the prototype used by Aven et al. In this arrangement, before bringing the system to room temperature, the molten material is separated from the crystal by quickly inverting the ampoule, which allows transfer of the liquid through the constriction into the upper portion. For example, the authors report successful extraction of noble impurities (Cu, Ag, Au) in CdTe after annealing in molten Cd for 48 h at 700°C.

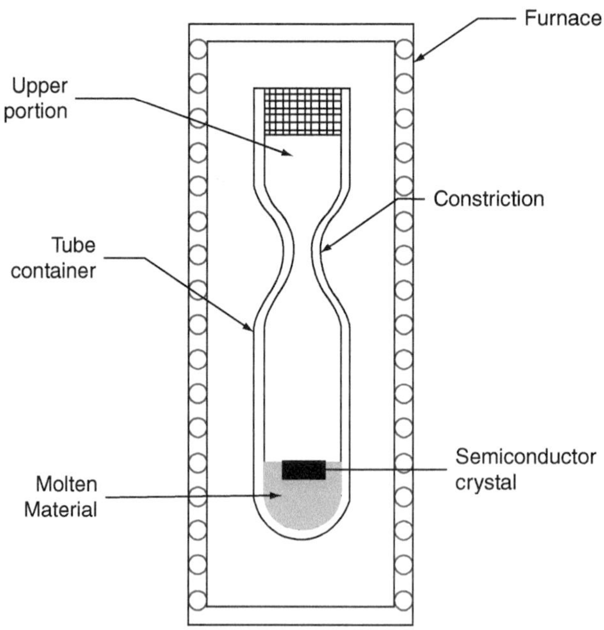

Figure 19 Schematic of a system identical to that used by Aven et al. in the early sixties to purify II-VI semiconductor crystals.

Various slightly modified, complex and improved versions followed Aven's original invention. Mead and McCaldin [72] developed a complex system that required two different baths of molten material, placed in opened crucibles and under a flow of an inert gas. The semiconducting wafers are first annealed for impurity extraction in the first bath made of one of the molten materials described in Aven's invention. The wafers, placed in a carrier, are removed from the first bath and quickly placed in a second one of a low melting point material such as gallium or indium, the temperature of which is slightly above the melting point of the material in the first bath. This allows dissolution of the material from the first bath that remained on the wafer. Subsequently, the wafers are removed from the bath and the low melting point material may be wiped off their surfaces at reasonably low temperature. The authors applied this technique on Al:ZnS crystals that were annealed at 800 °C in Zn then at 420 °C in In.

In the liquid diffusion technique developed by Tregilgas [73], a thin layer of the extracting material predeposited on the surface of the wafer is transformed into liquid droplets during annealing and impurities from the wafer segregate into the droplets by diffusion. The authors used Cd as the extracting material on CdZnTe and CdTe substrates.

4.2. By solid phase diffusion

In this technique developed by Dudoff et al. [74], a sacrificial layer of high purity HgTe (or HgCdTe) is formed on a finely polished CdZnTe substrates by liquid phase epitaxy (LPE), by any vapor transport techniques such as molecular beam epitaxy (MBE), metal organic chemical vapor deposition (MOCVD), photo-assisted molecular beam epitaxy (PAMBE), or by isothermal vapor phase epitaxy (ISOVPE) from the Te-rich bulk alloy [75, 76].[3] For impurity extraction, this assembly is subsequently annealed for a few hours to a few days, at a temperature between 300 and 450 °C in an overpressure (~1600 Torr) of mercury to prevent loss of Hg from the layer by sublimation. After annealing and cool down to room temperature, the sacrificial layer along with a contiguous fraction (~100 μm thick) of the substrate is removed by polishing. During the annealing step, fast diffusing impurities migrate from the substrate into the layer. The process may be repeated until the levels of the remaining impurities in the CdZnTe substrate are satisfactory.

[3] Synthesis of the Te-rich bulk alloy can be done by reacting the components in evacuated and sealed thick-wall ampoules at a temperature 10 to 20 °C above the melting point of the selected composition. ISOVPE can be carried out at a temperature in the 530 – 575 °C range. In the case ISOVPE is used for formation of the layer, the impurity extraction may simply be performed in the same ampoule as one of the steps of the same process by lowering the temperature of system to the desired value.

The success of this process is contingent to the sacrificial layer fulfilling some requirements. Preferably, its composition is selected so as to maximize impurity segregation therein. Indeed, the impurities will migrate toward the layer if their segregation coefficient is higher therein. According to the authors [74], this is observed for $Hg_{1-x}Cd_xTe$ layers of Cd composition $x \leq 0.2$, because of their high Hg content. The ratio of the thickness of the substrate to that of the layer may be in the 55:1 to 400:1 range.

5. CONCLUSION

CdZnTe substrates produced from bulk crystals grown from the melt or from the solution usually contain relatively large densities of structural defects such as second phase microparticles that make them unsuitable for growth of HgCdTe epi-layers. These second phase particles are essentially precipitates and inclusions of tellurium or cadmium. Precipitates and inclusions differ by their size, mode of generation and are result of crystallization from nonstoichiometric melts. In vacuum, a CdZnTe melt will always be Te-rich due to Cd evaporation. Precipitates are smaller than inclusions; they occur at low temperature during the cool down due to retrograde solubility, when the crystal's temperature crosses the solidus. Entrapment of Te (or Cd) rich liquid droplets at an unstable crystallization interface produces inclusions. Dislocations associated with these particles as well as the impurities they contain adversely affect device performance, as they propagate into the epi-layer during epitaxy.

Formation of Te second phase microparticles can be prevented through control of melt stoichiometry, which can be achieved by applying *in situ* passive or dynamic control of Cd evaporation techniques. The latter method has proved to be the most effective way for controlling the emergence of Te particles.

Postgrowth annealing provides another option for elimination of the second phase particles. Te precipitates and inclusions can indeed be eliminated by annealing Te-rich wafers in Cd atmosphere. While elimination of Cd particles can be achieved via annealing in Te atmosphere, this process however promotes formation of Te precipitates that may fortunately be eliminated by a subsequent annealing in Cd atmosphere.

To clean up impurities released from the precipitates and inclusions during postgrowth annealing, a substrate purification step is required. This can be achieved by promoting impurity diffusion from the CdZnTe substrate into an impurity-gettering liquid phase or a solid phase compound of higher impurity segregation coefficients. It was demonstrated that molten Cd or a Te-rich $Hg_{1-x}Cd_xTe$ ($x \leq 0.2$) layer produced excellent results.

REFERENCES

[1] M. Bruder, H.-J. Schwarz, R. Schmitt, H. Maier, M.-O. Möeller, J. Cryst. Growth 101 (1990) 266–269.

[2] Yu.M. Ivanov, V.V. Artemov, V.M. Kanevsky, A.N. Polyakov, V.S. Chudakov, E.M. Pashaev, R.A. Senin, Eur. Phys. J. Appl. Phys. 27 (2004) 371–374.

[3] L.O. Bubulac, W.F. Tennant, D.D. Edwall, E.R. Gertner, J.C. Robinson, J. Vac. Sci. Technol. A3 (1) (1985) 163–170.

[4] R.S. List, J. Vac, Sci. Technol. B10 (4) (1992) 1651–1657.

[5] S.M. Johnson, D.R. Rhiger, J.P. Rosbeck, J.M. Peterson, S.M. Taylor, M.E. Boyd, J. Vac. Sci. Technol. B10 (4) (1992) 1499–1506.

[6] G.R. Woolhouse, T.J. Magee, H.A. Kawavoshi, C.S.H. Leung, R.D. Ormond, J. Vac. Sci. Technol A3 (1) (1985) 83–88.

[7] S. Kumar, A.K. Kapoor, A. Nagpal, S. Sharma, D. Verma, A. Kumar, R. Raman, P. K. Basu, J. Cryst. Growth 297 (2006) 311–316.

[8] Y. Chang, C.H. Grein, J. Zhao, S. Sivanathan, C.Z. Wang, T. Aoki, D.J. Smith, P. Wijewarnasuriya, V. Nathan, J. Appl. Phys. 100 (2006) 114316–114321.

[9] H.R. Vydyanath, J.A. Ellsworth, J.B. Parkinson, J.J. Kennedy, B. Dean, C.J. Johnson, G.T. Neugebauer, J. Sepich, P.-K. Liao, J. Electron. Mater. 22 (8) (1993) 1073–1080.

[10] S. Sen, C.S. Liang, D.R. Rhiger, J.E. Stannard, H.F. Arlinghaus, J. Electron. Mater. 25 (8) (1996) 1188–1195.

[11] J. Shen, D.K. Aidun, L. Regel, W.R. Wilcox, J. Cryst. Growth 132 (1993) 250–260.

[12] P. Rudolph, M. Neubert, M. Muhlberg, J. Cryst. Growth 128 (1993) 582–587.

[13] P. Rudolph, Prog. Cryst. Growth Charact. 29 (1994) 275–381.

[14] P. Rudolph, A. Engel, I. Schentke, A. Grochocki, J. Cryst. Growth 147 (1995) 297–304.

[15] H.G. Brion, C. Mewes, I. Hahn, U. Schäufele, J. Cryst. Growth 134 (1993) 281–286.

[16] E. Belas, M. Bugár1, R. Grill1, P. Horodyský1, R. Feš1, J. Franc1, P. Moravec1, Z. Matěj, P. Höschl1, Elimination of Inclusions in CdZnTe Substrates by Post-growth Annealing, II-VI workshop, October, 2006, New Port Beach, CA, USA.

[17] J.L. Pautrat, N. Magna, J.P. Faurie, J. Appl. Phys. 53 (12) (1982) 8668–8677.

[18] R.S. Rai, S. Mahajan, S. McDevitt, C.J. Johnson, J. Vac. Sci. Technol. B9(3) (1991) 1892–1896.

[19] A. Nouruzi-Khorasani, P.S. Dobson, J. Cryst. Growth 92 (1988) 208–214.

[20] H.F. Schaake, J.H. Tregilgas, J. Electron. Mater. 12 (1983) 931–936.

[21] H.N. Jayatirtha, D.O. Henderson, A. Burger, M.P. Volz, Appl. Phys. Lett. 62(6) (1993) 573–575.

[22] J. Zhu, X. Zhang, B. Li, J. Chu, Infrared Phys. Technol. 40 (1999) 411–415.

[23] S.H. Shin, J. Bajaj, L.A. Moudy, D.T. Cheung, Appl. Phys. Lett 43(1) (1983) 68–70.

[24] K. Zanio, Semiconductors and Semimetals Volume 13 – Cadmium Telluride, Academic Press, New York, 1978, pp. 48–121.

[25] R. Korenstein, R.J. Olson, D. Lee, P.K. Liao, C.A. Castro, J. Electron. Mater. 24 (1995) 511–514.

[26] J.P. Tower, S.P. Tobin, M. Kestigian, P.W. Norton, A.B. Bollong, H.F. Shaake, C.K. Ard, J. Electron. Mater. 24 (1995) 497–504.

[27] S. Bernardi, Mater. Sci. Eng. B28 (1994) 21–28.

[28] E. Weiss, O. Klin, E. Benory, E. Kedar, Y. Juravel, J. Electron. Mater 30 (6) (2001) 756–761.

[29] P. Rudolph, U. Rinas, K. Jacobs, J. Cryst. Growth 138 (1994) 249–254.

[30] J.K. Butler, E. Doty, B. Apotovsky, J. Lajzerowicz, L. Verger, Mater. Sci. Eng. B16 (1993) 291–296.

[31] M. Azoulay, S. Rotter, G. Gafni, R. Tenne, M. Roth, J. Cryst. Growth 117 (1992) 276–280.

[32] T. Asahi, O. Oda, Y. Taniguchi, A. Koyama, J. Cryst. Growth 149 (1995) 23–29.

[33] C. Szeles, S.E. Cameron, J.-O. Ndap, W.C Chalmers, IEEE Trans. Nucl. Sci. 49 (2002) 2535–2540.
[34] J.-O. Ndap, W.C. Chalmers, K.L. Echard, C. Szeles, IEEE Nuclear Science Symposium, October 19th–24th, 2003 Portland, OR, USA, Un published work.
[35] G. Brandt, H. Ennen, R. Moritz, W. Rothemund, J. Cryst. Growth 101 (1990) 226–231.
[36] P. Rudolph, H. Schröter, U. Zimmermann, R. Boyn, Adv. Mater. Opt. Electron. 3 (1994) 289–296.
[37] Yu.M. Ivanov, J. Cryst. Growth 194 (1998) 309–316.
[38] K.Y. Lay, D. Nichols, S.M.C. Devitt, B.E. Dean, C.J. Johnson, J. Cryst. Growth 86 (1988) 118–126.
[39] L. Carlson, C.N. Ahlquist, J. Appl. Phys. 43 (1972) 2529–2536.
[40] A. Zappettini, T. Görög, M. Zha, L. Zanotti, G. Zucalli, C. Paorici, J. Cryst. Growth 214/215 (2000) 14–18.
[41] N. Armani, C. Ferrari, G. Salviati, F. Bissoli, M. Zha, L. Zanotti, Mater. Sci. Eng. B91-92 (2002) 353–357.
[42] E. Raiskin, J.F. Butler, IEEE Trans. Nucl. Sci. 35 (1988) 81–85.
[43] C. Szeles, S.E. Cameron, S.A. Soldner, J.-O. Ndap, M.D Reed, J. Electron. Matter 33 (6) (2004) 742–751.
[44] S.B. Trivedi, H. Wiedemeier, J. Electrochem. Soc. 134 (1987) 3199–3201.
[45] J.H. Greenberg, V.N. Guskov, M. Fiederle, K.-W. Benz, J. Electron. Mat. V33 (6) (2004) 719–723.
[46] J.H. Greenberg, V.N. Guskov, J. Cryst. Growth 289 (2006) 552–558.
[47] J. Greenberg, Thermo-dynamic Basis of Crystal Growth: P-T-X Phase Equilibrium and Non-Stoichiometry, Springer-Verlag, Berlin, 2002, pp. 96–103.
[48] K. Peters, A. Wenzel, P. Rudolph, Cryst. Res. Technol. 25 (1990) 1107–1116.
[49] R. Fan, R.F. Brebrick, J. Phys. Chem. Solids 57 (4) (1996) 443–450.
[50] D. Bensahel, M. Dupuy, J.C. Pfister, J. Cryst. Growth 47 (1979) 727–732.
[51] P. Rudolph, Cryst. Res. Technol. 38 (7-8) (2003) 542–554.
[52] U. Becker, P. Rudolph, R. Boyn, M. Wienecke, I. Utke, Phys. State Solidi (a) 120 (1990) 653–660.
[53] H.F Schaake, J.H. Tregilgas, J.D. Beck, M.A. Kinch, B.E. Gnade, J. Vac. Sci. Technol. A 3 (1985) 143–149.
[54] W.G. Pfann, Trans. Met. Soc. AIME 194 (1952) 747–753.
[55] R.T. Delves, B.R. Pamplin (Eds.), Crystal Growth, Pergamon, Oxford, 1974, pp. 40–103.
[56] W.A. Tiller, K.A. Jackson, J.W. Rutter, B. Chalmers, Acta Metallurg. 1 (1953) 428–433.
[57] W.N. Mullins, R.F. Sekerka, J. Appl. Phys. 35 (1964) 444–451.
[58] K.A. Jackson, Kinetic Processes, Wiley-VCH Verlag GmbH & Co, KGaA weinheim, 2004, pp. 143–153.
[59] R.J. Dinger, I.L. Fowler, Rev. Phys. Appl. 12 (1977) 135–140.
[60] N.V. Sochinskii, E. Diéguez, U. Pal, J. Piqueras, P. Fernández, F. Agulló-Rueda, Semicond. Sci. Technol. 10 (1995) 870–875.
[61] R.U. Barz, P. Gille, J. Cryst. Growth 149 (1995) 196–200.
[62] L.R. Shiozawa, J.M. Jost, G.A. Sullivan, Research on improved II-VI crystals, Final technical report ARL, 68-0153, August, 1968.
[63] Yellin, Szapiro, J. Cryst. Growth 73 (1985) 77–82.
[64] HSC Chemistry software version 4.1 by Outotec Research (formerly known as Outokumpu Research) Oy, Ori, Finland.
[65] Greenberg, Thermo-dynamic Basis of Crystal Growth: P-T-X Phase Equilibrium and Non-stoichiometry, Springer-Verlag, Berlin, 2002, pp. 123–132.
[66] J.H. Greenberg, J. Cryst. Growth 197 (1999) 406–412.
[67] K. Yoshino, M. Yoneta, K. Ohmori, H. Saito, M. Ohishi, T. Yabe, J. Electron Mater. 33 (6) (2004) 579–582.

[68] J.O. Ndap, unpublished work.
[69] L. Yujie, M. Guoli, Z. Xiaona, J. Wanqi, J. Electron Mater 31 (8) (2002) 834–840.
[70] L. Guoqian, Z. Xiaolu, H. Hui, J. Wanqi, Semicond. Sci. Technol. 21 (2006) 392–396.
[71] M. Aven, Preparation of II-VI Semiconducting Compounds by Solvent Extraction, US patent 3146204, August 25, 1974.
[72] C.A. Mead, J.O. McCaldin, Method for Processing Semiconductors, US patent 3650823, March 21, 1972.
[73] J.H. Tregilgas, Method of Impurity Gettering, US patent 6143630, November 7, 2000.
[74] G.K. Dudoff, K.A. Harris, L.M. Mohnkern, R.J. Williams, R.W. Yanka, T.H. Meyers, Method of Purifying Substrate From Unwanted heavy Metals, US Patent No 5454885, October 3, 1995.
[75] P. Becla, J. Lagowski, H.C. Gatos, H. Ruda, J. Electrochem. Soc. 128 (5) (1981) 1171–1173.
[76] J.G. Fleming, D.A. Stevenson, J. Vac. Sci. Technol. A 5 (6) (1987) 3383–3385.

CHAPTER VC

Point Defects

CHAPTER VC1

Theoretical Calculation of Point Defect Formation Energies in CdTe

M.A. Berding

1. INTRODUCTION

There have been significant advances over the last 40 years in our ability to calculate the ground state electronic properties of molecules and solids. Many of the advanced computational methods are based on density functional theory [1]. There are many excellent reviews of recent developments in density functional theory, highlighting both successes and limitations of the theory and the reader is referred to them for an in-depth discussion, see for example Ref. [2].

Most calculations of the properties of crystalline solids and their associated defect, impurity, and dopant properties are based on the local-density approximation (LDA) to the density functional theory. In the LDA, the inhomogeneous exchange-correlation energy functional in the real system is approximated at each location by that of a homogeneous electron gas with the same density [3]. Recent advancements to the LDA included development of improved forms for the function describing the correlation energy and extensions to the LDA to include generalized gradient corrections.

The most reliable ab initio calculations of native point defect properties in CdTe are based on the LDA. But reliable prediction of the defect densities requires additional computations: it is the defect *free energy* that determines the defect concentrations in CdTe as a function of equilibration temperature. Furthermore, most native defects occur in CdTe in both their neutral and one or more ionized states, and thus the net density defects will depend on the Fermi level during equilibration.

SRI International, 333 Ravenswood Avenue, Menlo Park, CA 94025, USA

In this chapter the recent theoretical work on native defects in CdTe is discussed. The focus is on the calculations based on *ab initio* work, which in many cases confirmed the understanding of the defect structure in CdTe determined from experiments [4–6].

2. FORMATION ENERGIES

Defect formation energies in crystalline solids such as CdTe are calculated by taking the difference in total energies of a system with and without the defect of interest. One complication that arises is that although isolated defect energies are desired in most cases, the CdTe host in which the defect is imbedded must be spatially extended to properly describe the crystalline binding properties. For crystalline solids, this is done by using periodic boundary conditions. A similar approach is used for defect calculations, where the defect is imbedded in a "supercell" which is periodically repeated. Supercells containing >128 atoms are typically used in most present-day calculations, and relaxation within the supercell about the defect site are calculated self-consistently.

There are three classes of native point defects occurring in CdTe: vacancies, antisites, and interstitials. Because CdTe is a binary compound, two versions of each vacancy and antisite defects are possible, one involving the cadmium atom and one involving the tellurium atom. For interstitials, there are both Cd and Te interstitials, but there are several different interstitial sites at which interstitial defects are typically located. A listing of native defects which are found in any significant density in CdTe is shown in Table 1. Notation for defects is a primary symbol referring to the defect identity (V refers to a vacancy) and a subscript referring to the defect site (I refers to an interstitial).

In addition to native point defects, defect complex must also be considered. Entropy disfavors the formation of defect complexes. While the

Table 1 Native point defects and defect complexes in CdTe and their ionization states

Defect	Ionization state (eV) [8]
V_{Cd}	$a1 = 0.2, a2 = 0.8$
Te_{Cd}	$d1 = 0, d2 = 0.4$
Te_I	Donor
V_{Te}	$d1 = 0.4, d2 = 0.5$
Cd_I	$d1 = 0, d2 = 0.2$
Te_{Cd}-V_{Cd}	Neutral only

*a*1 and *a*2 refer to first and second acceptor levels, and similarly for *d*1 and *d*2.

number of possible defect complexes is large, only in a very small subset of these defect complexes is the binding energy high enough to lead to their formation in significant quantities. In CdTe the only defect complex was predicted to occur in the material, the Te_{Cd}-V_{Cd} complex [7].

The *ab initio* formation energies of native point defects were calculated by several groups. Berding [7, 8] used the full-potential linearized muffin-tin orbital method [9] within the LDA to calculate the energies of all native point defect and all binary defect complexes composed of the dominant native point defects. Because of computational limitations, very small supercells (32-atomic sites per supercell) were used. Wei and Zhang [10] used the linearized augmented plane wave method [11] within the LDA to calculate the defects and doping trends in CdTe; one significant difference in their work compared to that of Berding is in the assignment of the ionization states of several defects. In a more recent work [12], the open-source VASP code [13] was used to calculate properties of defects in CdTe, with a goal of understanding their role in compensation of shallow acceptor states. More details on the general calculational approach to defect energy in CdTe can be found in a recent review [14].

3. ELECTRONIC EXCITATION ENERGIES

The native defects in CdTe can occur in both neutral and in most cases in one or more ionization states. In the aforementioned references on the calculations of the ground state formation defect energies in CdTe, the ionization states were also calculated using their various implementations of the LDA. While various procedures have been developed to deduce the excitation energies, because of the LDA is a ground state theory, the accuracy of the *ab initio* prediction of the ionization states is suspect. Recently, calculations capable of more accurately describing excited states of a system, for example those based on the GW approximation [15], have been developed, but their computational complexities limit them to systems with few atoms.

In Table 1 we summarize the ionization levels determined by Berding [8]. In a recent paper [16] both calculated and experimentally deduced ionization energies for all of the primary point defects in CdTe are tabulated. Their table shows a great variation among both calculated and experimentally deduced ionization energies, emphasizing the difficulty in definitively determining ionization energies of specific defects both experimentally and theoretically. One significant difference in the work by Berding and the later work by Wei et al. and Du et al. is in the assignment of the tellurium antisite, a dominant defect in the material, as a negative-U defect, that is one in which the first donor state is unstable.

4. DEFECT FREE ENERGIES

To calculate the density of native defects in CdTe, the free energy associated with the defects must be calculated. The zinc-blend phase of CdTe can exist over an extended range of chemical potentials for cadmium and tellurium. One way to visualize this is in the phase diagram of the partial pressure of cadmium or tellurium as a function of $1/T$, as shown in Fig. 1.

Because of differences used in reporting the defect formation energies in the various LDA calculations, details on how defect formation energies are mapped to the phase diagram differ. The common feature is that a reference chemical potential of the constituent species must be specified and bound by the boundaries of the phase diagram.

An important contribution to the defect free energy comes from changes in the vibrational entropy of the lattice. Berding et al. [17] developed an approach for calculating this contribution to the free energy using a classic elastic model. This contribution to the free energy can be significant when comparing a vacancy, which involves removal of an oscillator from the lattice, and an interstitial, which involves the addition of an oscillator to the lattice.

The configurational free energy is typically treated within the quasi-chemical formalism from which defect concentrations are calculated. Additional contributions coming from relaxation degeneracies must also be taken into consideration and can differ depending on the symmetry of the lattice relaxation around a defect. When referencing to the gas phase,

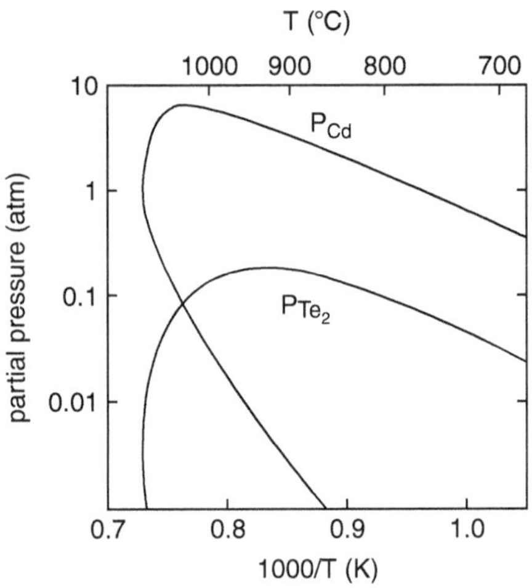

Figure 1 *P-T* phase diagram of CdTe. Defect densities depend on at which *P* and *T* the materials are equilibrated.

as was done by Berding et al. [17], the free energy of the vapor phase species must also be calculated.

Just as the chemical potential for each of the constituent species in CdTe must be specified to determine the native point defect equilibrium concentration, the chemical potential, or Fermi level, for the electrons and holes must be calculated. The Fermi level depends on the temperature and the doping level in the system, as well as on the native defect concentrations, and must be determined self-consistently for the desired equilibration conditions. Because extended valence and conduction band states can also accommodate holes and electrons, respectively, a model of these states must be included in the calculation of the Fermi level. This is discussed briefly by Berding et al. [17]. Franc et al. [18] made some important improvements to that work by including higher lying CdTe bands that are important when calculating the Fermi level at high temperature where equilibration occurs.

5. PREDICTION OF NATIVE POINT DEFECT DENSITIES IN CdTe

Densities of native point defects are typically calculated using the quasichemical formalism, for example, see Ref. [19]. This formalism is generally applicable when the defect densities are sufficiently low ($<10^{21}$ cm^{3} or so). Typically the chemical potential of the cadmium (or tellurium) is specified by the equilibration conditions (e.g., the temperature and the partial pressure of one of the elemental constituents). Depending on the equilibration conditions, the Fermi level is either specified (e.g., by a majority dopant or by an elevated temperature in which the material is "intrinsic" and determined by the density of states in the valence and conduction bands) or must be determined self-consistently (e.g., when it is determined by the density of the native defects themselves). In the quasichemical formalism, defect complexes must be specified as a separate defect; for example, the Te_{Cd}-V_{Cd} complex is an additional mechanism by which tellurium antisites and cadmium vacancies can be introduced into the lattice during equilibration.

In Fig. 2, the predicted native defect and free-carrier concentrations at 700°C are shown as a function of cadmium partial pressure within the existence region of the compound [20]. Also shown are the carrier concentrations at room temperature, assuming the net concentration of native defects are frozen from the 700°C equilibration conditions, and only the Fermi level is allowed to re-equilibrate upon cooling. As shown in Fig. 2, the primary native defect predicted under tellurium-rich conditions is the cadmium vacancy, in agreement with previous work. This work also predicts a high density of Te_{Cd} and a significant fraction of Te_{Cd}-V_{Cd} complexes. Under cadmium-rich conditions, Cd_I is predicted to be the

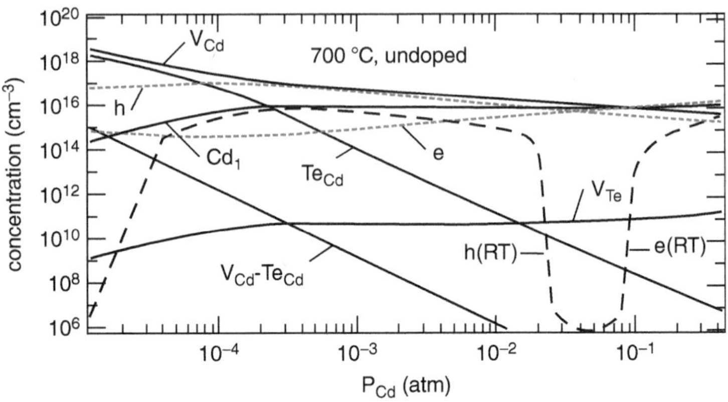

Figure 2 Predicted native point defect densities as a function of cadmium partial pressure in the existence region of CdTe. h and e are electrons and holes and RT refers to room temperature.

dominant defect. The native defect densities shown in Fig. 2 include all ionization states of that defect.

The work by the group at Charles University in Prague, see for instance Ref. [21], has taken an approach to predicting the native defects in the material by using input from the *ab initio* calculations, but by varying several of the energies to better match their extensive experimental data. This approach has met with great success in both explaining the experimental data, with a foundation in theory.

6. NATIVE DEFECTS AND THEIR RELATIONSHIP TO DOPING

Much of the work done on modeling native defects in CdTe has been aimed at in understand the doping properties of this material. Unlike other wide-gap II-VI compounds, CdTe can be doped by p- and n-type, although the maximum hole concentrations that can be achieved is $\sim 10^{15}$ cm^{-3}. The topic of doping and its relationship to defects was recently reviewed by Zhang [14] and earlier by Chadi [22]. One of the issues that must be addressed is complexes that form between the dopant atom and native defects, forming for inactive complexes.

7. FUTURE CHALLENGES

The enormous growth in computation power over the last several decades has permitted the calculation of defect properties in CdTe with some degree of accuracy. The most challenging aspect in the current state-of-the-art is in the prediction of the ionization states of the defects.

When comparing theory with experiments, defect equilibrium in the theoretical calculations is assumed. Because of the slow diffusion rates, it is assumed that the defect structure established under high temperature growth conditions is frozen-in upon cooling, but this may not always be the case. For example, when tellurium inclusions are in the material, it can provide reservoirs where local re-equilibration can take place. The next challenge for *ab initio* calculations is in modeling diffusion rates in CdTe.

REFERENCES

[1] P. Hohenberg, W. Kohn, Phys. Rev. 136 (1964) B864–B871.
[2] J. Seminario (Ed.), Recent Developments and Applications of Modern Density Functional Theory, Elsevier Science, 1996.
[3] W. Kohn, J. Sham, Phys. Rev. 140 (1965) A1133–A1138.
[4] R. Brebrick, R. Fang, J. Phys. Chem. Solids 57 (1996) 451–460.
[5] P. Höschl, R. Grill, J. Franc, P. Moravec, E. Belas, Mater. Sci. Eng. B16 (1993) 215–218.
[6] P. Fochuk, O. Korovyanko, O. Panchuk, J. Cryst. Growth 197 (1999) 603–606.
[7] M. Berding, Appl. Phys. Lett. 74 (1999) 552–554.
[8] M. Berding, Phys. Rev. B 60 (1999) 8943–8950.
[9] O. Andersen, O. Jepsen, D. Glotzel, in: F. Bassani (Ed.), Highlights of Condensed Matter Theory, North-Holland, Amsterdam, 1985.
[10] S. Wei, S. Zhang, Phys. Rev. B 66 (2002) 155211.
[11] S. Wei, H. Krakauer, Phys. Rev. Lett. 55 (1985) 1200–1203.
[12] M. Du, H. Takenaka, D. Singh, Phys. Rev. B 77 (2008) 094122–094126.
[13] G. Kresse, J. Furthmuller, Phys. Rev. B 54 (1996) 11169–11186.
[14] S. Zhang, J. Phys.: Condens. Matter. 14 (2002) R881–R903.
[15] L. Hedin, Phys. Rev. 139 (1965) A796–A823.
[16] V. Kosyak, A. Opanasyuk, Semi. Phys. Quant. Elect. Opto. 10 (2007) 95–102.
[17] M. Berding, M. van Schilfgaarde, A. Sher, Phys. Rev. B 50 (1994) 1519–1534.
[18] J. Franc, R. Grill, L. Turjanska, P. Höschl, E. Belas, P. Moravec, J. Appl. Phys. 89 (2001) 786–788.
[19] F. Kröger, The Chemistry of Imperfect Crystals, John Wiley, New York, 1964.
[20] M. Berding, unpublished.
[21] R. Grill, J. Franc, P. Höschl, I. Turkevych, E. Belas, P. Moravec, M. Fiederle, K. Benz, IEEE Trans. Nucl. Sci. 49 (2002) 1270–1274.
[22] D. Chadi, Annu. Rev. Mater. Sci. 24 (1994) 45–62.

CHAPTER VC2

Experimental identification of intrinsic point defects

CHAPTER VC2A

Characterization of Intrinsic Defect Levels in CdTe and CdZnTe

Kelvin G. Lynn* and **Kelly A. Jones**[†]

1. INTRODUCTION

In this chapter, experimental and theoretical results are reviewed which show that high-resistivity $Cd_{(1-x)}Zn_xTe$ (CZT) and CdTe can be obtained from high-purity undoped materials under certain growth conditions. Intrinsic defects are responsible for the compensation. It has been shown that by *sufficiently reducing the Cd vacancy concentration* and by complexing with other deep intrinsic defects, this reduction can be used to compensate the material [1, 2]. Intrinsic defects with electronic levels near the middle of the bandgap are believed to be present in all high-purity CdTe/CZT material and are independent of Zn composition up to ~15% (CdTe/CZT), growth technique and thermal history. The donor levels follow the conduction band with Zn addition [3]. The concentrations of these levels can vary by orders of magnitude depending on the growth, composition conditions and postprocessing. These native intrinsic deep levels can provide compensation in high-purity CdTe/CZT, however, they limit any application requiring an active depth of more than a few millimeters. Even with the addition of a Group III or Group VII for self-compensation, one of the levels is not removed which has the largest trapping cross section limiting the depletion depth. These can be reduced in concentration by careful postprocessing, however, the ability to realize the control of the levels for thick samples is at best challenging. This can be averted by the addition of an appropriate second doping element to modify or passivate these complexes without decreasing the resistivity of the host material. This idea has not been used in other

*Center for Materials Research, Washington State University, Pullman, Washington, USA
[†]Mechanical and Materials Engineering, Washington State University, Pullman, Washington, USA

compound semiconductors but represents a new idea for compensation in compound semiconductors. Presently these deep levels have been tentatively identified as a complex of a single or a double Cd vacancy bound to a Te antisite. The second vacancy as a di-vacancy or possibly a vacancy bound to a jog (vacancy-like) of an edge dislocation. The vacancy concentration must be low enough that deep level defects successfully compensate the material to yield a resistivity $>10^9$ Ω cm. This material can be grown in excess Te and have been produced by Cd over pressure either during growth or subsequent careful postprocessing. Considering current growth techniques, the lowest concentration of impurities in grown material is $\sim10^{15}$ cm^{-3}, which is much larger than the 10^8 cm^{-3} required for intrinsic material. The high resistivity of these crystals is due to electrical compensation. The presence of these deep levels relaxes the extremely tight purity requirements for obtaining high resistivity, impurity levels 10^{14}-10^{15} cm^{-3} which are achievable with current growth methods. The trapping of electrons at either one of the intrinsic deep levels limits the active volume or the depletion depth of the detectors using standard Schottky contacts [4, 5]. These deep level defects can be modified by doping using an appropriate second doping element which modifies the properties of these complexes. This new complex is more thermally stable with a lower trapping cross section [6–8] allowing for large volumes and fully active detectors.

Normally, high-purity undoped CdTe/CZT is low resistivity; however, it can be high resistivity when there is a significant presence of intrinsic defects, and their complexes to pin the Fermi level if there is a *small concentration of Cd vacancies*. These detectors do not fully deplete for more than a few millimeters under bias and are not useful for large volume radiation detectors. A more reliable and frequently used technique for producing high-resistivity CdTe crystals is to grow it from a slightly Te-rich melt, doped with either a group III (Al, Ga, In) or group VII (Cl, Br) element. However, these doping elements render the material high resistivity but still do not work for large depletion depth detectors. In any growth technique, incorporation of defects like (1) cadmium vacancy (V_{Cd}), (2) Te antisite, (3) intrinsic complexes (4) dislocation, and (5) residual impurities, especially hydrogen is inevitable. In the authors' opinion, this has led to a great confusion in the area as sometimes small amounts of impurities or intrinsic levels, less than 10 ppb, could modify the properties [9]. One should realize that these can form a binding between the various levels, which can form complexes during the cool down which can be partly due to the retrograde solubility. These defects introduce localized defect levels in the bandgap that act as traps and recombination centers for mobile charge carriers in CdTe or CZT crystals. Although some of the defects in CdTe/CZT have been studied to a considerable depth, the $0/-$ ionization level of A-center and donors are well established. The position of the first $0/-$ and second $-/-2$ ionization levels of the Cd-vacancy is still debated [10–13]. Both of these defects can be characterized as relatively shallow acceptors in CdTe/CZT. Isolated Cd vacancies cannot explain the pinning of the Fermi level to the middle of the bandgap in CdTe/CZT. To explain the high resistivity in CdTe/CZT, a compensation mechanism invoking deep level defects has been investigated [3, 14–17]. The explanation, which follows, provides strong evidence that these levels are well founded within this model as the Te antisite complexes with a Cd vacancy or a Cd di-vacancy which is not so different from the isolated Cd-vacancy except with an extra Te in a Cd position. In TEES experiments, V_{Cd} related defects were reported at 0.21 eV and $E_c - 0.73$ eV [18]. The latter was found to be responsible for the degrading performance of the detectors. This is consistent with samples taken from the same ingot that were associated with not fully active detectors [18]. Hage-Ali and Siffert suggested [19] that the single ionized level of the V_{Cd} was at $E_v + 0.38$ eV and stated that a "well established doubly charged Cd vacancy" at E_c −0.6-0.7 eV. Fiederle et al. using PICTS signal found that undoped CZT is dominated by deep level traps around 0.8 and 0.6 eV [20]. It was suggested that the Te antisite can act as a deep level to compensate excess electron carriers [20]. Tessaro and Mascher [21] performed positron lifetime measurements on material characterized by WSU and discovered that the positron lifetime in the alloys were larger than expected for the annihilation to be caused by a single vacancy. They suggested neutral di-vacancy concentrations in the mid-10^{16} cm^{-3} [21] and would be consistent with the present conclusions if bound to the antisite and was neutral or negatively charged. This interpretation of a single or a double Cd vacancy bound to a Te antisite allows one to resolve these discrepancies associated with the assignments of the Cd vacancy and as well the cross section differences, shown in Table 1. These more recent results are supported by various experiments and independent theoretical calculations [24, 35]. This chapter reviews the literature and will highlight the discoveries that have drawn to the conclusions.

2. CHARACTERIZATION OF VARIOUS DEFECT LEVELS IN CdTe/CZT

Understanding the intrinsic defects of pure CdTe/CZT is absolutely necessary in determining a well-founded approach for electrical compensation which has now mainly focused on the Cd vacancy with group III or group VII elements. This model needs to be expanded to produce and resolve the different levels that have been assigned experimentally. The isolated vacancy is supported by theoretical calculations as shallow acceptor levels are near the valance band [2]. Du et al. [24] calculated using the density functional theory (DFT) within the local density approximation (LDA). The (0/−) and (−/−2) acceptor levels for the Cd vacancy were located at 0.18 and 0.25 eV above the valence band [24]. Our work

Table 1 Ionization energies of native defects

Defect	Thermal ionization, *E* (eV)	Photoionization, *E* (eV)	Method	References
V_{Cd}	$E_v < 0.47$		Photo-EPR	[12]
V_{Cd}		Acceptor: 0.78	PICTS	[22]
V_{Cd}	$E_v + 0.2, 0.8$		Theory	[23]
V_{Cd}	$E_v + 0.14$, 0.4, 0.76		DLTS, PICT	[14]
V_{Cd}^{-}	$E_v + 0.38$		TSC	[19]
V_{Cd}^{-2}	$E_c - 0.70$		TSC	[19]
V_{Cd}^{-}	$E_v + 0.18$		EDFT theory	[24]
V_{Cd}^{-2}	$E_v + 0.26$		EDFT theory	[24]
V_{Cd}	$E_v + 0.1$		Theory	[25]
V_{Cd}^{-}	$E_v + 0.1$		TEES	[2]
V_{Cd}^{-2}	$E_v + 0.24$		TEES	[2]
V_{Cd}	$E_v + 0.43$		TEES	[13]
V_{Cd}	$E_v + 0.21$; $E_c - 0.73$		TEES	[18]
V_{Cd}	$E_v + 0.325$		Galvano-magnetic	[57]
V_{Cd}^{-2}	$E_v + 0.43$		PICTS	[20]
V_{Cd}^{-2}		<0.4	Photo-EPR	[27]
V_{Cd} complex	$E_v +$ (0.35-0.41)		TSC, TEES, PICT	[28]
Te_{Cd}	$E_c - 0.2$		Photo-EPR	[29]
Te_{cd}	$E_c - 0.2$, 0.4		Theory	[23]
Te_{Cd}	$E_c - 0.4$ [03]		TEES	[2]
Te_I	$E_c - 1.28$, 1.48		EDFT theory	[24]
Te_I	Donor		Theory	[23]
Te_{cd} complex	$E_v + 0.32$		DLTS, PICT	[14]
$Te_{Cd}^{-2} + V_{Cd}^{-2}$	$E_v + 0.69$		EDFT theory	[24]

(*continued*)

Table 1 (*continued*)

Defect	Thermal ionization, *E* (eV)	Photoionization, *E* (eV)	Method	References
$Te_{Cd} + V_{Cd}$	$E_v + 0.7^a$, 0.79		TEES	[2]
$Te_{Cd} + 2V_{Cd}$	$E_v + 0.46$, 0.7^a		TEES	[2]
$V_{Te}^{+/0}$	$E_c - 1.4$		EPR	[10]
V_{Te}	$E_c - 1.4$	Donor: 1.4	Photo-EPR	[29]
V_{Te}	$E_c - 1.1$		DLTS, PICT	[14]
V_{Te}	$E_c - 0.4$, 0.5		Theory	[23]
V_{Te}	$E_v + (0.19\text{-}0.22)$?		TSC, TEES, PICT	[28]
V_{Te}		Donor 1.1	PL	[26, 31]
Cd_I	$E_c - 0.64$		DLTS, PICT	[14]
Cd_I	$E_c - 0.54$		PICT	[30]
Cd_I	$E_c - 0.56$		TSC	[19]
Cd_I	$E_c - 0.5$		Theory	[25]
Cd_I	$E_c - 0$, 0.2		Theory	[23]
Cd_I	$E_c - 0.71$		EDFT theory	[24]
Stress related	$E_v + (0.46\text{-}0.50)$		TSC, TEES, PICT	[28]
Zn related	0.6?		TSC, TEES, PICT	[28]
O_{Te}-H complex	$E_c - 0.73$		EDFT theory	[24]

[a] $E_v + 0.7$ defect level was suggested to be either $Te_{Cd} + V_{Cd}$ or $Te_{Cd} + 2V_{Cd}$ [2].

has been done over the last 12 years to understand how high-purity grown material under excess Te can be very high resistivity but does not function as a large volume radiation detector. Low levels of impurities can greatly affect the material electrical properties. Significant numbers of intrinsic and extrinsic defect levels have been identified in CdTe and CZT samples grown by different methods either unprocessed or postprocessed. Regardless of the fabrication method, high-purity material exhibits similar deep levels. The intrinsic deep levels can be acceptors or donors depending on the position of the Fermi level. For example, one can find an ingot that will change its resistivity by orders of magnitude by only an mm or two along the growth axis.

The aim of this chapter is to understand and characterize the various defect levels in CdTe/CZT in as grown and as well in postgrowth processed conditions. The thermal ionization energy (E_{th}) and trapping cross section (σth) are the signatures of the defect levels. Each is reviewed or estimated by using thermal stimulated current (TSC) and thermoelectric effect spectroscopy (TEES). These results were compared to other earlier studies [36–39] and theoretical energy levels of intrinsic and some extrinsic defects in CdTe were predicted by first-principle band structure calculations [40]. Based on Soundararajan et al. [2] a comparison and interpretation of the defects levels are given in Fig. 1. These defect levels determine the compensation mechanism in CdTe/CZT and reveals insight into material properties.

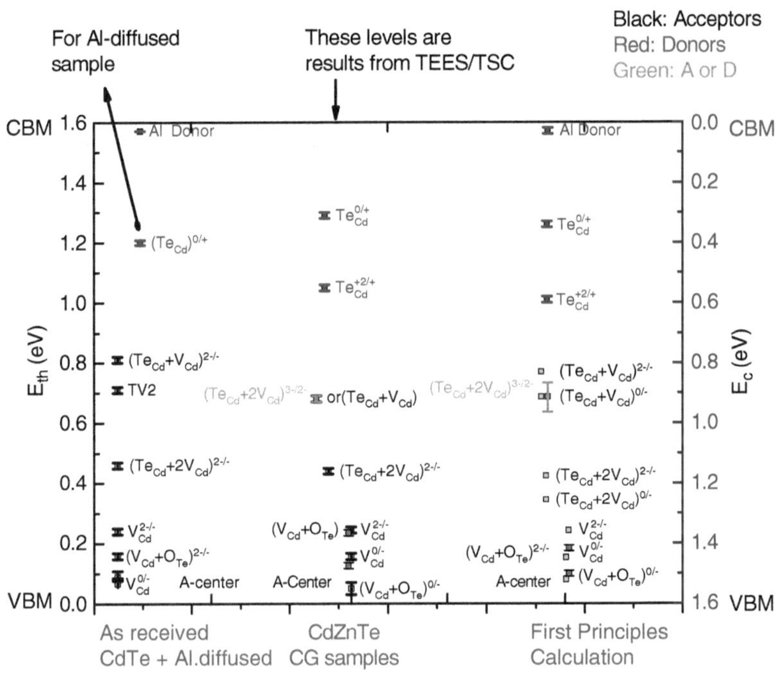

Figure 1 The thermal ionization energy of various intrinsic defects and defect complexes calculated experimentally in pure CdTe (sample A1), and CdTe with Aluminum diffused (sample A2a), and theoretically calculated using the first-principles band structure methods [2].

By using the material as a radiation detector, it can clearly demonstrate the electronic defects within the bandgap. While high-resistivity material is essential for a working radiation detector, this is not a sufficient requirement. The charge carrier drift lifetime has to be on the order of several to tens of microseconds, as well as the mobility-lifetime ($\mu\tau$) value for electrons and holes should be within a factor of a 1000 for use in high count rate applications ($>10^6$ counts per second). The mean drift lifetime is shortened by the density and cross section of the traps present in the material. Typically, defect levels of less than 0.2 eV play a minor role in the drift length as the trapping decay time of ~0.1 μs without an applied electric field at room temperature. Carriers that recombine with the opposite charge will show a decrease in the induced charge. These properties are measured in mobility-lifetime measurements. Furthermore, if the depletion depth does not extend over the full volume of the complete detector it will lead to unphysical results. This phenomenon is also related to polarization of the detector (collapse of the electric field) which has hampered CdTe/CZT for widespread use as an ambient radiation detector for almost 50 years. Trapped charge will generate an electric field that counteracts the external field from the applied bias thus making the field not linear with the thickness. The efficiency (full depletion) of the detector suffers. The lack of depletion can be observed by measuring the capacitance of the detector. McGregor et al. [41] showed for materials with a high bulk resistivity a lower frequency is needed to examine the capacitance. Figure 2 shows capacitance results for two samples: one depletes when a bias is applied; and the other does not when the deep levels are present. Both samples have high resistivity but only the low capacitance sample functions as an efficient detector. This is similar to that observed in GaAs detectors [41]. The capacitance was measured at a 10 kHz frequency with an accuracy of 5×10^{-15} F. A fully depleting detector will show proportional behavior to C, $W \alpha 1/C^2$ where W is the depletion width. The lack of depletion depth is a serious problem for larger detector crystals. Raytheon [4] reported on detectors with excellent energy resolution of less than 5% at the 122 keV line of ^{57}Co with a small active volume. WSU also published a similar result for a 1 cm^2 area 2 mm thick sample with an active depth of ~0.5 mm [42]. For radiation detector applications, it is essential that the compensating deep levels do not significantly affect the depletion and/or charge transport in the material especially for single carrier devices.

An attempt was made to review all of the published results from other researchers when undoped samples were discussed or measured and suggested intrinsic defects. To provide further support, five samples will be discussed for their purity and subjected to a variety of tests. Glow discharge mass spectroscopy (GDMS) was performed on adjacent samples in the following results. Table 2 shows high purity for five of the

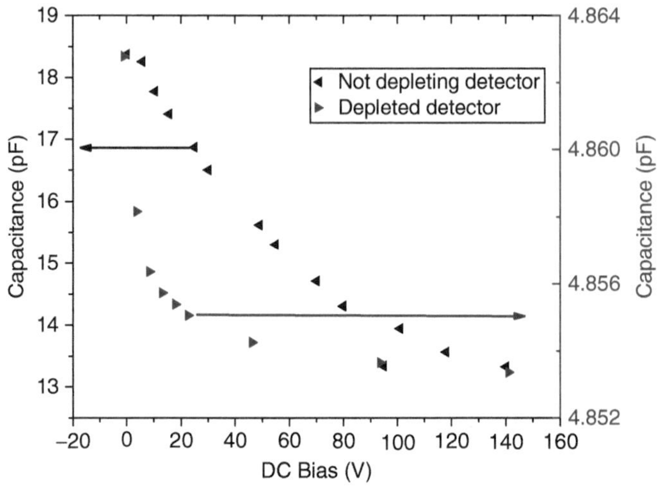

Figure 2 Capacitance-bias curve for two samples. In a nondepleting sample (left pointing symbols, left scale), the capacitance is much higher than for a depleted sample (right pointing symbols, right scale). The samples are of equal area and thickness. *Note*: the capacitance is close to a factor of 3 different between the two samples.

Table 2 GDMS data ppb atomic of electrically active impurity atoms for the measured samples (ND: below detection limit)

Sample	Li	Na	O	Mg	Al	Cu	Ni
CdTe	ND	3	91	18	13	2	ND
CZT	ND	ND	76	78	18	ND	ND
VBOC CZT	450	8	530	ND	25	10	ND
VB CZT	12	120	130	430	40	12	5
Al-doped HPB CdTe	ND	12	93	19	1000	3	25
Sample	Cl	Ga	Si	N	K	P	Zn
CdTe	2	1	5	55	ND	2	12 ppb
CZT	ND	ND	2	ND	ND	ND	10%
VBOC CZT	7	ND	2	35	25	0.7	10%
VB CZT	ND	ND	3	420	42	8	8%
Al-doped HPB CdTe	37	ND	8	12	4	ND	17 ppb

samples; high pressure Bridgman (HPB) CdTe, HPB CdZnTe, vertical Bridgman with overpressure Cd (VBOC), vertical Bridgman (VB) grown CdZnTe, and an Al-doped HPB CdTe sample. Each sample exhibits high resistivity from current versus voltage measurements taken from −1 to 1 V. These results are shown in Table 3. The results of the radiation detector performance measurements are shown in Fig. 3 for five samples studied.

Table 3 Resistivity

Sample	Resistivity (Ω cm)
Pure HPB CT	$1.06(7) \times 10^{10}$
Pure HPB CZT	$2.21(13) \times 10^{9}$
Pure VBOC CZT	$2.7(04) \times 10^{10}$
Al-doped HPB CT	$1.66(1) \times 10^{8}$
Commercial detector	$1.78(1) \times 10^{10}$

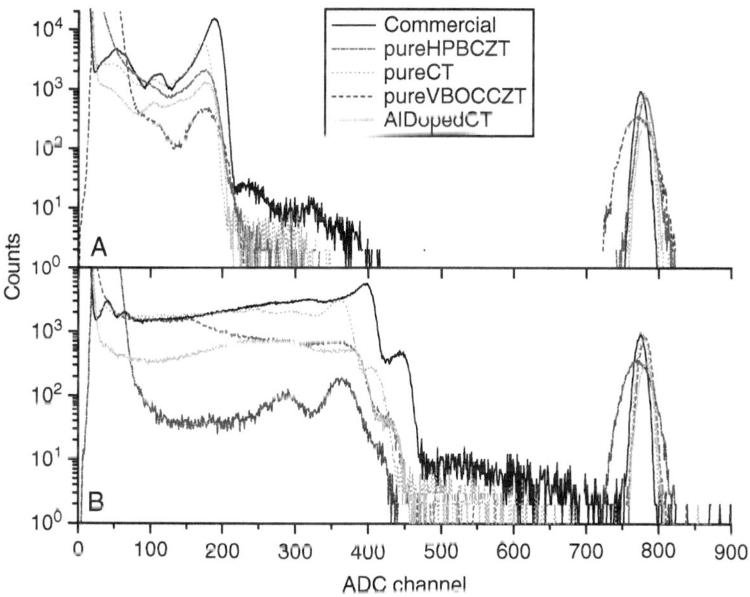

Figure 3 Pulse height spectra for all the samples (A) for the 59.5 keV Am^{241} gamma line and (B) for the Co^{57} source. Electronic test pulses accumulate near channel 800 and the width of these curves are associated with capacitance noise. The measurement conditions are identical for all samples.

Table 4 FWHM of the pulser peak in the pulse height spectra. Large width corresponds to small depletion depths

	Pulser FWHM in channels	
Sample	^{241}Am spectrum	^{57}Co spectrum
Pure HPB CdTe	14.51 ± 0.07	14.47 ± 0.08
Pure HPB CZT	17.07 ± 0.15	15.17 ± 0.09
Pure VBOC CZT	35.4 ± 0.2	35.0 ± 0.3
Al-doped HPB CdTe	16.97 ± 0.19	16.64 ± 0.18
Commercial detector	13.76 ± 0.05	14.10 ± 0.08

Figure 3A shows the detector response to 59.5 keV photons from ^{241}Am and Fig. 3B to 122 and 136 keV photons from ^{57}Co. All pulse height spectra were accumulated under identical source-detector geometries, amplifier settings, and bias voltage and collected for the same live time. Dramatic differences in the detector response were visible. Table 4 displays the pulser FWHMs for Fig. 3A and B. The 59.5 keV ^{241}Am line was well resolved by the pure VBOC CZT sample while its response to the higher 122 keV from ^{57}Co exhibited pronounced low energy tailing and exhibited different intensity escape peaks. The pure CZT samples from HPB and Al-doped CdTe growths produced a 59.5 keV peak but could not resolve the 122 keV photo peak. The data for pure CdTe and CZT suggest small depletion depths. There was not enough active detector volume to completely absorb the 122 keV photons. The electronic pulser width is also directly related to the capacitance of the planar detector (depletion width) [43]. These samples were all close to the same thickness and had similar leakage currents. The lack of depletion indicates the electric field did not extend across the full detector, due to deep levels.

Fiederle et al. [44] and later Krsmanovic et al. [1] developed a numerical model to describe the compensative behavior of CZT grown by physical vapor transport using a model of Neumark [9]. This work did not include lack of depletion in these calculations. It is not clear exactly how to model the field dependence of the trapping cross sections. Calculations were adapted in this study to include deep acceptors and deep donor levels. An excess of shallow acceptors (N_a) over shallow donors (N_d), $\Delta N_s = (N_a - N_d) > 0$ can be compensated by $N_{dd} > N_s$, to produce high resistivity. Similarly, an excess of donors can be compensated by deep acceptors if $N_{da} > \Delta N_s$. Figure 4 shows the calculated electrical resistivity of CdTe and $Cd_{1-x}Zn_xTe$ ($x = 0.1$) as a function of deep donor concentration N_{dd} for a net shallow acceptor excess of $\Delta N_s = 8.8 \times 10^{15}$ cm^{-3}

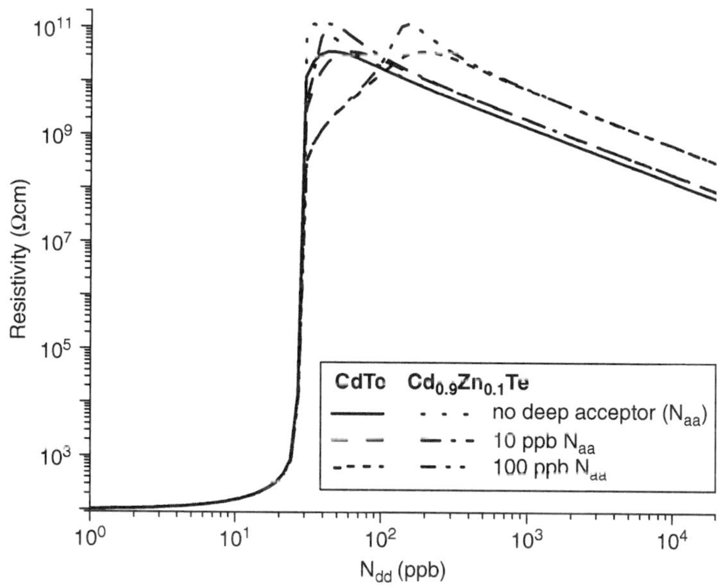

Figure 4 Resistivity versus the concentration of deep donors at 0.76 and 0.82 eV is shown for CT and CZT with 10% Zn, respectively. In addition to a deep donor a deep acceptor N_{aa} is present with two choices of concentrations. The Neumark condition shows that very small changes in deep donor concentration are responsible for large increase in resistivity of the pure material. When the concentration of deep donors exceeds the number of uncompensated acceptors (10 and 100 ppb are used in this model calculation with 0.7 eV), the resistivity rises sharply. The model suggests that further improvements in resistivity are possible if the concentration of deep acceptors is raised to 100 ppb. However, this may cause adverse effects for the mobility-lifetime product.

(300 ppb). Results are shown for three different deep acceptor concentrations $\Delta N_{aa} = 0, 10, 100$ ppb ($0, 3 \times 10^{14}$ cm^{-3}, 3×10^{15} cm^{-3}). Since all but the VBOC sample shows n-type conductivity at room temperature, the electrical compensation of the samples is due to a deep donor. The measured resistivity of the studied samples was between $1.06 \times 10^{9}$and 5.57×10^{9} Ωcm shown in Table 3 which corresponds to a deep donor concentration range of 200 (5.9×10^{15} cm^{-3}) to 600 ppb (1.8×10^{16} cm^{-3}) from Fig. 4. No impurities were found in the samples at such a high concentration, from Table 2. It is suggested that the deep donor level of an intrinsic defect is responsible for the compensation.

To identify and resolve the predicted and theoretical deep levels, DLTS, TSC, and TEES measurements were performed on similar samples as above. Lang first introduced DLTS in the 1970s and it has become a well-established technique for observing deep levels in semiconductors [14, 45–48]. While TEES can observe deep levels in high-resistivity material, it fails with low-resistivity material because the large number of carriers at high temperature dominates any signal from a deep level. In contrast, standard DLTS requires a low-resistivity sample in which there are sufficient carriers to completely fill a trap during a bias pulse. In this way, the two techniques were used in a complementary fashion; TEES was used for high-resistivity samples and DLTS was used for low-resistivity samples for the same material.

In the standard DLTS technique, measuring capacitance transients from a reversed biased diode after a bias pulse monitors the charge state of the deep levels. This pulse temporally collapses the depleted region allowing carriers to fill the traps. An Arrhenius plot of $\ln(\tau_e T^2)$ versus $1/k_b T$ yields a straight line whose slope gives the thermal activation energy of the level. The emission time constant at a single temperature is determined by performing a thermal scan which produces a spectrum peak when the experimental rate window equals the emission time [47]. Although majority and minority carrier traps can be distinguished by the polarity of the capacitance transient, only majority carrier traps are observed when Schottky barriers are used to create the required diode.

Low-resistivity, high-purity samples were studied where the Cd vacancy concentration was higher than intrinsic deep levels. Similar to the high-resistivity samples, two-deep levels were detected and are shown in Fig. 5. One energy level was estimated to be at ~0.79 eV. Another study by Cavallini et al. [17] demonstrated the presence of a similar deep level by DLTS a value of a hole trap 0.75 eV above the valence band. They also found electron trap at 0.79 eV and suggested this was related to the same defect.

TSC and TEES measurements were performed on many samples; however, the results for only four CdTe samples are shown in Fig. 6. The four samples and postgrowth processing conditions are listed in Table 5. The GDMS results for the samples are in Table 6. These spectroscopy measurements generate a small thermoelectric current as a function of the sample temperature at a constant heating rate. A small temperature gradient across the sample is the driving force for the current in these experiments. The peaks occur when previously charged traps thermally release their charge. The measurement is fundamentally polarity sensitive, thus allowing the distinction of electrons from holes with ohmic contacts. Positive currents indicate hole transport and negative currents indicate electron transport, except near the Fermi level. A deep donor level can appear as a hole trap rather than an electron donor state because it lies below the Fermi level or *visa versa*. "Normal" donors with shallow levels are ionized until the sample is cooled. Photo excitation will fill these levels with electrons. In this case, however, the trap holds on to electrons until thermal excitation removes the electron into the conduction band.

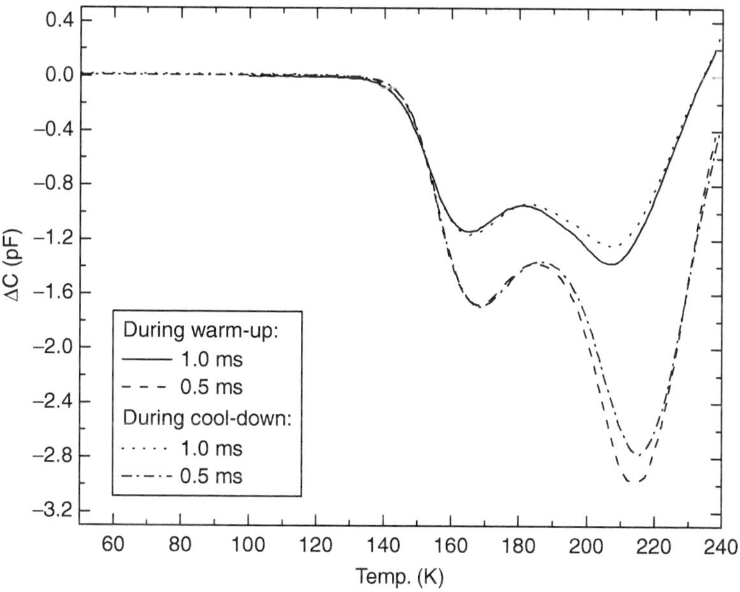

Figure 5 Deep level transient spectroscopy data (DLTS) for a low resistivity high-purity CZT sample with insufficient deep levels to pin the Fermi level in the middle of the gap (note the two-deep levels).

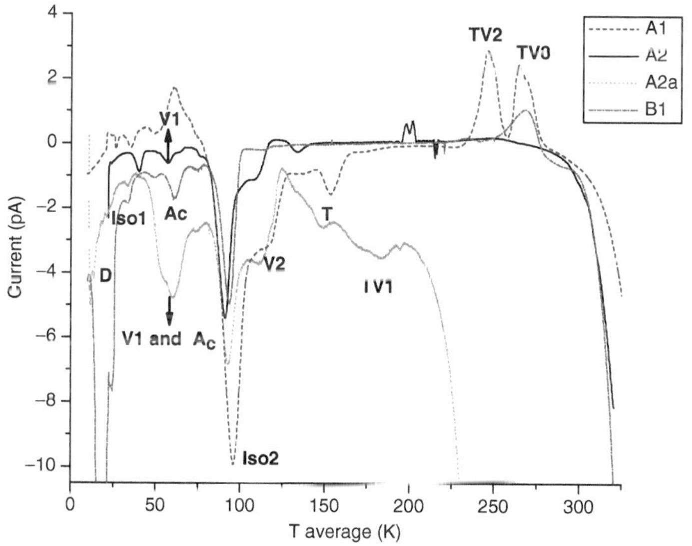

Figure 6 The TEES and TSC data of samples A1 (as-grown high-purity CdTe), B1 (as grown, Al-doped CdTe). A2 was grown and A2a (Al diffused CdTe, annealed at 225 °C for 24 h). We have observed 10 different defect levels: D, Iso1, V1, Ac, Iso2, V2, T, TV1, TV2, and TV3.

Table 5 CdTe samples with their respective growth/postgrowth processing conditions

CdTe sample	Processing condition	Aim of study
A1 (reference sample)	Sample from ingot A (As grown, high-purity undoped CdTe), close to the heel	To characterize the intrinsic defect levels in the high-purity CdTe sample
A2	Sample from Ingot A (adjacent to and taken 1 mm below the position of sample A1)	To compare the effects of before and after Al-diffusion on the intrinsic defect levels in high-purity CdTe
A2a	Aluminum was sputtered on the entire sample A2, then the sample was annealed at ~225 °C for 24 h in Argon	To study the effect of low-temperature Al diffusion on the deep-level defects in high-purity CdTe
B1	Al-doped CdTe	The effect of Al doping on deep level defects in CdTe

Table 6 GDMS data ppb atomic of electrically active impurity atoms for the measured samples (ND: below detection limit)

Sample	Mg	Na	O	Mg	Al	Cu	Ni
HP CdTe "A"	20	27	180	20	39	4	2
HP CdTe "B" Al doped	18	54	282	18	1300	6	24
Sample	S	Ca	Cr	Fe	Se	Sn	In
HP CdTe "A"	14	ND	ND	ND	85	ND	ND
HP CdTe "B" Al doped	52	10	ND	35	ND	ND	ND
Sample	Cl	Ga	Si	K	P	Zn	
HP CdTe "A"	9	ND	5	ND	ND	76	
HP CdTe "B" Al doped	120	ND	9	54	ND	16	

The TEES and TSC data of samples A1 (as grown high-purity CdTe), B1 (as grown, Al-doped CdTe). A2 was grown and A2a (Al diffused CdTe, annealed at 225°C for 24 h) are shown in Fig. 6. Ten different defect levels have been observed: D, Iso1, V1, Ac, Iso2, V2, T, TV1, TV2, and TV3. These levels are associated with D is a donor, and Iso1 and Iso2 are a vacancy complex with oxygen, V1 and V2 are the two levels of the Cd vacancy, Ac is an A-center, and T is antisite. The TV1-TV3* are the antisite complexes. The antisite (T), Te_{Cd} consisted of a group of five neighboring Te atoms and are considered the birth of a precipitate. Some defect levels were specific to one sample and some were found to occur in more than one sample. For example, there are other transitions on the T site but are not consistently seen under various conditions. The thermal ionization energies of the defect levels in samples A1 were calculated using the initial rise method where an exponential function was fitted to the initial slope of the current peaks, as given in Eq. (1).

$$\ln\left(\frac{I}{I_0}\right) = E_{\mathrm{th}}\left(\frac{1}{k_{\mathrm{B}}T_0} - \frac{1}{k_{\mathrm{B}}T}\right) \tag{1}$$

where, I_0 and I are the currents at temperatures T_0 and T, respectively, and k_B is the Boltzmann's constant.

The thermal ionization energy (E_{th}) and trapping cross-section (σ) of all the observed traps levels in sample A1, A2, A2a (A2a A1 diffused, annealed at 225°C in Ar for 24 hrs), as well as for some of the observed traps in sample B1 (Al-doped) were determined by using the variable heating rate method using Equation 2. This equation assumes a negligible re-trapping of the liberated charge carriers, and a high recombination rate that is a slowly varying function of temperature [48].

$$E_{th} = k_B T_m \ln\left(\frac{N_c k_B T_m^2 \nu\sigma}{\beta E_{th}}\right) \tag{2}$$

which can be rewritten as:

$$\ln\left(\frac{T_m^2}{\beta}\right) - \left(\frac{E_{th}}{k_B T_m}\right) \quad \ln\left(\frac{k_B N_c \nu\sigma}{E_{th}}\right) \tag{3}$$

where: T_m (K) is the maximum temperature of the peak, N_c is the density of carriers; β is the heating rate (K/sec); ν is the thermal velocity of the charge carriers, k_B is Boltzmann constant.

$$\sigma = \exp(-Y_{Intercept})\left(\frac{E_{th}}{N_c \nu k_B}\right) \tag{4}$$

In the variable heating rate method, the TEES or TSC measurement is repeated at various heating rates and the $\ln(T_{m^2}/\beta)$ quantity plotted as the function of $1/k_BT_m$ where T_m is the temperature of the current peak maximum. The slope of this curve according to Equation 3 is the thermal ionization energy of the defect level. Once the thermal ionization energy of the defect level is determined, Equation 4 can be used to estimate the capture cross section. The ionization energies of defect levels in the samples A1, A2a, and B1 were determined by the initial slope method and variable heating method.

Energy level, D is a shallow donor due to Al. TEES data in Fig. 6, from A2a and B1 exhibited this level and from the GDMS results in Table 6 and post treatment of A2a from Table 5 explains this level. The A-center level is also in samples B1 and A2a. The A-center (V_{Cd} + Al_{Cd}) has been determined in previous studies, specifically by TEES and has a thermal ionization energy $E_{Ac} = 0.13[01]$ eV above the VBM and the capture cross section was found to be $\sigma_{Ac} = 5.45[1.85] \times 10^{-16}$ cm^2. Similar energies have been previously found by Castaldini et al. [11]. V1 was determined by the A2 spectrum as the 0/+ ionization of the V_{Cd} with $E_{V1} = 0.1[04]$ eV and $\sigma_{V1} = 7.4[1.0] \times 10^{-16}$ cm^2 consistent with point defect cross section. V2 the −/−2 ionization state of the V_{Cd} at 115 K has an $E_{V2} = 0.24[01]$ eV and $\sigma_{V2} = 1.85[05] \times 10^{-16}$ cm^2. The defect energy level assignment for both ionizations of the V_{Cd} is still well undefined from TEES and other contributors [10–13]. The multiple defect levels associated with the V_{Cd} has led to many possible assignments all related in the present interpretation to complexes paired with the V_{Cd}. Emanuelsson et al. and Szeles et al. also observed −1/−2 ionization state of V_{Cd} by EPR ~0.47 eV and TEES ~0.43 eV, respectively [12, 13]. Emanuelsson et al. [12] reported the identification of the Cd vacancy in CdTe by electron paramagnetic resonance (EPR). It was found that the defect has a trigonal symmetry and suggested Jahn-Teller distortion. This suggested that the level observed was the transition of the second ionized states of V_{Cd} and the optical excitation was positioned less than 0.47 eV above the valance band. These results can be reinterpreted by binding the V_{Cd} to the antisite which results in better agreement with the increased cross section determined by the experimental data. It should be noted in deformed samples a broad acceptor level occurs around 135 K which seemed consistent with dislocations being clearly the most likely candidate.

Iso1 level was determined $E_{Iso1} = 0.06[05]$ eV with a trapping cross section $\sigma_{Iso1} = 3.6[07] \times 10^{-16}$ cm^2. This would be the first ionization transition of the V_{Cd} + O_{Te} pair. GDMS supported the possibility of oxygen as all samples contain >90 ppb O. Similar results have been found for shallow acceptor levels associated with oxygen [39, 40]. It should be noted that levels of oxygen determined by GDMS have significant background issues and are sometimes not reported. These particular samples were of the lowest levels of oxygen ever processed. First, principle calculations also support V_{Cd} + O_{Te} pair with a predicted energy level of 0.08 eV [36]. Other complexes involving hydrogen, oxygen, and V_{Cd}

have been discussed in Du et al. and are consistent with our experimental results [24]. It is clear that hydrogen could play a significant role passivating acceptor defects such as the V_{Cd} or in Te clusters. It is also clear that the exact role of hydrogen is not understood; however, in growth experiments at WSU, it was evident that the concentrations of the shallow acceptors levels were reduced when a background of hydrogen gas was used in the ampoule during growth. An optical study and electrical measurement on deformed hydrogenated material found that neutralization of the Cd vacancies was associated with an increase of hydrogen [49]. Du et al. [24] suggested that the O_{Te}-H complex may be a deep level that exhibits amphoteric character and reasonably high concentrations. It is noteworthy that HPB consistently exhibits the lowest oxygen concentration in GDMS analysis (Table 2), as it is usually grown with some hydrogen to reduce the oxygen content. The hydrogen could react with oxygen in the melt and be removed from the material by making H_2O or forming CO or CO_2 with the graphite crucible. This also supports the role of hydrogen in reducing or passivating the amount of V_{Cd} defects in the material. Results on oxygen doping in CdTe thin films [50] have also been found to be in agreement with our results of shallow acceptors. Hsu et al. [50] found that by increasing oxygen doping that the resistivity decreased exponentially due to an increasing shallow acceptor.

The trap Iso2 was found to occur in all the CdTe samples. This defect level had an ionization energy of $E_{Iso2} = 0.19[05]$ eV and a trap cross section $\sigma_{Iso2} = 6.3[2.0] \times 10^{-16}$ cm^2. This level was associated with the $-/-2$ ionization level of the ($V_{Cd} + O_{Te}$) defect pair [20] and has a cross section consistent with a point-like defect.

One level of T was determined to be $E_T = 0.4[03]$ eV and $\sigma_T = 4.9[12] \times 10^{-13}$ cm^2. This level was assigned as the Te_{Cd} which is predicted by Berding [37]. Berding's calculations indicate Te_{Cd} is a double donor with the levels 0.4 and 0.2 eV below the conduction band edge and can be present in significant concentrations in CdTe if the crystal is solidified or annealed under Te rich conditions. Crystals grown for radiation detector applications are typically grown in a Te rich mixture [51]. More than 90% of the Te excess is expected to form Te precipitates during the cool down of the crystals [52]. Te_{Cd} consists of a group of five neighboring Te atoms which might be considered as the early stage of a growing Te precipitate and appears to complex with a vacancy in the early stages. Recent combined EPR and modulated photocurrent suggested that the level was $E_c - 0.20$ eV for the $+/+2$ ionization state of Te_{Cd} and the $0/+$ ionization state of V_{Te} as a donor level at $E_v + 0.2$ eV [29]. These results were supported by positron lifetime measurements. It should be noted that V_{Te} would not expect to play a role since it has high formation energy under Te-rich conditions. It was stated that the EPR photo ionization threshold at 1.4 eV is a direct transition between the valence band and the Te_{Cd} level [29]. At low Cd partial pressures (Te rich), the three most predominant defects are first V_{Cd} followed by the Te_{Cd}, and the complex V_{Cd}-Te_{Cd} (TV2) suggesting a strong binding energy, consistent with our findings [24]. Fiederle et al. [53] also mentioned that a combination of a V_{Cd} and a Te antisite as a possible intrinsic defect, however, also included Ge or Sn as possible candidates. The deep levels observed all showed very large cross sections with values of 10^{-9} to 10^{-12} cm^2 [53]. The stabilized Te_{Cd} was calculated by Du et al. [24] with its level calculated to ~0.4 eV above E_v, shown in Fig. 7B. The Te atom is displaced from the Cd site and moves to the interstitial site. This is significantly lower energy than the second ionized level when the Te is sitting in the Cd site, which was around 1.3 eV above E_v. It was also found that it could have a −2 state (Te_{Cd}^{-2}) and could bind with a Cd vacancy, consistent with TV2 assignment to follow. The complex $Te_{Cd}^{-2} + V_{Cd}^{-2}$ was found to have a binding energy of ~0.47 eV. There is a large relaxation when it combines with the V_{Cd}^{-2} considering the Cd is missing in the nearest neighbor, shown in Fig. 7C. This would be a deep acceptor in the mid-gap around 0.69 eV. Figure 7 displays (A) Te split interstitial and (B) model of the complexes associated with Te_{Cd} without proper relaxation. This work lends further support that the Te antisite alone cannot pin the Fermi level and suggests a possible mid-gap complex of $Te_{Cd} + V_{Cd}$ pair.

Figure 8A presents the TEES data on sample A1 for the level TV2 at various heating rates. This was done by cooling the sample down to 200 K and illuminating it followed by the TEES at various heating rates. In this method, the TV2 deep level (at 245 ± 1 K) was isolated. The energy and capture cross section of this level was calculated from the Arrhenius plot

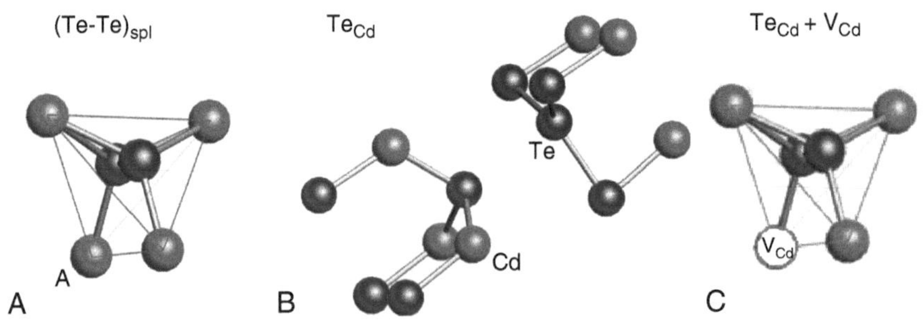

Figure 7 Structures of (A) $(Te\text{-}Te)_{spl}$ and (B) Te_{Cd} and a vacancy complexes at neutral charge states where the transition was calculated to be (0/−2) is near the middle of the gap. There is a large relaxation when it combines with the V_{Cd}^{-2} considering the Cd is missing in the nearest neighbor as shown schematically in (C). This would be a deep acceptor if there was suppression of the Cd vacancies. The four Cd atoms vertices of a tetrahedron around the Te site in (A) are connected by thin gray lines to guide the eye [21]. Copyright American Institute of Physics.

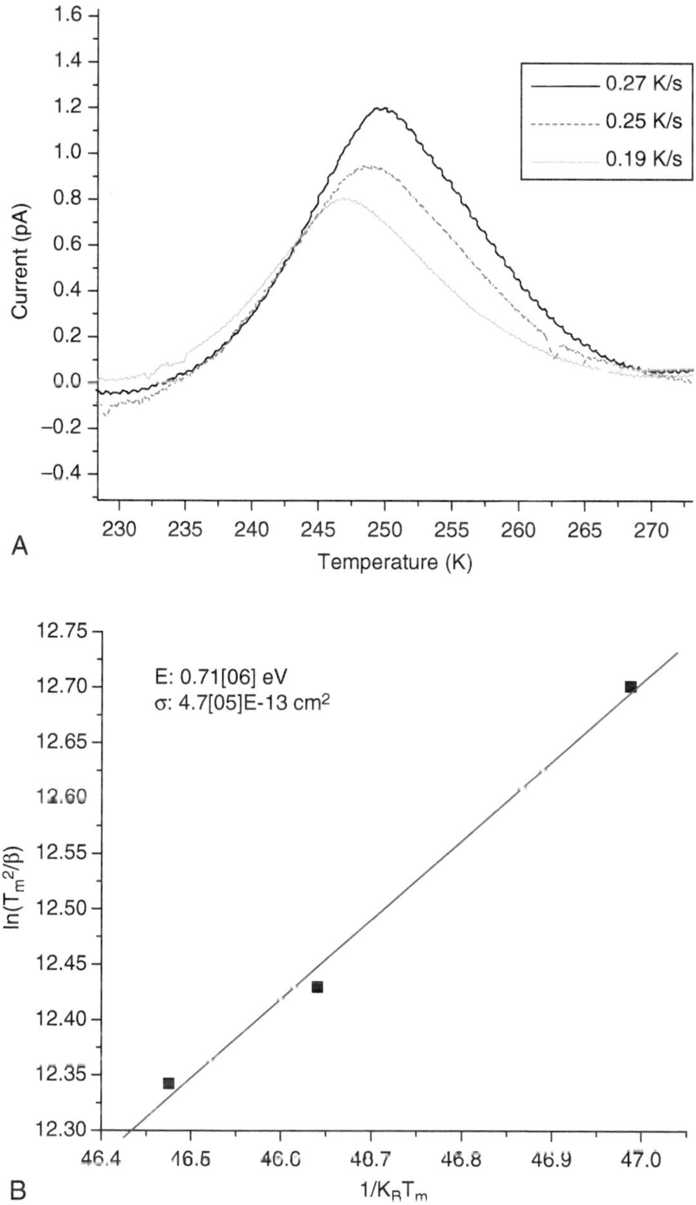

Figure 8 (A) Variable heating rate TEES data on defect level TV2 in sample A1. (B) Arrhenius plot showing the thermal ionization energy and the trap cross section of defect level TV2.

shown in Fig. 8B, $E_{TV2} = 0.7[06]$ eV and $\sigma_{TV2} = 4.7[05] \times 10^{-13}$ cm^2. Similar results were obtained on an earlier study [1] on this high-purity CdTe sample in which the deep level TV2 at (245 $\pm$ 1 K) was attributed to the $-/-2$ ionization level of the cadmium vacancy coupled to an antisite. Since the measured thermoelectric current is positive, the peaks are the result of hole emission. This was consistent with the presence of Al doping in sample B1, A2a, and Ref. [1]. Clearly, the addition of Al reduces the concentration of this defect level and would affect V_{Cd}. It is noteworthy that the TV2 peak has appeared with both positive and negative currents in our experiments, suggesting that the TV2 defect is amphoteric. The thermal ionization energy of the $-/-2$ V_{Cd} acceptor level has been calculated to be 0.24 eV, which is much smaller than the experimental ionization energy of the TV2 deep level, 0.71 eV. Also interesting is the increase of the level T (Te_{Cd}) visible in B1 and A2a from Fig. 6. The authors now suggest the TV2 could be the transition to one of the higher charge state levels of the Tellurium antisite-cadmium vacancy pair ($Te_{Cd} + V_{Cd}$). This conclusion is also supported by values obtained from first principles total energy and band structure calculations [24] and our results. In a separate paper Wei and Zhang [36], found a 1.24 eV energy for a neutral Te_{Cd} and 0.34 for the (+/0) for the donor transition level. In the review by Meyer and Stadler [54], they presented the first ionized state for the Cd vacancy near or possibly in the valence band and the doubly ionized state around 0.2 eV. Further, it was stated that the energy level at 0.78 eV for the doubly ionized state of the Cd vacancy was in contradiction to their EPR results [54].

In a similar method, the second-deep level possibly TV3 (at 260 $\pm$ 5 K) was isolated by cooling the sample to 255 K and illuminating it at 255 K followed by a TEES at variable heating rates. The thermal ionization energy and the trapping cross section of the deep levels were extracted to be $E_{TV3} = 0.79[06]$ eV and $\sigma_{TV3} = 1.58[01] \times 10^{-12}$ cm^2. TV3 is assigned to be the transition to one of the higher charge state levels of the Tellurium antisite-double Cd vacancy pair ($Te_{Cd} + 2V_{Cd}$) or $-/0$ ionization state of ($Te_{Cd} + V_{Cd}$). This complex would be consistent with the larger cross section contrary to the $-/-2$ ionization state of V_{Cd} cross section. When the TV3 peak is mainly present and then Al diffusion at 225°C is performed, the sample can become n- or p-type. The Al will perform one of either two roles. First, in the case for A2a shown in Fig. 6, the sample becomes more n-type, due to the defect $Te_{Cd} + V_{Cd}$ complex. The Al fills the single vacancy and creates a donor site and Te_{Cd} rendering the material n-type. The T level (Te_{Cd}) is visible in Fig. 6 but not clear enough to identify the level in the A2a spectrum. This is due to the difficulty to fill all the levels, as A2a has become low resistivity. The second case for assignment of the $Te_{Cd} + 2V_{Cd}$ complex is when the Al fills one of the

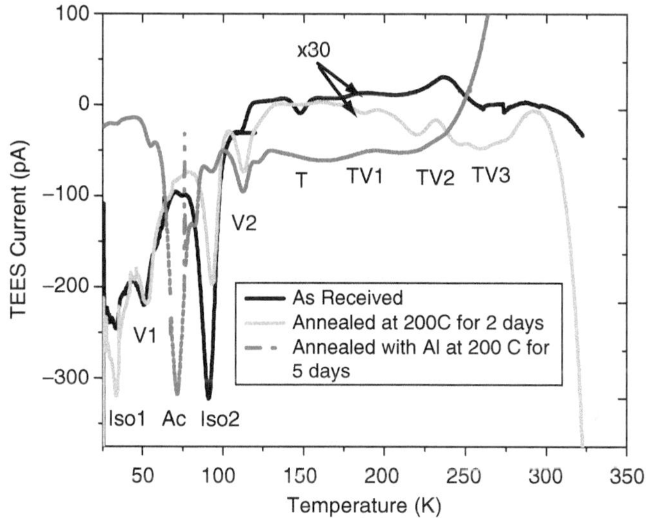

Figure 9 TEES spectra from sample near A1 as grown; control, after annealed 200 °C, after Al sputtered and annealed 200 °C 5 days.

two vacancies creating A-centers. This will result in p-type conductivity which is strong evidence for this assignment. A similar experiment supporting this conclusion of TV3 with two Cd vacancies pair with a Te antisite was performed on a third sample near A1. Three TEES spectra of this sample are shown in Fig. 9. The sample was processed by annealing in Ar at 200 °C for 2 days and remeasured to be sure the defect complexes were not changing with the low-temperature anneal. Next, the sample was sputtered with Al (fully encapsulating the sample to eliminate the loss of Cd) and annealed in Ar at 200 °C for 5 days. The annealing time was extended to allow for the Aluminum metal to diffuse into the sample filling the Cd vacancies. By extending the annealing time of the Al coated sample to 5 days instead of 1 (A2a) demonstrates the changing from n- to p-type shown in Fig. 9. Changing defect levels of the Al annealed sample displays a prominent Ac level rendering the material p-type. The level TV3 is a deep acceptor associated level of ($Te_{Cd} + 2V_{Cd}$) pair or ($Te_{Cd} + V_{Cd}$). It was observed that Al doping during growth did not compensate this deep defect level in as grown material. With aluminum doping, the second-deep level TV2 was reduced, however, the third-deep level TV3 was still present and can be seen in Fig. 6 which provides support to the di-vacancy pair. It is believed that the level TV3 may actually be both levels. The authors have previously assigned as ($Te_{Cd} + 2V_{Cd}$) and ($Te_{Cd} + V_{Cd}$). In many of the TEES spectrums, there is a second peak associated with TV3 as seen in Fig. 6 A1 and Fig. 8 control and Al sputtered with anneal. Defect level TV3* would be identified in these examples. The position of TV3 and TV3* in the TEES spectra was similar due to the relative thermal ionization energies and size of the cross section. To date, the only mechanism to potentially identify between the two defects is the Al diffusion experiment. If the sample contains more single V_{Cd} complexes, the Al sputtered anneal will render the sample n-type and double V_{Cd}, respectively, p-type. Transmission electron microscope studies show that Te clusters can also be neighbored by voids. The formation of secondary phases due to the retrograde solubility provides supporting evidence that complexes are present in grown material [55]. TV1 was determined from the A2a spectrum using the variable heating method. The ionization energy and the trapping cross-sectional area was found, $E_{TV1} = 0.46[03]$ eV and $\sigma_{TV1} = 1.1[01] \times 10^{-14}$ cm^2. This level is assigned to the $0/-$ transition of the ($Te_{Cd} + 2V_{Cd}$) complex and would be in reasonable agreement of the model of Du et al. [24] with the transition ($0/-2$).

Table 7 shows the thermal ionization energy (E_{th}) and trapping cross section (σth) of the observed defect levels in samples A1, A2, A2a, and B1. The thermal ionization energies of various defects and defect complexes in pure and separately doped CdTe and CZT were theoretically calculated using the first principles band structure method [39]. This result is shown in Fig. 1 [2].

Table 7 Thermal ionization energy and thermal trapping cross section of various defect levels in the CdTe samples studied

Trap	Tavg$_{max}$ ± 1 K	Energy (eV)	Trap cross section (cm^2)	Defect level
1	18	–	–	D: Al Donor
2	33	0.06[05]	$3.6[07] \times 10^{-16}$	Iso1: $(V_{Cd} + O_{Te})^{-/0}$ $(V_{Cd} + H)$
3	48	0.1[04]	$7.4[1.0] \times 10^{-16}$	V1: $V_{cd}{}^{-/0}$
4	60	0.13[01]	$5.45[1.85] \times 10^{-16}$	Ac: $(V_{Cd} + Al_{Cd})^{-/0}$
5	92	0.19[05]	$6.3[2.0] \times 10^{-16}$	Iso2: $(V_{Cd} + O_{Te})^{-2/-}$
6	115	0.24[01]	$5.4[3.6] \times 10^{-16}$	V2: $VCd^{-2/-}$
7	147	0.41[03]	$3.6[1.3] \times 10^{-13}$	T: $(Te_{Cd})^{0/+a}$
8	180	0.46[03]	$1.1[01] \times 10^{-14}$	TV1: $(Te_{Cd} + 2V_{Cd})^{-2/0}$
9	245	0.70[06]	$4.7[1.45] \times 10^{-13}$	TV2: $[Te_{Cd} + V_{cd}]^{-2/-}$
10	260	0.79[06]	$1.58[01] \times 10^{-12}$	TV3: $(Te_{Cd} + V_{Cd})^{-/0}$ $(Te_{Cd} + 2V_{Cd})^{+/+2}$

[a]This can be a donor or acceptor $E_c - 0.4$ eV or $E_v + 0.43$ eV from theory [56].

3. CONCLUSIONS

Results show that semiinsulating CdTe and CZT are compensated by intrinsic defects that satisfy the Neumark model. Detector, capacitance and TEES data support that the depletion region width depends on the concentration of Te antisites associated with Cd vacancies. Fully depleted detectors show almost none of these deep levels. Previously reported results indicated that high purity, high-resistivity CdTe/CZT was due to isolated Cd vacancies, which is in conflict with EPR results. The V_{Cd} alone cannot explain the pinning of the Fermi level in the middle of the gap as discussed. The isolated Te antisite, as supported by experimental work and theoretical papers, does not appear to be supported in this mid-gap level. The Neumark model explains a compensation mechanism that accurately describes the dramatic changes in resistivity of high-purity CdTe/CZT. The fabrication profile of CdTe/CZT material will affect the intrinsic defect levels associated with the material. This chapter identified possible deep level defects that render the high-purity materials characteristics. A full set of defect levels and estimates of the cross sections have been listed and compared with previous literature. Many of the experiments have observed different defect levels in as grown, and postgrowth processed CdTe/CZT samples using EPR, PAS, PICTS, DLTS, TEES, and TSC techniques. Some of the defect levels were common to all samples; however, there were some other transitions that have not been included in this work but would be consistent with other transitions of these complexes. The thermal ionization energies and trapping cross sections were estimated for each of the defect levels. The ionization energy values obtained were compared to the theoretical calculations of the transition energy levels of various intrinsic and defect complexes in CdTe/CZT. Based on this comparison, trends observed in other experiments and treatment of the samples, an interpretation of the different defect levels observed experimentally was suggested but clearly is not the definitive answer but is hopefully emerging to a better understanding of the intrinsic defects in CdTe and CZT.

ACKNOWLEDGMENTS

The authors thank the US Department of Energy, NA-22, Contracts DE-FG52-06/27497/A000 and DE-FG52-08NA28769 for their financial support of this research. In particular, KGL wishes to acknowledge Robert Triboulet for much needed prodding, critical suggestions, and corrections. I would also like to thank Roger Saunders for careful reading of this chapter. The authors also thank Raji Soundararajan and Santosh Swain at the Center for Materials Research, Dr Mary Bliss at Pacific Northwest National Laboratory, Csaba Szeles at eV Products, and Dr Su-Huai Wei at the National Renewable Energy Laboratory for their research contributions.

REFERENCES

[1] N. Krsmanovic, K.G. Lynn, M.H. Weber, R. Tjossem, Th. Gessmann, Phy. Rev. B 62 (2000) 16279–16282.
[2] R. Soundararajan, K.G. Lynn, S. Awadallah, C. Szeles, S.-H. Wei, J. Electron. Mater. 35 (2006) 1333–1340.
[3] Cs. Szeles, Y.Y. Shan, K.G. Lynn, E.E. Eissler, Nucl. Inst. Meth. Phys. Res. A 380 (1996) 148–152.
[4] D.S. McGregor, R.A. Rojeski, G.F. Knoll, F.L. Terry, J. East, Y. Eisen, J. Appl. Phys. 75 (1994) 7910–7915.
[5] W.J. Hamilton, D.R. Rhiger, S. Sen, M.H. Kalisher, K. James, C.P. Reid, V. Gerrish, C.O. Baccash, IEEE Trans. Nucl. Sci. 41 (1994) 989–992.
[6] US patent (7067008).
[7] US patent (7192481).
[8] WIPO Patent Application WO/2008/054840.
[9] G.F. Neumark, Phys. Rev. B 26 (1982) 2250–2252.
[10] D.M. Hofmann, W. Stadler, P. Christman, B.K. Meyer, Nucl. Instrum. Methods Phys. Res. A 380 (1996) 117–120.
[11] A. Castaldini, A. Cavallini, B. Fraboni, et al., Appl. Phys. Lett. 69 (1996) 3510–3512.
[12] P. Emanuelsson, P. Omling, B.K. Meyer, M. Wienecke, M. Schenk, Phys. Rev. B 47 (1993) 15578–15580.
[13] C. Szeles, Y.Y. Shan, K.G. Lynn, A.R. Moodenbaugh, E.E. Eissler, Phys. Rev. B 55 (1997) 6945–6949.
[14] A. Castaldini, A. Cavallini, B. Fraboni, P. Fernandez, J. Piqueras, J. Appl. Phys. 83 (1998) 2121–2126.
[15] L.C. Islett, P.K. Raychaudhuri, J. Appl. Phys. 55 (1984) 3605–3612.
[16] E.Y Yee, R.B. James, R.W. Olsen, H. Hermon, J. Electron. Mater. 28 (1999) 766–773.
[17] A. Cavallini, B. Fraboni, W. Dusi, IEEE Trans. Nucl. Sci. 52 (2005) 1964–1967.
[18] Z.C. Huang, E. Eissler C.R. Wie, Nucl. Instrum. Methods Phys. Res. B 100 (1995) 507–510.
[19] M. Hage-Ali, P. Siffert, Nucl. Instrum. Methods Phys. Res. A 322 (1992) 313–323.
[20] M. Fiederle, V. Babentsov, J. Franc, A. Fauler, J.P. Konrath, Cryst. Res. Technol. 38 (2003) 588–597.
[21] G. Tessaro, P. Mascher, J. Cryst. Growth 197 (1999) 581–585.
[22] W. Stadler, D.M. Hofmann, B.K. Meyer, R. Krause-Rehberg, A. Polity, Th. Abgarjan, M. Salk, K.W. Benz, M. Azoulay, Acta Phys. Pol. A 88 (1995) 921–930.
[23] M. Hage-Ali, P. Siffert, in: T.E. Schlesinger, R.B. James (Eds.), Semiconductors and Semimetals, vol. 43, Academic Press, New York, 1995, p. 219.
[24] M.-H. Du, H. Takenaka, D.J. Singh, J. Appl. Phys. 104 (2008) 093521–093529.
[25] M.A. Berding, Appl. Phys. Lett. 74 (1999) 552–554.
[26] C. Barnett Davis, D.D. Allred, A. Reyes-Mena, J. González-Hernández, O. González, B.C. Hess, W.P. Allred, Phys. Rev. B 47 (1993) 13363–13369.
[27] Y.C. Chang, R.B. James, in: R.B. James, R.C. Sherato (Eds.), Hard X-Ray, Gamma-Ray and Neutron Detector Physics and Applications, vol. 3768, SPIE, Bellingham, WA, Phys. Rev B, 1999, p. 381.
[28] P. Fougeres, Thesis, Louis Pasteur University, Strasbourg, 1998.
[29] D. Verstraeten, C. Longeaud, A. Ben Mahmoud, H.J. von Bardeleben, J.C. Launay, O. Viraphong, Ph.C. Lemaire, Semicond. Sci. Technol. 18 (2003) 919–926.
[30] M. Fiederle, D. Ebling, C. Eiche, D.M. Hofmann, M. Salk, W. Stadler, K.W. Benz, B.K. Meyer, J. Cryst. Growth 138 (1994) 529–533.
[31] W. Stadler, D.M. Hofmann, H.C. Alt, T. Muschik, B.K. Meyer, E. Weigel, G. Müller-Vogt, M. Salk, E. Rupp, K.W. Benz, Phys. Rev. B, 51 (1995) 10619–10630.

[32] C. Corbel, L. Baroux, F.M. Kiessling, C. Gely-Sykes, R. Triboulet, Mater. Sci. Eng. B 16 (1993) 134–138.
[33] T.E. Schlesinger, J.E. Toney, H. Yoon, E.Y. Lee, B.A. Brunett, L. Franks, R.B. James, Mater. Sci. Eng. 32 (2001) 103–189.
[34] P. Fougeres, P. Siffert, M. Hageali, J.M. Koebel, R. Regal, Nucl. Instrum. Methods Phys. Res. A 428 (1999) 38–44.
[35] M.-H. Du, H. Takenaka, D.J. Singh, Phys. Rev. B 77 (2008) 094122–094126.
[36] S.-H. Wei, S.B. Zhang, Phys. Rev. B 66 (2002) 155211–155220.
[37] M.A. Berding, Phy. Rev. B 60 (1999) 8943–8950.
[38] S.A. Awadalla, A.W. Hunt, K.G. Lynn, H. Glass, C. Szeles, S.-H. Wei, Phys. Rev. B 69 (2004) 075210–075213.
[39] Y. Marfaing, Thin Solid Films 387 (2001) 123–128.
[40] S.-H. Wei, S.B. Zhang, Phys. State Solidi (b) 229 (2002) 305–310.
[41] D.S. McGregor, G.F. Knoll, Y. Eisen, R. Brake, IEEE Trans. Nucl. Sci. 39 (1992) 1226–1236.
[42] K.G. Lynn, M. Weber, H.L. Glass, J.P. Flint, Cs. Szeles, MRS Symp. Proc. 487 (1998) 319–328.
[43] V. Radeka, IEEE Trans. Nucl. Sci. 21 (1974) 51–64.
[44] M. Fiederle, C. Eiche, M. Salk, R. Schwarz, K.W. Benz, W. Stadler, D.M. Hofmann, B.K. Meyer, J. Appl. Phys. 84 (1998) 6689–6692.
[45] D.V. Lang in: Topics in Applied Physics, vol. 37, Springer-Verlag, Berlin, 1979, pp. 93–133.
[46] D. Kwon, R.J. Kaplar, S.A. Ringel, A.A. Allerman, S.R. Kurtz, E.D. Jones, Appl. Phys. Lett. 74 (1999) 2830–2832.
[47] D.V. Lang, J. Appl. Phys. 45 (1974) 3023–3032.
[48] R.H. Bube, Photoconductivity in Solids, Wiley, New York, 1960, p. 242.
[49] F. Lmai, M. Brihi, Z. Takkouk, K. Guergouri, F. Bouzerara, M. Hage-Ali, J. Appl. Phys. 103 (2008) 084504–084508.
[50] T.M. Hsu, R.J. Jih, P.C. Lin, H.Y Ueng, Y.J. Hsu, H.L. Hwang, J. Appl. Phys. 59 (1986) 3607–3609.
[51] E. Raiskin, J.F. Butler, IEEE Trans. Nucl. Sci. 35 (1988) 81–84.
[52] P. Rudolph, M. Neubert, M. Muhlberg, J. Cryst. Growth 128 (1993) 582–587.
[53] M. Fiederle, A. Fauler, J. Konrath, V. Babentsov, J. Franc, R.B. James, IEEE Trans. Nucl. Sci. 51 (2004) 1864–1868.
[54] B.K. Meyer, W. Stadler, J. Cryst. Growth 16 (1996) 119–127.
[55] M. Duff, SVNL and Chuck Henager PNNL, Private Communication.
[56] S.-H. Wei, RNEL, Private Communication.
[57] I. Turkevych, R. Grill, J. Franc, P. Hoschl, E. Belas, P. Moravec, M. Fiederle, K.W. Benz, Cryst. Res. Technol. 38 (2003) 288–296.

CHAPTER VC2B

Experimental Identification of the Point Defects

P. Fochuk and **O. Panchuk**

1. INTRODUCTION

The development of advanced CdTe devices needs an improvement of our knowledge of its PD structure. In spite of the fact that there are a lot of publications devoted to this problem the situation is not clear even with native PDs. Especially it is complicated in Te-rich part of the phase Cd-Te diagram. Up to now it is not easy to make "sure" choice between V_{Cd}, Te_{Cd}, and Te_i (the designation of the point defects (PDs) corresponds to similar used by Kröger [1]) as dominant species in this area. In this chapter, we divided all native PDs in four groups and consider information in chronological order: donors-acceptors-neutral PDs-antisites. It should be noticed that especially few published experimental results concern the last two groups.

Perfect or ideal is a crystal when for all is atoms exist neighboring and distant order and they take the right position in the ideal crystalline lattice that is typical for this structure. Such situation is theoretically possible only at 0 K. All real crystals, which are in use, differ from ideal ones due to the atom displacements. These crystal structure deviations determine the "structural-sensitive" semiconductor properties.

The process of PDs formation is reversible and at high temperature they can exist in the crystal in thermodynamic equilibrium or not. That equilibrium can be described by the help of the quasichemical defect formation reactions (QCDR) what was at first time proposed by Kröger and Vink [1, 2]. These reactions take place due to defect possibility to migrate inside the crystal.

Chernivtsi National University, 2 vul. Kotsiubinskoho, Chernivtsi 58012, Ukraine

To describe the PD equilibrium in the crystal two main approaches usually are explored. The statistic-thermodynamic approach is based on the full energy distribution function for any rationally chosen defect crystal model. From this function one can get the formula for the full crystal energy and minimizing it to find the equilibrium conditions.

2. CHARGED PDs

2.1. Donor native PDs

Nobel was the first who described the entire CdTe PD structure in details using Hall effect measurements at 300 K [3]. On the basis of obtained results and using Brouwers electro-neutrality equation approximation method [4] he calculated different QCDR constants. On his mind at Cd saturation the dominant native-charged PDs were electrons and compensating these centers—Cd_i^+ PDs, at Te saturation—correspondingly holes and Cd vacancies (V_{Cd}^-). Concentrations are shown in the Table 1 according to Nobel [3].

Constructing approximated PD models by use of free carrier density experimental data Nobel could obtain the values of different PDs concentrations. Comparing the expressions for these values at the boundary of two approximation regions he was able to calculate the equilibrium constants governing the respective QCDRs. Using similar data obtained at different temperatures, he calculated the QCDR enthalpies.

The weakness of Nobel's PDs theory was the use of low temperature (300 K) electrical measurements instead of high temperature ones. His samples at first were annealed at high temperature under defined Cd vapor pressure and then quenched. Nobel assumed that during very fast quenching the high-temperature defect equilibrium (HTDE) should be retained. Thus he used room temperature Hall measurements to obtain high-temperature free electron density values. Nevertheless it should be admitted, that the Nobel's QCDRs thermodynamic data [1, 3], see Table 3, are close enough to later obtained more precise values.

The obtained by Nobel slope value in the free electron density versus Cd vapor pressure dependencies (logarithmic scale) equal to $\gamma = ½$ (Eq. (1)) issued in the conclusion that Cd_i^+ is the main native positive-charged PD at P(Cd, max).

Table 1 Content of some PD (cm^{-3}) in CdTe [3]

Temperature (K)	P_{Cd}	$[e^-]$	$[h^+]$	$[Cd_i^+]$	$[V_{Cd}^-]$
973	Max.	8×10^{16}	5×10^{14}	8×10^{16}	2×10^{14}
	Min.	5×10^{14}	3×10^{17}	10^{14}	3×10^{17}

$$[e^-] = A \times P(Cd)^\gamma, \tag{1}$$

The location of the Cd_i^+ energetic level in the gap ($E_d = E_C-(0.01\text{-}0.02)$ eV) was determined experimentally by Nobel [3] and later by Segall et al. [5]. They measured the Hall effect both in high purity and doped CdTe samples. Cd_i^+ as the dominant donor PD at high temperature was reported in conductivity measurements by Matveev et al. [6] though more sure conclusions follow from Hall effect measurements [7]. As was first showed by Whelan and Shaw [8] in direct high-temperature conductivity measurements in carefully purified CdTe samples at 760-1220 K in Eq. (1) the slope γ in reality equals to 1/3.

According to the QCDR theory [1] the $\gamma = 1/3$ value is realized only if the native donors are doubly charged: Cd_i^{2+} or V_{Te}^{2+}. Also the same authors defined the incorporation energy of the double-charged native donors into the CdTe lattice $-(1.65 \pm 0.15)$ eV and the ionization energy of the singly ionized intrinsic donor $-\leq 0.24$ eV.

Later the domination of double-charged native donors above 773 K was confirmed by direct determination of electron density at HTDE by Zanio et al. [9] and Smith [10]. Rud' and Sanin [11] observed also the $\gamma = ⅓$ slope in conductivity measurements. As in Ref. [8] he determined the incorporation energy of double-charged native donors (Cd_i^{2+}) into the CdTe lattice $-(1.1 \pm 0.15)$ eV by temperature measurements of conductivity and estimated the second ionization energy of this PD $-\leq 0.21$ eV. This value (1.1 eV) is essentially less than that proposed by Nobel (2.28 eV) and Whelan (1.65 eV) [8]. Its use in PD structure calculations gives poor coincidence with experiment. The reason is in the necessity to use Hall data only instead of conductivity (Fig. 1).

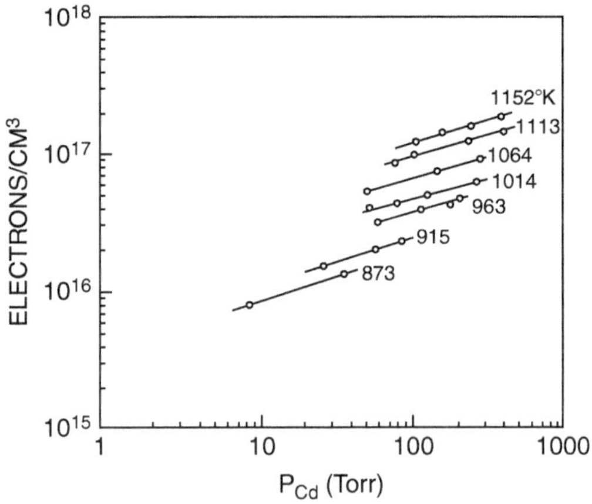

Figure 1 Dependence of electron density $[e^-]$ on P_{Cd} in CdTe [10]. The all isotherms slope is equal to 1/3.

Table 2 Relation $[Cd_i^{2+}]/[V_{Te}^{2+}]$ at different temperatures [12]

Temperature (K)	1170	1070	970
$[Cd_i^{2+}]/[V_{Te}^{2+}]$	3:1	1.3:1	1:2

The subsequent Hall effect experiments at HTDE performed by Chern in undoped CdTe [12], confirmed the results obtained in [9–11]. He also observed the $\gamma = 1/3$ slope (Eq. (1)) in the whole investigated temperature range (870-1150 K). Therefore, he suggested that under the Cd saturation as the dominant double-charged donor PD can act both Cd_i^{2+} and V_{Te}^{2+}. Using the results of a component self-diffusion study in CdTe [13] Chern defined the conditions where every of these PDs becomes apparent (Table 2): V_{Te}^{2+} dominates below 800 °C, and Cd_i^{2+}—above 800 °C.

Later Fochuk et al. [14] performed high temperature (500-1200 K) *in situ* Hall effect measurements at HTDE under well-defined *P*(Cd) in CdTe single crystals grown by different techniques. In all samples and at all investigated temperatures above 770 K the γ value in Eq. (1) was equal to ⅓ that corroborates with previous results [9–12]. To verify Chern's suggestion concerning different donor domination temperature areas, these authors annealed CdTe samples at 970-1170 K at maximal *P*(Cd) with following quenching into cold water. After quenching the free electron density $[e^-]$, which is typical at high temperature in undoped CdTe, reduced more than in 20-30 times, whereas after annealing at lower *T* values and subsequent quenching this reducing was lower in times only, not in decades. For explanation it was taken into account the much higher Cd interstitials' mobility, than that of Te vacancies, which are essential in the decay of the donor PDs solid solution at sample cooling. During the quenching the solubility of interstitial Cd in CdTe decreases and the Cd_i^{2+} quickly migrate into precipitates:

$$nCd_i^{2+} + 2ne^- \leftrightarrow (Cd_i^{o})_n \tag{2}$$

Therefore, the $[e^-]$ sharply falls. So, it was supposed that at high temperature ($\geq\sim$970 K) dominate Cd interstitials and below this temperature—the Te vacancy, whose precipitation at temperature lowering is not so intensive as for Cd_I species due to significantly lower diffusivity of Te atoms in CdTe [13].

The regions where every of these PD prevails were defined from the temperature dependencies of the free electron density measured at constant *P*(Cd) (0.001, 0.01, 0.1, and 1 atm) [14]. Two different slopes (Fig. 2) at every constant *P*(Cd) were observed: smaller at low temperature and larger at higher temperature. They can be attributed to the two-mentioned donor PDs. Their formation enthalpies were determined

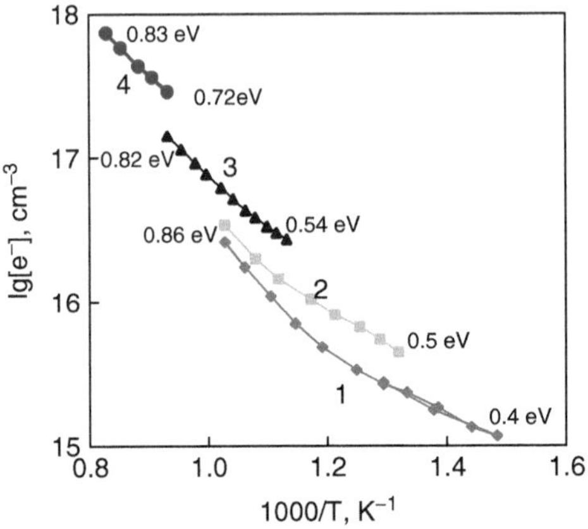

Figure 2 Temperature dependence of the electron concentration in undoped CdTe at constant *P*(Cd). 1—0.001 atm, 2—0.01 atm, 3—0.1 atm, 4—1 atm [14]. The activation energy (in eV) is shown for the beginning and the end of every curve.

from temperature dependencies of electron concentration (Fig. 2) using the Arrhenius equation ($\Delta H_{V\,{}^{2+}_{Te}} = 1.3$ eV; $\Delta H_{Cd\,{}^{2+}_{i}} = 2.5$ eV). It was shown that V^{2+}_{Te} evidently dominates at low temperatures (under some 900 K) whether Cd^{2+}_i begins to prevail at $T > 930$ K.

In his review devoted to the PD structure in II-VI compounds, commenting optical investigations' results, Taguchi cites the ionization energy for $V^{+}_{Te} - E_C - 0.018$ eV that was obtained by him earlier (reference in the signature of Table 2.3 in Ref. [15]).

Meyer et al. [16, 17], studying CdTe crystals, doped by III and VII group elements, and annealed in vacuum or argon gas (650-1050 K), established by EPR that V^{+}_{Te} has a cubic symmetry and the unpaired electrons are uniformly distributed among all four neighboring Cd atoms. However, the use of undoped CdTe in such experiments of highest purity seems to be preferred.

Ye and Chen [18] studied the location of deep levels in the gap by spectroscopy investigations. The $E_C - 0.61$ eV level he found in undoped CdTe was attributed to Cd^{2+}_i. This value differs from $E_d \approx E_C - 0.2$ eV that was proposed by other authors [8, 11] and is commonly used by many scientists in PD structure calculations.

Allen analyzed the validity of the values of V_{Te} energetic level location in the gap obtained by different spectroscopic methods [19]. The conclusion was made that the V^{o}_{Te}/V^{+}_{Te} level is located, most probably only at 0.2 eV above the valence band (citation in Ref. [16]) and the $[V^{+}_{Te}] \approx 10^{15}$ cm^{-3} at photoexitation.

2.2. Acceptor native PDs

To acceptor native, PDs belong Cd vacancies and Te interstitials.

Lorenz [20–21] investigated the optical absorption and Hall effect measurements in Te-saturated CdTe crystals and revealed a double-charged acceptor with an $E_C - 0.06$ eV level. He attributed it to the V_{Cd}^{2-}. However, the position of this PD near the middle of the gap looks more reliable.

Ivanov studied local centers in CdTe by electroabsorption and determined the position of the V_{Cd}^{-} level in the gap as $E_V + 0.05$ eV [22]. By PL investigations in CdTe crystals Agrinskaya et al. [23] found acceptor levels near $E_V + 0.06$ eV and attributed it to V_{Cd}^{-}, too. The PL measurements in CdTe, doped by Cl, allowed her to determine the ionization energy of native acceptor (V_{Cd}^{-}) as $E_V + 0.07$ eV [24]. Both values are close to $E_V + (0.05\text{-}0.06)$ eV that is mostly used as the first V_{Cd} ionization energy.

Taguchi and Ray [15] pointed out the next values of PD energetic levels location in the gap using the data of optical measurements: $V_{Cd}^{-} - E_V + 0.06$ eV, $V_{Cd}^{2-} - E_V + \sim 0.6$ eV (see Table 2.3 in Ref. [15]).

The latter PD was identified by Emmanuelson et al. [25] by EPR studies at 25 K in samples that were annealed at 1020 K under Te saturation during 5 h. The defect possesses trigonal symmetry, and the hole, trapped by this PD, is strongly localized at one of four neighboring Te atoms. Photo-EPR measurements demonstrated that the acceptor level $2-/-$ is located in the gap lower than $E_V + 0.47$ eV. It is necessary to note that Bardeleben et al. [26] are in some doubt about this value.

Jasinskaite et al. [27] using high-temperature Hall effect investigations in CdTe⟨Cl⟩ crystals determined the V_{Cd}^{2-} enthalpy formation ($E_{VCd} = 1.1$ eB) according to the reaction:

$$1/2\mathrm{Te}_2(\mathrm{g}) \leftrightarrow \mathrm{Te}_{\mathrm{Te}}^{\mathrm{o}} + \mathrm{V}_{\mathrm{Cd}}^{2-} + 2\mathrm{h}^{+} \tag{3}$$

Szeles et al. [28] revealed a deep level at $E_V + 0.43$ eV in $Cd_{1-x}Zn_xTe$ ($x = 0.12$) crystals, grown by the high-pressure Bridgman (HPB) method and annealed under Argon at 860 K during 4 h, using TEES and TSC measurements. He supposed that this level can belong to the Cd or Zn vacancy but the final conclusion is absent.

Using PICTS investigations, Fiederle found in CdTe several levels with different energies [29]. The $E_V + 0.43$ eV level was attributed to V_{Cd}^{2-}, while $E_C - 0.54$ eV—to the Cd_i^{2+}. The first value is in agreement with other authors' data [16], but the second is in 2-2.5 times larger.

Positron annihilation spectroscopy was used to study Cd vacancies in CdTe, too, since they can interact with positrons [30–32]. Both the average lifetime of positrons ($\tau \approx 285$ ps) and its temperature dependency were established. Because it is an indirect method of $[V_{Cd}]$ determination it did not allow up to now to get the real [V_{Cd}] values or any thermodynamic parameters.

Investigating CdTe crystals, grown under different deviation from stoichiometry, Rudolph et al. [33] concluded about Cd vacancies domination under Te saturation and Te_i^+—under Cd saturation. If the first suggestion is generally accepted, the second one is proposed by this author only and looks as a misprint because in Ref. [34] he considered Te_i^- as one of the PD existing at Te excess.

Separately it is possible to discuss the experimental data for Te interstitial PD. The fullest references according to it have been gathered by Chern et al. [12]. He supposed that Te can produce two acceptor PD in CdTe matrix: the first is Te_i as a single-charged acceptor and the second Te_i as a double-charged acceptor (according to Lorenz [20, 21]). The latter was not demonstrated by later publications therefore we did not put it in the Table 3. Lyahovitskaya et al. [35] suggested that Te interstitials are donors. In our mind, it is more reliable to take into account the Chern's supposition [12] about Te interstitials as a single acceptor PD Te_i^-. We could not find any information about its experimental investigation. However, Te_i^- can play an important role at Te excess in the CdTe lattice.

2.3. Neutral PDs

Both Cd and Te vacancies or interstitials are possible native neutral PDs in the CdTe lattice. Up to now no one reliable method was proposed for their direct investigation. Among indirect methods precise determination of the lattice period and measurements of prevalent component mass after temperature treatment was proposed for this purpose [36].

Kharif studied the Cd side of the homogeneity region boundary using Cd mass determination that educes from crystals, saturated by Cd and annealed at different temperatures and *P*(Cd) [36, 37]. He found that at constant temperature the concentration of native neutral PD is proportional to $P_{Cd}^{1/3}$ whereas the Cd solubility in CdTe varies from 10^{19} to 2×10^{18} cm^{-3} (at 800 and 1300 K, correspondingly). The main PDs responsible for this phenomenon are proposed to be associates of Cd interstitials and Cd divacancies ($Cd_i^o V_{Cd}^o V_{Cd}^o$), ($Cd_i^+ V_{Cd}^- V_{Cd}^o$), ($Cd_i^{2+} V_{Cd}^- V_{Cd}^-)^o$, ($Cd_i^{2+} V_{Cd}^{2-} V_{Cd}^o)^o$. However, it is not simple to understand both the stable existence of Cd interstitials close to a Cd vacancy and the nature of attraction force between the two (nominally neutral) PD species.

2.4. Antisite PDs

Antisite PDs are meant by native atoms which are located in places that belong to another component. In the ionized state they can form the following centers: Te_{Cd}^+, Te_{Cd}^{2+}, Cd_{Te}^-. Although another assumption can be found, too: Cd_{Te}^+ and Te_{Cd}^- [38, 39].

Table 3 The location of PD energetic levels in the gap determined by different technique in experiments

PD	Method of investigation	Level (eV)	Year	Authors	References	Notes
Cd_i^+	Hall effect	$E_C - 0.02$	1959	Nobel	[3]	
	Hall effect	$E_C - 0.01$	1963	Segall	[5]	
Cd_i^{2+}	HT conductivity	$\leq E_C - 0.24$	1968	Whelan	[8]	or $[V_{Te}^{2+}]$
	HT conductivity	$\leq E_C - 0.21$	1971	Rud′	[11]	
	DLTS	$E_C - 0.61$	1990	Ye	[18]	
	PICTS	$E_C - 0.54$	1994	Fiederle	[29]	
V_{Te}^+	–	$E_C - 0.018$	1983	Taguchi	[15]	
	EPR	$E_V+0.2$	1992	Meyer	[16]	
V_{Te}^{2+}	HT conductivity	$\leq E_C - 0.24$	1968	Whelan	[8]	or $[Cd_i^{2+}]$
V_{Cd}^-	Electroabsorption	$E_V+0.05$	1971	Ivanov	[22]	
	PL	$E_V+0.06$	1971	Agrinskaya	[23]	
	PL	$E_V+0.07$	1987	Agrinskaya	[24]	
	Optical measurements	$E_V+0.06$	1983	Taguchi	[15]	
	Admittance spectroscopy	$E_V+0.61$	2001	Gilmore	[65]	
V_{Cd}^{2-}	Hall effect	$E_C - 0.06$	1963	Lorenz	[20]	
	Optical measurements	$E_V+0.6$	1983	Taguchi	[15]	
	EPR	$\leq E_V+0.47$	1993	Emmanuelson	[25]	
	TEES, TSC	$E_V+0.43$	1997	Szeles	[28]	
	PICTS	$E_V+0.43$	1994	Fiederle	[29]	
	TEES	$E_V+0.735$	2000	Krsmanovich	[62]	

(*continued*)

Table 3 *(continued)*

PD	Method of investigation	Level (eV)	Year	Authors	References	Notes
Te_i^-	–	E_V+0.15	1959	Nobel	[4]	
			1975	Chern	[12]	
Te_{Cd}^+	–	$E_C - 0.56$	1986	Maksimovskij	[39]	
	TDCM, PICTS	$E_C - 0.75$	1998	Fiederle	[52]	
	Hall effect	$E_C - 0.01$	2001	Chu	[55]	
Te_{Cd}^{2+}	Hall effect	$E_C - 0.75$	2001	Chu	[55]	
	Modulated photoconductivity, EPR	$E_C - 0.20$	2003	Verstraeten	[61]	
	TEES	E_V+0.743	2000	Krsmanovich	[62]	

TDCM, time-dependent charge measurements; PICTS, photo-induced current transient spectroscopy; DLTS, deep level transient spectroscopy; EPR, electron paramagnetic resonance; TSC, thermostimulated current.

At first, based on experimental results, idea about existence in sufficient quantities of antisite PDs in CdTe was proposed by Martynov and Kobeleva [40] and later supported by Maksimovskij and Kobeleva [41]. The authors suggested that the revealed $E_d = E_c - 0.56$ eV donor level belongs to Te^+_{Cd} and the hole density values in the Te homogeneity region boundary are in good agreement with Chern's calculations [12] taking into account the results in Ref. [40]. At the same time it was assumed that the Te^+_{Cd} content is much higher than that of $[Te_i]$ and $[V_{Cd}]$ and strongly depends on P_{Cd}:

$$[Te^+_{Cd}] = 1.3 \times 10^{15} \exp(-6.45 \times 10^4/T) \times P^2_{Cd} \quad (4)$$

However, it is not clear how the $[h^+]$ experimental values were obtained because in undoped CdTe at high-temperature measurements the samples usually reveal n-type conductivity.

A great influence on the CdTe PD structure conceptions had Berding's models. Berding was the first who executed *ab initio* theoretical calculations and concluded the possible existence of Te_{Cd} PDs in CdTe [42]. Her results suggested that Te_{Cd} dominates in CdTe at low P(Cd) because Te_{Cd} has a lower formation enthalpy than V_{Cd} (2.29 comparing to 4.70 eV). Based on these assumptions, the hypothesis of many scientists, that explain CdTe properties with the regard for Te_{Cd}, are based up to now. In her latest works [43, 44] Berding moves this defect on the second place (after V_{Cd}) at 970 K and high Te vapor pressure. However, at Cd saturation on the second position after Cd^{2+}_i another antisite PD appears—Cd_{Te}. The concentration of both antisite PDs is sufficient and generally exceeds the content of native PDs. Elementary evaluations of the logarithmic $[e^-]$ versus P(Cd) slope result in a +1 value, whereas in reality it equals to +1/3.

Similar *ab initio* calculations of PD concentrations were performed also by Chen et al. [45, 46]. However, there is an essential disagreement between these calculations and high-temperature measurements. In Ref. [47], Brebrick used Wienecke's data [48] and obtained the next results: in Cd saturated CdTe dominates V^{2+}_{Te} ($E_C - 0.03$ eV); at Te excess—Te^+_{Cd} ($E_C - 0.05$ eV) or Te^-_i. These results coincide with Ref. [10] but disagree with Ref. [49]. However, there are very few experimental results that directly can confirm Berding's calculations. Also they are in variance with high-temperature Hall effect investigations results [48]. Even if one supposes that the two dominant PDs in the Te saturation part of the CdTe T-x diagram compensate one another $[Te^+_{Cd}] = [V^+_{Cd}]$, the hole density has to depend on the Te vapor pressure [50], which is not the case in experiments (Jasinskaite only reported such unconfirmed later dependence in undoped CdTe after additional purification in the Cd + CdTe melt [51]). It is supposed that the Te^+_{Cd} donor PD compensates the acceptor ones Te^-_i

and/or V_{Cd}^{-} what should explain the hole density independence upon P_{Te2}. However, this assumption is not in accord with calculations based on the Krögers theory [1]. Indeed, computer simulations of the CdTe native PD structure versus P_{Cd} value show that at P_{Te2} rising the logarithmic slope for both Te_i^{-} and V_{Cd}^{-} PDs is equal to $+1/4$, whereas for the hypotetic Te_{Cd}^{+} this value equals to $+1/2$ [50]. This signifies that the latter donor PD can not efficiently compensate the former acceptor PDs (that could occur at conditions of identic slopes versus P_{Te2} and close formation enthalpies of the native donor and acceptor PDs what seems unlikely). Anyway the occurrence of the antistructural PD Te_{Cd}^{+} is rather questionable.

On the basis of deep level spectroscopy and conductivity measurements in CdTe⟨Cl⟩ crystals, Fiederle et al. [52] assumed the existence of a deep donor level ($E_C - 0.75$ eV) in the middle of the gap attributing it to Te_{Cd}. In that case only and taking into account the chlorine segregation coefficient it will be possible to explain such high resistivity as $\rho \geq 10^8$ Ω cm (Fig. 3). However, not in all cases the $E_C - 0.75$ eV level is observed and there are other phenomena responsible for high resistivity.

Matveev and Terent'ev [53, 54] explained the nature of the p-n transition that appears at Te vapor pressure reducing during annealing, by the Te_{Cd} formation, resulting in [V_{Cd}] diminishing and misbalances the precise self-compensation mechanism in CdTe⟨Cl⟩. Such supposition does not correspond to the QCDR theory.

The origin of the $E_C - 0.01$ eV level that appears at CdTe and CdZnTe crystals growth, Chu et al. [55] explains by the existence of singly ionized

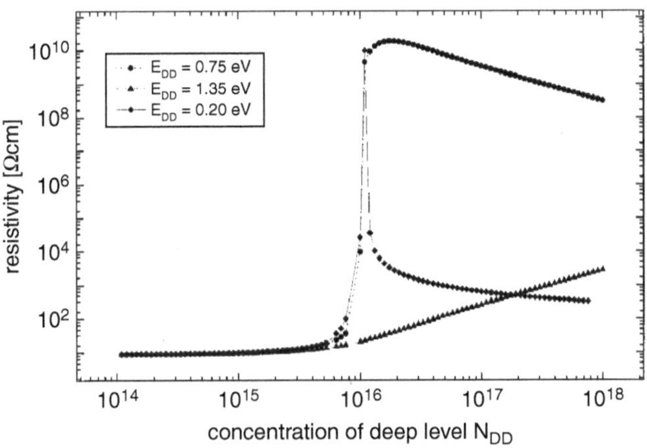

Figure 3 CdTe resistivity dependency at 300 K on deep donor content and its location in the gap [52].

Te_{Cd} donor. In the same time for the $E_C - 0.75$ eV level should be responsible the double-charged Te_{Cd} donor. It was assumed that the growth conditions (Te excess) do not favor to "classical" shallow donor formation—Cd_i^{2+} and V_{Te}^{2+}, therefore, the revealed levels ($E_C - 0.01$ and $E_C - 0.75$ eV) belong to Te_{Cd}^{+} *i* to Te_{Cd}^{2+}. Such suggestion can be considered as very relative.

A comprehensive study of the defect structure based on high-temperature galvanomagnetic measurements and the shape of *T-x* projection of the *P-T-x* phase diagram was reported in a couple of papers [56–60]. The authors conclude that all experimental data taken into account can be described with a good precision by a defect model involving Cd_i, V_{Cd}, and Te_{Cd} PDs only. The electrical properties of Te_{Cd} could not be deduced within their approach and variant defect models are analyzed [60].

By means of EPR and modulated photoconductivity Verstraeten et al. [61] in as-grown and annealed CdTe⟨V⟩ crystals determined the +/2+ level for the antisite $Te_{Cd}^{2+} - E_C - 0.20 \pm 0.01$ eV. This value is much less than proposed by Fiederle et al. [52] for the first ionization level.

The influence of the two native PDs on self-compensation processes in semiinsulating CdTe and CdZnTe crystals, grown by the classical Bridgman and HPB method was revealed by TEES investigations [62]. One of them—with a $E_V + (0.735 \pm 0.005)$ eV thermalization energy—was attributed to V_{Cd}^{2-}, the second—to Te_{Cd} $\{E_V+(0.743 \pm 0.005)\}$ eV, that together with the Cd vacancy forms an associate.

In Refs. [63, 64], Babentsov, on the basis of experimental and calculated data, determined the full energy dependency on the Fermi level position for eight native PD. He supposes that for formation and transformation of V_{Cd} are responsible two reactions:

$$Cd_{Cd}^{0} \leftrightarrow V_{Cd} + Cd_i \quad (5)$$

$$(Te_{Te}^{0}) + V_{Cd} \leftrightarrow V_{Te} + Te_{Cd} \quad (6)$$

The authors suggest that the most likely deep native PD, which can stabilize the Fermi level near the middle of the gap, is Te_{Cd}. Unfortunately, the choice of selection for calculating defect ionization energies among many possible is not argued enough, and they differ in several times.

A very large value for V_{Cd}^{-} first ionization energy (0.61 eV) was determined by Gilmore et al. [65] from admittance spectroscopy of CdTe crystals.

There are several reviews [15, 19, 66–70] where the PD structure in II-VI compounds is characterized. However, there are few data only concerning the experimental determination of the thermodynamic parameters of native PD in CdTe.

At present one can observe significant discrepancies between concerning the positions of native PD levels in the CdTe gap (Table 3). Some values differ from others in several times. That does not allow predicting the real CdTe PD structure both at high and at low temperatures. Besides, many authors assumed the existence of large quantities of antisite PDs in CdTe what was not confirmed by direct experiments up to now.

The electrical properties of the CdTe crystals are determined both by the PDs ionization energy (given by the location of the PD level in the gap) and the concentration of the atomic PDs in the lattice. The PDs content depends on their formation energy. Table 4 shows the available literature thermodynamic parameters (formation entropy and enthalpy).

Conventionally one can consider the location of Cd_i^+ and V_{Cd}^- in the gap $E_C-0.01$ and $\sim E_V + 0.06$ eV, respectively. For Cd_i^{2+} the value $\sim E_C-0.2$ eV seems more reliable instead of $\sim E_C-(0.5\text{-}0.6)$ eV. The data for V_{Te}^+, V_{Te}^{2+}, Te_{Cd}^+, and Te_{Cd}^{2+} are contradictory. The double-charged Cd vacancy can be attributed to very deep acceptors with $E_V + (0.43\text{-}0.75)$ eV. The difficulties in interpretation of experimentally obtained PD ionization values consist in attributing them to respective PDs.

At medium P(Cd) values, where CdTe congruent sublimation occurs, HTDE measurement is difficult to perform due to sample instability. On the other hand, at minimal P(Cd) values, that is under Te saturation, the PD structure is rather unclear, including the formation constants of the main PDs—V_{Cd}^- and Te_i^-, (K_{11}, K_{12}) which should be corrected.

The formation constants, shown in the Table 4, allow to obtain a good agreement with experimental results mainly for the Cd-rich part of the Cd-Te phase diagram. The part of them concerning the native donors (K_8, K_7) was specified experimentally in recent time [14], the others by mathematical optimization methods using the full electro-neutrality condition.

The main problem in native PDs study is the absence of direct investigation methods. The presence of foreign impurities (including oxygen) in the purest CdTe crystals is not less than 10^{14}-10^{15} atm/cm^3 and leaves mainly one way to study them: the high-temperature Hall effect measurements, when the charged free carrier density is above this value. Then using experimental dependencies of $[e^-]$ on temperature and component pressure it is possible to adjust them with calculated models.

Using the native PD formation constants set (Table 4) the modeling of the PDs concentration temperature and Cd vapor pressure dependencies in undoped CdTe become possible [14]. Also in combination with the In incorporation into the CdTe lattice constants it was possible to explain the dependence of electron density in CdTe:In crystals on temperature, dopant content, and P(Cd) [71–73].

Table 4 Quasichemical reaction for native PDs and their formation thermodynamic parameters

No	QCDR reaction	Equilibrium constant[a]	$K_i^0 = \Delta S/k$	ΔH (eV)	References
1	$0 \leftrightarrow e^- + h^+$	$K_1 = [e^-][h^+]$	(a) 4.58×10^{40}	(a) 1.73	[4]
			(c) 8.24×10^{40}	(c) 1.32	[10]
			(d) 3×10^{40}	(d) 1.60	[14]
2	$Cd(v) + V_{Cd}^o \leftrightarrow Cd_{Cd}^o$	$K_{16} = [V_{Cd}^o]^{-1} P_{Cd}^{-1}$	(a) 5×10^{-27}	(a) −3.32	[4]
3	$Cd_{Cd}^o + e^- \leftrightarrow Cd(v) + V_{Cd}^-$	$K_{11} = [V_{Cd}^-][e^-]^{-1} P_{Cd}$	(a) 9.8×10^7	(a) 2.08	[12]
			(b) 1.6×10^5	(b) 1.54	[70]
			(c) 8×10^5	(c) 2.08	[14]
4	$Cd_{Cd}^o + 2e^- \leftrightarrow Cd(v) + V_{Cd}^{2-}$	$K_{10} = [V_{Cd}^{2-}][e^-]^{-2} P_{Cd}$	(a) 2.4×10^{-14}	(a) 0.88	[12]
			(b) 2×10^{-15}	(b) 1.28	[14]
5	$Cd(v) \leftrightarrow Cd_i^o$	$K_{18} = [Cd_i^o] P_{Cd}^{-1}$	1×10^{10}	0.95	[4]
6	$Cd(v) \leftrightarrow Cd_i^+ + e^-$	$K_{17} = [Cd_i^+][e^-] P_{Cd}^{-1}$	1×10^{30}	−0.81	[4]
7	$Cd(v) \leftrightarrow Cd_i^{2+} + 2e^-$	$K_9 = [Cd_i^{2+}][e^-]^2 P_{Cd}^{-1}$	(a) 6.16×10^{61}	(a) 2.28	[12]
			(b) 1×10^{62}	(b) 2.5	[14]
8	$Cd(v) \leftrightarrow Cd_{Cd}^o + V_{Te}^{2+} + 2e^-$	$K_8 = [V_{Te}^{2+}][e^-]^2 P_{Cd}^{-1}$	(a) 7.78×10^{57}	(a) 1.47	[12]
			(b) 7.76×10^{57}	(b) 1.07	[70]
			(c) 3×10^{57}	(c) 1.3	[14]
9	$CdTe + e^- \leftrightarrow Cd(v) + Te_i^-$	$K_{12} = [Te_i^-][e^-]^{-1} P_{Cd}$	(a) 3.95×10^3	(a) 1.19	[12]
			(b) 2×10^3	(b) 1.00	[70]
			(c) 4×10^2	(c) 1.00	[14]

[a]Constant numbers correspond to ones denoted in Ref. [14].

REFERENCES

[1] F.A. Kröger, The Chemistry of Imperfect Crystals, North Holland, New York, Amsterdam, 1964.
[2] F.A. Kröger, H.J. Vink, Solid State Physics, New York, Academic Press, vol. 3, 1956, pp. 307–345.
[3] D. Nobel, Phil. Res. Rep. 14 (1959) 361–399, 430–492.
[4] G. Brouwer, Phil. Res. Rep. 9 (1954) 366–377.
[5] B. Segall, M.R. Lorenz, R.E. Halsted, Phys. Rev. 129 (1963) 2471–2481.
[6] O. Matveev, Yu. Rud', K. Sanin, Physika i Tekhnika Poluprovodn. (rus.). 3 (1969) 924–926.
[7] P. Fochuk, Bull. Chernivtsi Univ. Chem. (Ukrainian) Chernivtsi 51 (1999) 28–35.
[8] R.C. Whelan, D. Shaw, Phys. State Solidi 29 (1968) 145–52.
[9] K. Zanio, in: R.K. Willardson, A.C. Beer (Eds), Cadmium Telluride, Semiconductors and Semimetals, vol. 13, Academic Press, New York, 1978.
[10] F.T. Smith, Metal. Trans. 1 (1970) 617–621.
[11] Yu. Rud', K. Sanin, Physika i Tekhnika Poluprovodn. (rus.). 5 (1971) 284–292.
[12] S.S. Chern, H.R. Vydyanath, F.A. Kröger, J. Solid State Chem. 14 (1975) 33–43.
[13] S.S. Chern, F.A. Kröger, J. Solid State Chem. 14 (1975) 44–51.
[14] P. Fochuk, R. Grill, O. Panchuk, J. Electron. Mater. 35 (2006) 1354–1359.
[15] T. Taguchi, B. Ray, Progr. Cryst. Growth Ch. 6 (1983) 103–162.
[16] B.K. Meyer, P. Omling, E. Weigel, G. Müller-Vogt, Phys. Rev. B. 46 (1992) 15135–15138.
[17] B.K. Meyer, W. Stadler, J. Cryst. Growth 161 (1996) 119–127.
[18] C. Ye, J. Chen, J. Appl. Phys. 67 (1990) 2475–2481.
[19] W.J. Allen, Semicond. Sci. Technol. 10 (1995) 1049–1064.
[20] M.R. Lorenz, B. Segall, Phys. Lett. 7 (1963) 18–20.
[21] M.R. Lorenz, B. Segall, H.H. Woodbury, Phys. Rev. 134 (1964) 751–760.
[22] V. Ivanov, V. Stopachinskij, V. Chapnin, Physika i Tekhnika Poluprovodn. (rus.). 5(1971) 101–105.
[23] N. Agrinskaya, E. Arkad'eva, O. Matveev, Physika i Tekhnika Poluprovodn. (rus.). 5(1971) 863–868.
[24] N. Agrinskaya, O. Matveev, Physika i Tekhnika Poluprovodn. (rus.). 4 (1987) 676–679.
[25] P. Emanuelsson, P. Omling, B.K. Meyer, M. Wienecke, M. Schenk, Phys. Rev. B 47 (1993) 15578–15580.
[26] H.J. Bardeleben, T. Arnoux, J. Launay, J. Cryst. Growth 197 (1999) 718-723.
[27] R. Jasinskaite, A. Sakalas, A. Martinaitis, P. Höschl, Phys. State Solidi A 82 (1984) K199.
[28] C. Szeles, Y.Y. Shan, K.G. Lynn, A.R. Moodenbaugh, Phys. Rev. B 55 (1997) 6945–6949.
[29] M. Fiederle, D. Ebling, C. Eiche, D.M. Hoffman, M. Salk, W. Stadler, K.W. Benz, B.K. Meyer, J. Cryst. Growth 138 (1994) 529–534.
[30] R. Krause-Rehberg, H. Leipner, T. Abgarjan, A. Polity, J. Appl. Phys. 66 (1998) 599–614.
[31] G. Tessaro, P. Masher, J. Cryst. Growth 197 (1999) 581–585.
[32] G. Tessaro, P. Mascher, Mat. Sci. Forum 258-263 (1978) 1335–1340.
[33] P. Rudolph, M. Neubert, M. Mühlberg, J. Cryst. Growth 128 (1993) 582–587.
[34] P. Rudolph, Progress. Cryst. Growth Ch. 29 (1994) 275–381.
[35] V. Lyahovitskaya, L. Chernyak, J. Greenberg, L. Kaplan, C. David, J. Cryst. Growth 197 (1999) 106–109.
[36] Ya. Kharif, N. Kudryashov, T. Strunilina, Neorgan. Mater. (rus.). 23 (1987) 1140–1143.
[37] Ya. Kharif, T. Strunilina, Strunilina T., P. Kovtunenko, Neorgan. Mater. (rus.). 25 (1989) 571–577.
[38] A. Gilmore, Studies of the basic electronic properties of CdTe thin films and completed CdTe/CdS solar cells, Ph.D. Thesis, Colorado Mines School, Colorado USA, 2002.
[39] R. Kremer, W. Leight, J. Cryst. Growth 86 (1988) 490–496.

[40] V.N. Martynov and S. Kobeleva, Kristallografia (in Russian), 28 (1983): 394–397.
[41] S. Maksimovskij, S. Kobeleva, Neorgan. Mater. (rus.), 22 (1986) 922–925.
[42] M. Berding, M. Schiffgaarde, A. Paxton, A. Sher, J. Vac. Sci. Technol. 8 (1990) 1103–1107.
[43] M. Berding, A. Sher, M. Schiffgaarde, J. Electron. Mater. 24 (1995) 1127–1135.
[44] M. Berding, Phys. Rev. B 60 (1999) 8943–8950.
[45] Q. Chen, M. Hillert, B. Sundman, J. Electron. Mater. 27 (1998) 145–158.
[46] Q. Chen, M. Hillert, J. Alloys Comp. 245 (1996) 125–131.
[47] R. Brebrick, R. Fang, J. Phys. Chem. Solids 57 (1996) 451–460.
[48] M. Wienecke, H. Berger, M. Schenk, Mater. Sci. Eng. 16 (1993) 219–222.
[49] J. Greenberg, J. Cryst. Growth 161 (1993) 1–11.
[50] O. Panchuk, P. Fochuk, O. Korovyanko, L. Shcherbak, Ye. Verzhak, A. Fatuyev, Phys. State Solid Chem. 3 (2006) 821–824.
[51] R. Jasinskaite, A. Martinaitis, A. Sakalas, O. Panchuk, Solid State Commun. 58 (1986) 681–682.
[52] M. Fiederle, C. Eiche, M. Salk, R. Schwarz, K.W. Benz J. Appl. Phys. 84 (1998) 6689–6692.
[53] O. Matveev, A. Terent'ev, Physika Poluprovodn. (rus.)., 32 (1998) 159–163.
[54] O. Matveev, A. Terent'ev, Physika Poluprovodn. (rus.)., 34 (2000) 1316–1321.
[55] M. Chu, S. Terterian, D. Ting, C. Wang, H. Gurgenian, S. Mesropian, J. Appl. Phys. 79 (2001) 2728–2780.
[56] J. Franc, P. Hoschl, R. Grill, L. Turjanska, E. Belas, P. Moravec, J. Electron. Mater. 30 (2001) 595–602.
[57] R. Grill, J. Franc, P. Hoschl, I. Turkevych, E. Belas, P. Moravec, M. Fiederle, K.W. Benz, IEEE. Trans. Nucl. Sci. 49 (2002) 1270–1274.
[58] R. Grill, J. Franc, I. Turkevych, P. Hoschl, E. Belas, P. Moravec, Phys. State Solidi 2 (2005) 1489–1494.
[59] R. Grill, J. Franc, I. Turkevych, P. Hoschl, E. Belas, P. Moravec, J. Electron. Mater. 34 (2005) 939–943.
[60] R. Grill, J. Franc, P. Hoschl, I. Turkevych, E. Belas, P. Moravec, IEEE. Trans. Nucl. Sci. 52 (2005) 1925–1931.
[61] D. Verstraeten, C. Longeaud, A.B. Mahmoud, H.J. von Bardeleben, J.C. Launay, O. Viraphong, Ph.C. Lemaire, Semicond. Sci. Technol. 18 (2003) 919–926.
[62] N. Krsmanovic, K.G. Lynn, M.H. Weber, R. Tjossem, Th. Gessmann, Phys. Rev. B 62 (2000) 279–282.
[63] V. Babentsov, V. Corregidor, K. Benz, M. Fiederle, T. Feltgen, E. Dieguez, Nucl. Instr. Methods A 458 (2001) 85–89.
[64] V.N. Babentsov, L.V. Gorban, L. Rashkovetskii, E.A. Salkov, Opt. Spektrosk. (rus.), 68 (1990) 820–821.
[65] A.S. Gilmore, V. Kaydanov, T.R. Ohno, NCPV Review Meeting, Denver, USA, 2001, 4 p.
[66] K. Zanio, in: R.K. Willardson, A.C. Beer (Eds.), Cadmium Telluride, Semiconductors and Semimetals, vol. 13, Academic Press, New York, 1978.
[67] G.D. Watkins, J. Cryst. Growth 159 (1996) 338–344.
[68] G.F. Neumark, Mater. Sci. Eng. R 21 (1997) 1–46.
[69] R. Grill, A. Zappettini, Progress. Cryst. Growth Ch. 48/49 (2004) 209–244.
[70] V.M. Glazov, L.M. Pavlova, Neorgan. Mater. (in russ.) 30 (1994) 629–634.
[71] P. Fochuk, O. Korovyanko, O. Panchuk, J. Cryst. Growth 197 (1999) 603–606.
[72] P. Fochuk, O. Korovyanko, I. Turkevych, O. Panchuk, P. Siffert, J. Cryst. Growth 207 (1999) 273–277.
[73] P. Fochuk, O. Panchuk, P. Feychuk, L. Shcherbak, A. Savitskyi, O. Parfenyuk, M. Hage-Ali, P. Siffert, Nucl. Instr. Methods A458 (2001) 104–112.

CHAPTER **VI**

Doping

O. Panchuk and **P. Fochuk**

Contents

1. PRELIMINARY REMARKS

The doping of CdTe was initially discussed in Zanio's monograph [1]. Subsequently, a large number of publications appeared and many new dopants were investigated. Generally "doped CdTe", that is CdTe

Chernivtsi National University, 2 vul. Kotsiubinskoho, Chernivtsi 58012, Ukraine

CDTE and Related Compounds
DOI: 10.1016/B978-0-08-046409-1.00006-X

containing foreign atoms is a broad concept. Therefore, the topic will be restricted here to the behaviour of specially introduced impurities, or dopants, into CdTe single crystals. In this context "the behaviour of dopants" principally means the chemical properties of these impurities, in other words the interaction of foreign atoms with the CdTe lattice, particularly their segregation in the ingot during crystallization and the diffusivity and solubility of the dopant atoms. Many physical properties have been investigated in doped CdTe, but only the results that directly refer to the state of an impurity atom, as a foreign point defect in the lattice, are considered in the main text. Such an analysis is restricted to the high-temperature electrical behaviour in the range 600–1200 K, when the high-temperature equilibrium is established. However, in some cases electrical properties at low-temperature, 4–400 K, and, for example optical spectra or EPR measurements are relevant for the clarification of the state of the dopant atoms. Furthermore, these restrictions exclude consideration of non-crystalline CdTe layers, as well as nanoparticles, ternary alloys of the $Cd_xZn_{1-x}Te$ type and binary or more complicated doping processes.

The results concerning the chemical behaviour of the dopants are discussed in the main text. Mostly information about the sample preparation, the investigation methods and the main results are given. In some cases where abundant information exist, figures and generalizing tables are included for the convenience of the reader.

Generally, the results obtained by different investigators are listed chronologically in the bibliography given at the end of this chapter. In addition a supplementary bibliography with the symbol "D" is available on the CD joined to the book. This list includes references to publications that do not directly refer to the topics of the chapter, but may be useful for readers who wish to obtain additional and deeper information. In the CD bibliography list the publication reference format includes the paper title, the source, and the papers are in addition classified according to both the Elements Group and the corresponding topics. The titles of papers not published in English, for example in Russian, have been translated into English.

In the main text the notations listed in Table 1 with their meaning are used.

2. IMPURITIES IN CdTe

As it is not possible to obtain absolutely pure crystals of CdTe, impurities are always present. These impurities can be characterized as either *residual* or *uncontrolled* (Section 2.1). However, *dopants* (Section 2.2) are impurities which are specially introduced to modify the physical or chemical properties of the material.

Table 1 List of notations

Notations	Meaning
CdTe:F	CdTe crystal containing the impurity F
C_L	Initial dopant concentration in the liquid (L) before crystallization
C_S(F)	Concentration of dopant (F) in the solid (S)
$C_S(g = 0)$	Dopant concentration at the beginning of the ingot ($g = 0$)
c.s.	CdTe congruent sublimation (full stoichiometry conditions)
D_{eff}(F)	Effective diffusion coefficient of the dopant F
DSC	Dopant self-compensation
F	General notation for a dopant ("Foreign atom")
FPD	Foreign point defect (impurity or dopant)
g	Relative position of the sample in the ingot (end of ingot: $g = 1$)
HTDE	High-temperature defect equilibrium
k_{eff}(F)	Effective dopant segregation coefficient at given conditions
k_{eq}(F)	Equilibrium dopant segregation coefficient (calculated at $\upsilon = 0$)
MS	Mass spectrometry
v	Crystal growth rate
NPD	Native point defect
ODMR	Optically detected magnetic resonance
PA	Positron annihilation
PAC(S)	Perturbed angular correlation (spectroscopy)
P(Cd, c.s.)	Cd vapour pressure at congruent sublimation and given temperature
P(Cd, max)	Maximal Cd vapour pressure at given temperature
P(Cd, min)	Minimal Cd vapour pressure at given temperature
PICTS	Photo-induced current transient spectroscopy
PL	Photoluminescence
QCDR	Quasichemical defect reaction
RT	Room temperature (around 300 K)
S(F)	Dopant solubility (in the solid)
SC	Self-compensation
T	Temperature (K)
THM	Traveller heating method (of crystal growth)
TSC	Thermostimulated current method (of electrical measurements)

2.1. Residual impurities

Efforts are usually made to keep the amount of residual impurities as low as possible to ensure a minimal effect on the crystals electrical properties. This level must be lower than that of the native point defects (NPD), which is about 10^{13}–10^{15} cm^{-3} in CdTe crystals at RT. To ensure the minimum content of uncontrolled impurities various sophisticated procedures have been proposed, but the most effective is combinations of repeated vacuum distillation and zone refining [2, 3]. Such methods resulted in impurity contents as low as 10^{15}–10^{16} cm^{-3}, depending on the nature of the impurities. This effectively means that there are only 1–2 impurity atoms in 10^7–10^6 CdTe "molecules", which generally satisfies most of the purposes. At RT the average NPD content, which corresponds to the thermodynamical PD equilibrium, is far lower, however, it cannot be achieved in practice. At high temperatures the HTDE is established. It is caused by both an effective mass-exchange between the solid CdTe and the vapour phase and the activation of diffusion processes in the crystal. With even slow cooling to lower temperatures, below about 600 K, these high-temperatures equilibria are, step by step, totally "frozen in". Therefore, as the temperature is lowered the PD content is kept constant, which is close to the above respective minimum concentrations. As result, it is only in very highly purified crystals that the electrical properties at RT are defined by native PDs and not by residual impurities. The principal uncontrolled impurities in CdTe are Cu, Li, Na, K, Al, O, Cl, Si, Ga, Fe and P [4–7].

2.2. Dopants

As in the other II-VI compounds, has often not a single behaviour. In elemental semiconductors the dopant atoms are usually uniquely located at a substitional or an interstitial site in the lattice. Therefore, they exhibit, for example in Ge or Si crystals, either an acceptor (B, In) or a donor (P, As) electrical behaviour. In contrast to this situation, a series of dopants in CdTe reveals a peculiar amphoteric (e.g. "dual") behaviour, occupying either normal atomic or interstitial sites. This results, as in the case of Cu, in the formation of both interstitial Cu_i, a donor FPD, and substitutional Cu_{Cd}, an acceptor FPD. Note that here and in the following, the Kröger notation system for PDs in quasichemical defect reactions (QCDR) in solids [8] is used. Amphoteric dopant behaviour of the "Cd-site–Te-site" type has also been suggested. In another case the donor foreign atom in a Cd position forms associates with native PDs as in the reaction: $In_{Cd}^{+} + V_{Cd}^{2-} \leftrightarrow (In_{Cd}^{+}V_{Cd}^{2-})^{-}$. Other combinations are possible. The depths of the dopant donor/acceptor energy levels in the gap with respect to the reference band also greatly differ: from 0.01 (In donor) to some 0.8 eV (V donor).

When discussing the behaviour of dopants in CdTe, it should be borne in mind that the crucial problem, in the studies of impurity solubility, diffusion or segregation coefficients, is the method of evaluating the impurity content. In the majority of investigations the radiotracer method was found to be the most effective, mainly due to its sensivity (down to some 10^{13} at/cm^3 or even lower). Nevertheless atomic adsorption spectroscopy is successfully used as well as modern mass-spectrometry methods.

The general aspects of dopant behaviour are discussed in Section 3, and specific values for each particular impurity are given in Section 4.

3. GENERAL ASPECTS OF DOPANT BEHAVIOUR

3.1. Segregation

CdTe doping is mainly achieved throughout single crystal growth from a doped melt, the most common methods being vertical Bridgman or THM. During crystal growth the impurity is partly incorporated into the solid phase, which means that the dopant content in the melt changes during solidification. Of fundamental importance is the Pfann's theoretical prediction [2, 3] concerning the effective segregation coefficient $k_{\mathrm{eff}} = C_{\mathrm{S}}/C_{\mathrm{L}}$, where C_{L} is the initial dopant concentration in the melt and C_{S} is its content in the solid. Two cases are possible: k_{eff} can be greater or lower than unity. Usually the latter is realized, which means that during crystal growth the impurity is driven into the melt. This results in the dopant concentration in the solid increasing during crystallization from the start to the end of the process. When $k_{\mathrm{eff}} > 1$ the inverse situation occurs. An experiment consists in cutting the ingot in slices perpendicular to the direction of growth. Each slice is analysed to obtain the dopant concentration. The analysis data enable to construct a segregation curve along the Bridgman ingot (the total ingot length g is taken as unity). The curve represents the dependence

$$C_{\mathrm{S}} = C_{\mathrm{L}} k_{\mathrm{eff}} (1 - g)^{k-1} \tag{1}$$

from which the value of k_{eff} can be determined. In general the value depends on a number of technological parameters, such as the method of crystal growth, zone growth or Bridgman crystallization, the rate of growth, the initial dopant content in the melt, the ampoule diameter and the temperature gradient across the crystallization front etc... Therefore, for some elements the dependence (Eq. (1)) does not correctly represent the segregation curve and additional calculations are needed [9]. If one has to use a k_{eff} value listed in the literature it is recommended to pay attention to the dopant analysis method used in the original work, the most accurate one being the radiotracer method.

Knowledge of the k_{eff} value is important for the evaluation of the dopant concentration in a specific slice as the dopant content in the ingot slices can significantly vary from the bottom to the top of the ingot even up to a tenfold variation. This is not always accepted by some investigators. Equation (1) allows the appropriate dopant concentration to be calculated for any slice knowing the C_L, k_{eff} and g values. For a given dopant F the equilibrium $k_{eq}(F)$ is usually significant as it corresponds to the solid–liquid boundary. In practice this is the k_{eff} value at $\upsilon = 0$. It cannot be obtained directly by experimentation but it can be determined by graphical analysis of the experimental data.

Most dopants in CdTe are characterized by k_{eff} values lower that unity, usually in the 10^{-1}–10^{-3} range. In contrast some dopants, Zn and Mn, whose the tellurides form continuous solid solutions with CdTe, possess k_{eff} values close to unity.

In general, the k_{eff} value gives only approximate information concerning the solubility of dopants in CdTe. As mentioned above, the segregation coefficients most often cited in the literature are approximate values only, as in most publications the C_L value, the growth rate or even the crystallization method are not indicated. Below we make reference to all the available information concerning the growth procedure.

3.2. Diffusion

General aspects of diffusion have been considered in a series of monographs, for example [10]. There are usually several diffusion mechanisms simultaneously in play: therefore, the process is characterized by an "effective diffusion coefficient" D_{eff}. The temperature dependence of the diffusion coefficient is an essential factor of the behaviour of foreign atoms in the crystal for two reasons. The first is that it gives evidence of doping saturation, whatever from the vapour, the liquid or the solid phase. The use of the simplest equation

$$x^2 = 4D_{eff}t \qquad (2)$$

enables one to calculate the time t necessary to uniformly saturate a crystal with a minimal "$x/2$" dimension provided the diffusion coefficient $D_{eff}(F)$ at the specific temperature is known. The second reason is that the D_{eff} value extrapolated to RT allows an appreciation of the dopant solid solution stability in the crystal. In practice it can be assumed that if the D values, extrapolated from high temperature to RT, are higher than some specific value, the probability of dopant solid solution decay at RT is significant and that such a material is unstable. Some exemplary diffusion data are shown in Table 2.

It can be seen in Table 2 that the saturation time (the time necessary for reaching the maximum solubility according to the solidus boundary at the given thermodynamic conditions) is experimentally acceptable only for

Table 2 Effective diffusion coefficients (cm^2/s) of some elements in CdTe at 523 and 1073 K

Dopant	Diffusion coefficient	D_{eff} value (cm^2/s) 523 K	1073 K	Saturation time[a]
Cu	$D_{eff} = 3.7 \exp(-0.67\ eV/kT)$ [11]	1.3×10^{-10}	1.38×10^{-7}	5 h
In	$D_{eff} = 3.3 \times 10^{-4} \exp(-1.1\ eV/kT)$ [12]	8×10^{-15}	2.2×10^{-9}	13.1 days
Ge	$D_{eff} = 1.64 \times 10^{-3} \exp(-2.0\ eV/kT)$ [13]	4.1×10^{-23}	4.6×10^{-13}	689 years

[a] 2 mm thick sample at 1073 K.

Cu and In. This is the surest way to investigate the CdTe–F solidus at different temperatures and stoichiometric conditions. However, the Cu–CdTe solid solution breaks down even at room temperature. The D_{eff} values listed in Table 1 were obtained by extrapolation of the diffusion coefficients to lower temperatures (except for Cu). The temperature of 573 K is, more or less, that at which atom movement in the crystal is virtually stopped if D_{eff} is less than 10^{-15}–10^{-16} cm^2/s. In this case it can be expected that the solid solution (and its electric properties) should not decay at temperatures close to RT. If the extrapolated D_{eff} value of a dopant at 573 K is above about (10^{-12}–10^{-13}) cm^2/s, the stability of the point defect system is doubtful as this is the case with Cu.

Dopants diffuse in CdTe crystals mostly by the classical vacancy, intersititial or interstitialcy mechanisms. To date more complicated mechanisms have been considered as rather exceptional. In general atoms of the IIA–VIIA groups of the periodic table, for example Zn, Ga, In, Tl, Ge, Sn, Sb, Cl, diffuse by the vacancy mechanism with significant activation energies (above 1 eV). On the other hand both the atoms of all B-Groups and the IA elements mostly diffuse by the interstitial mechanism with smaller activation energies, substantially below 1 eV. Significant diffusion coefficients at RT are possessed mainly by the IA Group and practically all the IB Group elements.

3.3. Solubility

The degree of solubility of a dopant is a crucial characteristic as it controls the theoretical maximum carrier density in the doped crystal. In general, the dopant solubility is illustrated on the micro-phase diagram. The dopant solubility mainly depends on the stoichiometric ratio in the crystal, which is determined by the Cd and Te vapour pressures in the ampoule during the crystal growth and annealing procedures. The microphase diagram referred to above is only a part of the three-dimensional *P–T–x* diagram of the triple Cd–Te–F system as it is difficult to represent the full diagram. Therefore, it is normal to use sections of the three-dimensional diagram with *P* constant. In such presentations the dopant solubility *S*(F) versus *T* graph is a curve with retrograde dependence. This means that *S*(F) initially increases with *T* reaching a maximum usually at about 970–1170 K, depending on the nature of the dopant, and then decreases to zero towards 1365 K, when *pure* CdTe starts to melt. In real conditions, crystallization of the doped melt, containing some 0.001–0.1% of the dopant, begins at a somewhat lower temperature, minus 0.01–0.1 K, but this is insignificant.

As CdTe has an atomic lattice, only a low metallic solubility is predicted. This corroborates the experiments where the maximal observed *S*(F) values do not exceed $\sim 10^{20}$ at/cm^3, or ~0.5–1.0 mol%, with favourable stoichiometric conditions and *temperature.*

In practice CdTe crystals are used near room temperature. Therefore, the high-temperature measurements of *S*(F) values are only indicative. The retrograde solubility phenomenon, as the temperature reduces towards RT, results in a decreasing solubility. When the precipitation of excess dopant occurs, the doping effect is partly lost. The dopant diffusion coefficient at the temperature of use determines the probability of decay. If the former is low, the doped crystal can maintain for years as a supersatured nonequilibrium solid solution which is thermodynamically unstable. In reality this situation occurs, because it is impossible to cool the ingot or sample so slowly that the crystal-dopant system could retain its equilibrium state. It is assumed that in such systems the interaction between native and foreign PDs and the vapour phase is "frozen-in" at temperatures lower than 500–600 K.

3.4. Electrical behaviour of dopants

As stated above, the dopants introduce either shallow or deep energy levels in the gap. The former are usually revealed by electrical measurements in the low-temperature range (4–400 K), whereas the latter are mostly determined from optical studies or Hall and conductivity data in conditions of HTDE, when even the deepest levels become activated. However, in the latter conditions an essential contribution to the electrical properties is made by the native PDs. Their density is defined by the deviation from stoichiometry which is determined by the thermodynamical conditions in the HTDE. Thus the electrical properties are defined by the interaction between native and foreign PDs and their respective contributions. This was shown by Kröger and Vink [14] and Brebrick [15].

Mandel's [16] analysis of wide-gap semiconductors (including II–VI semiconductors and CdTe), gives theoretical calculations showing that the introduction of electrically active dopant atoms in wide-gap semiconductors gives rise to oppositely charged native PDs. This means that doping CdTe by In, which locates in Cd sites, does not diminish the concentration of Cd vacancies, a common and complete misunderstanding not rarely discussed in the literature. On the contrary, their content rises as they are negatively charged and compensate for the In_{Cd}^{+} PDs. This follows from the mass-action law, to which the QCDR model of Kröger [8] also conforms. However, at least in some II–VI compounds, including CdTe, two types of compensation can be distinguished: (a) self-compensation (SC), i.e. "simple" compensation of the charged FPD by an oppositely charged NPD, and (b) dopant self-compensation (DSC), when the charged FPD is compensated by an associate (or complex) containing both this FPD and the respective NPD (a typical situation in CdTe heavily doped by In or Cl, see Sections 4.3.4. and 4.7.4). Of course it must be considered that for any dopant concentration both the SC and DSC processes coexist, their ratio diminishing with an increase in dopant content (see also Ref. [17]).

It is well known that electrical, photoelectrical and optical studies usually reveal many different levels in the gap (sometimes very close one to another and thus forming bands). Therefore, the attribution of some energy levels to certain FPD is generally complicated, which is due to energy interactions between different defects and the possibility of energy shift. This is a physical problem. Therefore, here, the energy values obtained by different authors are simply listed without specific comments.

General considerations concerning the influence of doping on the electrical properties of CdTe crystals can be found in Ref. [18].

Some reviews on compensation phenomena in II–VI compounds are available [19–25].

The electrical and other most important physical properties of doped crystals are considered in Part I ("Physics") of that book. Furthermore on the supplementary bibliographic CD the D-references concerning different physical properties of doped CdTe crystals are listed, most having been published in the former Soviet Union.

4. DOPANTS IN CdTe

4.1. Group I elements

4.1.1. Segregation (I)

Copper (Cu) In 1959 de Nobel [26, 27] found from Bridgman crystallization that $k_{eff}(Cu) \approx 0.5$. Later in 1968, Vanyukov [28], from normal freezing crystallization and using spectral analysis, determined the value $k_{eff}(Cu) = 0.151$. This chapter gave an equation for calculating the effective $k_{eff}(Cu)$ versus the growth rate v; however, the v symbol is erroneously lacking.

Later Cornet and co-authors [29] found $k_{eff}(Cu) \approx 0.2$, using the $_{64}Cu$ radiotracer in ingots grown by zone refining (width of the molten zone: 20–25 mm and $v = 5$–50 mm/h).

In 1973 Mykhailov et al. [30] determined the value $k_{eff}(Cu) = 0.46$, using spectral analysis on crystals grown by normal freezing ($v =$ 14.4 mm/h), whereas in another study [31] it was stated that $k_{eq}(Cu) = 2.5\times10^{-3}$, which corresponds to $k_{eff}(Cu) = 0.61$, 0.35 and 0.24 at the v values of 19.8, 10.8 and 5 mm/h, respectively.

Later Vanyukov and co-authors [32] investigated the segregation of Cu in CdTe grown by vertical zone melting at different growth rates (C_L data are not given). Unfortunately the $k_{eff}(Cu)$ values given in the article mistakenly diminish with increasing v, as confirmed by one of the authors [33]. The corrected dependencies are $k_{eff}(Cu) = 0.137$, 0.145 and 0.27 at growth rates of 10, 16 and 34 mm/h, respectively.

In Ref. [34] the Cu segregation in the zone refining process applied to the $CdCl_2$–CdTe (30%) melt was investigated. At temperatures in the range 613–803 K the determined $k_{eff}(Cu)$ values range between 0.135 and 0.166.

As it can be seen, the data reported by different authors gives substantially different $k_{eff}(Cu)$ values, possibly due to the relatively low analysis resolution of the the spectral method and different and not specified C_L values. Indeed summarizing the various data, related to various crystallization experiments and growth rates, $k_{eff}(Cu)$ values in the 0.137–0.61 range (median 0.374) are obtained.

Silver (Ag) Vanyukov [28] found with normal freezing the value $k_{eff}(Ag) = 0.066$. As with Cu the $k_{eff}(Ag) = f(v)$ dependence is unusable.

In 1973 Mykhailov and co-authors [30, 31] found that $k_{eff}(Ag) = 0.36$, 0.11 and 0.08 at the growth rates of 43.2, 14.4 and 10.8 mm/h, respectively. From these data using the $k_{eff}(Ag) = f(v)$ relationship [35] the equilibrium value $k_{eq}(Ag) = 0.066$ was calculated.

In 1975 for the first time Ag segregation was studied by the radiotracer method [36]. In Bridgman-grown ingots at the constant growth rate, v, of 3 mm/h, it was determined that varying the ingot annealing (cooling) times resulted in different Ag distributions along the ingot until to practically obtain an uniform distribution because of the rapid dopant diffusion (Fig. 1).

This indicates the impracticality of the $k_{eff}(Ag)$ determination as its value is very dependent on growth time, temperature gradient along the ingot, length and diameter etc. Nevertheless further attempts to determine $k_{eff}(Ag)$ were made.

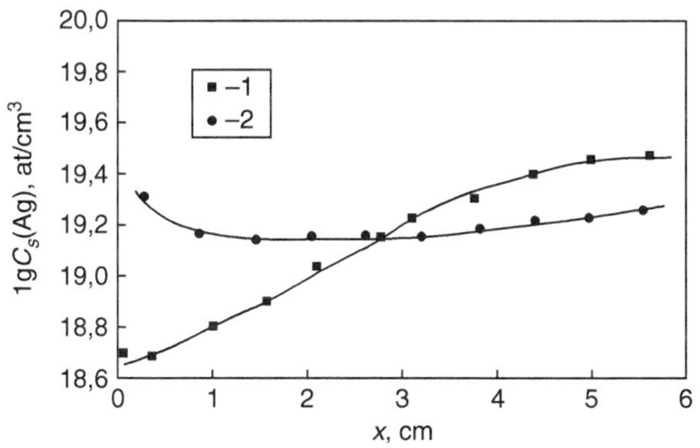

Figure 1 Ag content in CdTe versus ingot length: 1 – normal distribution after Bridgman growth; 2 – Ag concentration levelling after 50 h ingot annealing at 1120 K [36].

In 1976 Vanyukov and co-authors investigated [32] the vertical zone melting process and determined that at 10, 16 and 34 mm/h the k_{eff}(Ag) values were 0.12, 0.21 and 0.36, respectively.

Substantially lower values were obtained by Zimmermann and co-authors [37]: by using mass spectrometry and AAS analysis they found in Bridgman-grown ingots at $C_L = 2\times10^{18}$ at/cm^3 a C_S/C_L ratio resulting in k_{eff}(Ag) $= 0.016$.

In 1986 the Ag segregation in CdTe was studied once again [38]. Using the radiotracer method the k_{eff}(Ag) value was found to be 0.2 (both C_L and v are not specified).

It can be concluded that the spreading over the available data is considerable (0.016–0.36), most likely due to the rapid diffusion rate of Ag.

Gold (Au) Vanyukov [28] found in 1968 by spectral analysis that, with normal freezing, k_{eq}(Au) $= 0.056$. The formula for calculating k_{eff}(Au) at different v values is unusable.

Later in 1975, Mykhailov et al. [31] determined with spectral analysis the same equilibrium value from the following experimental data: for v values of 43.2, 19.8 and 14.4 cm/h, k_{eff}(Au) is equal to 0.43, 0.17, 0.09, respectivley.

The segregation coefficients of IA elements determined by different authors are given in Table 3.

The data concerning Cu and Ag are uncertain, as the mobility of both Cu and Ag in CdTe is very high (see Section 4.1.2) and the diffusion of these atoms in the ingot during crystal growth can severely affect the dopant segregation in the cooled ingot. In this situation the preparation of Cu- or Ag-doped samples at a particular concentration by use of the above-mentioned segregation coefficients is not advisable. A preferable doping alternative is the sample saturation by a well-defined dopant amount from the outer phase.

Table 3 Segregation coefficients of IA elements in CdTe

Dopant	Growth method	Growth rate (mm/h)	$k_{eff}(F)$	Analysis method	Authors	References
Cu	Bridgman	No data	0.5	Spectrochemical	de Nobel	[26, 27]
Cu	Normal freezing	No data	0.151	Spectrochemical	Vanyukov	[28]
Cu	Zone refining	5–50	0.2	Radiotracer	Cornet	[29]
Cu	Normal freezing	14.4	0.46	Spectrochemical	Mykhailov	[30]
Cu	Idem	19.8	0.61	Spectrochemical	Mykhailov	[31]
Cu	Idem	10.8	0.35	Spectrochemical	Mykhailov	[31]
Cu	Idem	5.0	0.24	Spectrochemical	Mykhailov	[31]
Cu	Zone refining	10	0.137	Spectrochemical	Vanyukov	[32]
Cu	Idem	16	0.145	Spectrochemical	Vanyukov	[32]
Cu	Idem	34	0.27	Spectrochemical	Vanyukov	[32]
Cu	Idem	No data	0.135–0.166	Spectrochemical	Polistansky	[34]
Ag	Normal freezing	No data	0.066	Spectrochemical	Vanyukov	[28]
Ag	Normal freezing	43.2	0.36	Spectrochemical	Mykhailov	[30, 31]
Ag	Normal freezing	14.4	0.11	Spectrochemical	Mykhailov	[30, 31]
Ag	Normal freezing	10.8	0.08	Spectrochemical	Mykhailov	[30, 31]
Ag	Zone refining	10	0.12	Spectrochemical	Vanyukov	[36]
Ag	Zone refining	16	0.21	Spectrochemical	Vanyukov	[36]
Ag	Zone refining	34	0.36	Spectrochemical	Vanyukov	[36]
Ag	Bridgman	No data	0.016	MS and AAS	Zimmermann	[37]
Ag	Normal freezing	No data	0.2	Radiotracer	Isshiki	[38]
Au	Normal freezing	No data	0.056	Spectrochemical	Vanyukov	[28]
Au	Normal freezing	43.2	0.43	Spectrochemical	Mykhailov	[31]
Au	Normal freezing	19.8	0.17	Spectrochemical	Mykhailov	[31]
Au	Normal freezing	14.4	0.09	Spectrochemical	Mykhailov	[31]

4.1.2. Diffusion (I)

Lithium (Li) Svob and Grattepain [39] studied the diffusion of lithium in CdTe at 300 K as a function of the stoichiometric conditions, defined by preliminary treatments in Cd or Te vapour. The diffusion profiles can be described by two different diffusion mechanisms, fast and slow. The interference between the two streams results in the ultimate profile. The authors assumed that the fast diffusion mechanism is related to interstitial Li atoms whereas the slow one implies substitutional (Li_{Cd}) atoms. The obtained diffusion coefficient of the two forms at RT differ significantly: D_{eff}(Li) (substitutional) $= 2 \times 10^{-14}$ cm^2/s at Cd saturation and 1.5×10^{-11} cm^2/s at Te saturation (greater V_{Cd} content), whereas D_{eff}(Li) (interstitial) $\approx 10^{-10}$ cm^2/s at 300 K and does not depend on the sample stoichiometry. Nevertheless for Li_{Cd} diffusion, the obtained D value should corroborate the respective value for Cd-self-diffusion. The latter was estimated to 10^{-26}–10^{-29} cm^2/s at RT in the literature, which means that the nature of the slow Li diffusion could not be unambiguously confirmed.

Copper (Cu) Woodbury and Aven [11] first investigated the diffusivity of Cu in CdTe using the radiotracer method in the range 97–300 °C. They obtained the temperature dependence D_{eff}(Cu) $= 3.7 \times 10^{-4}$ exp(-0.67 eV/kT) cm^2/s, indicating an interstitial diffusion mechanism, though a combined interstitial-substitutional mechanism was not excluded.

Later Panchuk and co-authors [40] found by use of the $_{64}$Cu radiotracer that Cu migrates in CdTe in the 523–653 K range mainly by the interstitial mechanism with D_{eff}(Cu) $= 9.57 \times 10^{-4}$ exp(-0.70 eV/kT) cm^2/s. This result seemed to be confirmed by the absence of stoichiometry influence (sample annealing in Cd or Te atmosphere) on the Cu diffusion.

Slightly different results were obtained by Jones and co-authors [41] who also used the radiotracer method: D_{eff}(Cu) $= 6.65 \times 10^{-3}$ exp(-0.57 eV/kT) cm^2/s. No difference induced by a vacuum treatment, Cd or Te vapour atmosphere was found. Diffusion probably occurs via the $(Cu_i^+V_{Cd}^{2-})^-$ complex. Both fast and slow diffusion components were revealed in the diffusion profiles.

Silver (Ag) As early as 1962 Čermak studied the diffusion of silver in CdTe [42]. The investigation method was simple: the time dependent movement of the boundary between CdTe ⟨Ag⟩ and undoped CdTe was studied by means of a microscope. Nevertheless the obtained diffusion relation D_{eff}(Ag) $= 164$ exp(-0.64 eV/kT) cm^2/s contains the 0.64 eV activation energy, close to the value obtained later using more precise diffusion experiments.

Later Hrytsiv [43] investigated Ag diffusion by the radiotracer method. In the 473–753 K temperature range the T dependence is given by the relation $D_{eff}(Ag) = 3.7 \times 10^{-3} \exp(-0.65\ eV/kT)\ cm^2/s$. The value does not depend upon the crystals stoichiometry (preliminary annealing under Cd or Te vapour pressure). Hrytsiv supposed that an interstitial diffusion mechanism was involved.

Internal drift effects on the diffusion of Ag in CdTe were recently studied in [44]. By the use of the $_{111}Ag$ radiotracer the anomalous diffusion profiles of Ag in CdTe were investigated. The observed profiles were explained by the presence of two Ag species (interstitial and substitutional) and their interaction with the electric field caused by the distribution of both native and foreign charged point defects.

Gold (Au) Teramoto and Takayanagi [45] were the first to study Au diffusion in CdTe. They obtained the relation $D_{eff}(Au) = 67 \exp(-2\ eV/kT)\ cm^2/s$. The high diffusion activation energy seems to indicate a vacancy diffusion mechanism.

The He-ions backscattering technique was employed by Hage-Ali et al. [46] to investigate the diffusion of gold from thin (2500–25,000 nm) films. The following diffusion relation was established: $D_{eff}(Au) = 9 \times 10^{-3} \exp(-1.7\ eV/kT)\ cm^2/s$ which gives $D_{Au}(1173\ K) = 4.3 \times 10^{-10}\ cm^2/s$, which is three orders of magnitude lower than the data of Teramoto. It is evident that the avalaible diffusion data for Au are not reliable and additional investigations are necessary.

Later Akutagawa and co-authors [47] investigated the Au diffusion coefficient at 1173 K by the He-ions backscattering technique. In two experiments they determined that $D_{eff}(Au)$ lies in the 10^{-6}–$10^{-8}\ cm^2/s$ range, which is close to the $10^{-7}\ cm^2/s$ value given by Teramoto for the same temperature.

4.1.3. Solubility (I)

Copper (Cu) According to Woodbury [11] the Cu solubility in CdTe versus T exponential dependence is linear in the 433–583 K range. The stoichiometric ratios in the crystal were not fixed in these experiments. At 433 K the Cu solubility was $5 \times 10^{16}\ at/cm^3$ and at 583 K it was $3 \times 10^{18}\ at/cm^3$.

From the surface concentration values in diffusion experiments (slow component) Jones and co-authors [41] obtained the relationship $S(Cu) = 1.56 \times 10^{23} \exp(-0.55\ eV/kT)\ at/cm^3$.

As Cu is a rapid diffusant in CdTe its solubility was investigated by Hrytsiv [43] using the method of full saturation (radiotracer method) from a surface deposited metallic layer. The Cu solubility has a retrograde character with a maximum value of $1.1 \times 10^{20}\ at/cm^3$ at 1070 K. The solubility curve in the 843–1273 K temperature range can be approximated by the equation $S(\text{Cu}) = 2.4 \times 10^{22} \exp(-0.46\,\text{eV}/kT)$ at/cm^3. This value was obtained in saturation experiments without control of the components vapour pressure. However, experiments at 773 and 973 K under maximal Cd or Te vapour pressures show that these stoichiometric conditions do not influence the Cu solubility.

Silver (Ag) The solubility of Ag in n-CdTe obtained from the Ag-radiotracer in saturation experiments exhibits a retrograde dependence with temperature as found by Panchuk et al. [17, 48]. The maximal solubility value (3.4×10^{19} at/cm^3) is reached at 1066 K in samples pre-annealed under maximal P_{Cd} (Fig. 2).

The part of the solubility curve in the 933–1066 K range is given by: $S(\text{Ag})) = 1.3 \times 10^{27} \exp(-1.54\,\text{eV}/kT)$ at/cm^3. Experiments under low P_{Cd} values resulted in higher solubility values, indicating an amphoteric behaviour due to partial solubility of Ag in Cd sites, the remainder forming the silver interstitial defects Ag_i.

Gold (Au) The results of He-ions backscattering techniques on CdTe ⟨Au⟩ indicated [47] that at 1073–1173 K Au atoms are located in both interstitial and substitional positions, the latter being preferred (in a 4:1 ratio). The absolute Au solubility values depends on the sample stoichiometry: at 1073 K and $P_{\text{Cd}} = 50$ Pa it equals some $2 \times \cdot 10^{19}$ at/cm^3, whereas at $P_{\text{Cd}} = 10^5$ Pa the value is 3×10^{18} at/cm^3. This indicates a preferred substitional location in Cd sites for Au atoms.

As shown by Hrytsiv [43], the Au solubility in CdTe obtained by radiotracer analysis is lower than that of Cu and Au. The data were

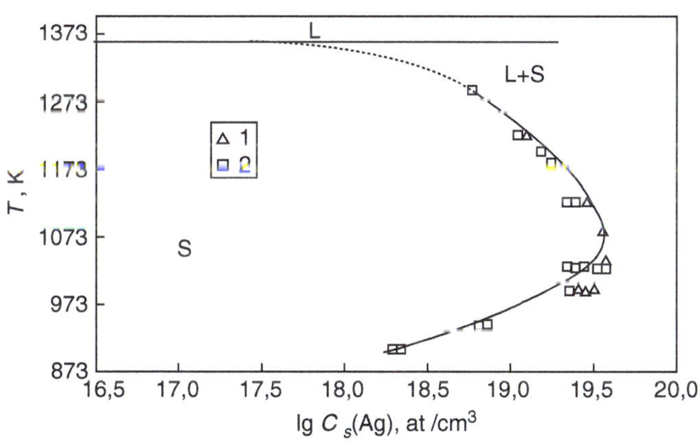

Figure 2 Solubility of Ag in CdTe at Cd saturation. Data obtained: 1 – by global Ag content measurements, 2 – by sample sectioning [48].

obtained in part from saturation experiments and in part from diffusion profiles, and a satisfactory agreement was observed. The retrograde solubility curve showed a maximum of 1.2×10^{18} at/cm^3 at 1220 K and P(Cd, max). The low-temperature part of the curve can be described by the relation: $S(\text{Au}) = 6.2 \times 10^{20} \exp(-0.59\ \text{eV}/kT)$ at/cm^3 [17]. By lowering P_{Cd} a rise of an order of magnitude of the S(Au) value was achieved, which corresponds to Akutagawa's experiments. This provides evidence of Au atoms being located in Cd sites at these temperatures.

4.1.4. Point defect electrical behaviour (I)

Lithium (Li) Vul and Chapnin [49] introduced Li by gaseous diffusion into CdTe crystals and observed the acceptor behaviour of the dopant, but the electrical characteristics of the sample changed with time.

Arkadyeva and co-authors [50] found that Li forms an approximately equal quantity of donor (interstital) and acceptor (substitutional) FPD in CdTe, the activation energy of the Li_{Cd} acceptor being 0.05 eV.

Such dual (amphoteric) behaviour of Li in CdTe was confirmed by Restle and co-authors [51] who found that the relationship between the two forms changed with T.

Copper (Cu) De Nobel [26, 27] doped CdTe with Cu and ascertained its electrical behaviour as that of an acceptor with a level at 0.33 ± 0.02 eV above the valence band. De Nobel was also the first to build up an approximated PD defect diagram in to interpret the P(Cd) free-carrier dependence determined by electrical measurements both in pure and Cu-doped material. The latter results are not sufficiently certain as the electrical properties of the CdTe samples were not investigated at HTDE but at RT after quenching from 873 to 1173 K. It is evident that even with rapid cooling the HTDE cannot be retained due to recombination processes and dopant solid solution decay etc.

Rud' and Sanin [52] investigated the conductivity of both undoped and Cu-doped CdTe at HTDE and confirmed the acceptor behaviour of Cu.

The estimation of de Nobel for the Cu level in the gap was corrected by Chamonal et al. [53] who determined its energy position at $E_{\text{V}} + 146$ meV. However, Hage-Ali claimed [54] that Cu is directly or indirectly correlated with two defect bands at 0.10–0.20 and 0.3–0.4 eV which present a fine structure with many peaks in TSC and PICTS.

The electrical properties of CdTe:Cu were studied at HTDE in [55, 56]; however, the free-carrier density was found not to differ from the values in undoped material even for $C_{\text{S}}(\text{Cu}) = 1 \times 10^{19}$ at/cm^3, which is still in the solidus area, evidently due to the strong self-compensation between donor and acceptor Cu FPDs.

Biglari et al. [57–59] found that the $E_v + 0.35$ eV level in the gap results from the Cu_{Cd} acceptor, whereas the $E_v + 0.15$ eV level arises from complexes between the former and, possibly, the non-controlled Cl_{Te} donor.

Silver (Ag) The acceptor behaviour of Ag introduced by diffusion in CdTe was investigated by Chamonal et al [53, 60]: a $E_V + 0.108$ eV acceptor level was identified by modern techniques.

Gold (Au) De Nobel [26, 27] first investigated the acceptor behaviour of Au in CdTe and pointed the corresponding level at $E_v + 0.33 \pm 0.02$ eV.

Later Molva et al. [61] determined the Au acceptor level in the CdTe gap more precisely as $E_V + 0.236$ eV.

Summary of Group I elements From diffusion, solubility and electrical investigations, it follows that the Group I dopants in CdTe mostly manifest amphoteric bevaviour, occupying both Cd atomic sites and interstitial positions. As the interstitial and substitutional FPD formation energies are obviously different, the relation $[F_i]/[F_{Cd}]$ changes with temperature. Taking into account that F_i are donors, whereas F_{Cd} are acceptors in CdTe, the free-carrier density provided by the dopant is defined by the difference between the contents of the two species. Evidently the actual carrier density is also influenced by the ionized NPD. This explains why electrical measurements in doped CdTe crystals reported by different investigators differ so much: the material was prepared at different temperatures, cooled at different rates and the real dopant content in the solid solution is different due to the decay of the latter. Thus it seems that some pertinent information concerning the $[F_i]/[F_{Cd}]$ relationship could be obtained in conditions of high-temperature defect equilibrium (HTDE), that is at temperatures above 600–800 K. However, the sole study done under these conditions concerned Cu which gave evidence of strong self-comensation processes. This can indicate that, in these conditions, the Cu donor and acceptor foreign PDs compensate each another, probably by forming associates of the $(Cu_{Cd}^- \, Cu_i^+)^-$ type. Unfortunately similar HTDE electrical measurements as a function of stoichiometry (e.g. of the P_{Cd} value) have not yet been obtained.

Summarizing, there is evidence that the location of the Group I atoms occurs both in atomic sites and in interstitial positions. The nature of the mutual compensation processes at HTDE is not yet known. However, the Group I dopants, due to their high diffusivity, resulting in decay of their solid solutions with CdTe, and their double dopant character are not well suited for controlled doping intended to produce CdTe material with prescribed electrical properties.

Note: For complementary information see "Bibliography on CD".

4.2. Group II elements

4.2.1. Segregation (II)

Beryllium (Be) Polistansky et al. [34] studied the segregation of Be with the zone crystallization method and obtained: $k_{eff}(Be) = 0.05$ and 0.175 at $v = 10$ and 26 mm/h growth rate, respectively. The equilibrium value was $k_{eq}(Be) = 0.029$ at $v = 0$.

Magnesium (Mg) Vanyukov et al. [28] stated that for normal crystallization $k_{eq}(Mg) = 2.223$. Unfortunately the formula giving the dependence of $k_{eff}(Mg)$ with v is unusable. However, it is indicated in Ref. [32] that the $k_{eff}(F)$ values exceed unity at $v = 10$, 16 and 34 mm/h. Woodbury and Lewandowki [62] being not aware of Vanyukov's investigations gave $k_{eff}(Mg) = 1$ for the zone melting process $\{C_L(Mg) = 0.2$–50% (atomic); $v = 5$–30 mm/h$\}$.

The Vanyukov's data were modified by Mykhailov et al. [31] who reported $k_{eff}(Mg)$ values of 1.4, 1.69 and 1.9 for $v = 10.8$, 19.8 and 79.2 cm/h, respectively.

Zinc (Zn) Vanyukov et al. [32] reported that $k_{eff}(Zn)$ equals unity at $v = 10$, 16 and 34 mm/h. These data were corrected in Ref. [34] using the zone crystallization technique which resulted in $k_{eff}(Zn) = 1.14$ at $v = 26$ mm/h. A close value was reported in Ref. [17] whereas in $Cd_{0.98}Zn_{0.02}Te$ a $k_{eff}(Zn)$ value of 1.16 was given [63].

4.2.2. Diffusion (II)

Zinc (Zn) Aslam et al. [64] found in 1992 that $D_{eff}(Zn)$ rises from 10^{-11} cm^2/s at 673 K to 10^{-9} cm^2/s at 1173 K.

In [63] Zn was in-diffused from the gaseous phase using a separate Zn radiotracer source in the 770–1170 K temperature range. The diffusion coefficient is $D_{eff}(Zn) = 1.39 \times 10^{-9} \exp(-0.08\ eV/kT)$ cm^2/s. The very low diffusion activation energy is questionable; nevertheless using this relation at $T = 1170$ K leads to $D_{eff}(Zn) = 6.3 \times 10^{-10}$ cm^2/s, which value slightly differs from the 1×10^{-9} cm^2/s result mentioned in Ref. [64]. These values are generally close to the Cd self-diffusion coefficients [65].

Clark and co-authors [66] assumed a complicated diffusion mechanism for Zn into CdTe which includes rapid in-diffusion followed by interdiffusion. They could not determine a diffusion coefficient as the diffusion process was time-dependent.

Mercury (Hg) Jones and co-authors [67, 68] found two-component profiles for the diffusion of Hg into CdTe. A function which gives a good fit with the experimental results was derived.

Summary of Group II elements No data concerning the solubility of Group II elements in CdTe were found. Generally all the indications are that these elements only form substitutional solid solutions in CdTe so that they do not directly influence the free-carrier density. However, changes are possible due to alteration of the band gap width.

Note: For complementary information see "Bibliography on CD".

4.3. Group III elements

4.3.1. Segregation (III)

Aluminium (Al) Vanyukov [28] found that k_{eq}(Al) equals 3.616 for a normal CdTe crystallization process. Using the same technique Mykhailov et al. [31] reported k_{eq}(Al) values of 1.35, 1.87 and 2.5 at $v = 10.8$, 43.2 and 79.2 mm/h, respectively, all data obtained by spectral analysis. In Ref. [32] a k_{eff}(Al) value >1 for v in the 10–34 mm/h range was given.

Gallium (Ga) It was reported in Ref. [32], from spectral analysis, that k_{eff}(Ga) = 0.25, 0.41 and 0.50 at v = 10, 16 and 34 mm/h, respectively. Later by use of the radiotracer technique [69] it was found that in Bridgman-grown CdTe (v = 3 mm/h) k_{eff}(Ga) equals 0.135 for C_L in the 3×10^{17}–3×10^{19} at/cm^3 range. Modelling of the Ga distribution along the ingot by use of the k_{eff} (Ga) values in Pfann's equation gives unsatisfactory results. This can be avoided by the use of a "reduced ingot length" G_x [9].

Indium (In) Mizuma et al. first found that for zone crystallization k_{eff}(In) = 0.07 [70]. Later Yokozawa et al. reported [71] that for Bridgman crystallization k_{eff}(In) = 0.11 at v = 10 mm/h and $C_L = (1–50) \times 10^{18}$ at/cm^3 with significant data spread.

In Ref. [72], the In segregation was first studied as a function of stoichiometry at C_L(In) = 8.9×10^{18} at/cm^3. It was found by spectrochemical emission that in Bridgman grown CdTe k_{eff}(In) = 8.5×10^{-3} and 4.9×10^{-1} for $P_{Cd} = 5.2 \times 10^5$ Pa and 0.8×10^5 Pa, respectively, which indicates that the In atoms are located in the Cd sites.

A value close to that of Yokozawa was found by Panchuk et al. [36] from the radiotracer method using Bridgman-grown CdTe crystals with v = 3 mm/h and C_L(In) = 2.6×10^{18} at/cm^3: k_{eff}(In) = 1.7×10^{-2}. Later these data were completed to cover the $(0.5–2.6) \times 10^{18}$ at/cm^3 C_L range [73].

Feichuk [74] found that at low In content in the Bridgman melt (up to 4×10^{17} at/cm^3) k_{eff}(In) = 0.04 at v = 3 mm/h, whereas for a higher In content in the melt the k_{eff}(In) diminishes (Fig. 3).

Essentially higher k_{eff}(In) values were reported by Vanyukov et al [32] with the zone crystallization process though at higher growth rates, but the C_L values were not indicated. Using spectral analysis they found

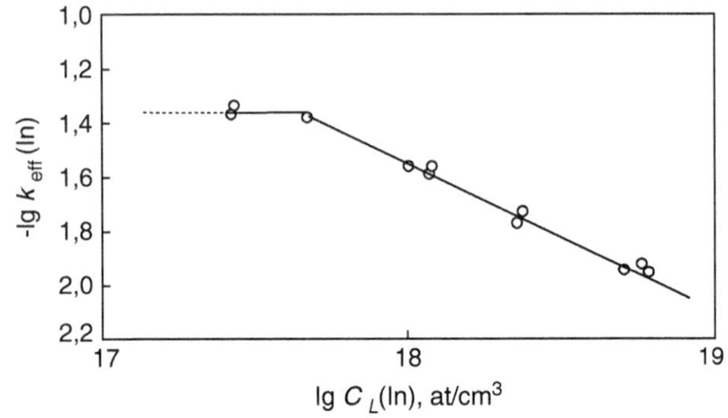

Figure 3 Dependence of the segregation coefficient k_{eff}(In) versus C_L(In) in the melt for a Bridgman CdTe crystal grown at $v = 3$ mm/h [74].

$k_{eff}(\mathrm{In}) = 0.135$, 0.152 and 0.162 at $v = 10$, 16 and 34 mm/h, respectively. The equilibrium k_{eq}(In) is calculated to be 0.132.

Weigel et al. [75] reported for THM growth (from Te solution) that $k_{eff}(\mathrm{In}) = 0.02 \pm 15\%$ for C_L(In) in the range (3×10^{15})–(1×10^{17}) at/cm^3. These data were obtained using the high-resolution AAS analysis method.

Among all the different data those of Feichuk [74] seem to coincide at best with other results {k_{eff}(In) decreasing with C_L rise} and therefore can be recommended as the most reliable (Table 4).

Thallium (Tl) The segregation of Tl in CdTe was studied using the radiotracer method in Refs. [74, 76]. The dopant segregates strongly in CdTe; the k_{eff}(Tl) values are low and poorly reproducible probably due to the large difference between the atomic radii of Cd and Tl. Feichuk [74] found by use of the radiotracer method that $k_{eff}(\mathrm{Tl}) \approx 3 \times 10^{-4}$ in Bridgman grown ingots with C_L in the (3×10^{17})–(5×10^{19}) at/cm^3 range.

As shown in Ref. [17] the temperature dependence is given by $k_{eq}(\mathrm{Tl}) = 5.5 \times 10^{-3} \exp(-0.35\ \mathrm{eV}/kT)$.

Scandium (Sc) In Ref. [34] for a zone crystallization process the value $k_{eff}(\mathrm{Sc}) = 3.3$ (at $v = 26$ mm/h) was reported.

Yttrium (Y) Using the zone crystallization method Polistansky et al. [34] reported $k_{eff}(\mathrm{Y}) = 0.022$ at $v = 40$ mm/h by applying the spectral analysis method.

Lanthanum (Ln) k_{eff}(La) for zone crystallization process was claimed to be equal to 0.05 at $v = 10$ mm/h [34].

Table 4 Segregation coefficient of In in CdTe

No	Growth method	Growth rate (mm/h)	k_{eff}(In)	Analysis method	Authors	Refererences
1	Bridgman	No data	0.07	Spectrochem.	Mizumi	[70]
2	Bridgman	10	0.11	Radiotracer	Yokozawa	[71]
3	Normal freezing	No data	8.5×10^{-3}	Spectrochem.	Lorenz	[72]
4	Normal freezing	No data	4.9×10^{-1}	Spectrochem.	Lorenz	[72]
5	Bridgman	3	1.7×10^{-2}	Radiotracer	Panchuk	[37, 73]
6	Bridgman	3	0.04	Radiotracer	Feichuk	[74]
7	Zone crystalliz.	10–34	0.135–0.162	Spectrochem.	Vanyukov	[32]
8	THM from Te solution	No data	0.02±15%	Atomic absorption	Weigel	[75]

Notes: No2: $C_L = (1\text{–}50) \times 10^{13}$ at/cm^3; No 3: $P_{Cd} = 3.2 \times 10^5$ Pa; No 4: $P_{Cd} = 0.8 \times 10^5$ Pa; No 3–4: $C_L(In) = 8.9 \times 10^{18}$ at/cm^3; No 5: $C_L(In) = 2.6 \times 10^{18}$ at/cm^3; No 7: $K_{eg}(In) = 0.132$; No 8: $C_L(In) = 3 \times 10^{15}\text{–}1 \times 10^{17}$ at/cm^3.

4.3.2. Diffusion (III)

Gallium (Ga) Feychuk [74] investigated the diffusion of Ga in CdTe by the radiotracer method. In experiments under Cd saturation it was found that the diffusion temperature dependence is described by $D_{\mathrm{eff}}(\mathrm{Ga}) = 1.75 \times 10^{-2} \exp[(-1.78 \pm 0.42)\ \mathrm{eV}/kT]\ \mathrm{cm}^2/\mathrm{s}$. A Cd vacancy diffusion mechanism was postulated.

A comprehensive study of the diffusion process was performed by Blackmore et al. [77]. By parallel use of the radiotracer and SIMS methods it was found that Ga atoms move in the CdTe lattice by two diffusion mechanisms. For the first the diffusion coefficient is $D_{\mathrm{eff}}(\mathrm{Ga}) = 3.1 \times 10^{-3} \exp(-1.52\ \mathrm{eV}/kT)\ \mathrm{cm}^2/\mathrm{s}$ and diffusion is hindered by increasing $P(\mathrm{Cd})$. For the second mechanism $D_{\mathrm{eff}}(\mathrm{Ga}) = 5.9 \times 10^{-2} \exp(-1.56\ \mathrm{eV}/kT)\ \mathrm{cm}^2/\mathrm{s}$ and is independent of the $P(\mathrm{Cd})$ value.

Indium (In) Kato and Takayanagi [78] first investigated In diffusion in Bridgman-grown CdTe samples in the 723–1273 K temperature range in vacuum. They obtained, by the use of the p–n transition method, the following diffusion coefficient: $D_{\mathrm{eff}}(\mathrm{In}) = 0.041 \exp[(-1.62 \pm 0{,}15]\ \mathrm{eV}/kT)\ \mathrm{cm}^2/\mathrm{s}$.

These results were roughly confirmed in 1966 by radiotracer investigations [79] though it is known that the p–n transition method is poorly suited for such experiments due to the large contribution of NPDs to the type and density of free carriers. Chern and Kröger [80] considered different In diffusion mechanisms in CdTe – by Cd vacancies and/or associates. The In diffusion coefficients were studied under Te vapour pressure in a very narrow (928–973 K) temperature range, which resulted in the relation: $D_{\mathrm{eff}}(\mathrm{In}) = 9 \times 10^{-3} \exp(-1.34\ \mathrm{eV}/kT)\ \mathrm{cm}^2/\mathrm{s}$.

Later Feichuk et al. [12] used the radiotracer technique (source: metallic In) under different stoichiometric conditions in the 873–1173 K range and obtained the following relations: at $P(\mathrm{Cd, c.s.})$ $D_{\mathrm{eff}}(\mathrm{In}) = 3.29 \times 10^{-4} \exp(-1.1\ \mathrm{eV}/kT)\ \mathrm{cm}^2/\mathrm{s}$ and at $P(\mathrm{Cd, max})$ $D_{\mathrm{eff}}(\mathrm{In}) = 1.71 \times 10^{-3} \exp(-1.5\ \mathrm{eV}/kT)\ \mathrm{cm}^2/\mathrm{s}$. These data indicate a significant contribution of substitional In to the diffusion mechanism. Generally In diffusion in CdTe is relatively fast.

In 1983 Watson and Shaw published a comprehensive work [81] on In diffusion for defined stoichiometry deviations and using variable sources (alloys of radioactive In with different amounts of Cd or Te). In the 773–1073 K range with an In/Te source the following relationship was given: $D_{\mathrm{eff}}(\mathrm{In}) = 3.22 \times 10^{-4} \exp(-1.13\ \mathrm{eV}/kT)\ \mathrm{cm}^2/\mathrm{s}$, which is identical to that obtained in Ref. [12]. However, with the use of an In/Cd source significant differences appear: the relation is $D_{\mathrm{eff}}(\mathrm{In}) = 1.44 \pm 2.98 \times 10^{-4} \exp(-1.82\ \mathrm{eV}/kT)\ \mathrm{cm}^2/\mathrm{s}$. The component pressure and dopant activity values resulting from specific dopant/component sources influence in a general way the diffusivity results. However, these dependences are

Table 5 Indium diffusion parameters in CdTe: $D_{eff} = D^0 \exp[-E_{act}/kT]$ cm^2/s

Stoichiometry conditions	Temperature range (K)	Experiment method	D^0 (cm^2/s)	E_{act} (eV)	Reference
Vacuum	723–1273	1*	4.1×10^{-2}	1.62 ± 0.15	[78]
Vacuum	723–1273	2*	8×10^{-2}	1.62	[79]
Te saturation	928–973	2*	9×10^{-3}	1.34	[80]
P(Cd, c.s.)	873–1173	2*	3.29×10^{-4}	1.1	[12]
Cd Saturation	873–1173	2*	1.71×10^{-3}	1.5	[12]
Te Saturation	773–1073	2*	3.22×10^{-4}	1.13	[81]
Cd Saturation	773–1073	2*	$(1.44 \pm 2.98) \times 10^{-4}$	1.82	[81]

Notes: 1*: p–n Transition method; 2*: Radiotracer method.

rarely used in diffusion investigations. However, higher diffusion activation energies at Cd saturation given in both works support a diffusion mechanism involving Cd vacancies. Table 5 gives data about In diffusion in CdTe.

Thallium (Tl) Panchuk et al. [76] studied Tl diffusion in CdTe by use of the Tl radiotracer and obtained the expression $D_{eff}(\text{Tl}) = 4.09 \times 10^{-4} \exp(-1.13\ \text{eV}/kT)$ cm^2/s at P(Cd, c.s.) and $D_{eff}(\text{In}) = 4.07 \times 10^{-4} \exp(-1.33\ \text{eV}/kT)$ cm^2/s at P(Cd, max) in the 873–1173 K temperature range.

4.3.3. Solubility (III)

Gallium (Ga) Fochuk et al. [82] used the radiotracer method in Ga solubility investigations and found that at P(Cd, max) the Ga solubility has a retrograde character with a maximum value of 1×10^{20} at/cm^3 at 1073 K. In Ref. [17] it was determined that the temperature dependence of Ga solubility in the 973–1073 temperature range was $S(\text{Ga}) = 2.95 \times 10^{25} \exp(-1.12\ \text{eV}/kT)$ at/cm^3 at P(Cd, max).

Indium (In) As early as 1962 Yokozawa and Teramoto [83] used the radiotracer method in In diffusion experiments without P(Cd) control. They found that the In surface concentration (i.e. the In solubility) was equal to 1.6×10^{19} at/cm^3 at 1173 K.

The In solubility at different temperatures and stoichiometry deviations was studied by the full sample saturation method using the

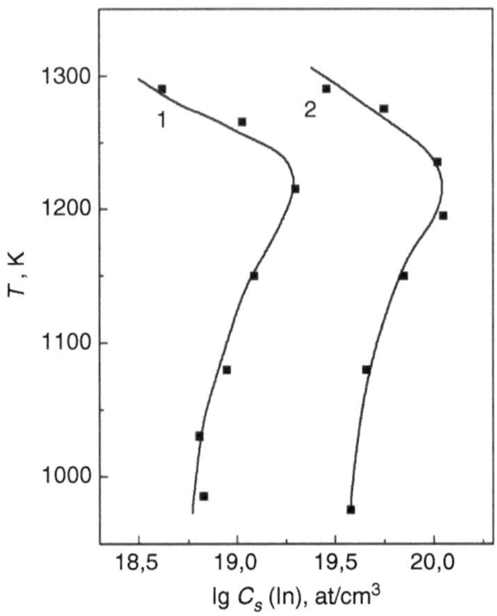

Figure 4 Temperature dependence of In solubility in CdTe at different stoichiometric conditions: 1 – at *P*(Cd, max), 2 – at *P*(Cd, c.s.) [84].

radiotracer method [84]. The increase of In solubility with decreasing *P*(Cd) indicates that In atoms occupy the Cd site positions (Fig. 4).

The maximum *S*(In) values at 1173 K are 1.6×10^{19} at/cm^3 at *P*(Cd, max) and 1.0×10^{20} at/cm^3 at *P*(Cd, c.s.). A numerical *S*(In)–*T* dependence is given in Ref. [17]: $S(\mathrm{In}) = 1.6 \times 10^{22} \exp(0.54\ \mathrm{eV}/kT)$ at/cm^3 at *P*(Cd, c.s.) and $S(\mathrm{In}) = 5.93 \times 10^{21} \exp(-0.62\ \mathrm{eV}/kT)$ at/cm^3 at*P*(Cd, max). These relationships are valid across the 873–1173 K temperature range.

The results obtained in the study [81] using different In sources (see 2.3.2 for In) are generally close to those published in Ref. [85]: maximal In solubilities at Te saturation/congruent CdTe sublimation on the order of 1–2 $\times 10^{20}$ at/cm^3 and close to 10^{18} at/cm^3 at *P*(Cd, max). This unambiguously confirms that In atoms are located in the Cd sites.

Thallium (Tl) Panchuk et al. [76] observed with the use of the $_{204}$Tl radiotracer a retrograde temperature solubility dependence: *S*(Tl) has a maximal value at *P*(Cd, c.s.) and 1170 K equal to 1×10^{18} at/cm^3, whereas at *P*(Cd,max) and 1170 K it is equal to 6.3×10^{17} at/cm^3. Lower solubility values at smaller V_{Cd} content under Cd saturation indicates that Tl atoms are located in the Cd sites. Later a numerical *S*(Tl) versus *T* dependence was proposed in Ref. [17] as $S(\mathrm{Tl}) = 1 \times 10^{22} \exp(-0.97\ \mathrm{eV}/kT)$ and $S(\mathrm{Tl}) = 1.6 \times 10^{21} \exp(-0.69\ \mathrm{eV}/kT)$ at/cm^3 at *P*(Cd, c.s.) and *P*(Cd, max), respectively, which is valid across the 873–1173 K temperature range.

4.3.4. Point defect electrical behaviour (III)

Gallium (Ga) CdTe ⟨Ga⟩ crystals were investigated much later than the In-doped material and therefore, the same DSC model (at high Ga content) was adopted to interpret the free electron density values measured at HTDE [17, 74]. The main FPDs are the Ga donor on the Cd site and the negatively charged Ga-vacancy associate (A centre, see below for the corresponding In complex).

In Ref. [85] a shallow donor level ($E_{Ga} = 0.068$ eV) and the Ga-related A-centre were identified by PICTS and PL measurements in CdTe ⟨Ga⟩.

Babentsov et al. [86] grew semi-insulating CdTe single crystals doped with Ga from the vapour phase. It was shown that the semi-insulating behaviour throughout the ingot is due to the compensation of shallow impurities by deep level centres. From the low-temperature PL spectra it was concluded that the shallow Ga_{Cd} donors are partly compensated by the A -centre, the other compensating centres being assumed to be (Ga_{Cd}–Cd_{Te}) complexes or residual acceptors (Na_{Cd} and/or Cu_{Cd}).

Later the thermodynamic constants governing the formation of the A-centre were calculated from experimental data by computer simulations of the NPD/FPD content at different T, stoichiometric ratios and C_S(Ga) values with satisfactory results [87, 88].

Indium (In) De Nobel first investigated [26, 27] the electrical properties of CdTe ⟨In⟩ crystals in 1959, determining the donor behaviour of In in CdTe ($E_c - 0.01$ eV) and building a model of In DSC by the $(In^{+}_{Cd}V^{2-}_{Cd})^{-}$ associate or A-centre (Fig. 5).

This idea was repeatedly used for the interpretation of electrical measurements results in CdTe-doped with different dopants, mostly Groups III and VII elements.

Höschl and Kubalkova [89] explained the free electron density limitations in CdTe ⟨In⟩ in terms of self-compensation in 1972.

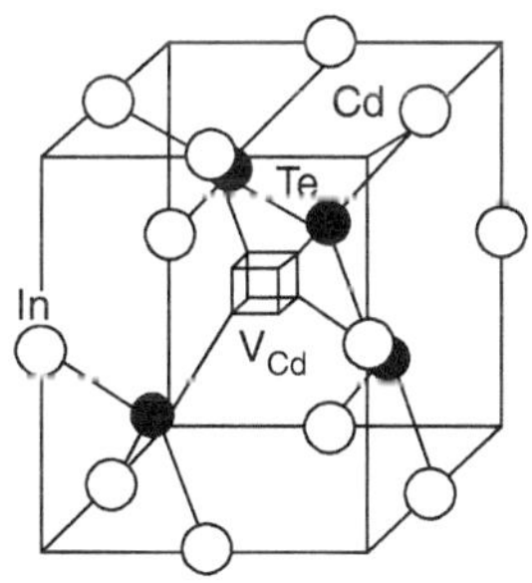

Figure 5 Atomic structure of the $(In^{+}_{Cd}V^{2-}_{Cd})^{-}$ associate (A-centre).

A comprehensive study of In-doped CdTe using direct high-temperature electric measurements was made by Chern from the Kröger's group [90, 91]. As he had no information concerning the In solubility in CdTe at HTDE, Chern limited his investigations to samples containing up to 3.6×10^{18} at/cm^3 In. This study resulted in a model of In DSC mechanism, the highest achieved free electron density being $\sim 2 \times 10^{18}$ cm^{-3}. Computer modelling was applied for the first time to the theoretical calculation of both the NPD and FPD contents; however, only the PD concentrations versus *P*(Cd) dependencies were considered.

Zanio et al. [92] used In for obtaining compensated high-resistivity CdTe. The binding energy of the A-centre in CdTe:In was found to be 0.15 eV [93]. Stadler et al. [94] investigated CdTe ⟨In⟩ by ODMR investigations which resulted in the confirmation of the A-centre occurrence in the crystals.

Shcherbak et al. [56, 95] carried out Hall effect measurements at HTDE (873–1173 K) in CdTe doped with In up to 1.5×10^{19} at/cm^3, in different stoichiometric conditions. The results are interpreted as DSC of the In donor by the A^- associate.

Ostheimer et al. [96] investigated In-doped CdTe by PACS. In samples with C_S(In) up to 10^{18} at/cm^3 only the compensating A-centre was identified, whereas at higher In concentrations another signal was observed attributed to the possible existence of a neutral $(In^+_{Cd}V^{2-}_{Cd}In^+_{Cd})^0$ associate. A similar associate – $(V^{2-}_{Cd}2D^+)^0$ – was mentioned earlier by Hage-Ali and Siffert [54] who related it to the $E_c - 0.06$ eV level in the gap. Being neutral this associate does not influence directly the free-carrier density and it is difficult to detect it from electrical measurements, though it indirectly shifts the Fermi level upward. Such a centre is easier to inspect in other II-VI semiconductors. The above cited $C_S(In) = 10^{18}$ at/cm^3 limit between the existing ranges of the different associates seems to be uncertain, as the PACS measurements were performed at *T* and *P*(Cd) values outside the conditions ensuring In solubilities as high as 10^{19}–10^{20} at/cm^3. In this situation it cannot be excluded that In solution decay occurs, with the initiation of precipitates of unknown origin (possibly In + Te), which could influence the results. Other PACS investigations of In-doped CdTe were published in Refs. [97–99]. In Ref. [97] it was reported that this method revealed, in addition to the A-centre, different In-related complexes, including those containing interstitial or antistructure components. Other methods for the identification of some proposed associates, for example In_{Cd}–V_{Te}, are desirable.

Fochuk et al. [73, 100] investigated the Hall effect at HTDE in heavily In-doped CdTe with regard to the temperature dependence of In solubility. Their study resulted in the construction of a PD structure model. Its use enabled to calculate several free-carrier density temperature dependencies at HTDE, taking into consideration both the stoichiometric conditions and the In content in the material. This modelling provided for the first time not only linear, but also turning point graphs, in specific thermodynamic conditions, with good agreement with experimental data (Figs. 6 and 7).

A characteristic luminescence line at 1.5842 eV due to emission from excitons bounded to compensating defects was observed in high-resistivity In-doped CdTe and ascribed to A^- complexes [101]. This defect is responsible for self-compensation through the balance between the In donors and the defect complex acceptors. Magneto-optical measurements also supported this self-compensation mechanism.

Grill et al. [102] carried out a theoretical study of the defect structure of donor-doped Te-saturated CdTe in the 473–1173 K temperature range. The authors concluded that a proper thermal treatment can be conveniently used for the optimization of room temperature electrical properties and for the preparation of semi-insulating detector grade material.

Positron annihilation spectroscopy was successfully used for the detection of A-centres in CdTe:In by Gely-Sykes et al. [103, 104], and

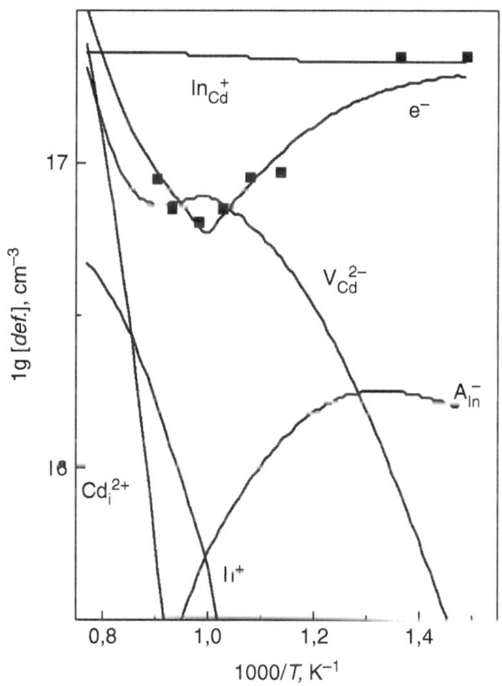

Figure 6 Point defect concentrations versus reciprocal temperature in CdTe with 1×10^{19} at/cm^3 In at $P(Cd) = 10^{-3}$ atm (points are experimental results, lines are theoretical modelling) [100].

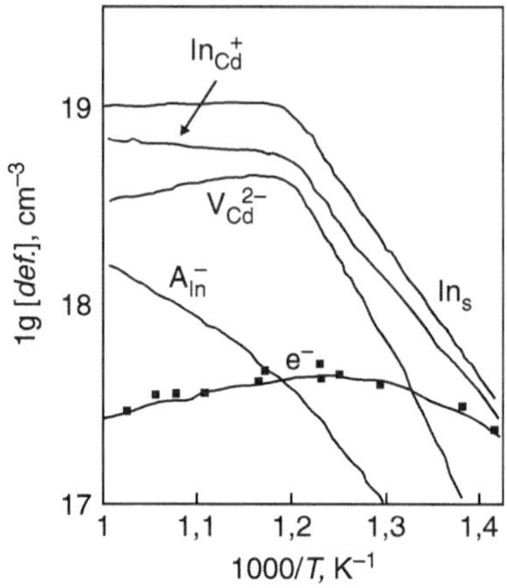

Figure 7 Point defect concentrations versus reciprocal temperature in CdTe with 2.2×10^{17} at/cm^3 In at $P(\text{Cd}) = 10^{-3}$ atm (points are experimental results, lines are theoretical modelling) [100].

also in Ref. [105]. These results were similar to the PAC data and supported the models of the A-centre discusssed above.

Another concept concerning the self-compensation of shallow donors in $A^{II}B^{VI}$ semiconductors was developed by Chadi et al. [106–108]. Using first-principles pseudopotential calculations the authors investigated the stability of the substitutional locations besides displaced atomic configurations for some dopants (D) in wide band gap II–VI semiconductors. They predicted the stability of the negatively charged so-called DX centres with large lattice relaxations for Group III atoms in some II–VI semiconductors. Though CdTe is not specifically mentioned, the appearance of a DX centre in that material is possible. The mechanism consists in the reaction $2D^0 \rightarrow DX^- + D^+$ with breaking of a D_{Cd}–Te_{Te} bond in heavily In-doped CdTe (Fig. 8).

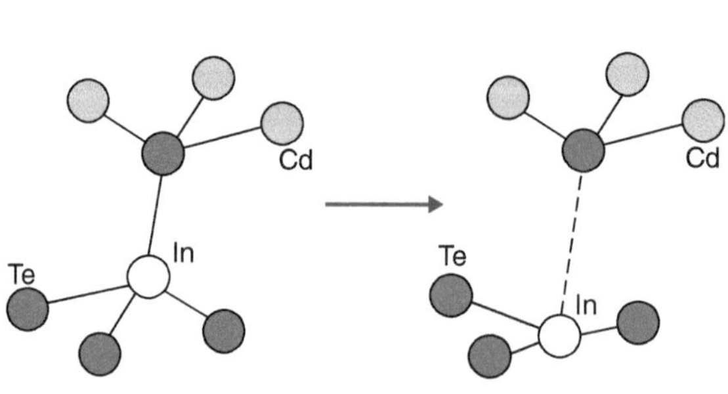

Figure 8 Atomic position of the In_{Cd} donor in the CdTe lattice (left) and with the broken In–Te bond in the DX-centre (right).

However, an unambiguous confirmation of DX centres occurence in CdTe has not yet been obtained. Neumark [23] noted: "Chadi's calculations were approximate and thus are not conclusive, although interesting". It was assumed in Refs. [109–111] that the persistent photoconductivity and EPR observed in In-doped CdTe is related to the DX centre model.

Thallium (Tl) The electrical behaviour of Tl in CdTe was studied in Refs. [74, 112] at HTDE. Conductivity measurements data indicated a diminishing of the indirectly calculated free electron density compared to non-doped material which led to conclude to the acceptor behaviour of the Tl_{Cd} centre and to a higher stability of Tl(I) compared to Tl(III). However, further investigations of the Hall effect [113] provided electron density values indicating that the previous results [74, 112] were rather erroneous. This is probably due to the calculation of $[e^-]$ from conductivity measurements by the use of carrier mobility values coming from other sources, and to the low solubility of the Tl dopant in CdTe. In these experiments [113] a weak donor effect of the Tl_{Cd} FPD was observed, probably due to the limited Tl solubility.

Summary of Group III elements Among these elements, In is the most used dopant (besides halogens) as a donor in CdTe. In and Ga segregation are not strong: the $k_{eff}(F)$ values lie approximately in the 10^{-2} area, which allows sufficiently doped material to be obtained by conventional crystal growth methods (Bridgman and THM). For the other elements the situation is as follows: Al has a large $k_{eff}(Al)$ value, whereas that of Tl is low, probably due to a greater atomic radius compared to Cd. CdTe doping by In and even Ga is more convenient to put into practice by diffusion saturation through the gaseous phase due to sufficiently high $D_{eff}(F)$ values at about 1100–1200 K. The advantage of such a doping method is the ability to obtain uniform doping of relatively small samples with annealing for 50–200 h. At the same time it is possible to choose the doping level, using published data concerning the $S(F)$ dependencies upon T and $P(Cd)$. The Ga and In solubility values are high, reaching limits of 10^{19}–10^{-20} at/cm^3. A $P(Cd)$ decrease results in increasing dopant solubility indicating that the dopants are predominantly located in Cd sites. As the diffusivity of In and Ga do not reach that of Cu or Ag at moderate T, the solid solutions obtained by saturation of In or Ga in CdTe are relatively stable, though their partial decay is possible. This problem is not resolved yet. It is significant that not only the common A-centre, but also a series of other minority associates, was observed in CdTe doped by IIIA elements. The donor activity of In and Ga atoms has limits due to the above SC (DSC) processes. The nature of the latter is not fully understood. On one hand the actual atomic configuration is evidently dopant concentration dependent (A or DX centres), but on the other, the mechanism to obtain high-resistivity material by almost full self-compensation should include different residual ionized impurities and/or minor NPDs while satisfying the full electro-neutrality condition. In practice high-resistivity CdTe:In is commonly obtained by THM growth from molten Te solutions at definite dopant concentrations and thermodynamic conditions, the required NPDs content being assured by the Te excess.

Note: For complementary information see "Bibliography on CD".

4.4. Group IV elements

4.4.1. Segregation (IV)

Carbon (C) Chibani et al. [114, 115] investigated CdTe:C(Si) crystals grown by THM from Te melt and gave the estimation: $k_{eff}(C) = (7 \pm 1) \times 10^{-3}$ with $C_L = 3.3 \times 10^{19}$ at/cm^3. The $C_S(C)$ was ambiguous, as well as that of Si.

Silicon (Si) In CdTe grown by zone crystallization [32] at $v = 10$–34 mm/h, a $k_{eq}(Si)$ above unity was obtained by spectral analysis.

In Ref. [34] with the same technique the value $k_{eff}(Si) = 1.05$ at $v = 26$ mm/h was determined. In THM-grown crystals from Te melts [114, 115] with $C_L(Si) = 9 \times 10^{19}$ at/cm^3, $C_S(Si)$ rose with g indicating a $k_{eff}(Si)$ value below unity.

Germanium (Ge) The first study of the segregation of Ge in Bridgman grown crystals ($v = 3$ mm/h) was by Shcherbak [116], using the $_{71}$Ge radiotracer. The $C_L(Ge)$ range was (8×10^{17})–(4×10^{19}) at/cm^3. Up to $C_L(Ge) = 1 \times 10^{19}$ at/cm^3, the $k_{eff}(Ge)$ value equals 3.2×10^{-3}, but it falls at higher $C_L(Ge)$ values.

In Ref. [117] the phase segregation in CdTe:Ge was investigated by different methods. The precipitates were mostly Ge + Te alloys.

Tin (Sn) Vanyukov [28] first found that for normal crystallization and with the use of spectral analysis, $k_{eq}(Sn) = 0.056$. Owing to errors in the $k_{eff}(Sn) - v$ numerical dependence the formula given by the author cannot be used for calculating $k_{eff}(Sn)$ at different growth rates.

Woodbury [62] investigated the segregation of Sn by using the $_{113}$Sn radiotracer in crystals grown by zone melting {$v = 5$–30 mm/h, C_LSn) in the (4×10^{12})–(5×10^{14}) at/cm^3 range}. It was found later that such low dopant contents do not affect the crystal electrical properties. In these conditions $k_{eff}(Sn) = 0.025$.

Mykhailov et al. [31] found that for the normal freezing process with $C_L(Sn) = 0.0114\%$ and $v = 5$, 10.8 and 19.8 mm/h, the respective $k_{eff}(Sn)$ values are 0.07, 0.12, and 0.18. Theoretical calculations [35] allowed to evaluate $k_{eq}(Sn) = 0.056$.

Later [32], in CdTe grown by zone crystallization with $v = 10, 16$ and 34 mm/h the respective k_{eq}(Sn) values of 0.025, 0.028 and 0.032 were obtained, which are close to the Woodbury's results.

Shcherbak [116] investigated the Sn segregation by the radiotracer method in Bridgman-grown CdTe ($v = 3$ mm/h). For C_L(Sn) in the (3×10^{17})–(1×10^{18}) at/cm^3 range she obtained $k_{eff}(\text{Sn}) = 8.9 \times 10^{-2}$. With C_L(Sn) increasing up to 1×10^{19} at/cm^3, k_{eff}(Sn) decreases to 1.9×10^{-2}.

Lead (Pb) Using normal crystallization Vanyukov [28] found by spectral analysis that $k_{eq}(\text{Pb}) = 0.037$. In another of his works [32] it is indicated that for zone melting at growth rates of 10, 16 and 34 mm/h, $k_{eff}(\text{Sn}) =$ 0.053, 0.076 and 0.096, respectively.

In Ref. [30] Mykhailov investigated the normal CdTe crystallization process with $C_L(\text{Pb}) = 1.13 \times 10^{18}$ at/cm^3. At $v = 5$, 10.8 and 19.8 mm/h the corrected k_{eff} (Pb) values were 0.07, 0.17 and 0.47, respectively. The value $k_{eq}(\text{Pb}) = 0.037$ was found by use of the relationship given in Ref. [35].

Titanium (Ti) In CdTe grown by zone melting the k_{eff}(Ti) values obtained from spectral analysis are 0.063 (at $v = 4$ mm/h) and 0.1 (at $v = 10$ mm/h) [34]. The calculated k_{eq}(Ti) value is 0.031.

4.4.2. Diffusion (IV)

Germanium (Ge) Ge diffusivity was studied in Ref. [13] at 903–1203 K by the radiotracer method in Bridgman grown crystals using different stoichiometric conditions. The $D_{eff}(\text{Ge}) = f(T)$ relations are: $D_{eff}(\text{Ge}) = 1.64 \times 10^{-3} \exp(-2.07\ \text{eV}/kT)$ cm^2/s and $D_{eff}(\text{Ge}) = 1.28 \times 10^{-8} \exp(-1.03\ \text{eV}/kT)$ cm^2/s at P(Cd, c.s.) and P(Cd, max), respectively.

Tin (Sn) The diffusion of Sn in CdTe was investigated by using the radiotracer method in Bridgman grown crystals with different stoichiometric ratios in the 1020–1190 K temperature range [118]. The following relations were obtained: $D_{eff}(\text{Sn}) = 8.3 \times 10^{-2} \exp(-2.2\ \text{eV}/kT)$ cm^2/s and $D_{eff}(\text{Sn}) = 6.9 \times 10^{-11} \exp(-0.38\ \text{eV}/kT)$ cm^2/s at P(Cd, c.s.) and P(Cd, max), respectively.

4.4.3. Solubility (IV)

Carbon (C) and Silicon (Si) Taking into account the small ($\sim 10^{-3}$) segregation coefficients of carbon and silicon in CdTe one can suppose that these dopants have low solubility even at high temperatures, probably below $(5\text{–}8) \times 10^{16}$ at/cm^3 [114, 115].

Germanium (Ge) As the diffusivity of Ge in CdTe is very low, the sample saturation method is inefficient and the solubility values were obtained from the Ge surface concentrations in diffusion profiles [13]. The $_{71}$Ge radiotracer analysis was used in Bridgman grown crystals. The experiments were performed in the 900–1190 K temperature range. It was found that in P(Cd, max) conditions the maximum Ge solubility of 4×10^{18} at/cm^3 was reached at 1170 K, whereas in P(Cd, c.s.) conditions the maximum value of 4×10^{19} at/cm^3 was reached at 950 K. In the 907–1093 K temperature range the S(Ge)–T dependence is given by 1.9×10^{23} exp $(-1.02\ \mathrm{eV}/kT)$ at/cm^3 at P(Cd, max) [17]. Later it was found that these values are rather unreliable due to the presence of dopant precipitates in the CdTe:Ge single crystals revealed by metallographic examination. Indirect data indicate that the Ge solubility close to RT does not significantly exceed $(5\text{–}7) \times 10^{16}$ at/cm^3.

Tin (Sn) Panchuk et al. [119] investigated with radiotracer analysis the temperature and P(Cd) dependencies of Sn solubility in CdTe. The solubility versus temperature dependence has a retrograde character with a maximum value of 5.6×10^{19} at/cm^3 at 1190 K and P(Cd, max), whereas at P(Cd, c.s.) the maximum value equals only 2.2×10^{19} at/cm^3 at 1070 K. The S(Sn)–T relationships are: $S(\mathrm{Sn}) = 1.35 \times 10^{26} \exp(-1.53\ \mathrm{eV}/kT)$ at/cm^3 at P(Cd, max) and $S(\mathrm{Sn}) = 1.93 \times 10^{24} \exp(-1.03\ \mathrm{eV}/kT)$ at/cm^3 at P(Cd, c.s.) [17, 116]. At RT the same remark as that made for Ge seems to be valid for Sn solubility.

4.4.4. Point defect electrical behaviour (IV)

Carbon (C) and Silicon (Si) Chibani et al. [114, 115] found that, in THM grown CdTe crystals, doping with $(3\text{–}9) \times 10^3$ at/cm^3 C or Si causes the appearance, in some samples, of supplementary bands (0.16–0.20 and 0.3–0.4 eV) in TSC measurements. However, there is only a weak correlation between the C or Si content and the electrical characteristics so that definite conclusions about the FPD of these dopants (whose solubility seems not to exceed $\sim 10^{16}$ at/cm^3) cannot be made.

Parfenyuk and co-authors [120] investigated CdTe:Si crystals grown from a CdTe melt with $C_L(\mathrm{Si}) = (2 \times 10^{18})\text{–}(5 \times 10^{19})$ at/cm^3 by the Bridgman–Stockbarger method. Though most samples of the ingot were highly resistive, electrical and PL studies indicated that Si impurities do not show any compensating or stabilizing effect in CdTe. Tentative high-temperature measurements of electrical properties in massively doped CdTe:Si at Cd saturation resulted in an insignificant rise in the density of free electrons. This can be caused by low Si solubility in CdTe even at HTDE [121].

Germanium (Ge) In the early 1970s it was found [122–125] that CdTe:Ge crystals are highly resistive ($\rho \approx 10^8\text{–}10^9\ \Omega$ cm) at RT and possess p-type conductivity, the latter being controlled by an acceptor level at E_v + (0.65–0.68) eV. The high resistivity region suddenly appears on reaching

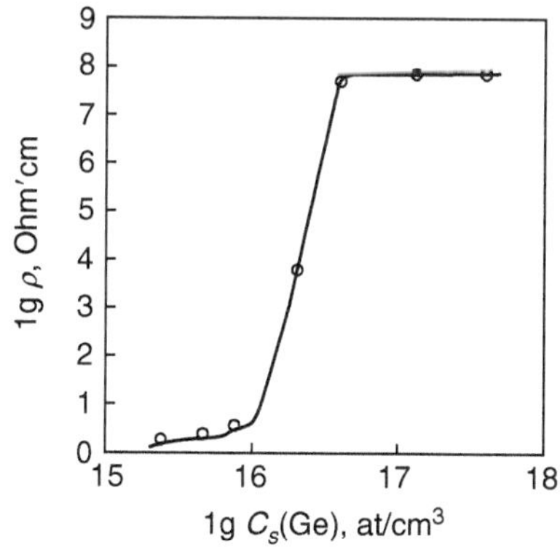

Figure 9 Variation of CdTe resistivity with Ge dopant content [17].

a Ge content (determined by radiotracer techniques) of some (2–3) $\times$ 10^{16} at/cm^3 (Fig. 9).

It is worth mentionning that in 1975 Kröger supposed that the high resistivity of CdTe:Ge could be due to self-compensation between different forms of the substitutional Ge dopant [126]. This view was later adopted in Refs. [13, 127]. The authors assumed that the compensating level corresponds to the $(Ge^{+}_{Cd}V^{2-}_{Cd})^{-}$ associate acting as an acceptor. The S(Ge)–P(Cd) dependence was interpreted as indicating the amphoteric behaviour of Ge in CdTe (with the presumed existence of both the Ge_{Cd} donor and Ge_{Te} acceptor FPDs). However, subsequent investigations revealed that the S(Ge) results were noticeably affected by the presence of dopant precipitates, so that the amphoteric model for Ge seems not to be soundly based.

This conclusion corroborates more recent studies leading to the sole donor behaviour of CdTe:Ge [128–130].

However, new theoretical calculations [131] indicated the possibility of Ge being located in Te sites with the dopant acting as an acceptor under conditions of Cd vapour saturation.

The existence of the E_v + (0.65–0.69) level was confirmed more than once by electrical [132, 133] as well as by magneto-optical investigations [134, 135].

In Ref. [136] the $Ge^{2+/3+}$ level was located around 0.6 eV above the valence band. Later Fiederle et al. [133] assumed that the above level is due to a deep Ge^{2+}/Ge^{3+} centre acting in the band as a recombination level. The authors concluded that, besides Ge, uncontrolled Fe and Cu FPDs also act as recombination or trapping centres.

In Ref. [137] the known electric, photoelectric and magnetic properties of CdTe:Ge and CdTe:Sn and their stability after high-temperature annealing under well-defined P(Cd) values were summarized. The analysis of recombination (r-centres) showed that their concentration is comparable with both that of the dissolved atoms and of the V_{Cd} defects. The established low electron-capture cross section values on r-centres and the respective ones for holes seem to support the supposition that these centres can be identified as $(Ge_{Cd}^{+}\ V_{Cd}^{2-})^{-}$ associates.

In one of the most recent publications [138] precise electrical transport measurements showed that at C_S(Ge) exceeding 5.6×10^{17} at/cm^3, the hole conductivity changes to n-type behaviour. A very similar model was proposed in Ref. [139].

Nykonyuk et al. [140] returned to the model of amphoteric behaviour of CdTe:Ge in order to explain the measured electrical and photoelectric properties of CdTe:Ge crystals after heat treatment. This view on the nature of the acceptor/donor levels created by the Ge FPD is not yet generally accepted.

Tin (Sn) The electrical properties of Sn-doped CdTe were first invesigated by Parfenyuk et al. [141]. Later studies [142] showed, that with increasing amounts of Sn, the crystal resistivity at 300 K sharply rose to $\approx 10^9$ Ω cm for C_S(Sn) $\approx$ (6–8) $\times$ 10^{16} at/cm^3 (value obtained by the radiotracer analysis). Another significant feature of the semi-insulating material is its rather unusual capability to retain its high resistivity state after a 2-h thermal treatment under P(Cd, max) followed with rapid cooling, even at the very high temperature of 1037 K.

As in the case of Ge, the same reasons suggest that the supposed amphoteric behaviour of Sn [119] is not well founded, as the Sn FPD rather acts as a deep donor [128], though recent theoretical models [131] indicate the possible existence of the acceptor Sn_{Te} in Sn-doped CdTe.

Rzepka et al. [136] reported the $Sn^{2+/3+}$ level at $E_V + 0.8$ eV. This explains the deep donor behaviour of Sn in CdTe. Measurements of electrical, photoelectric and magnetic properties of CdTe:Sn [137] indicated that the recombination processes are defined by both fast (s-) and slow (r-) mechanisms.

The photoelectric properties of CdTe:Sn crystals were studied in Ref. [143]. Compared to CdTe:Ge the former material is of higher resistance, the dark and non-equilibrium conductivity being due to electrons in both cases.

Franc et al. [144] investigated the defect structure of high-resistivity Sn-doped CdTe by a number of optical, photoelectrical and electrical methods. A model of the energy levels dominating the recombination processes in the material was elaborated, where the role of Sn as well as that of native PDs were discussed.

Grill et al. [145] studied theoretically and experimentally CdTe:Sn samples at HTDE (770–1270 K) under defined stoichiometric conditions. The theoretical defect structure analysis considered that three Sn-related defects exist in the material; a deep donor Sn_{Cd}, a shallow donor Sn_i, and the neutral complex $(Sn_{Cd}V_{Cd})$.

In a more recent publication [138] the C_S(Sn) range, in which both n- or p-type conductivity was observed, was defined more precisely and a theoretical analysis of the compensation mechanism in the material was proposed.

Franc et al. [146] investigated four CdTe ingots with a gradual increase in Sn concentration. It was confirmed that Sn strongly influences the resistivity and photoconductivity of the material. The Sn concentration must be higher than the total concentration of residual acceptors to achieve strong compensation. The middle-gap donor level pins the Fermi level. In total, six electron traps and three hole traps were identified in the band gap by several complementary techniques.

Lead (Pb) Savitskiy et al. [147] first investigated the electrical properties of Pb-doped CdTe. They found that such a doping resulted in p-type conductivity as well as in a specific resistivity increase, though not so pronounced as with Ge or Sn doping. As Ge or Sn, Pb is assumed to act as a deep donor leading to a compensating effect.

In Ref. [136] the $Pb^{2+/3+}$ energy level was located at $E_V + 0.4$ eV. The influence of Pb doping on different CdTe properties was studied in the publications [148, 149].

Recently in Ref. [138] the electrical and photoelectric properties of CdTe:Pb were thoroughly investigated and it was shown why this material is always of p-type conductivity.

Titanium (Ti) The investigation of the optical and photoelectric properties of CdTe:Ti crystals pointed to the existence of acceptor complexes of the V_{Cd}-donor type, Ti being the donor.

This assumption was confirmed by other methods, EPR and optical spectroscopy, in Ref. [150]. The authors found that in bulk-doped CdTe:Ti is present as Ti_{Cd}^{3+}.

Summary of Group IV elements The properties of CdTe, doped with Group IVA elements, especially Ge and Sn, were studied by a number of researchers because of the high resistivity of the doped crystals. The segregation of carbon and silicon in CdTe was reported as being almost absent as the k_{eff}(F) values were close to unity, whereas the segregation of Ge, Sn and Pb was significant, which complicates the obtention of even a moderately doped material. The diffusivities of these dopants are small, which should ensure the stability of the doped crystals and of their electrical properties close to room temperature. However, this would hinder the possibility of obtaining doped crystals by the diffusion saturation method. Moreover, the dopants solubilities are rather moderate even at high temperature, which impedes the synthesis of CdTe:F crystals with C_S(F) exceeding $\sim 1 \times 10^{17}$ at/cm^3. Bridgman-grown Ge-doped crystals with $C_S(F) \approx (4\text{–}9) \times 10^{16}$ at/cm^3 are usually p-type (n-type as an exception), while those doped by Sn demonstrate an inverse behaviour. Pb-doped crystals are always of p-type conductivity. The nature of the FPDs determining the hole conductivity in CdTe doped by these impurities has not been conclusively determined although most researchers suspect the respective A-centres. Besides electron conductivity seems to arise from individual FPDs acting as deep donors. The high resistivity of properly doped CdTe could be interesting for the fabrication of detectors, but the presence of deep levels in the gap acting as carrier traps results in low carrier lifetimes which cause the detecting properties to deteriorate.

Note: For additional information see bibliography on the CD.

4.5. Group V elements

4.5.1. Segregation (V)

Arsenic (As) In zone-grown CdTe ingots $k_{eff}(As) = 0.15$, 0.17 and 0.22 at growth rates of 10, 16 and 34 mm/h, respectively [32]. Moreover $k_{eq}(As) = 0.25$ as determined by spectral analysis.

Antimony (Sb) As early as 1966 Lorenz and Blum [72] prepared CdTe:Sb ingots by the horizontal crystallization technique under controlled stoichiometry conditions. Using emission spectrochemical analysis they found $k_{eff}(Sb) = 0.2$ at $P(Cd) = 5.2 \times 10^5$ Pa and $k_{eff}(Sb) = 0.012$ at $P(Cd) = 0.08 \times 10^5$ Pa. The $C_L(Sb)$ value was always 8.9×10^{18} at/cm^3. It was concluded that a P(Cd) rise increases the V_{Te} content and facilitates the incorporation of Sb atoms in the Te sites.

Vanyukov et al. [32] determined by spectral analysis k_{eff}(Sb) values in crystals grown by zone crystallization close to those of Lorenz: $k_{eff}(Sb) =$ 0.015, 0.025 and 0.028 at growth rates of 10, 16 and 34 mm/h, respectively. k_{eq}(Sb) was claimed to be 0.015. Later in Ref. [151], using the radiotracer technique in Bridgman-grown ingots, Fochuk et al. found $k_{eff}(Sb) = 4.2 \times 10^{-2}$ at $C_L(Sb) = (3 \times 10^{17})\text{–}(1 \times 10^{19})$ at/cm^3 and $v = 3$ mm/h.

Bismuth (Bi) Woodbury and Lewandowksi [62] found by the use of the radiotracer method on zone melted ingots ($v = 5\text{–}30$ mm/h) that the distribution profiles were not normal and could not be fitted with a standard curve. However, k_{eff}(Bi) appeared to be less than unity.

Vanadium (V) Polistanski [34] found that in zone melted crystals $k_{eff}(V) =$ 0.043 at $v = 10$ mm/h, whereas $k_{eff}(V) = 0.08$ at $v = 26$ and 40 mm/h. The equilibrium value k_{eq}(V) is 0.031.

4.5.2. Diffusion (V)

Phosphorus (P) Hall and Woodbury [152] were the first to use the radiotracer method to study the diffusion of phosphorus into CdTe with different stoichiometric ratios. It was found that the diffusion rate decreased with increasing P_{Cd} which indicates that a Te vacancy mechanism is involved. Although the T-dependence is not given, the P diffusivity is relatively high: at $P_{Cd} = 4 \times 10^{-2}$ Pa and 1223 K, $D_{eff}(P) = 4 \times 10^{-9}$ cm^2/s.

Cross-sectional TEM observations in CdTe:P allowed Loginov et al. to estimate a relatively low $D_{eff}(P) \approx 5 \times 10^{-12}$ cm^2/s at 873 K [153] which does not seem to agree with previous data.

Hoonnivathjana et al. [154] using the radiotracer method found that the stoichiometry ratio does not influence the $D_{eff}(P)$ value (873–1173 K). The T dependence of the diffusion coefficient was not given, but the diffusion activation energy was evaluated to 1.99 ± 0.10 eV.

Bismuth (Bi) Bi was in-diffused into THM and Bridgman-grown CdTe crystals at 573–723 K [47]. The metal penetration profiles were obtained by ion microprobe analysis. The pre-exponential term of the diffusion equation could not be defined, but the Bi diffusion activation energy was determined to be equal to 0.5 eV.

4.5.3. Solubility (V)

Phosphorus (P) In addition to the diffusivity the solubility of P in CdTe was investigated by Hall and Woodbury [152]. It increased with P_{Cd} in accordance with the incorporation model of P in Te sites. S(P) seems to be high: 9×10^{19} at/cm^3 at $P_{Cd} = 8 \times 10^5$ Pa and 1173 K as determined using the radiotracer method.

The phosphorus dissolution energy was evaluated to 1.30 ± 0.10 eV in Ref. [154], whereas the P solubility reached 4×10^{18} at/cm^3 at 1253 K. This value is essentially lower than in the former study, a possible explanation being, that in this study, the sample stoichiometry was not precisely defined.

4.5.4. Point defect electrical behaviour (V)

Nitrogen (N) Molva et al. [155] found from PL studies that the acceptor ionization energy of N in CdTe was 56 meV.

Phophorus (P) The electrical behaviour of P and As introduced in CdTe by diffusion was first investigated in 1964 by Morehead and Mandel [156]. The authors only determined the acceptor behaviour of these dopants, which implies that the dopant atoms are located in Te sites.

Later the localized vibrational modes, related to phosphorus in CdTe, were experimentally observed [157] and it was concluded that this dopant can go to either the substitutional or the interstitial sites.

A comprehensive study of the behaviour of P in CdTe was made by Selim and Kröger [158]. From RT Hall measurements carried out on P-doped crystals cooled from HTDE established with atmospheres of well-defined Cd and Te_2 pressures, they arrived at a defect model for P-doped CdTe. The temperature dependence of the Hall effect over the range 77–335 K showed that P gives rise to a shallow acceptor level at E_v + 0.035 eV. The acceptor centres are probably P_{Te} and P_i. The number of holes per P atom was close to one at low dopant contents after annealing at medium *P*(Cd). The ratio decreased with increasing P concentration as well as by annealing at low or high *P*(Cd), which indicates self-compensation. At high *P*(Cd) self-compensation involves the Cd_i^{2+} native defects. At low *P*(Cd) it involves P_{Cd}^{3+}. At high P concentrations the neutral associates $(P_{Cd}P_i)^0$ and $(P_{Cd}2P_i)^0$ are formed.

The same model accounts for the different chemical forms of phosphorus in CdTe considered in Ref. [159]. The authors theoretically studied the behaviour of P by a Green's function technique and they concluded that phosphorus is located interstitially rather than substitutionally.

An acceptor ionization energy of 68.2 meV related to phosphorus was determined from PL studies in Ref. [155].

Agrinskaya et al. [160] measured the optical and electrical properties of CdTe:P crystals. The P_{Te} acceptor showed a 0.06 eV activation energy, which agreed with the data given by Molva [155]. However, after annealing at 773 K a new E_v + 1.2 eV level appeared, probably due to the P_{Cd} donor or to a complex between P_{Te} and the Cd_i donors.

Arsenic (As) Morehead and Mandel [156] determined that the acceptor behaviour of As in CdTe was due to the existence of As_{Te} centres.

The energy level position of the acceptor centre was put at E_v + 92 meV by Molva et al. [155] from PL investigations.

The first high-temperature electrical measurements in CdTe:As crystals were detailed in Ref. [161]. In different stoichiometric conditions a substantial lowering of the free electron density values compared to the undoped material was observed, which gave support to the concept of the acceptor behaviour of As.

Antimony (Sb) In Bridgman grown crystals doped by Sb [162] the dependence of conductivity on temperature was investigated. Two acceptors levels were found at 0.262 and 0.068 eV, the former probably corresponding to the Sb_{Te} acceptor.

Later Nykonyuk et al. [163] thoroughly investigated the temperature dependencies of the Hall coefficient, charge-carrier mobility and photoconductivity under intrinsic-optical excitation in CdTe:Sb samples. It was concluded that, besides the Sb_{Te} acceptors, Sb_{Cd} donors and the associates $(Sb_{Te}Sb_{Cd})$ were most likely also present in the doped crystals. Although

C_L(Sb) varied in the (1×10^{17}–1×10^{19}) at/cm^3 range, the Sb_{Te} acceptors content did not exceed some 5×10^{16} cm^{-3}, due to the relative small k_{eff} (Sb) value. The corresponding ionization energy is 0.28 ± 0.01 eV, close to the value obtained by Iwamura [162].

Recently high-temperature Hall effect and PL measurements were performed [164]. It was confirmed that in CdTe:Sb crystals the dopant forms Sb_{Te} FPDs, which act as acceptors ($E_a = E_V + 0.29$ eV) up to 700–800 K in both Cd and Te vapour atmospheres. At higher temperatures the samples converted to n-type conductivity due either to Cd_i donors at high P(Cd) or to Sb_{Cd} donors under prevailing $P(Te_2)$.

Bismuth (Bi) In Ref. [165] an investigation of CdTe:Bi crystals with C_L(Bi) in the 10^{17}–10^{19} at/cm^3 range revealed that Bi_2Te_3 precipitates were formed, which confirmed that the Bi atoms trap Te atoms from the CdTe host lattice. At moderate Bi content the crystals were semi-insulating at RT, whereas at higher dopant concentrations resistivities of 1×10^5 Ω cm were measured. Photosensitivity studies showed an evolution from typical conductivity to shallow acceptor related conduction as the Bi concentration was increased. In the publications [166, 167] it was shown that Bi behaved as both the presumed Bi_{Te} acceptor ($E_V + 0.3$ eV) and a Bi_{Cd} donor ($E_V + 0.71$ eV).

Vanadium (V) In Ref. [168] from EPR and optical studies the position of the V donor level was located at $E_c - 0.76$ eV.

Joerger et al. [85] conducted PICTS studies and observed transitions from different charge states ($V_{Cd}^{+2/+3}$) of the vanadium donor.

In Ref. [169] a main electron trap at 0.95 eV, connected with V doping, is proposed as the main deep level involved in the photorefractive effect of CdTe:V.

EPR and optical spectroscopy on V-doped bulk CdTe were performed in Ref. [150] and only the V_{Cd}^{+3} ion could be definitely detected.

Summary of Group V dopants In general the behaviour of Group V elements in CdTe is poorly understood in comparison with Group III and VII dopants. The segregation of P is not known, whereas the As segregation is reported as being low. However, the doping procedure by these elements is not easy due to their high volatility at the melting point of CdTe and therefore, Sb and Bi are better suited for doping purposes. Only the diffusivity of Sb and Bi was investigated and low diffusion coefficients were found which results in high electric/optical stability of the doped crystals. However, this excludes the possibility of controlled and uniform doping by diffusion saturation. Evidence exists that the Group VA dopant atoms occupy predominantly Te sites in CdTe, acting as rather shallow acceptors with ionization energies in the

60–100 meV range, excepting for Sb (280 meV), whereas the location of Bi atoms is at present unclear. It is possible that the formation of Bi_{Cd} donors occurs. There are indications that phosphorus forms both acceptor FPD (substitutional P_{Te} or interstitial P_i) and donor FPD (P_{Cd}). Among the Group VB elements vanadium was studied by several authors and its behaviour as a donor in Cd sites was detected with the respective energy level at $E_c - 0.76$ eV, which explains the high resistivity of CdTe:V crystals.

Note: For additional information see "Bibliography on CD".

4.6. Group VI elements

4.6.1. Segregation (VI)

Selenium (Se) Fochuk et al. observed [170] in Bridgman-grown CdTe that with $C_L(\mathrm{Se}) = (4.6 \times 10^{17})$–$(7.9 \times 10^{18})$ at/cm^3 and at 3 mm/h growth rate the $C_S(\mathrm{Se})$ values along the ingot length obtained by radiotracer analysis showed only slight changes. The calculated $k_{\mathrm{eff}}(\mathrm{Se})$ value is 0.54 ± 0.15.

Chromium (Cr) In Ref. [32] for zone crystallization at $v = 10$ mm/h, and using spectral analysis, the value $k_{\mathrm{eff}}(\mathrm{Cr}) = 0.02$ was determined. The initial dopant concentration in the melt was not given.

Molybdenum (Mo) Polistanski at al. [34] found from spectral analysis of zone crystallized CdTe at $v = 10$ mm/h that $k_{\mathrm{eff}}(\mathrm{Mo}) = 0.073$.

Tungsten (W) The W segregation in CdTe was studied in Ref. [62]. The W distribution is not normal and $k_{\mathrm{eff}}(\mathrm{W})$ could not be evaluated, but zone refining seemed to be effective.

4.6.2. Diffusion (VI)

Oxygen (O) In Ref. [171] the oxygen diffusion was investigated in n- and p-CdTe by mass spectrometry. The diffusion coefficients at temperatures below 923 K are $D_{\mathrm{eff}}(\mathrm{O}) = 2 \times 10^{-9} \exp(-0.83\ \mathrm{eV}/kT)$ cm^2/s and $D_{\mathrm{eff}}(\mathrm{O}) = 5.6 \times 10^{-9} \exp(-1.27\ \mathrm{eV}/kT)$ cm^2/s in n- and p-CdTe, respectively. Above 923 K the oxygen diffusivity does not depend on the conductivity type: $D_{\mathrm{eff}}(\mathrm{O}) = 6 \times 10^{-10} \exp(-2.88\ \mathrm{eV}/\mathrm{k}T)$ cm^2/s.

Selenium (Se) By the use of the radiotracer method Kato et al. [172] found that in Bridgman grown p-CdTe crystals $D_{\mathrm{eff}}(\mathrm{Se}) = 1.7 \times 10^{-4} \exp(-1.35\ \mathrm{eV}/\mathrm{k}T)$ cm^2/s. It was assumed that diffusion occurred through Te vacancies.

Much later in Ref. [173] the value $D_{\mathrm{eff}}(\mathrm{Se}) = 1.10^{-5}$ cm^2/s at 1073 K was obtained. The corresponding coefficient calculated using Kato's equation [172] is very different: 7.53×10^{-11} cm^2/s.

4.6.3. Solubility (VI)

No information concerning the solubility of Group VI elements in CdTe has been found.

4.6.4. Point defect electrical behaviour (VI)

Oxygen (O) In Ref. [174] the properties of isoelectronic oxygen in II-VI semiconductors were studied by photoluminescence measurements. It was found that oxygen in CdTe, CdS and ZnS can act as acceptor and the respective energy levels in CdTe, CdSe and ZnS are shallower than those of typical acceptors such as Na. Two roles of oxygen in II–VI compounds were postulated acting either as an acceptor or as a trap, and they are classified according to the ionicity of the compound.

Chromium (Cr) EPR studies of CdTe:Cr were performed in Ref. [175]. The analysis of the spectra showed that the ionization degree of the incorporated Cr atoms is +3.

Summary of Group VI elements The available information concerning the behaviour of Group VI elements is rather limited. This can be explained by the relatively large or even complete solubility of their chalcogenides in quasibinary systems with CdTe. Only the segregation of Se was studied and and it was found to be low. The available data on diffusion coefficients varied considerably and should be reinvestigated. The same remark applies for oxygen diffusion, where the published results do not have a simple explanation. The later element is an acceptor in CdTe, acting probably as an interstitial FPD, possibly as an uncontrolled impurity due to its prevalence in the atmosphere. The Group VIB metallic dopants (Cr, Mo, W) have not been extensively investigated. It is known that they segregate relatively strongly in CdTe, which complicates the doping procedures in crystal growth. Their location in the lattice has not been identified, although it can be supposed that these atoms, being metallic, should rather substitute for Cd and mainly act as deep donors.

Note: For additional information see "Bibliography on CD".

4.7. Group VII elements

4.7.1. Segregation (VII)

Chlorine (Cl) In Ref. [37] Bridgman-grown CdTe single crystals were directly doped by elemental chlorine. Its content in the solid was measured by SIMS. With $C_L(Cl) = 9.5 \times 10^{18}$ at/cm^3, the value $k_{eff}(Cl) = 0.37 \pm 0.2$ was obtained.

Iodine (I) In Ref. [38] with CdTe grown by normal crystallization the neutron activation method led to $k_{eff}(I) = 0.22$.

Manganese (Mn) In Ref. [32] it was found that, with CdTe grown by normal crystallization at $v = 10–34$ mm/h, $k_{eff}(Mn)$ was equal to unity. This is very close to the result of Ref. [34] where the value $k_{eff}(Mn) = 1.005$ at $v = 26$ mm/h was given for ingots produced by zone crystallization.

4.7.2. Diffusion (VII)

Chlorine (Cl) In Ref. [176] the diffusion of Cl in CdTe was investigated for different stoichiometric conditions from P(Cd, max) to P(Cd, min) using the radiotracer method and no major differences were detected (Fig. 10). The Cl diffusion coefficient is $D_{eff}(Cl) = (0.071) \exp[-(1.6 \pm 0.07)\ eV/kT]\ cm^2/s$. The diffusion mechanism is probably assured by neutral defects of the $(V_{Cd}\ V_{Te})^0$ type.

In Ref. [177] the jump frequency of different Cl isotopes was theoretically analysed. It was concluded that Cl diffusion does not occur through a simple vacancy mechanism.

Jones et al. [178] studied Cl diffusion in the 573–973 K range from a $CdCl_2$ source under P(Cd, max). The diffusion profiles were found to be composed of four parts, the fastest component gave diffusion coefficients values that agreed with previously published results.

Bromine (Br) The diffusion of Br into CdTe at 293–593 K was analysed by the use of SIMS measurements which revealed four component diffusion profiles [179]. The corresponding D^0 values ranged from 2×10^{-12} to 8.2×10^{-15} cm^2/s. The diffusion activation energies were between 0.14 and 0.26 eV. These results strongly differ from the respective values obtained for Cl.

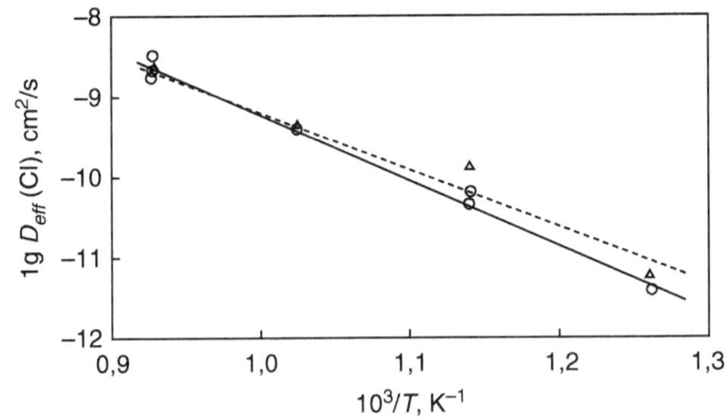

Figure 10 Temperature dependence of Cl diffusion in CdTe at: P(Cd, max) – circles and P(Cd, min) – triangles [176].

Iodine (I) Malzbender et al. [180] measured the diffusion of iodine from an elemental source. The I radiotracer profiles at 293 K consisted of four parts, the fastest corresponding to a $D_{eff}(I)$ value of $\sim 2 \times 10^{-14}$ cm^2/s.

In another study [181] the same authors gave the following relationship for the fastest diffusion component: $D_{eff}(I) = (7 \pm 3) \times 10^{-11} \exp[-(0.21 \pm 0.05)\ \text{eV}/kT]$ cm^2/s.

Manganese (Mn) The diffusivity of Mn in CdTe was measured between 773 and 1073 K under saturated Cd or Te vapour pressure [182]. The variation of $D_{eff}(Mn)$ with $P(Cd)$ was measured at 873 K. The diffusion coefficient is $D_{eff}(Mn) = (22.5 \pm 3.30) \exp[-(2.35 \pm 0.09)\ \text{eV}/kT]$ cm^2/s for saturated Te vapour, and $D_{eff}(Mn) = (1.12 \pm 9.12) \times 10^3 \exp[-(2.76 \pm 0.18)\ \text{eV}/kT]$ cm^2/s under P(Cd, max). High diffusion energy values point to a vacancy diffusion mechanism.

4.7.3. Solubility (VII)

Bromine (Br) Experiments by Jones et al. on the diffusion of Br into CdTe [178] showed that the surface bromine concentration was equal to 5×10^{19} at/cm^3, and that the Br solubility was independent of $P(Cd)$.

4.7.4. Point defect electrical behaviour (VII)

Chlorine is a shallow donor in CdTe related to the formation of the Cl_{Te} FPD. The idea of self-compensation of the halogen donors (D^+_{Te}) by the $(D^+_{Te}V^{2-}_{Cd})^-$ associates corresponds to the formation of A-centres in CdTe ⟨In⟩ and thus to DSC (see Section 4.3.4). Such compensation allows high-resistive CdTe:Cl crystals to be obtained [183]. The $E_c - (0.02\text{–}0.03)$ eV energy level was found in this material and attributed to the individual Cl_{Te} donor [184]. The main difference between halogen- and In-containing associates consists in the ionized donor atom being the nearest neighbour of the negatively charged Cd vacancy in CdTe:Cl, whereas in CdTe:In the In atom is separated by an intermediate Te atom.

SC in CdTe doped by Cl, Br or I was identified in electrical, optical, PA, PL, EPR, PAC studies conducted by many investigators on high-resistive detector grade CdTe, see Refs. [54, 185–200].

In Ref. [201] a theoretical compensation model was built and relationships for the calculation of the concentrations of various PDs were obtained.

In 1977 Agrinskaya and Matveev [202] performed electrical measurements on CdTe crystals doped with $C_S(Cl) = 1 \times 10^{17}\text{–}2 \times 10^{18}$ at/cm^3 and found a series of energy levels in the $E_v + (0.15\text{–}0.60)$ eV range. The lowest activation energy corresponds to the $V_{Cd}\text{–}Cl_{Te}$ associate, whereas the Fermi level pinning towards the middle of the gap is due to the generation of additional Cd vacancies during annealing in vacuum.

Castaldini et al. reported in Ref. [203] that the $E_v + 0.12$ eV trap level corresponds to the $(D^+_{Te}V^{2-}_{Cd})^-$ complex.

However, alternative compensation mechanisms were proposed. A comprehensive study of SC in CdTe:Cl was made by Matveev and Terentev [204]. The self-compensation of charged point defects in CdTe: Cl was investigated down to the lowest limit of the free-carrier densities (n_i, p_i) over the entire range of component vapour pressures in equilibrium with the crystal during annealing. Under control of $P(Te_2)$, the free electron density is observed to increase from 10^7 to 10^{14} cm^{-3} as $P(Te_2)$ rises from the value corresponding to P(Cd, max) up to Te saturation. This result is attributed to the formation of a donor Te_{Cd} antistructural defect. The concentration of cadmium vacancies is thus lowered to the point where the mechanism of exact self-compensation of CdTe:Cl is disrupted, and low-resistivity n-type crystals are obtained.

Another compensation model in CdTe:Cl was proposed in Refs. [205, 206] for explaning the results of TSC capacity studies. It was suggested that Cl atoms pairs are formed in the CdTe lattice: one of the atoms (donor) lies in the Te site and the other one occupies one of two distinct interstitial positions. This complex between two Cl species (Cl_{Te}–Cl_i) can bind one electron, the activation energy of which depends on the position of Cl_i in the lattice.

As in the case of In dopant, DX centres are assumed to be present in CdTe:Cl [207]. The difference with In is caused by a new type of lattice instability involving two broken bonds.

Chlorine-doped CdTe is mostly used for detector fabrication. The influence of the growth conditions on the detector properties are considered in many publications, for instance [208, 209].

Among the elements of the VII B Group, it is known that in CdTe Mn forms a deep donor (0/+) level in the valence band [210].

Summary of Group VII elements The available data show that the segregation of halogen atoms in CdTe is generally not strong, which facilitates doping by Cd halogenides during crystal growth. Though the solubility values are known only for Br they seem to be sufficiently high for the other halogens to achieve the required electrical properties. In most diffusion studies high activation energies are reported indicating a significant contribution of the vacancy diffusion mechanism, although low energies were sometimes found. Low diffusion coefficients at moderate temperatures ensure a relatively high stability of the halogen atoms in CdTe solid solutions in these conditions, which is essential for detector fabrication. Occupying predominantly Te sites in the lattice, the halogen atoms act as shallow donors in CdTe. They are usually strongly compensated by halogen-containing A-centres in Te-rich material. In such a material an excess of Cd vacancies was found in PA studies, therefore, the suggestions that interstitial Cl atoms contribute to compensation need to be verified. An alternative compensation mechanism is postulated due to the formation of DX centres as in the case of In. Anyway the compensated high-resistivity CdTe:Cl material is widely used for detector fabrication.

Note: For complementary information see "Bibliography on CD".

4.8. Group VIII elements

4.8.1. Segregation (VIII)

Iron (Fe) Slack and Galginaitis [211] produced CdTe single crystals doped by FeTe using vertical zone crystallization at a growth rate of 5 mm/h with $C_L(\mathrm{Fe}) = 1 \times 10^{20}$ at/cm^3 and calculated $k_{\mathrm{eff}}(\mathrm{Fe}) = 0.3$. In Ref. [28], it was found by spectral analysis, that for normal crystallization, $k_{\mathrm{eq}}(\mathrm{Fe}) = 0.15$.

Mykhailov et al. [31] performed Bridgman crystallization at v values of 14, 30 or 55 mm/h. From spectral analysis they obtained the following $k_{\mathrm{eff}}(\mathrm{Fe})$ values: 0.28, 0.48 and 0.67, respectively.

Woodbury and Lewandowski [62] investigated the $_{59}\mathrm{Fe}$ radiotracer segregation in zone melting process with $C_L(\mathrm{Fe}) = 10^{14}$–$10^{18}$ at/cm^3 and $v = 5$–30 mm/h. $k_{\mathrm{eff}}(\mathrm{Fe})$ values in the range 0.47–0.58 were found.

Data compatible with Ref. [31] were obtained by Vanyukov et al. [32] using zone crystallization. At $v = 10$, 16, and 34 mm/h the $k_{\mathrm{eff}}(\mathrm{Fe})$ values were 0.36, 0.41 and 0.54, respectively, whereas the equilibrium value $k_{\mathrm{eq}}(\mathrm{Fe})$ was 0.29.

Cobalt (Co) In Ref. [211] the value $k_{\mathrm{eff}}(\mathrm{Co}) = 0.3$ was obtained for vertically crystallized CdTe at a growth rate of 0.5 mm/h.

Later Woodbury [11, 62] found $k_{\mathrm{eff}}(\mathrm{Co}) = 0.27$ under conditions mostly similar to those used in the above case of Fe.

Issik et al. [38] found, using the radiotracer method, $k_{\mathrm{eff}}(\mathrm{Co}) = 0.3$ in Bridgman-doped crystals.

CdTe zone crystallization studies [32], supported by spectral chemical analysis, resulted in $k_{\mathrm{eff}}(\mathrm{Co}) = 0.19$, 0.31 for $v = 0.42$ at 10, 16 and 34 mm/h, respectively.

Nickel (Ni) Vanyukov [28] first found that, with normal CdTe crystallization, $k_{\mathrm{eq}}(\mathrm{Ni}) = 0.082$. Unfortunately the $k_{\mathrm{eff}}(\mathrm{Ni})$ versus "v" dependence is unusable owing to errors in the formula.

Consistent data are given in Ref. [31] for normal crystallization: at $v = 5$, 10.8 and 19.8 mm/h the values $k_{\mathrm{eff}}(\mathrm{Ni}) = 0.11$, 0.19 and 0.29 were obtained. $k_{\mathrm{eq}}(\mathrm{Ni})$ calculated at $v = 0$ is equal to 0.082.

Investigation of zone crystallization in Ref. [32] led to $k_{\mathrm{eff}}(\mathrm{Ni}) = 0.05$, 0.065 and 0.07 at growth rates of 10, 16 and 34 mm/h, respectively.

Palladium (Pd) Polistansky et al. [34] found that, in Bridgman grown CdTe crystals, k_{eff} (Pd) was equal to 0.025 and 0.037 at growth rates of 26 and 40 mm/h, respectively. The equilibrium k_{eq}(Pd) value was 0.017.

4.8.2. Diffusion (VIII)

Iron (Fe) In the study of Vul et al. [212] Fe was diffused into n- or p-CdTe under controlled Cd vapour pressure. D_{eff}(Fe) at 1173 K was estimated at 4×10^{-8} cm^2/s, but the Fe analysis method is not specified. The authors assumed that Fe atoms diffuse by the interstitial mechanism.

The diffusion of Fe from a separate Fe source was studied by the radiotracer method in the temperature range 900–1173 K [213]. The stoichiometry was controlled by the P(Cd) value in a two-zone ampoule. The diffusion was independent of P(Cd), suggesting a probable interstitial mechanism, which is confirmed by the diffusion coefficient: $D_{eff}(\text{Fe}) = 1.16 \times 10^{-5} \exp(-0.77\ \text{eV}/kT)$ cm^2/s. This results in $D_{eff}(\text{Fe}) = 1 \times 10^{-8}$ cm^2/s at 1173 K and indicates that full saturation of a 1–1.5 mm thick CdTe sample can be achieved in a time of 70–156 h.

Cobalt (Co) The diffusion of a Co radiotracer was studied in Bridgman grown CdTe samples [17]. D_{eff}(Co) does not depend on P(Cd) and its temperature dependence is given by $D_{eff}(\text{Co}) = 3.8 \times 10^{-8} \exp(-0.75\ \text{eV}/kT)$ cm^2/s. Thus Co atoms diffuse significantly more slowly than Fe and the saturation procedure cannot be used to obtain uniformly doped CdTe:Co samples.

4.8.3. Solubility (VIII)

Iron (Fe) In Ref. [212] CdTe:Fe samples obtained by Fe diffusion from a separate Fe source under controlled Cd vapour pressure were studied. It is claimed that it was possible to attain Fe concentrations as high as 1×10^{20} at/cm^3, but the analysis method was not specified.

In Ref. [213] CdTe samples were uniformly saturated by Fe radiotracer atoms under a specified Cd vapour pressure in the 773–1273 K temperature range. The saturation uniformity was proved by sectioning techniques. The obtained Fe solubility dependence on temperature has a retrograde character with an maximum at 1250 K. Below that temperature the Fe solubility dependence is given by the relation: $S(\text{Fe}) = 3.28 \times 10^{23} \exp(-1.05\ \text{eV}/kT)$ at/cm^3. The Fe diffusion activation energy is relatively low and is independent of the stoichiometry, probably owing to an interstitial mechanism. Nevertheless the Fe solubility depends on the P(Cd) value. This indicates a possible temperature dependent redistribution of Fe atoms between Cd sites and interstitial positions in the lattice. Another specific feature of the behaviour of Fe atoms in CdTe is the time-dependent character of the saturation process [17]. As the diffusion of CdTe:Fe is relatively rapid, full saturation of the 1–1.5 mm thick

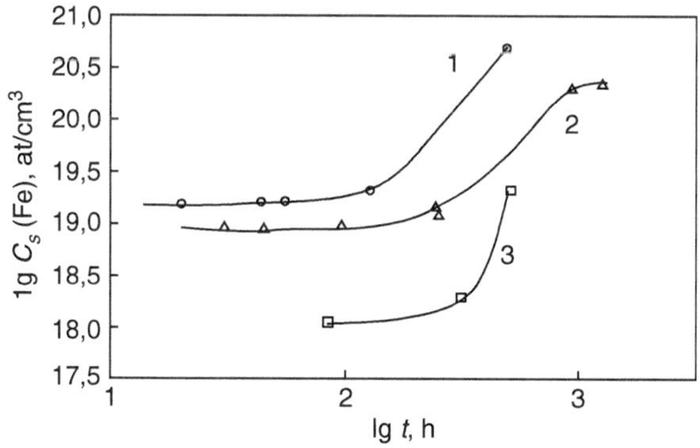

Figure 11 Time dependence of CdTe saturation with Fe. Respective temperatures of sample and Cd vapour source: 1 – 1230 and 1220 K, 2 – 1170 and 1120 K, 3 – 1230 and 1140 K [17].

samples used was achieved in a few dozen hours. Prolonged saturation times of 100–150 h resulted in the same S(Fe) values. However, if the saturation time exceeded 200–250 h, C_S(Fe) began to rise reaching at some temperatures a value of more than (1–2) $\times$ 10^{20} atoms/cm^3 (Fig. 11).

Metallographic and autoradiographic experiments at 300 K do not indicate the presence of second phase inclusions. The nature of the mentioned effect is currently not clear.

4.8.4. Point defect electrical behaviour (VIII)

Iron (Fe) Vul et al. [212] claimed that in CdTe:Fe no levels which could be attributed to the Fe donor were observed. In a series of investigations using different methods [214–217] the $Fe^{2+/3+}$ donor level was identically found to be located at E_c – (1.40–1.45) eV.

Cobalt (Co) In Ref. [218] it was reported from luminescence investigations that the ionization energy of the Co dopant is equal to 0.44 eV

Nickel (Ni) Kaufmann [219] found using electron spin resonance that an isolated ionized Ni acceptor Ni_{Cd} ($3d^9$) existed in CdTe:Ni single crystals, corresponding to the Ni^{3+} state.

Summary of Group VIII elements The Fe diffusion is independent of P(Cd) which indicates that at HTDE dopant diffusion occurs via Fe interstitials over a wide temperature range. However, the study of Fe solubility provides evidence of a clear stoichiometry dependence. Therefore, in crystals doped at high T both Fe_{Cd} and Fe_i FPDs are present. Nevertheless even at 1173 K, electrical measurements on Fe saturated CdTe in

P(Cd, max) conditions do not reveal any difference between the free-carrier density in doped and undoped material. This seems somewhat strange, as Fe diffusion and solubility experiments indicate the presence of interstitial Fe atoms, which should ionize relatively easily in contrast to the Fe_{Cd} centre which is known as a deep donor. Taking into account the pecularities of Fe dissolution in CdTe, as reported in the section "Solubility", it can be concluded that the behaviour of Fe in CdTe requires additional investigations.

Note: For additional information see "Bibliography on CD".

REFERENCES

[1] K. Zanio, R.K. Willardson, A.C. Beer (Eds.), Cadmium Telluride, Semiconductors and Semimetals, vol. 13, Academic Press, New York, 1978.
[2] W.G. Pfann, Zone Melting, Wiley, New York, 1959.
[3] W.G. Pfann, J. Metals 4 (1952) 747–750.
[4] J.M. Francou, K. Saminadayar, J.L. Pautrat, Phys. Rev. D 19 (1990) 12035–12046.
[5] R. Triboulet, A. Aoudia, A. Lusson, J. Electron. Mater. 24 (1995) 1061–1065.
[6] B. Yang, T. Miki, Y. Doumae, M. Isshiki, J. Cryst. Growth 179 (1997) 410–414.
[7] B. Yang, Y. Ishikawa, T. Miki, Y. Doumae, T. Tomizono, M. Isshiki, J. Cryst. Growth 159 (1996) 171–174.
[8] F.A. Kröger, The Chemistry of Imperfect Crystals, North-Holland, Amsterdam, 1964.
[9] R.I. Rozin, O.L. Kreinin, M.P. Shashkovskaya, Proc. Acad. Sci. USSR Chem Ser. (in Russian) 6 (1971) 1313–1316.
[10] D. Shaw, Atomic Diffusion in Semiconductors, Plenum Press, London, 1973.
[11] H.H. Woodbury, M. Aven, J. Appl.Phys. 39 (1968) 5485–5488.
[12] P. Feichuk, O. Panchuk, A. Savitskiy, Inorg. Mater. 15 (1979) 1390–1392.
[13] O. Panchuk, L. Shcherbak, Inorg. Mater. 15 (1979) 1049–1052.
[14] F.A. Kröger, H.J. Vink, Solid State Physics, vol. 3, Academic Press, New York, 1956, pp. 307–345.
[15] R.F. Brebrick, J. Phys. Chem. Solids 18 (1961) 116–128.
[16] G. Mandel, Phys. Rev. A 134 (1961) 116–128.
[17] O. Panchuk, Controlled Doping of $A^{II}B^{VI}$ Semiconductors (in Russian), Doctoral Thesis., Moscow University, Moscow, 1988.
[18] R.O. Bell, F. Wald, P. Siffert, A. Cornet (Eds.), Proceedings of the international Symposium on Cadmium Telluride – A material for gamma-ray detectors, 1971, pp. 6–15.
[19] Y. Marfaing, Rev. Phys. Appl. 12 (1977) 211–217.
[20] Y. Marfaing, Progr. Cryst. Growth Character. Mater. 4 (1981) 317–343.
[21] J.L. Pautrat, J.M. Francou, N. Magnea, E. Molva, J. Cryst. Growth 82 (1985) 194–204.
[22] Y. Marfaing, J. Cryst. Growth 161 (1996) 205–211.
[23] G.F. Neumark, Mater. Sci. Eng. R 21 (1997) 1–46.
[24] U.V. Desnica, Progr. Cryst. Growth Character. Mater. 36 (1998) 291–357.
[25] U.V. Desnica, I.D. Desnica-Frankovic, Vacuum 61 (2001) 361–365.
[26] D. Nobel, Philips Res. Rep. 14 (1959) 361–399.
[27] D. Nobel, Philips Res. Rep. 14 (1959) 430–492.
[28] A.V. Vanyukov, Nauchnye Trudy Moskovsk. Instituta Stali i Splavov (in Russian) 52 (1968) 244–261.
[29] A. Cornet, P. Siffert, A. Cocne, J. Cryst. Growth 7 (1970) 329–332.

[30] V.A. Mykhailov, V.V. Golubev, B.S. Krasulina, in Khalkohenidy Zinka, Kadmiya, Rtuti (in Russian), Moscow, Moskovskiy Institut Stali i Splavov 73 (1973) 36–39.
[31] V.A. Mykhailov, S.S. Tykhonov, I.B. Semenova, B.S. Krasulina, in Khalkohenidy Zinka, Kadmiya, Rtuti (in Russian), Moscow, Moskovskiy Institut Stali i Splavov 73 (1973) 39–42.
[32] A.V. Vanyukov, Yu.M. Ivanov, A.V. Namm, N.G. Sedelnikov, B.S. Krasulina, O.F. Kravchenko, Elektronnaya Tekhnika: Materialy (in Russian) 6 (1976) 72–77.
[33] Yu M. Ivanov, private communication, (2007).
[34] D.G. Polistansky, B.S. Krasulina, Nauchnye Trudy Moskovsk. Instituta Stali i Splavov (in Russian) 146 (1983) 35–36.
[35] I.A. Burton, R.C. Prim, W.P. Slichter, J. Phys.Chem.Solids 21 (1953) 1987–1990.
[36] O. Panchuk, V. Hrytsiv, L. Shcherbak, P. Feichuk, D. Belotskiy, in Legirovannye poluprovodniki (in Russian), Moscow, Nauka 1975, pp. 89–92.
[37] H. Zimmermann, R. Boyn, C. Albers, K.W. Benz, J. Cryst. Growth 128 (1993) 593–598.
[38] M. Isshiki, M. Sato, K. Masumoto, J. Cryst. Growth 78 (1986) 58–62.
[39] L. Svob, C. Grattepain, J. Solid State Chem. 20 (1977) 297–303.
[40] O. Panchuk, V. Hrytsiv, D. Belotskyi, Inorgan. Mater. 11 (1975) 1510–1512.
[41] E.D. Jones, N.M. Stewart, J.B. Mullin, J. Cryst. Growth 117 (1992) 244–248.
[42] K. Čermak, Českoslov. Časop. Fiz. (in Czech) A12 (1962) 84–85.
[43] V. Hrytsiv, Investigation of Cu, Ag and Au behavior in Cadmium Telluride (in Russian), Ph.D. Thesis., Chernovtsy University, Chernovtsy (Ukraine), 1975.
[44] H. Wolf, F. Wagner, T. Wichert, R. Grill, E. Belas, J. Electron. Mater. 35 (2006) 1350–1353.
[45] I. Teramoto, S. Takayanagi, J. Phys. Soc. Jpn. 17 (1962) 1137–1141.
[46] M. Hage-Ali, I.V. Mitchell, J.J. Grob, P. Siffert, Thin Solid Films 19 (1973) 409–418.
[47] T. Akutagawa, D. Turnbull, W.K. Chu, J.W. Mayer, J. Phys. Chem. Solids 36 (1975) 521–528.
[48] O. Panchuk, V. Hrytsiv, D. Belotskyi, Inorg. Mater. 10 (1974) 497–499.
[49] B.M. Vul, V.A. Chapnin, Sov. Phys. Solid State 8 (1966) 206–207.
[50] E.N. Arkadeva, O.A. Matveev, V.A. Sladkova, Sov. Phys. Semicond. 2 (1968) 1514–1517.
[51] M. Restle, K. Bharut-Ram, H. Quintel, C. Ronning, H. Hofsäss, J. Cryst. Growth 161 (1996) 168–171.
[52] Yu. Rud, K.V. Sanin, Sov. Phys. Semicond. 6 (1972) 645–647.
[53] J.P. Chamonal, E. Molva, J.L. Pautrat, Solid State Commun. 43 (1982) 801–805.
[54] M. Hage-Ali, P. Siffert, Nucl. Instrum. Methods A 322 (1992) 313–323.
[55] P. Fochuk, L. Shcherbak, P. Feichuk, O. Panchuk, CALPHAD 21 (1997) 469–473.
[56] P. Fochuk, O. Panchuk, V. Kocherhan, L. Yatsunyk, Inorg. Mater. 32 (1996) 1187–1189.
[57] B. Biglari, M. Samimi, M. Hage-Ali, Appl. Phys. Lett. A 43 (1987) 47–52.
[58] B. Biglari, M. Samimi, M. Hage-Ali, J. Cryst. Growth 89 (1988) 428–434.
[59] B. Biglari, M. Samimi, M. Hage-Ali, Nucl. Instrum. Methods A 283 (1989) 249–254.
[60] J.P. Chamonal, E. Molva, J.L. Pautrat, L. Revoil, J. Cryst. Growth 59 (1982) 297–300.
[61] E. Molva, J.M. Francou, J.L. Pautrat, K. Saminadayar, L.S. Dong, J. Appl. Phys. 56 (1984) 2241–2249.
[62] H.H. Woodbury, R.S. Lewandowski, J. Cryst. Growth 10 (1971) 6–12.
[63] L. Shcherbak, Optoelektronika i poluprovodnikovaya Tekhnika (in Russian) 31 (1996) 39–44.
[64] N. Aslam, E.D. Jones, T.C.Q. Noakes, J. Cryst. Growth 117 (1992) 249–253.
[65] D. Shaw, J. Cryst. Growth 86 (1988) 778–786.
[66] J.C. Clark, E.D. Jones, J.B. Mullin, A.W. Brinkman, Mater. Sci. Forum 182–184 (1995) 63–68.

[67] M.U. Ahmed, E.D. Jones, J.B. Mullin, N.M. Stewart, Mater. Sci. Forum 182–184 (1992) 51–56.
[68] E.D. Jones, V. Thambipillai, J.B. Mullin, J. Cryst. Growth 118 (1992) 1–6.
[69] O. Panchuk, P. Feichuk, L. Shcherbak, P. Fochuk, R. Boichuk, Inorg. Mater. 21 (1985) 974–975.
[70] K. Mizuma, O. Mikami, K. Ono, M. Kamiya, Electron. Commun. Lab. Technol. (in Japan) 10 (1961) 895–924.
[71] M. Yokodzawa, S. Otsuka, S. Takayanagi, Jpn. J. Appl. Phys. 4 (1965) 1018.
[72] M. Lorenz, S.E. Blum, J. Electrochem. Soc. 113 (1966) 559–563.
[73] P. Fochuk, O. Panchuk, P. Feichuk, L. Shcherbak, A. Savitskyi, O. Parfenyuk, M. Ilashchuk, M. Hage-Ali, P. Siffert, Nucl. Instrum. Methods A 458 (2001) 104–112.
[74] P. Feichuk, Study of the Mechanism of Group III Dopants Introduction into CdTe (in Russian), Ph.D. Thesis., Chernovtsy University, Chernovtsy (Ukraine), 1979.
[75] A. Weigel, B. Müller-Vogt, B. Steinbach, W. Wendl, W. Stadler, D. Hoffman, B.K. Meyer, Mater. Sci. Eng. B 16 (1993) 17–22.
[76] O. Panchuk, P. Feichuk, L. Shcherbak, Inorg. Mater. 22 (1986) 1438–1442.
[77] G.W. Blackmore, E.D. Jones, J.B. Mullin, N.M. Stewart, Mater. Sci. Eng. B 16 (1993) 186–190.
[78] H. Kato, S. Takayanagi, Jpn. J. Appl. Phys. 2 (1963) 250–251.
[79] M. Jokozawa, H. Kato, S. Takayanagi, Denki Kagaku (in Japan.) 43 (1966) 828–833.
[80] S.S. Chern, F.A. Kröger, Phys. Status Solidi A 25 (1974) 215–222.
[81] E. Watson, D. Shaw, J. Phys. C Solid State 16 (1983) 515–537.
[82] P. Fochuk, O. Panchuk, D. Belotskyi, U. Kazmiruk, The diffusion and solubility of Ga in CdTe (in Russian), Deposited in UkrNIINTI (Kiev, Ukraine) N 2090, Uk-85, pp. 1–11, 1985.
[83] M. Yokozawa, I. Teramoto, Nat. Technol. Rept. (in Japan) 8 (1962) 428–435.
[84] O. Panchuk, P. Feychuk, L. Shcherbak, Inorg. Mater. 19 (1983) 327–330.
[85] M. Joerger, M. Laasch, T. Kunz, M. Fiederle, J. Meinhardt, K.W. Benz, K. Scholz, W. Wendl, G. Müller-Vogt, Cryst. Res. Technol. 32 (1997) 1103–1113.
[86] V. Babentsov, V. Corregidor, J.L. Castano, M. Fiederle, T. Feltgen, K.W. Benz, E. Dieguez, Cryst. Res. Technol. 36 (2001) 5250542.
[87] P. Fochuk, O. Panchuk, O. Korovyanko, J. Alloys Comp. 371 (2004) 10–14.
[88] P. Fochuk, O. Korovyanko, I. Turkevych, O. Panchuk, Inorg. Mater. 38 (2002) 350–354.
[89] P. Höschl, S. Kubalkova, Ceskoslov. Časop. Fiz. (in Czech) B 22 (1972) 530–533.
[90] S.S. Chern, H.R. Vydyanath, F.A. Kröger, J. Solid State Chem. 14 (1975) 33–43.
[91] S.S. Chern, The Defect Structure and Tracer Self-Diffusion in CdTe, Ph.D. Thesis., University of Southern California, Los-Angeles, 1973.
[92] K. Zanio, H. Montano, D.F. Krajenbrink, Appl. Phys. Lett. 27 (1975) 159–162.
[93] J.W. Griffith, R. Lundquist, R. Platzer, J.A. Gardner, G. Karczewski, J.K. Furdyna, Mater. Sci. Forum 143–147 (1994) 405–410.
[94] W. Stadler, B.K. Meyer, D.M. Hoffmann, B. Kowalski, Mater. Sci. Forum 143–147 (1994) 399–404.
[95] L. Shcherbak, P. Feichuk, P. Fochuk, O. Panchuk, J. Cryst. Growth 161 (1996) 219–222.
[96] V. Ostheimer, T. Filz, J. Hamann, S. Lauer, D. Weber, H. Wolf, T. Wichert, Mater. Sci. Forum 258–263 (1997) 1341–1346.
[97] D. Wegner, B. Meyer, J. Phys. Condens. Matter 1 (1989) 5403–5410.
[98] U. Reislöhner, N. Achtziger, M. Rüb, M. Witthuhn, J. Cryst. Growth 159 (1996) 372–375.
[99] U. Reislöhner, N. Achtziger, W. Witthuhn, J. Cryst. Growth 197 (1999) 576–580.
[100] P. Fochuk, O. Korovyanko, I. Turkevych, O. Panchuk, P. Siffert, J. Cryst. Growth 207 (1999) 273–277.
[101] S. Seto, K. Suzuki, V.N. Abastillas, K. Inabe, J. Cryst. Growth 214–215 (2000) 974–978.

[102] R. Grill, P. Fochuk, J. Franc, B. Nahlovsky, P. Höschl, P. Moravec, Z. Zakharuk, Ye. Nykonyuk, O. Panchuk, Phys. Status Solidi B 243 (2006) 787–793.
[103] C. Gely, C. Corbel, R. Triboulet, Comp. Rend. Acad. Sci. Ser. II 309 (1989) 179–182.
[104] C. Gely-Sykes, C. Corbel, R. Triboulet, Solid State Commun. 80 (1991) 79–83.
[105] H. Kauppinen, L. Baroux, K. Saarinen, C. Corbel, P. Hautojarvi, J. Phys. Condens. Matter 9 (1997) 5495–5505.
[106] D.J. Chadi, J. Cryst. Growth 138 (1994) 295–300.
[107] D.J. Chadi, Phys. Rev. Lett. 72 (1995) 534–538.
[108] C.H. Park, D.J. Chadi, Phys. Rev. B 52 (1995) 11884–11890.
[109] S. Lany, H. Wolf, T. Wichert, Mater. Res. Soc. Proc. 763 (2003) 1.3.1–1.6.1.
[110] S. Lany, H. Wolf, T. Wichert, Phys. Rev. Lett. 92 (2004) 225504/225501–225504/225504.
[111] Z. Rivera-Alvarez, L. Hernandez, M. Becceril, A. Picos-Vega, O. Zelaya-Angel, R. Ramirez-Bon, J.R. Vargas-Garcia, Solid State Commun. 113 (2000) 621–624.
[112] Ye. Nykonyuk, O. Panchuk, P. Feichuk, N. Kuchma, L. Shcherbak, Inorg. Mater. 22 (1986) 1442–1446.
[113] O. Korovyanko, High-Temperature Defect Equilibrium in CdTe doped by Group III Elements (in Ukrainian), Ph.D. Thesis., Chernovtsy University, Chernovtsy, 2003.
[114] L. Chibani, M. Hage-Ali, J.A. Stoquert, J.P. Koebel, P. Siffert, Appl. Surf. Sci. 50 (1991) 177–180.
[115] L. Chibani, M. Hage-Ali, J.A. Stoquert, J.P. Koebel, P. Siffert, Mater. Sci. Eng. B 16 (1993) 202–206.
[116] L. Shcherbak, Investigation of the defect structure of CdTe, doped by Germanium and Tin (in Russian), Ph.D. Thesis., University of Chernovtsy, Chernovtsy, 1977.
[117] L. Shcherbak, P. Feichuk, O. Panchuk, Inorg. Mater. 34 (1998) 17–22.
[118] O. Panchuk, L. Shcherbak, P. Feichuk, A. Savitskyi, Inorgan. Mater. 14 (1978) 41–44.
[119] O. Panchuk, L. Shcherbak, P. Feichuk, A. Savitskiy, D. Belotskiy, Inorg. Mater. 14 (1978) 38–41.
[120] O. Parfenyuk, M. Ilashchuk, K. Ulyanitskii, P. Fochuk, O. Strilchuk, S. Krylyuk, D. Korbutyak, Semiconductors 40 (2006) 143–147.
[121] P. Fochuk, (private communication).
[122] V. Matlak, Ye. Nykonyuk, A. Savitskyi, K. Tovstyuk, Sov. Phys. Semicond. 6 (1972) 1760–1761.
[123] Ye. Nykonyuk, V. Matlak, R. Ivanchuk, A. Savitskyi, in "Fizychna Elektronika" (in Ukrainian), Lviv, Vyshcha Shkola, 1970, pp. 8–14.
[124] A. Savitskiy, Ye. Nykonyuk, V. Matlak, R. Ivanchuk, V. Zayachkivskiy, in Pytannya suchasnoho pryrodoznavstva (in Ukrainian), Lviv, Vyshcha Shkola, 1975, pp. 52–59.
[125] L. Shcherbak, Ye. Nykonyuk, O. Panchuk, A. Savitskiy, P. Feichuk, V. Matlak, Inorg. Mater. 13 (1977) 415–418.
[126] F.A. Kröger, Rev. Phys. Appl. 12 (1975) 205–210.
[127] O. Panchuk, L. Shcherbak, Ye. Nykonyuk, A. Savitskiy, Inorg. Mater. 16 (1980) 433–437.
[128] G. Brunthaler, W. Jantsch, U. Kaufmann, J. Schneider, Phys. Rev. B 31 (1985) 1239–1243.
[129] W. Jantsch, G. Brunthaler, Appl. Phys. Lett. 46 (1985) 665–668.
[130] W. Jantsch, G. Hendorfer, J. Cryst. Growth 101 (1990) 404–411.
[131] J.E. Jaffe, J. Appl. Phys. 99 (2006) 033704.
[132] C. Scharager, P. Siffert, P. Höschl, P. Moravec, M. Vaneček, Phys. Status Solidi (a) 66 (1981) 87–91.
[133] M. Fiederle, V. Babentsov, J. Franc, A. Fauler, K.W. Benz, R. James, E. Cross, J. Cryst. Growth 243 (2002) 77–86.
[134] B. Farid, Spectroscopic investigation and characterization of photorefractivity of the CdTe:Ge semiconductor (in French), Ph.D. Thesis., University D.Diderot, Paris, 2000.

[135] B. Farid, F. Ramaz, O. Panchuk, P. Fochuk, T. Arnoux, J.C. Launay, B. Briat, Opt. Mater. 18 (2001) 7–11.
[136] E. Rzepka, Y. Marfaing, M. Cuniot, R. Triboulet, Mater. Sci. Eng. B 16 (1993) 262–267.
[137] O. Panchuk, A. Savitskiy, P. Fochuk, Ye. Nykomyuk, O. Parfenyuk, L. Shcherbak, M. Ilashchuk, L. Yatsunyk, P. Feichuk, J. Cryst. Growth 197 (1999) 607–611.
[138] P. Gorley, O. Parfenyuk, M. Ilashchuk, I. Nikolayevych, Inorg. Mater. 41 (2005) 1266–1269.
[139] N. Sochinskii, M. Lozano, G. Pellegrini, M. Ullan, Nucl. Instrum. Methods A 568 (2006) 451–454.
[140] Ye. Nikonyuk, Z. Zakharuk, A. Rarenko, Ye. Rybak, M. Kuchma, V. Shlyakhoviy, P. Fochuk, Phys. Stat. Solidi C 4 (2006) 750–753.
[141] O. Parfenyuk, Ye. Nykonyuk, A. Savitskyi, M. Ilashchuk, V. Matlak, P. Pavlin, Sov. Phys. Semicond. 14 (1980) 1662–1664.
[142] O. Parfenyuk, A. Savitskiy, P. Pavlin, A. Albota, Izvestiya Vuzov SSSR, Fizika (in Russian) 4 (1986) 66–69.
[143] A. Savitskiy, O. Parfenyk, M. Ilashchuk, K. Ulyanitsky, A. Savchuk, P. Gorley, Opt. Mater. 18 (2001) 167–169.
[144] J. Franc, M. Fiederle, V. Babentsov, A. Fauler, K.W. Benz, R. James, J. Electron. Mater. 32 (2003) 772–777.
[145] R. Grill, I. Turkevych, J. Franc, E. Belas, P. Moravec, P. Höschl, Phys. Status Solidi C 4 (2004) 727–730.
[146] J. Franc, H. Elhadidy, V. Babentsov, A. Fauler, M. Fiederle, J. Mater. Res. 21 (2006) 1025–1032.
[147] A. Savitskiy, O. Parfenyuk, M. Ilashchuk, P. Pavlin, Inorg. Mater. 25 (1986) 1564–1567.
[148] P.A. Gorley, Y. Vorobiev, V. Makhniy, O. Parfenyuk, M. Ilashchuk, J. Gonzalez-Hernandez, P.P. Gorley, Mater. Sci. Eng. B 99 (2003) 584–587.
[149] A. Savitsky, O. Parfenyuk, M. Ilashchuk, A. Savchuk, S. Chupyra, Semiconductors 38 (2004) 499–504.
[150] H.-J. Schulz, J. Kreissl, Opt. Mater. 4 (1995) 202–209.
[151] P. Fochuk, O. Panchuk, D. Bilotskyi, Inorg. Mater. 22 (1986) 438–439.
[152] R.B. Hall, H.H. Woodbury, J Appl. Phys. 39 (1968) 5361–5365.
[153] Y.Y. Loginov, P.D. Brown, K. Durose, J. Woods, N. Thompson, A. Brinkman, A. Alnajjar, J. Cryst. Growth 117 (1992) 259–265.
[154] E. Hoonnivathjana, E.D. Jones, I.V. Viney, L.J. Duckers, J. Electron. Mater. 27 (1998) 610–614.
[155] E. Molva, K. Saminadayar, J.L. Pautrat, E. Ligeon, Solid State Commun. 48 (1983) 955–960.
[156] F.F. Morehead, G. Mandel, Phys. Lett. 10 (1964) 5–6.
[157] B.V. Dutt, W.G. Spitzer, J. Appl. Phys. (1977) 954–960.
[158] F.A. Selim, F.A. Kröger, J. Electrochem. Soc. 124 (1977) 401–408.
[159] K. Ramachandran, T.M. Haridsasan, J. Physiq. Colloq. 42 (1981) 679–681.
[160] N.V. Agrinskaya, O.A. Matveev, A.I. Terentev, V.V. Shashkova, Sov. Phys. Semicond. 23 (1989) 274–276.
[161] Yu. Obedzynska, P. Fochuk, Z. Zakharuk, O. Panchuk, Bull. Chernivtsi Univ.: Chem (in Ukrainian), Chernivtsi 306 (2006) 103–106.
[162] Y. Iwamura, S. Yamamori, Jpn. J. Appl. Phys. 24 (1985) 361–362.
[163] Ye. Nikonyuk, Z. Zakharuk, V. Shlyakhovyi, P. Fochuk, A. Rarenko, Semiconductors 35 (2001) 405–408.
[164] P. Fochuk, R. Grill, Ye. Nykonyuk, J. Krustok, N. Armani, Z. Zakharuk, M. Grossberg, O. Panchuk, IEEE Trans. Nucl. Sci. 54 (4-1) (2007) 763–768.
[165] S. Neretina, N.V. Sochinskii, P. Mascher, E. Saucedo, J. Chevallier, B.L. Sopori, M. Tabe, P. Kiesel, Mater. Res. Soc. Symp. Proc. 864 (2005) 567–572.

[166] E. Saucedo, O. Martinez, C.M. Ruiz, O. Vigil-Galan, I. Benito, L. Fornaro, N.V. Sochinskii, E. Dieguez, J. Cryst. Growth 291 (2006) 416–423.

[167] E. Saucedo, C.M. Ruiz, V. Bermudez, E. Dieguez, E. Gombia, A. Zapettini, A. Baraldi, N.V. Sochinskii, J. Appl. Phys. 100 (2006) 104901–104907.

[168] P. Christmann, J. Kreissl, D.M. Hofmann, B.K. Meyer, R. Schwarz, K.W. Benz, J. Cryst. Growth 161 (1996) 259–263.

[169] G. Bremond, A. Zerrai, G. Marrakchi, A. Aoudia, Y. Marfaing, R. Triboulet, M.C. Busch, J.M. Koebbel, M. Hage-Ali, P. Siffert, Opt. Mater. 4 (1995) 246–251.

[170] P. Fochuk, D. Belotskyi, O. Panchuk, R. Boichuk, Segregation of Se in CdTe (in Russian), Deposited at UkrNIINTI, 2091, Uk- 85, Kiev (Ukraine), (1985).

[171] F. Vodovatov, G. Indenbaum, A. Vanyukov, Sov. Phys. Semicond. 12 (1970) 17–19.

[172] H. Kato, M. Yokozawa, S. Takayanagi, Jpn. J. Appl. Phys. 4 (1965) 1019.

[173] D. Shaw, Semicond. Sci. Technol. 12 (1997) 1079–1091.

[174] K. Akimoto, H. Okuyama, M. Ikeda, Y. Mori, Appl. Phys. Lett. 60 (1992) 91–93.

[175] I. Stefaniuk, M. Bester, I.S. Virt, M. Kuzma, Acta Phys. Polonica A 108 (2005) 413–418.

[176] D. Shaw, E. Watson, J. Phys. C: Solid State 17 (1984) 4945–4950.

[177] K. Sadaiyandi, K. Ramachandran, Phys. Stat. Solidi B 160 (1990) K85–88.

[178] E.D. Jones, J. Malzbender, J.B. Mullin, N. Shaw, J. Phys. Condens. Matter 6 (1994) 7499–504.

[179] J. Malzbender, E.D. Jones, N. Shaw, J.B. Mullin, Mater. Sci. Forum 182–184 (1995) 57–62.

[180] J. Malzbender, E.D. Jones, N.N. Shaw, J.B. Mullin, Adv. Mater. Opt. Electron. 5 (1995) 157–161.

[181] E.D. Jones, J.C. Clark, J. Malzbender, J.B Mullin, N. Shaw, A.W. Brinkman, J. Electron. Mater. 24 (1995) 581–585.

[182] N.Y. Jamil, D. Shaw, Semicond. Sci. Technol. 10 (1995) 952–958.

[183] P. Siffert, A. Cornet, R. Stuck, R. Triboulet, Y. Marfaing, IEEE Trans. Nucl. Sci. 1 (1975) 221–225.

[184] G. Neu, Y. Marfaing, R. Triboulet, M. Escorne, Rev.Phys. Appl. 12 (1977) 263–266.

[185] N.V. Agrinskaya, E.N. Arkadeva, Nucl. Instrum. Methods A 283 (1989) 260–262.

[186] C. Canali, G. Ottaviani, R.O. Bell, F.V. Wald, J. Phys. Chem. Solids 35 (1974) 1405–1413.

[187] L. Chibani, M. Hage-Ali, P. Siffert, J. Cryst. Growth 161 (1996) 153–158.

[188] M. Hage-Ali, C. Scharager, J. Koebel, P. Siffert, Nucl. Instrum. Methods 176 (1980) 499–502.

[189] D.M. Hoffmann, P. Omling, H.G. Crimmeis, B.K. Meyer, K.W. Benz, D. Sinerius, Phys. Rev. B Condens. Matter 45 (1992) 6247–6250.

[190] P. Höschl, P. Moravec, J. Franc, E. Belas, R. Grill, Nucl. Instrum. Methods A 322 (1992) 371–374.

[191] P. Höschl, J. Franc, R. Grill, P. Moravec, E. Belas, Mater. Sci. Eng. B 16 (1993) 215–218.

[192] T. Ido, A. Hertel, R. Triboulet, Y. Marfaing, J. Phys. Chem. Solids 48 (1987) 536–539.

[193] K. Injae, P. Jinki, S. Dongyoon, K. Eunsung, C. Ilsoo, C. Gwangjae, S. Jonghyung, M. Byungdon, S. Jeongchil, K. Sunung, P. Mannjang, Ungyong Mulli (J. Korean Phys. Soc. in Korean) 11 (1998) 594–599.

[194] H. Kauppinen, L. Baroux, K. Saarinen, C. Corbel, P. Hautojarvi, J. Phys. C Condens. Matter 9 (1997) 5495.

[195] T.A. Kuhn, W. Ossau, A. Waag, R. Bicknell-Tassius, G. Landwehr, J. Cryst. Growth 117 (1992) 660–665.

[196] J. Lee, N.C. Giles, C.J. Summers, J. Appl. Phys. 77 (1995) 4544–4547.

[197] J. Lee, N.C. Giles, D. Rajaoce, C.J. Summers, Appl. Phys. 78 (1997) 5669–5670.

[198] M. Samimi, B. Biglari, M. Hage-Ali, J. Koebel, P. Siffert, Phys. Status Solidi A 100 (1987) 251–258.

[199] P. Siffert, B. Rabin, H.Y. Tabatai, R. Stuck, Nucl. Instrum. Methods A 150 (1978) 31–37.

[200] W. Stadler, D.M. Hofmann, B.K. Meyer, R. Krause-Rehberg, A. Polity, Th. Abgarjan, M. Salk, K.W. Benz, M. Azoulay, Acta Phys. Polonica A 88 (1995) 921–924.
[201] R. Stuck, A. Cornet, C. Scharager, P. Siffert, J. Phys. Chem. Solids 37 (1976) 989–997.
[202] N.V. Agrinskaya, O.A. Matveev, Rev. Phys. Appl. 12 (1977) 235–237.
[203] A. Castaldini, A. Cavallini, B. Fraboni, P. Fernandez, J. Piqueras, Appl. Phys. Lett. 69 (1996) 3510–3512.
[204] O.A. Matveev, A.I. Terentev, Semiconductors 32 (1998) 144–147.
[205] R. Legros, Y. Marfaing, R. Triboulet, Rev. Phys. Appl. 12 (1977) 245–248.
[206] R. Legros, Y. Marfaing, R. Triboulet, J. Phys. Chem. Solids 39 (1978) 179–184.
[207] D.J. Chadi, C.H. Park, Mater. Sci. Forum 196–201 (1995) 285–292.
[208] M. Hage-Ali, P. Siffert, Mater. Res. Soc. Sympos. Proc. 303 (1992) 479–486.
[209] Ch. Steer, L. Chibani, J.M. Koebel, M. Hage-Ali, P. Siffert, Mater. Res. Soc. Sympos. Proc. 302 (1992) 457–462.
[210] V.I. Sokolov, Sov. Phys. Solid State 29 (1987) 1061–1061–1065.
[211] G.A. Slack, S. Galginaitis, Phys. Rev. A 133 (1964) 253–268.
[212] B.M. Vul, V.S. Ivanov, V.A. Rukavishnikov, V.M. Salman, V.A. Chapnin, Sov. Phys. Semicond. 6 (1972) 1106–1109.
[213] O. Panchuk, R. Fesh, A. Savitskyi, L. Shcherbak, Inorg. Mater. 17 (1981) 1004–1006.
[214] G. Brunthaler, U. Kaufmann, J. Schneider, J. Appl. Phys. 56 (1984) 2974–2976.
[215] K. Lischka, G. Brunthaler, W. Jantsch, J. Cryst. Growth 72 (1985) 355–359.
[216] B.K. Meyer, H. Linke, P. Omling, M. Salk, K.W. Benz, Mater. Sci. Eng. B 16 (1993) 243–245.
[217] A.J. Szadkowski, J. Phys. Condens. Matter 2 (1990) 9853–9859.
[218] A. Radlinski, J. Phys. C 12 (1979) 4477–4482.
[219] U. Kaufmann, J. Windscheif, G. Brunthaler, J. Phys. C Solid State Phys. 17 (1984) 6169–6176.

CHAPTER VII

Impurity Compensation

Yves Marfaing

Contents

1. INTRODUCTION

Impurity compensation refers to the fact that the doping efficiency of a majority impurity is reduced by the presence of oppositely charged defect centres. This is a general situation in semiconductors because several types of impurities and defects usually coexist. However, it is common to distinguish between different cases, depending on the compensation degree.

Weak or moderate compensation applies to well-defined n-type or p-type semiconductors. It restricts the doping efficiency of the doping

GEMaC, CNRS/UVSQ, F-92195 Meudon Cedex, France

CDTE and Related Compounds
DOI: 10.1016/B978-0-08-046409-1.00007-1

impurity, within the solubility limit of the latter. The concentration and nature of the compensating defects could be closely linked to the doping impurity: this is the so-called self-compensation phenomenon, of frequent occurrence in II–VI compounds.

Strong compensation can lead to a semi-insulating state. This "pathological" situation is indeed observed in CdTe and has been the subject for numerous analyses and speculations. It also makes possible such important applications as optical modulators and radiation detectors.

In the following we will first consider the case of "compensated conductivity" which refers to n-type or p-type doped CdTe. Then we will specifically discuss the semi-insulating behaviour of CdTe. All these phenomena involve in some manner crystal native defects the properties of which will be critically examined. Finally we will briefly describe the behaviour of amphoteric impurities which could compensate themselves without the contribution of defects.

2. COMPENSATED CONDUCTIVITY

Preparation of doped CdTe and analysis of the related electrical conductivity have led to consider the implication of several types of defects in doping and compensation. We do not intend here to give a complete analysis of the doping processes in CdTe but rather to identify and describe the main compensating centres.

2.1. n-Type CdTe

Let us consider the doping behaviour of the donor impurity In_{Cd} incorporated in the total concentration $[In]_S$. Compensation is commonly attributed to two main defects: the cadmium vacancy V_{Cd} (double acceptor) and the single acceptor complex (V_{Cd}–In_{Cd}). Electrical neutrality and mass conservation are expressed by

$$n = [In^+] - [V_{Cd}^-] - 2V_{Cd}^{2-} - [(V_{Cd} - In_{Cd})^-], \quad (1)$$

$$[In]_S = [In^+] + [(V_{Cd} - In_{Cd})^-], \quad (2)$$

where n is the free electron concentration. Equation (2) implies that the indium donor and the acceptor complex are fully ionised in n-type CdTe at the doping temperature.

The role of cadmium vacancies was recognised very early [1–3] while the presence of donor–vacancy pairs was not postulated in some of these works [3].

Ionisation of the cadmium vacancy is described by the following relations (omitting the degeneracy factors):

$$[V_{Cd}^{-}] = [V_{Cd}^{o}]\exp\left(\frac{E_F - E_1}{k_B T}\right), \tag{3}$$

$$[V_{Cd}^{2-}] = [V_{Cd}^{o}]\exp\left(\frac{2E_F - E_1 - E_2}{k_B T}\right), \tag{4}$$

where k_B is Boltzmann's constant and T is the temperature. E_F is the Fermi level and E_1, E_2 are the first and second ionisation levels of the vacancy in the band gap. V_{Cd}^{o} designates the neutral cadmium vacancy the concentration of which is directly linked to the Cd chemical potential in the external phase in equilibrium with the solid. For a gaseous phase the pertinent quantity is the partial pressure P_{Cd}, hence

$$[V_{Cd}^{o}] = K_{CdV} P_{Cd}^{-1}. \tag{5}$$

One clearly sees from Eqs. (3) and (4) that the concentrations of negatively charged vacancies increase when the Fermi level goes up, i.e. for n-type doping: this represents the self-compensation effect. By expressing E_F as a function of the electron concentration n in a non-degenerate semiconductor one obtains the expressions commonly used in the analyses of the defect structure [4–8]

$$[V_{Cd}^{-}] = K_{nCV}^{-} K_{CdV} n P_{Cd}^{-1} \tag{6}$$

$$[V_{Cd}^{2-}] = K_{nCV}^{2-} K_{CdV} n^2 P_{Cd}^{-1} \tag{7}$$

K_{nCV}^{-} and K_{nCV}^{2-} are ionisation reaction constants of the form

$$K_{nCV}^{-} = N_c^{-1}\exp[(E_c - E_1)/k_B T],\ K_{nCV}^{2-} = N_c^{-2}\exp[(2E_c - E_1 - E_2)/k_B T], \tag{8}$$

where N_c is the effective density of states in the conduction band and E_c is the bottom energy of the latter. Finally the donor–vacancy pair concentration is simply expressed in terms of a pairing constant K_P

$$[(V_{Cd} - In_{Cd})^{-}] = K_P[V_{Cd}^{2-}][In^{+}]. \tag{9}$$

Combining the above relations leads to

$$[In]_S = \frac{n(1 + K_{nCV}^{-} K_{CdV} P_{Cd}^{-1} + 2n K_{nCV}^{2-} K_{CdV} P_{Cd}^{-1})(1 + n^2 K_P K_{nCV}^{2-} K_{CdV} P_{Cd}^{-1})}{1 - n^2 K_P K_{nCV}^{2-} K_{CdV} P_{Cd}^{-1}}. \tag{10}$$

A solution of this form was already given in Ref. [9]. It reveals the existence of a doping limit at high indium concentration: $n_{max} = (K_P K_{nCV}^{2-} K_{CdV})^{-1/2} P_{Cd}^{1/2}$ in agreement with experiments (Fig. 1) [10]. For a given total impurity content the resulting electron concentration increases with increasing P_{Cd} as shown in Fig. 2 [2]. The respective contributions of Cd vacancies and donor–vacancy pairs, as deduced from the defect model used in this work, are also shown. These observations support the

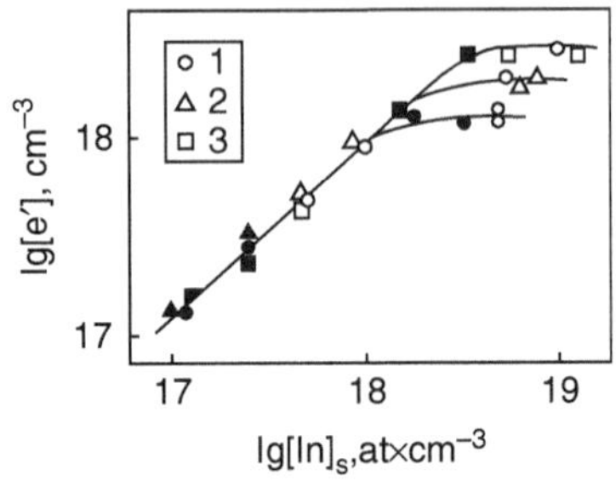

Figure 1 Electron concentration in CdTe at equilibrium temperature as a function of indium concentration. Temperatures: (1) 973 K; (2) 1073 K; (3) 1173 K (Reprinted with permission from Ref. [10]. Copyright [1996] by Elsevier).

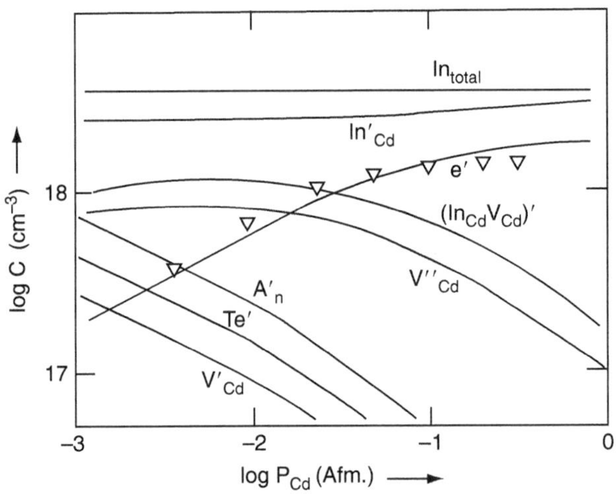

Figure 2 Defect concentration isotherms for CdTe doped with 3.6×10^{18} In cm^{-3} at 700 °C; ∇: experimental electron concentration (Reprinted with permission from Ref. [4]. Copyright [1975] by Elsevier).

compensation model presented here. Furthermore direct evidence of the involvement of vacancies and/or donor–vacancy pairs has been obtained from positron annihilation and perturbed angular-correlation experiments [11–15]. A good example shown in Fig. 3 is the correlation occurring between the electron concentration measured in heavily iodine-doped CdTe and an annihilation parameter related to the presence of Cd vacancies [12]. The detailed properties of the compensating defects will be discussed later on.

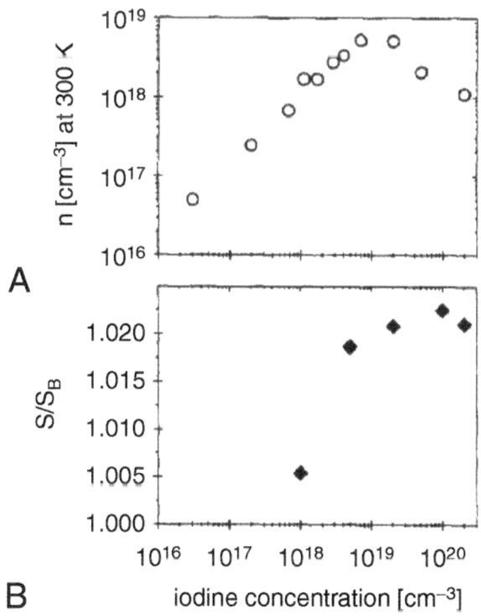

Figure 3 Electron concentration n at room temperature in CdTe and related positron annihilation parameter S/S_B as a function of iodine concentration (Reprinted with permission from Ref. [12]. Copyright [1996] by Elsevier).

2.2. p-Type CdTe

Doping with acceptor impurities is much less documented. Observation of p-type conductivity is hampered by the fact that most of the acceptors such as Cu_{Cd}, P_{Te} have large ionisation energies compared to the group III and group VI donors. Furthermore the compensation scheme does not appear as simple as on the n-type side.

The first compensating point defects to be considered are those related to Cd excess, Cd_i and V_{Te}, which are double donor centres. The latter one is now considered to be of negligible importance with respect to Cd_i from *ab initio* calculations [16, 17]. Doping conditions under Te excess is a natural way to reduce the compensating effect of Cd_i. However, another double donor defect then appears: the antisite Te_{Cd}. Such a defect was not considered in the early studies. Its importance is now supported from theoretical [16, 17] and experimental works [18, 19]. The neutrality equation should write

$$p - [A^-] + [V_{Cd}^-] + 2[V_{Cd}^{2-}] - [Cd_i^+] - 2[Cd_i^{2+}] - [Te_{Cd}^+] - 2[Te_{Cd}^{2+}], \quad (11)$$

where $[A^-]$ refers to an undefined ionised acceptor impurity A.

Ionisation of a donor defect D can be represented by formal relations like

$$[D^+] = [D^\circ](p/N_v)\exp\frac{E_1 - E_v}{k_B T} \equiv pK_{pD}^+[D^\circ], \quad (12)$$

$$[D^{2+}] = [D^\circ](p/N_v)^2\exp\frac{E_1 + E_2 - 2E_v}{k_B T} \equiv p^2K_{pD}^{2+}[D^\circ], \quad (13)$$

where E_1 and E_2 are the defect energy levels and N_v, E_v now refer to the valence band.

Equilibrium with an external gaseous phase is described by

$$[\mathrm{Cd_i^o}] = K_{\mathrm{CdI}} P_{\mathrm{Cd}}, \tag{14}$$

$$[\mathrm{Te_{Cd}^o}] = K_{\mathrm{TeA}} P_{\mathrm{Cd}}^{-2}. \tag{15}$$

The neutrality equation then takes the form

$$\begin{aligned} p = [\mathrm{A^-}] &+ p^{-1} K_g K_{\mathrm{CdV}} (K_{n\mathrm{CV}}^- + 2p^{-1} K_g K_{n\mathrm{CV}}^{2-}) P_{\mathrm{Cd}}^{-1} \\ &- p K_{\mathrm{CdI}} (K_{p\mathrm{CI}}^+ - 2p K_{p\mathrm{CI}}^{2+}) P_{\mathrm{Cd}} - p K_{\mathrm{TeA}} (K_{p\mathrm{TA}}^+ - 2p K_{p\mathrm{TA}}^{2+}) P_{\mathrm{Cd}}^{-2}, \end{aligned} \tag{16}$$

where p is the free hole concentration related to the electron concentration n by

$$np = N_c N_v \exp(-E_g/k_B T) \equiv K_g, \tag{17}$$

where E_g is the band gap energy. The two last terms of Eq. (16) involve the compensating donor centres. Their Cd pressure dependence is opposite, so that their respective contributions change across the existence region of CdTe. At low Cd pressure $\mathrm{Te_{Cd}}$ should be dominant over $\mathrm{Cd_i}$ and should compensate not only the acceptor impurity but also the cadmium vacancy the concentration of which is large in this domain. An example of such a behaviour in acceptor-doped CdTe is given by the theoretical results displayed in Fig. 4 [16]. The doping efficiency is maximum at low Cd pressure as expected. Near the Te-saturated limit there is an almost exact compensation between the two charged defects $\mathrm{V_{Cd}^-}$ and $\mathrm{Te_{Cd}^{2+}}$ so that $p \cong [\mathrm{A^-}]$ at the high temperature equilibrium. This result is

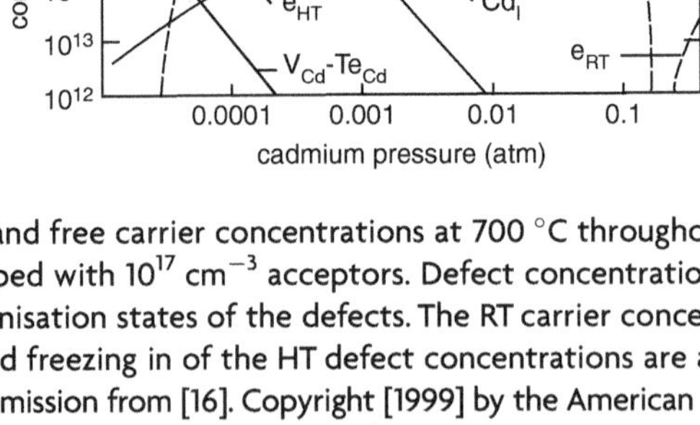

Figure 4 Defect and free carrier concentrations at 700 °C throughout the existence region of CdTe doped with 10^{17} cm^{-3} acceptors. Defect concentrations represent a sum over all the ionisation states of the defects. The RT carrier concentrations obtained after quenching and freezing in of the HT defect concentrations are also shown (Reprinted with permission from [16]. Copyright [1999] by the American Physical Society - http://link.aps.org/abstract/PRB/v60/p8943)

obtained for particular values of the defect parameters which will be now discussed.

2.3. Present status of defect properties

The above phenomenological presentation has allowed us to identify the main compensating defects brought into play. We now have to define more precisely the relevant thermodynamical and electronic parameters.

2.3.1. Cadmium vacancy

The vacancy parameters were usually derived from fits of electrical measurements according to a given defect model [2, 5, 7, 8, 20, 21]. This leads to determine reaction constants like the product $K^{2-}_{n\mathrm{CV}}K_{\mathrm{CdV}}$ (Eq. (7)). If the vacancy energy levels can be obtained from related or independent observations, the formation energy of the neutral vacancy can then be deduced. First, principles *ab initio* calculations appeared more recently [16, 17]. They enable one to evaluate both the neutral vacancy parameters and the ionisation energy levels. A comparison between all these data reveals a rather wide dispersion. Table 1 gathers the energy level values appearing in several works. The position of the second ionisation level is worth considering. The largest value around 0.6–0.7 eV has been favoured in the earlier works because it provides a simple explanation for the formation of semi-insulating CdTe (Section 3). However, more recent studies including direct identification from paramagnetic resonance [22] lead to retain a smaller value in the range 0.2–0.45 eV.

Thermodynamics of V_{Cd} is illustrated in Fig. 5 which displays the temperature dependence of the neutral vacancy concentration in Te-saturated CdTe as computed from several studies. For a given temperature, the variation range is at least of two orders of magnitude. This shows the large uncertainty which still subsists in the properties of the cadmium vacancy. However, it is possible to restrain this variation range by considering the estimations obtained from positron annihilation

Table 1 Ionisation energy levels of the Cd vacancy in CdTe

$E_1 - E_v$ (eV)	$E_2 - E_v$ (eV)	References
0.05–0.06	0.6–0.7	Exp. [2] 1964 [3, 4, 10, 20, 25]
	<0.47	Exp. [22] 1993
	0.43	Exp. [23] 1997
	0.23	Exp. [24] 1998
0.2	0.8	Theory [16] 1999
0.13	0.21	Theory [17] 2002
$0.12E_g$	$0.3E_g$	Exp. [21] 2002

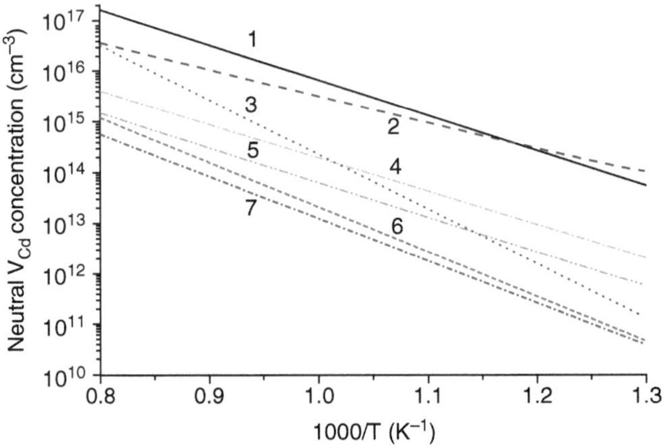

Figure 5 Concentration of the neutral Cd vacancy in Te-saturated CdTe as a function of reciprocal temperature computed from several studies: (1) Chern [4], (2) Berding [16], (3) Brebrick [20], (4) Brebrick [20], (5) Grill [21], (6) Wei [17], (7) Fochuk [7].

experiments. The concentrations of Cd vacancies obtained in these works for Te-rich crystals quenched to room temperature is around 1–2 $\times$ 10^{16} cm^{-3} [26–28]. This should be compared to the results of Fig. 6 which gives the total concentration of Cd vacancies (neutral and ionised) calculated from the above data (Table 1 and Fig. 5) for pure Te-saturated crystals in which $p-n=[V_{Cd}^{-}]+2[V_{Cd}^{2-}]$. One can conclude that the sets of

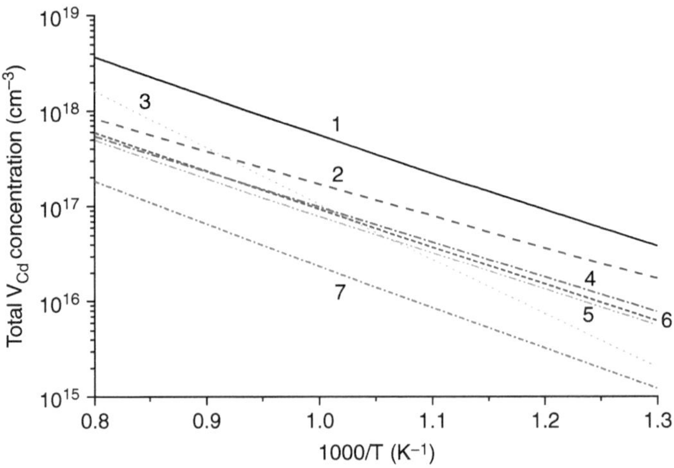

Figure 6 Total concentration of Cd vacancies in pure Te-saturated CdTe as a function of reciprocal temperature computed from several studies: (1) Chern [4], (2) Berding [16], (3) Brebrick [20], (4) Brebrick [20], (5) Grill [21], (6) Wei [17], (7) Fochuk [7].

parameters leading to the lowest values ($\sim 10^{16}$ cm^{-3} at 800–1000 K) have to be preferred.

2.3.2. Donor-cadmium vacancy complex

This defect, also called A-centre, has been studied from a long time by means of photoluminescence and deep centre dedicated techniques like DLTS, TSC, PICTS (e.g. [29–34]). A clear signature of the (Cl_{Te}–V_{Cd}) centre was obtained from EPR and ODMR experiments [35]. The related acceptor energy level is set at 0.12 eV above the valence band. For the (In_{Cd}–V_{Cd}) centre a slightly larger ionisation energy of 0.14 eV is determined [32]. The pairing constant of Eq. (9) takes the general form $K_P = K_{Po}$ $\exp(-E_B/k_BT)$, where E_B is the binding energy of the A-centre. Some values of K_{Po} and E_B extracted from different studies are collected in Table 2. A rather wide dispersion is here also observed.

An evaluation of the A-centre concentrations in Cl-doped CdTe was made from positron annihilation data [37]. It results in rather large concentrations of *A*-centres between 3.6×10^{16} and 1.3×10^{17} cm^{-3}, according to the doping value. Let us finally mention that other types of In-related defect complexes were detected using the perturbed angular-correlation method [38].

2.3.3. Tellurium antisite

The energy levels of this donor defect have not been experimentally determined in a direct way. The available data come from theoretical works and analyses of electrical measurements which assume somewhat arbitrary values. They are collected in Table 3. From the energies of formation given in these works one can calculate the concentrations of neutral Te_{Cd} in specified conditions, for example at the Te-rich phase boundary (Fig. 7). The results are not too much dispersed in the intermediate and low-temperature ranges, except in two works (2 and 5) which make use of similar formation energies.

As a final note for this section one can refer to the global optimised fit of the various high-temperature electron concentration measurements obtained by previous researchers which was recently published [40]. The defects Cd_i, V_{Cd}, Te_i, Te_{Cd}, V_{Te} were considered.

Table 2 Parameters of the reaction constant $K_P = K_{Po} \exp(-E_B/k_BT)$ associated with the Cd vacancy–donor pair formation in CdTe (Eq. (9))

K_{Po} (cm^3)	E_B (eV)	References
2.9×10^{-23}	−0.83	[4]
6×10^{-21}	−0.92	[7]
1.3×10^{-22}	−0.6	[36]

Table 3 Ionisation energy levels of the antisite Te_{Cd} in CdTe

$E_c - E_1$ (eV)	$E_c - E_2$ (eV)	References
0	0.4	Theory [16] 1999
0.34	0.59	Theory [17] 2002
0.05		Exp. [20] 1996
0	$0.5E_g$	Exp. [39] 2005

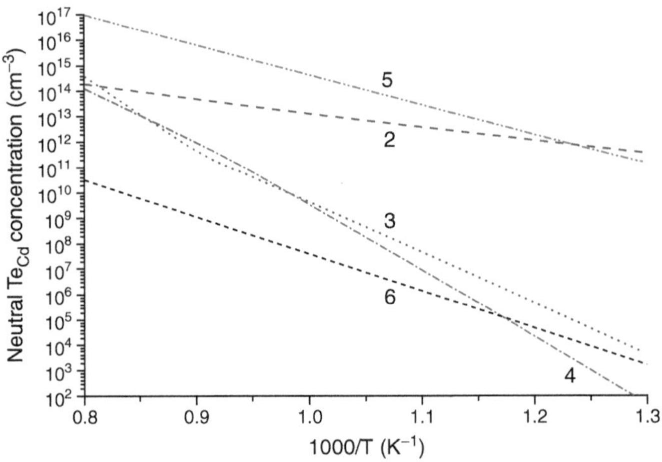

Figure 7 Concentration of the neutral Te antisite in Te-saturated CdTe as a function of reciprocal temperature computed from several studies: (2) Berding [16], (3) Brebrick [20], (4) Brebrick [20], (5) Grill [21], (6) Wei [17].

3. SEMI-INSULATING STATE

De Nobel first observed that donor-doped CdTe could exhibit a sudden conductivity drop at low P_{Cd} pressure as shown in Fig. 8 [1]. This trend logically points to the involvement of cadmium vacancies which were the basis of the early defect structure models [2–6]. On the other hand it is unrealistic to consider that only shallow donors and acceptors in strictly balanced concentrations are implied. Consequently the second ionisation level of V_{Cd} which was assumed in these works to be deep (Table 1) was invoked in favour of the Fermi level stabilisation at a midgap position [41]. The influence of a donor/acceptor deep level on electrical conductivity has been theoretically analysed in several studies [42, 43].

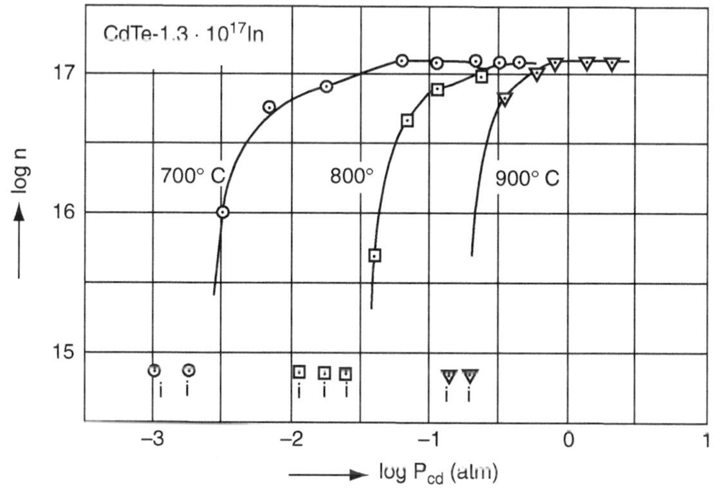

Figure 8 Room temperature electron concentration in CdTe doped with 1.3×10^{17} In cm^{-3} and annealed at several temperatures (700, 800 and 900 °C) in atmospheres of varying Cd pressure (Reprinted with permission from Ref. [1]. Copyright [1959] by Philips).

Figure 9 illustrates the case of a p-type semiconductor containing three defect states, shallow donors (N_D), shallow acceptors (N_{A1}) and deep acceptors (N_{A2}), respectively [42]. Strong compensation occurs when the following inequalities are satisfied

$$N_{A1} < N_D < (N_{A1} + N_{A2}). \tag{18}$$

The striking result is that the conductivity decreases by orders of magnitude for a slight variation in the overall compensation ratio. Physically this means that the Fermi level drops from a position close to the shallow impurity level to essentially pin at the deep level. This behaviour may appear for minority deep level states provided that $N_{A2} > N_D - N_{A1}$. These considerations can be extended in a straightforward way to the opposite case of a n-type semiconductor containing shallow donors (N_{D1}) and acceptors (N_A) where the deep state is a donor one (N_{D2}) [43]. The conditions for compensation are then

$$N_{D1} < N_A < (N_{D1} + N_{D2}). \tag{19}$$

In the framework of this general scheme, two main questions have now to be addressed. The first one is about the actual nature of the deep levels responsible of Fermi level pinning and the related defect structure model. The second one is relative to the application of these models to detector-grade semi-insulating CdTe, which is characterised by a quite low concentration of deep centres.

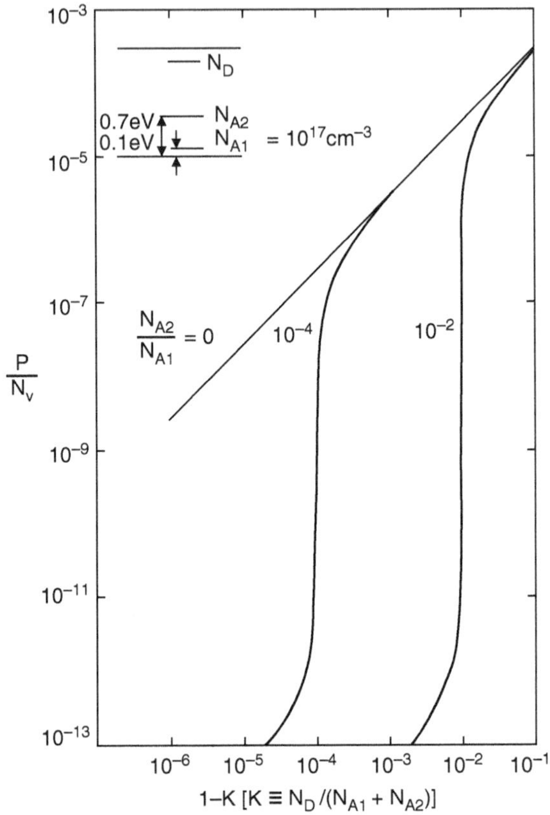

Figure 9 Normalised 300 K hole concentration (p/N_v) plotted versus the difference with unity ($1 - K$) of the compensation ratio $K = N_D/(N_{A1} + N_{A2})$ for different ratios of deep to shallow acceptor centres (N_{A2}/N_{A1}) (Reprinted with permission from Ref. [42]. Copyright [1982] by the American Physical Society - http://link.aps.org/abstract/PRB/v26/p2250).

3.1. Deep levels and related models of semi-insulating (SI) state

For the sake of completeness the category of deep metal impurities should be first mentioned. This includes some group IV elements and transition metals which are deep donors when substituted for Cd, apart from Ni which is a deep acceptor. The relevant energy levels given in various papers [44–47] are collected in Table 4.

Electrical properties of Ge- and Sn-doped CdTe grown from the melt have been published [48, 49]. These studies reveal the strong resistivity increase occurring above some doping concentration (Fig. 10). SI titanium- and vanadium-doped cadmium telluride crystals grown from the vapour phase were also reported [50]. Vanadium-doped CdTe has been extensively studied within the context of photorefractivity [51].

Table 4 Energy levels of group IV and transition metal elements substituted for Cd in CdTe (from Refs. [44–47])

	Ge	Sn	V	Ni
$E_c - E$ (eV)	0.95	0.85	0.93	0.92

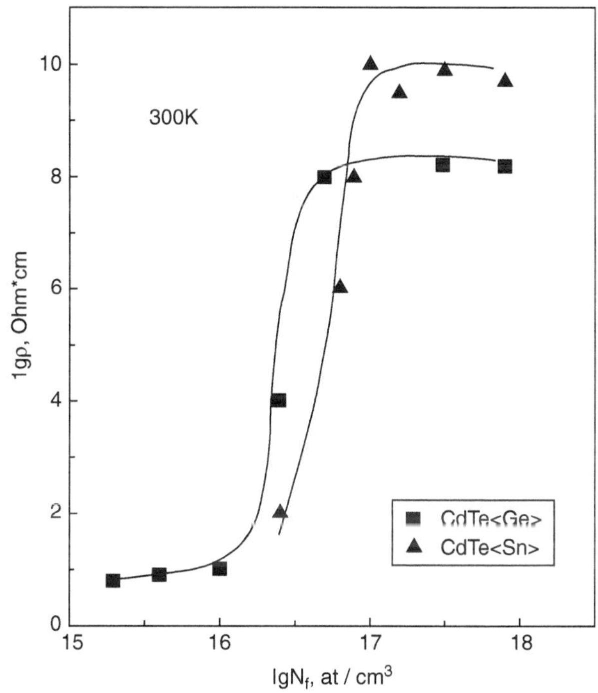

Figure 10 Room temperature resistivity of CdTe doped with Ge and Sn as a function of dopant concentration (Reprinted with permission from Ref. [48]. Copyright [1999] by Elsevier).

The associated deep centres introduced at the concentration of $\sim 10^{15}$–10^{16} cm^{-3} constitute the photo-active states required for the photorefractive effect. However, such concentration of deep traps is too high in regard to applications to radiation detectors. Other ways of preparing SI material where native defects play a significant role have now to be explored.

The de Nobel's experiments presented in Fig. 8 were analysed by Chern et al. [4] who came to the defect structure shown in Fig. 11. In the region of conductivity drop obtained at low Cd vapour pressure electrical neutrality is described by the relation

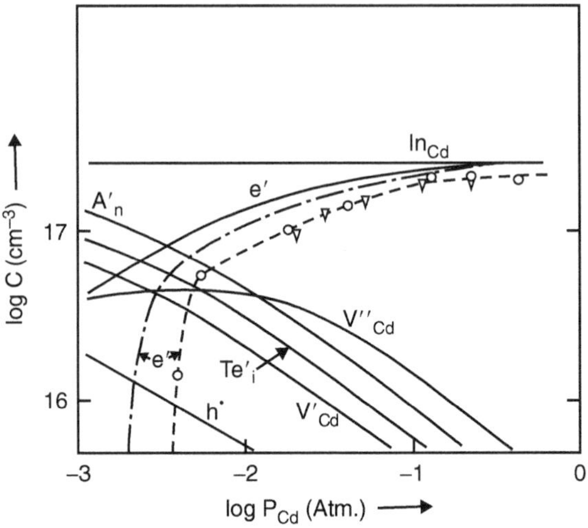

Figure 11 Defect concentration isotherms for CdTe doped with 2.7×10^{17} In cm^{-3} at 700 °C; ∇: experimental electron concentration. The RT carrier concentrations obtained after quenching and freezing in of the HT defect concentrations are also shown: −•−• calculated; -o– experiments by de Nobel [1] with 2×10^{17} In cm^{-3} (Reprinted with permission from Ref. [4]. Copyright [1975] by Elsevier).

$$n - p = [In^+_{Cd}] - [Te^-_i] - [V^-_{Cd}] - 2[V^{2-}_{Cd}]. \tag{20}$$

The defects are assumed to be frozen in upon quenching to room temperature. The semi-insulating state is then defined by the quantities shown in Table 5. D designates the incorporated donor and A is any acceptor other than V_{Cd} (here Te_i). $[V_{Cd}]_S$ is the total concentration of Cd vacancies. The ratio $[V^{2-}_{Cd}]/[V_{Cd}]_S = 0.44$ is the occupation factor of the vacancy second ionisation level set at $E_v + 0.65$ eV. The Fermi level is pinned on this level. A rather similar analysis was done for Cl-doped CdTe but including the donor–vacancy A-centre instead of Te_i in the neutrality Eq. (20) [52]. The Hall coefficient and the conductivity of the SI crystals were fitted in the temperature range 250–450 K. The fit led to the quantities presented in Table 5 and to the determination of the deep acceptor level $E_v + 0.69$ eV around which the Fermi level was pinned. This level was associated to the second ionisation energy of the Cd vacancy.

Another modelling of donor-doped CdTe was done by Berding [16] the results of which are shown in Fig. 12. A range of semi-insulating behaviour appears at P_{Cd} around 10^{-2} atm. Electrical neutrality then approximately writes

$$n - p \cong [D^+] - [V^-_{Cd}] - 2[V^{2-}_{Cd}]. \tag{21}$$

The concentration of antisites Te_{Cd} is negligible in this range. The concentrations of the frozen in defects in the SI state are given in Table 5.

Table 5 Defect properties of donor-doped semi-insulating CdTe annealed at different temperatures under low Cd vapour pressure

T (°C)	P_{Cd} (atm)	[D] (cm^{-3})	$[V_{Cd}]_S$ (cm^{-3})	$[V_{Cd}^{2-}]$ (cm^{-3})	$[Te_{Cd}]_S$ (cm^{-3})	$[Te_{Cd}^{+}]$ (cm^{-3})	$[A^{-}]$ (cm^{-3})	References
700	2×10^{-3}	2×10^{17}	9×10^{16}	4×10^{16}			7×10^{16}	[4]
450	8×10^{-10}	1×10^{17}	2.4×10^{14}	1.1×10^{14}			5×10^{16}	[52]
700	1×10^{-2}	1×10^{17}	9×10^{16}	1×10^{16}				[16]
700	1.3×10^{-5}	1×10^{15}	2.8×10^{17}	2.8×10^{17}	4.3×10^{17}	3×10^{17}	4×10^{14}	[39]

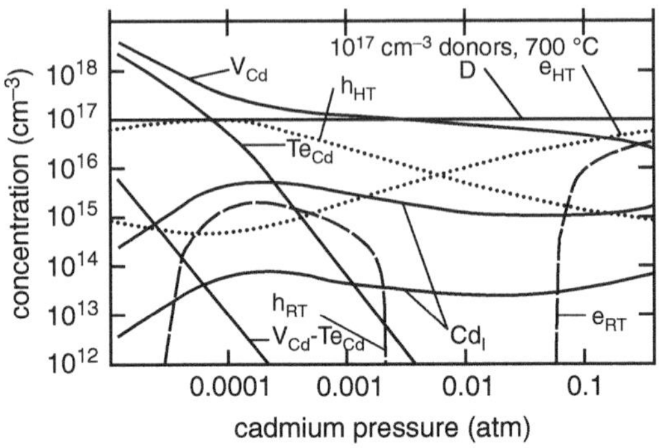

Figure 12 Defect and free carrier concentrations throughout the existence region of CdTe doped with 10^{17} cm^{-3} donors at 700 °C. Defect concentrations represent a sum over all the ionisation states of the defects. The RT carrier concentrations obtained after quenching and freezing in of the HT defect concentrations are also shown (Reprinted with permission from Ref. [16]. Copyright [1999] by the American Physical Society - http://link.aps.org/abstract/PRB/v60/p8943).

As in the previous examples the Fermi level at room temperature is pinned on the vacancy second ionisation level set at $E_v + 0.8$ eV.

More recent analyses have adopted a shallower second ionisation level of V_{Cd} at $0.3E_g$ above the valence band [39, 53]. The characteristics of CdTe doped with 10^{15} donors and annealed under Te-saturated conditions at various temperatures between 700 and 100 °C are displayed in Fig. 13 [39]. Two ranges of semi-insulating behaviour appear. Around 700 °C electrical neutrality is described by

$$p - n = 2[V_{Cd}^{2-}] - [Te_{Cd}^{+}] - 2[Te_{Cd}^{2+}] + [V_{Cd}^{-}] - [D^{+}] + [A^{-}]. \qquad (22)$$

The dominant defects are the two first in the second hand. A^- is the ionised donor vacancy complex $(D^+ - V_{Cd}^{2-})^-$. At room temperature the defects are distributed as indicated in Table 5. The Cd vacancies are fully ionised whereas the second ionisation level of Te_{Cd} set at $E_c - 0.5E_g$ is partially occupied. This pins the Fermi level around midgap whence the SI state. The low-temperature range of Fig. 13 corresponds to samples which are equilibrated from internal Te-rich sources, like Te precipitates. It will be discussed in the following section.

3.2. Application to detector-grade material

Achievement of semi-insulating CdTe is commonly accounted for by the presence of deep levels which pin the Fermi level around midgap. In the usual defect structure models presented above the concentration of such deep centres can reach values up to 10^{16}–10^{17} cm^{-3} (Table 5). These concentrations are highly detrimental to the operation of X- and γ-rays

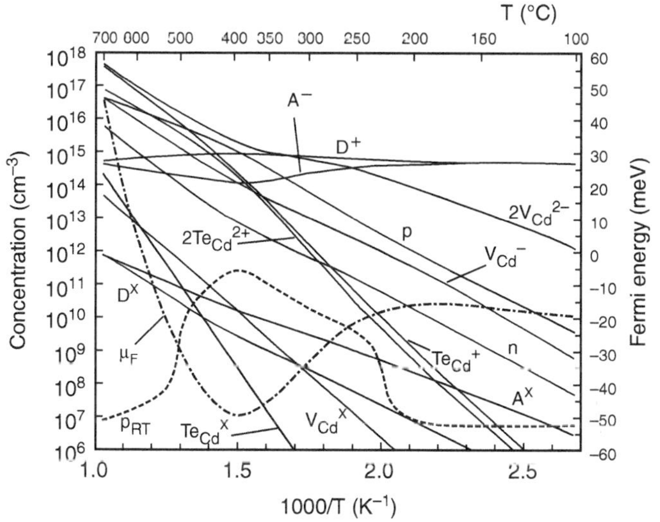

Figure 13 Defect and carrier concentrations in CdTe containing 10^{15} shallow donors cm^{-3} and a midgap trap density $N_{deep} = 10^{13}$ cm^{-3} as a function of annealing temperature in Te-saturated conditions. The Fermi energy μ_F is referred to the intrinsic level and depicted on the right axis. The RT hole concentration is shown by the dash line (Reprinted with permission from Ref. [39]. Copyright [2005] by the IEEE).

CdTe detectors [54, 55]. Long-carrier lifetimes of $\sim 10^{-6}$ s require a deep level concentration as low as 10^{11}–10^{13} cm^{-3}, depending upon the capture cross section of the traps. Assuming that these deep levels are involved in the compensation process, this concentration is also a measure of the difference $\Delta N = |N_D - N_A|$ between ionised shallow donor and acceptor centres. The absolute concentrations of these shallow states are thus bounded to about 10^{13}–10^{15} cm^{-3}. In addition a high electrical resistivity of the material in the 10^9–10^{11} Ω cm range is needed. The highest resistivities are more easily achievable in $Cd_{1-x}Zn_xTe$ (CZT) alloys, the band gap of which increases with Zn content. A value of x around 10% is commonly used. The defect properties are then not expected to be much different from those of CdTe described in the previous sections.

Crystal growth techniques for preparing SI material can have recourse to intentional donor doping or not, what deserves two distinct examinations. The first category includes the travelling heater method (THM), mainly used with chlorine doping [56], and the Bridgman methods, either in vertical [57] or horizontal configuration [58], often practised with In or Al doping. The purpose of donor doping at the level of a few 10^{17} cm^{-3} is to compensate the Cd vacancies which are generated during growth from Te-rich melts and are more generally related to the high Cd evaporation rate. The dependence of electrical resistivity and carrier lifetime of THM

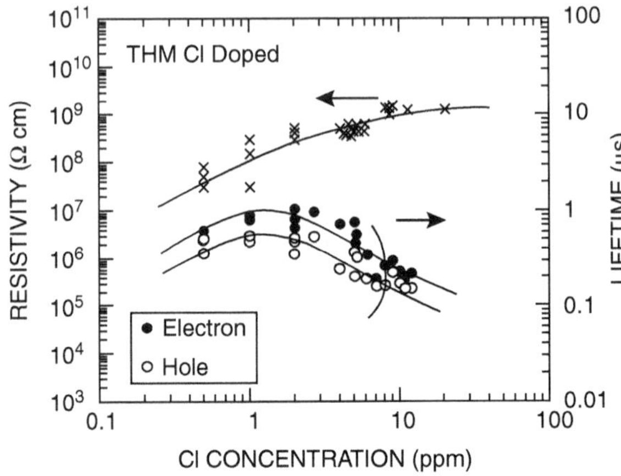

Figure 14 Resistivity and carrier lifetime in THM-grown CdTe as a function of Cl concentration (Reprinted with permission from Ref. [59]. Copyright [1993] by Elsevier).

grown crystals upon Cl concentration is presented in Fig. 14 [59]. As a general rule the room temperature resistivity is limited to around $10^9\ \Omega$ cm with p-type conductivity [60]. The hole carrier concentration is about $10^8\ cm^{-3}$ which puts the Fermi level at ~0.65 eV above the valence band. The increase in carrier lifetime shown in Fig. 14 also indicates that the concentration of deep traps decreases with increasing Cl content. Chlorine is thus able to passivate defects either punctual or extended. This is connected to the "self-purification" effect described in Ref. [61] and to the specific role of Cl in polycrystalline CdTe layers [62].

Chlorine doping has also been used in CdTe and CZT crystals grown by the horizontal Stockbarger method and subsequently annealed under a controlled Cd vapour pressure [63, 64]. In this way p-type as well as n-type semi-insulating crystals were obtained (Fig. 15).

The compensation models summarised by the data in Table 5 in principle apply to these donor-doped SI crystals but do not account for the low trap concentration required for detector-grade material. Besides we have to consider the changes occurring during the cooling sequence or any anneal which follows the growth process. Cooling or low-temperature anneals can be assumed to take place under Te-rich conditions – except specially defined Cd-rich treatments – because of the retrograde solubility of Te in CdTe. When thermodynamic equilibrium is assumed to be realised throughout a wide temperature range, changes in the defect structure such as those depicted in Fig. 13 are predicted [39]. It appears that below 200 °C the donor impurities are almost exactly compensated by the donor–vacancy pairs A^-. The concentration of the free Cd vacancies, which are the remaining native defects, decreases from $10^{14}\ cm^{-3}$ at

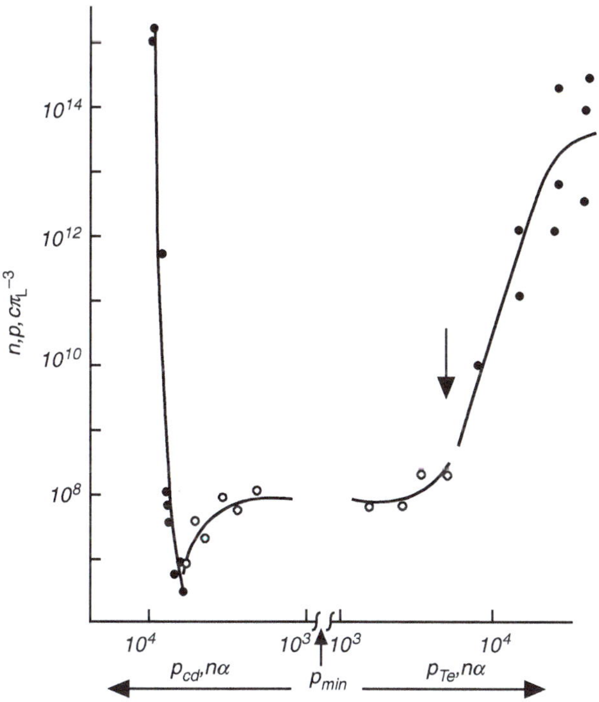

Figure 15 Room temperature free carrier density in CdTe doped with 2×10^{18} Cl cm^{-3} and annealed at 900 °C in atmospheres of varying Cd and Te pressure. The dark symbols correspond to *n* and the light symbols to *p* (Reprinted with permission from Ref. [63]. Copyright [1998] by Springer Science, MAIK "Nauka/Interperiodica" and the American Institute of Physics)

200 °C to 10^{13} cm^{-3} at 150 °C. A further reduction in the uncompensated charge

$$\Delta N = 2[V_{Cd}^{2-}] + [V_{Cd}^{-}] - [D^{+}] + [A^{-}] \tag{23}$$

can take place through the formation of neutral double-donor vacancy complexes according to the reaction

$$(D^{+} - V_{Cd}^{2-})^{-} + D^{+} \rightarrow (2D^{+} - V_{Cd}^{2-})^{\circ}. \tag{24}$$

The electronic structure of this double-donor vacancy complex was analysed by Bell [65]. Its local structure vas more recently investigated from magnetoluminescence spectroscopy conducted on In-doped CdTe [66, 67]. In the final compensated state the electrical neutrality equation is approximately reduced to $[D^{+}] \cong [(V_{Cd} - D)^{-}]$ and the majority of the Cd vacancies are included in the A-centres whence the relation between the total concentrations of donors and vacancies

$$[V_{Cd}]_S = \frac{1}{2}[D]_S. \tag{25}$$

This is the basic assumption formulated in the early chemical self-compensation model of Canali et al. [68].

The Fermi level finally pins on some deep level around midgap. A concentration of $N_{deep} = 10^{13}$ cm^{-3} was postulated for the computation of Fig. 13. Indeed deep levels at 0.6–0.8 eV above the valence band are commonly detected by various appropriate techniques in semi-insulating CdTe and CZT [69–71]. These deep levels are associated either to V_{Cd}, Te_{Cd} or to some uncontrolled transition metal. In the present case of p-type crystals another possibility is the interstitial Te_i. According to Ref. [17] Te_i is a negative-U double acceptor with the transition level $0/-2$ set at 0.57 eV above the valence band. The equilibrium concentration of Te_i^o is several orders of magnitude below that of V_{Cd}^o but some excess concentration could be generated during the cooling process leading to Te precipitates.

All the above discussion rests on the hypothesis of thermodynamic equilibrium achieved down to $T \approx 200$ °C. This can be checked as done in Ref. [39] by comparing the annealing time with the diffusion-limited characteristic time $t = L^2/D$, where D is the chemical diffusion coefficient and L is the defect migration range inversely related to the Te precipitates density. It appears that an anneal time of 10^4 s at 200 °C is compatible with a precipitate density $\geq 10^{10}$ cm^{-3}. Such a value falls in the range of precipitate density observed in transmission microscopy [72]. Let us mention in this context that the influence of the rate of ingot cooling upon the properties of Cl-doped CdTe has been investigated [73]. At the highest cooling rate of $v = 48$ K/h a semi-insulating crystal was only obtained for the lowest doping concentration used in this study $N(\mathrm{Cl}) = 2\text{–}4 \times 10^{18}$ cm^{-3}.

Semi-insulating CdTe crystals prepared without intentional doping have been grown by a number of techniques [71, 74–76]. The growth conditions are not always well defined and electrical resistivity is limited to $\sim 10^8$ Ω cm. Nowadays attention is focused on CZT crystals of the highest resistivity prepared by the high-pressure Bridgman method [77, 78]. They are grown from highly purified elements in open graphite crucibles under an over-pressure of an inert gas, typically argon, of about 100 atm to reduce the loss of volatile components. After growth, the ingot is slowly cooled down to room temperature at a rate of a few °C/h [79]. The crystals obtained in this way are n-type with a resistivity in the 10^{10}–10^{11} Ω cm range for $x_{Zn} \sim 0.1$. The Fermi level is then located at ~0.75 eV below the conduction band. The particular properties of this material could be associated to a growth process close to stoichiometry and to a low impurity content attested from photoluminescence measurements [77, 80]. The defect structure of near stoichiometric CdTe and CZT at high temperatures is dominated by two native defects, Cd_i and V_{Cd}, according to the models discussed previously. Strict compensation is predicted in a narrow P_{Cd} range only. In fact achievement of the semi-insulating state is obtained via the post-growth cooling stage. During temperature decrease the equilibrium concentration of Cd vacancies decreases more rapidly than the concentration of Cd interstitials as shown in temperature dependence studies [4, 16]. This leads to super-saturation of vacancies and finally to formation of Te precipitates. Internal equilibrium corresponding to a Te-rich material then prevails. The situation is rather similar to that analysed above for donor doping and illustrated in Fig. 13. However, in the present case the Fermi level is stabilised in the upper half of the band gap which could be related to excess residual donor impurities in this undoped material. The condition defined by Eq. (19) is fulfilled and a deep donor such as Te_{Cd} is a natural candidate for Fermi level pinning. As a matter of fact the deep level structure of undoped SI CZT is very rich. As many as 12 levels were detected by Castaldini et al. [81]. Several of them were encountered in other studies devoted to HPB grown crystals [25, 32, 82].

4. AMPHOTERIC IMPURITIES

In some conditions substitutional donor impurities can undergo lattice relaxation leading to the formation of negatively charged species, the so-called DX centre [83, 84]. This is a true self-compensation effect described by the reaction

$$2\,D^{0} \rightarrow DX^{-} + D^{+}, \tag{26}$$

where D^{o} and D^{+} denote the neutral and the ionised substitutional donor, respectively. The DX centre forms through a single bond breaking. The substitutional cation impurity or Cd next to a substitutional anion displaces along the ⟨111⟩ direction, breaking one bond and changing the local symmetry from T_d to C_{3v}. The atomic and electronic structure of DX centres associated with group III and VI donors have been theoretically studied in the framework of the density functional theory (DFT) [17, 84, 85]. A low formation energy for the DX configuration compared with the substitutional one implies the possible conversion of a charged shallow donor D^{+} to a negatively charged deep acceptor DX^{-}. The theoretical results are somewhat different depending on the authors. The calculation done in Ref. [84] shows that, in CdTe, Ga forms a DX-centre but Al and In do not. However, as the Zn concentration is increased in CZT, Al and In are predicted to become DX-centres. In addition, for CdTe, Cl, Br and I are not DX centres, but alloying with Zn is predicted to make two DX centres for Cl and one DX centre each for Br and I. A degenerate electron population due to strong doping can also induce the formation of a stable DX centre [85]. In contrast with the above predictions the calculation performed in Ref. [17] for CdTe indicate that DX centres are the more stable configuration in the case of Al_{Cd}, Ga_{Cd}, In_{Cd} and I_{Te}. Several experimental results should be mentioned in relation with this topic. Electron concentration up to 10^{19} cm^{-3} in MBE grown Al-doped CdTe was reported [86] showing that Al is the most efficient donor in CdTe and is not affected by a DX transition up to this concentration. On the other hand the electron concentration decrease observed at high doping level in I-doped CdTe (Fig. 3) can be interpreted in terms of compensation by a negatively charged DX centre [9]. Positron annihilation spectroscopy experiments showed positron trapping at DX centres in Cl-doped CdZnTe [87]. Moreover evidence for an In DX centre in CdTe was obtained from perturbed angular-correlation experiments [85].

In conclusion the formation of DX centres can account for some compensation in donor-doped CdTe and even more in CZT. However, the role of this phenomenon in achievement of the semi-insulating state appears to be limited.

A more classical case of amphoteric compensation concern impurities which are incorporated in both substitutional and interstitial sites, such as those of group I: Li, Na, Cu, Ag. These elements are acceptors when substituted for Cd and donors in interstitial positions. Transition between the two sites is described by a reaction like

$$\mathrm{Li}_{\mathrm{Cd}}^{-} + \mathrm{h}^{+} \leftrightarrow \mathrm{Li}_{\mathrm{i}}^{+} + \mathrm{V}_{\mathrm{Cd}}^{-}. \tag{27}$$

When the formation energies of the two impurity species are comparable they can coexist and compensate themselves. This is especially the case for Li and Na [17]. In an experimental study a Li-doped sample was given a low-temperature annealing (250 °C) under vacuum [88]. The density of holes decreased by three orders of magnitude and the Fermi level shifted into the band gap to $E_F \sim E_v + 0.25$ eV. This position could be related to the energy levels of Cd vacancy and shows that the shallow acceptor level of Li_{Cd} was compensated. At the limit of strong compensation $[\mathrm{Li}_{\mathrm{Cd}}^{-}] = [\mathrm{Li}_{\mathrm{i}}^{+}]$ and Eq. (27) indicates that p saturates to a value proportional to $[\mathrm{V}_{\mathrm{Cd}}^{-}]$.

5. CONCLUSION

Impurity compensation is a widespread phenomenon which affects conductive as well as high resistivity CdTe and CdZnTe crystals. As a general rule compensation occurs through the generation of native defects with a charge opposite to that of the doping impurities. Some of these defects are reasonably well identified in CdTe: interstitial Cd_i, vacancy V_{Cd}, antisite Te_{Cd}. However, knowledge of their electronic and thermodynamic properties still lacks reliability and accuracy. This situation limits the validity of the defect models and of the related predictions. Nevertheless compensation of donor impurities is qualitatively well described in terms of the double acceptor Cd vacancy and the donor–vacancy pairs. These defects play a major role in the achievement of donor-doped semi-insulating CdTe. Compensation of acceptor impurities is rather poorly understood. Relevant compensating donor defects could be Cd interstitials in near stoichiometric crystals and Te antisites in Te-rich material. More information about the Te antisite and possible associates with acceptor impurities is needed to master this side of the compensation process. In any case the final compensated state is determined through the post-growth cooling stage. Most of the impurity–defect complexes and precipitates are formed during this phase. As suggested in Ref. [89] the high defect mobilities in II–VI compounds and especially in CdTe could account for the high occurrence of defect–impurity interaction and consequently for the very pronounced self-compensation effects. This stresses the importance of defect kinetics in these phenomena which are usually analysed in the framework of equilibrium thermodynamics. Yet a remarkable outcome of all these theoretical and experimental studies is worth being underlined: that is the development of a detector-grade semi-insulating material characterised by a very low density of active localised centres. This achievement is both astonishing and encouraging in view of expected future progress regarding homogeneity and reproducibility of SI crystals.

REFERENCES

[1] D. de Nobel, Philips Res. Rep. 14 (1959) 361–399.
[2] F.A. Kröger, Chemistry of Imperfect Crystals, North-Holland, Amsterdam, 1964.
[3] G. Mandel, Phys. Rev. 134 (1964) A1073–A1079.
[4] S.S. Chern, H.R. Vydyanath, F.A. Kröger, J. Solid State Chem. 14 (1975) 33–43.
[5] P. Höschl, S. Kubalkova, Czech. J. Phys. B 22 (1972) 530–533.
[6] Y. Marfaing, Rev. Phys. Appl. 12 (1977) 211–217.
[7] P. Fochuk, O. Korovyanko, O. Panchuk, J. Cryst. Growth 197 (1999) 603–606.
[8] D.M. Freik, U.M. Pysklynets, L.I. Mezhylovska, Inorg. Mater. 41 (2005) 565–569.
[9] Y. Marfaing, Thin Solid Films 387 (2001) 123–128.
[10] L. Shcherbak, P. Feichouk, P. Fochouk, O. Panchouk, J. Cryst. Growth 161 (1996) 219–222.
[11] C. Gely-Sykes, C. Corbel, R. Triboulet, Solid State Commun. 80 (1991) 79–83.
[12] F. Fischer, A. Waag, L. Worschech, W. Ossau, S. Scholl, G. Landwehr, J. Mäkinen, P. Hautojärvi, C. Corbel, J. Cryst. Growth 161 (1996) 214–218.
[13] H. Kauppinen, L. Baroux, K. Saarinen, C. Corbel, P. Hautojärvi, J. Phys. Condens. Matter. 9 (1997) 5495–5505.
[14] R. Kalish, M. Deicher, G. Schatz, J. Appl. Phys. 53 (1982) 4793–4799.
[15] U. Reislöhner, N. Achtziger, W. Witthuhn, J. Cryst. Growth 197 (1999) 576–580.
[16] M.A. Berding, Phys. Rev. B 60 (1999) 8943–8960.
[17] S.-H. Wei, S.B. Zhang, Phys. Rev. B 66 (2002) 155211–1–10.
[18] O.A. Matveev, A.I. Terent'ev, Technol. Phys. Lett. 23 (1997) 140–141.

[19] M. Chu, S. Terterian, D. Ting, C.C. Wang, H.K. Gurgenian, S. Mesropian, Appl. Phys. Lett. 79 (2001) 2728–2730.
[20] R.F. Brebrick, R. Fang, J. Phys. Chem. Solids 57 (1996) 451–460.
[21] R. Grill, J. Franc, P. Höschl, I. Turkevych, E. Belas, P. Moravec, M. Fiederle, K.W. Benz, IEEE Trans. Nucl. Sci. 49 (2002) 1270–1274.
[22] P. Emanuelsson, P. Omling, B.K. Meyer, M. Wienecke, M. Schenk, Phys. Rev. B 47 (1993) 15578–15580.
[23] Cs. Szeles, Y.Y. Shan, K.G. Lynn, A.R. Moodenbaugh, E.E. Eissler, Phys. Rev. B 55 (1997) 6945–6949.
[24] U. Reislöhner, N. Achtziger, W. Witthuhn, J. Cryst. Growth 197 (1999) 576–580.
[25] N. Krsmanovic, K.G. Lynn, M.H. Weber, R. Tjossem, Th. Gessmann, Cs. Szeles, E.E. Eissler, J.P. Flint, H.L. Glass, Phys. Rev. B 62 (2000) R16279–R16282.
[26] C. Corbel, L. Baroux, F.M. Kiessling, C. Gely-Sykes, R. Triboulet, Mater. Sci. Eng. B 16 (1993) 134–138.
[27] H. Zimmermann, R. Boyn, P. Rudolph, J. Bollmann, A. Klimakow, R. Krause, Mater. Sci. Eng. B 16 (1993) 139–144.
[28] R. Krause-Rehberg, H. Zimmermann, A. Klimakow, Th. Drost, Phys. Status Solidi A134 (1992) K45–K47.
[29] N.V. Agrinskaya, E.N. Arkad'eva, O.A. Matveev, Sov. Phys. Semicond. 5 (1971) 767–772.
[30] C.E. Barnes, K. Zanio, J. Appl. Phys. 46 (1975) 3959–3964.
[31] T. Ido, A. Heurtel, R. Triboulet, Y. Marfaing, J. Phys. Chem. Solids 48 (1987) 781–790.
[32] W. Stadler, D.M. Hofmann, H.C. Alt, T. Muschik, B.K. Meyer, E. Weigel, G. Müller-Vogt, M. Salk, E. Rupp, K.W. Benz, Phys. Rev. B 51 (1995) 10619–10630.
[33] A. Zumbiehl, S. Mergui, M. Ayoub, M. Hage-Ali, A. Zerrai, K. Cherkaoui, G. Marrakchi, Y. Darici, Mater. Sci. Eng. B 71 (2000) 297–300.
[34] C. Eiche, D. Maier, D. Sinerius, J. Weese, K.W. Benz, J. Honerkamp, J. Appl. Phys. 74 (1993) 6667–6670.
[35] D.M. Hofmann, P. Omling, H.G. Grimmeiss, B.K. Meyer, K.W. Benz, D. Sinerius, Phys. Rev. B 45 (1992) 6247–6250.
[36] R. Grill, P. Fochuk, J. Franc, B. Nahlovskyy, P. Höschl, P. Moravec, Z. Zakharuk, Ye. Nykonyuk, O. Panchuk, Phys. Status Solidi B 243 (2006) 787–793.
[37] B.K. Meyer, W. Stadler, J. Cryst. Growth 161 (1996) 119–127.
[38] D. Wegner, E.A. Meyer, J. Phys. Condens. Matter. 1 (1989) 5403–5410.
[39] R. Grill, J. Franc, P. Höschl, I. Turkevych, E. Belas, P. Moravec, IEEE Trans. Nucl. Sci. 52 (2005) 1925–1931.
[40] L. Yujie, M. Guolic, J. Wanqi, J. Cryst. Growth 256 (2003) 266–275.
[41] F.A. Kröger, J. Phys. Chem. Solids 26 (1965) 1707–1715.
[42] G.F. Neumark, Phys. Rev. B 26 (1982) 2250–2252.
[43] M. Fiederle, C. Eiche, M. Salk, R. Schwarz, K.W. Benz, W. Stadler, D.M. Hofmann, B.K. Meyer, J. Appl. Phys. 84 (1998) 6689–6692.
[44] W. Jantsch, G. Hendorfer, J. Cryst. Growth 101 (1990) 404–413.
[45] P. Moravec, M. Hage-Ali, L. Chibani, P. Siffert, Mater. Sci. Eng. B 16 (1993) 223–227.
[46] E. Rzepka, Y. Marfaing, M. Cuniot, R. Triboulet, Mater. Sci. Eng. B 16 (1993) 262–267.
[47] D.M. Hofmann, W. Stadler, P. Christmann, B.K. Meyer, Nucl. Instrum. Methods Phys. Res. A 380 (1996) 117–120.
[48] O. Panchuk, A. Savitskiy, P. Fochuk, Ye. Nykonyuk, O. Parfenyuk, L. Shcherbak, M. Ilashchuk, L. Yatsunyk, P. Feychuk, J. Cryst. Growth 197 (1999) 607–611.
[49] P.N. Gorlei, O.A. Parfenyuk, M.I. Ilashchuk, I.V. Nikolaevich, Inorg. Mater. 41 (2005) 1266–1269.
[50] M. Fiederle, D. Ebling, C. Eiche, P. Hug, W. Joerger, M. Laasch, R. Schwarz, M. Salk, K.W. Benz, J. Cryst. Growth 146 (1995) 142–147.

[51] Y. Marfaing, J. Cryst. Growth 197 (1999) 707–717.
[52] P. Höschl, P. Moravec, J. Franc, E. Belas, R. Grill, Nucl. Instrum. Methods Phys. Res. A 322 (1992) 371–374.
[53] R. Grill, J. Franc, I. Turkevych, P. Höschl, E. Belas, P. Moravec, J. Electron. Mater. 24 (2005) 939–943.
[54] Cs. Szeles, Phys. Status Solidi B 241 (2004) 783–790.
[55] T.E. Schlesinger, J.E. Toney, H. Yoon, E.Y. Lee, B.A. Brunett, L. Franks, R.B. James, Mater. Sci. Eng. R 32 (2001) 103–189.
[56] M. Hage-Ali, P. Siffert, Nucl. Instrum. Methods Phys. Res. A 322 (1992) 313–323.
[57] M. Fiederle, A. Fauler, J. Konrath, V. Babentsov, J. Franc, R.B. James, IEEE Trans. Nucl. Sci. 51 (2004) 1864–1868.
[58] P. Cheuvart, U. El Hanany, D. Schneider, R. Triboulet, J. Cryst. Growth 101 (1990) 270–274.
[59] M. Ohmori, Y. Iwase, R. Ohno, Mater. Sci. Eng. B 16 (1993) 283–290.
[60] M. Fiederle, D. Ebling, C. Eiche, D.M. Hofmann, M. Salk, W. Stadler, K.W. Benz, B.K. Meyer, J. Cryst. Growth 138 (1994) 529–533.
[61] O.A. Matveev, E.N. Arkad'eva, L.A. Goncharov, Sov. Phys. Dokl. 20 (1975) 220–221.
[62] V. Consonni, N. Baier, S. Renet, A. Brambilla, G. Feuillet, Phys. Status Solidi C 3 (2006) 807–811.
[63] O.A. Matveev, A.I. Terent'ev, Semiconductors 32 (1998) 144–147.
[64] O.A. Matveev, A.I. Terent'ev, N.K. Zelenina, V.N. Guskov, V.E. Sedov, A.A. Tomasov, V.P. Karpenko, Semiconductors 39 (2005) 1034–1039.
[65] R.O. Bell, Solid State Commun. 16 (1975) 913–916.
[66] L. Worschech, W. Ossau, G. Landwehr, Phys. Rev. B 52 (1995) 13965–13974.
[67] S. Seto, K. Suzuki, V.N. Abastillas Jr., K. Inabe, J. Cryst. Growth 214/215 (2000) 974–978.
[68] C. Canali, G. Ottaviani, R.O Bell, F.V. Wald, J. Phys. Chem. Solids 35 (1974) 1405–1413.
[69] J. Franc, V. Babentsov, M. Fiederle, E. Belas, R. Grill, K.W. Benz, P. Höschl, IEEE Trans. Nucl. Sci. 51 (2004) 1176–1181.
[70] A. Cavallini, B. Fraboni, W. Dusi, IEEE Trans. Nucl. Sci. 52 (2005) 1964–1967.
[71] M. Zha, E. Gombia, F. Bissoli, A. Zappettini, L. Zanotti, Phys. Status Solidi B 229 (2002) 15–18.
[72] R.S. Rai, S. Mahajan, S. McDewitt, C.J. Johnson, J. Vac. Sci. Technol. B 9 (1991) 1892–1896.
[73] O.A. Matveev, A.I. Terent'ev, Semiconductors 34 (2000) 1264–1269.
[74] R. Triboulet, Y. Marfaing, A. Cornet, P. Siffert, J. Appl. Phys. 45 (1974) 2759–2765.
[75] P. Rudolph, S. Kawasaki, S. Yamashita, S. Yamamoto, y. Usuki, Y. Konagaya, S. Matada, T. Fukuda, J. Cryst. Growth 161 (1996) 28–33.
[76] M. Zha, A. Zappettini, F. Bissoli, L. Zanotti, V. Corregidor, E. Dieguez, J. Cryst. Growth 260 (2004) 291–297.
[77] F.P. Doty, J.F. Butler, J.F. Schetzina, K.A. Bowers, J. Vac. Sci. Technol. B 10 (1992) 1418–1422.
[78] J.F. Butler, F.P. Doty, B. Apotovsky, J. Lajzerowicz, L. Verger, Mater. Sci. Eng. B 16 (1993) 291–295.
[79] P. Fougères, M. Hage-Ali, J.M. Koebel, P. Siffert, S. Hassan, A. Lusson, R. Triboulet, G. Marrakchi, A. Zerrai, K. Cherkaoui, R. Adhiri, G. Brémond, O. Kaitasov, M.O. Ruault, J. Crestou, J. Cryst. Growth 184 (/185) (1998) 1313–1318.
[80] M. Fiederle, T. Feltgen, J. Meinhardt, M. Rogalla, K.W. Benz, J. Cryst. Growth 197 (1999) 635–640.
[81] A. Castaldini, A. Cavallini, B. Fraboni, P. Fernandez, J. Piqueras, J. Appl. Phys. 83 (1998) 2121–2126.
[82] A. Zerrai, K. Cherkaoui, G. Marrakchi, G. Brémond, P. Fougères, M. Hage-Ali, J.M. Koebel, P. Siffert, J. Cryst. Growth 197 (1999) 646–649.
[83] D.J. Chadi, K.J. Chang, Phys. Rev. Lett. 61 (1988) 873–876.

[84] C.H. Park, D.J. Chadi, Phys. Rev. B 52 (1995) 11884–11890.
[85] S. Lany, H. Wolf, T. Wichert, Phys. Rev. Lett. 92 (2004) 225504–1–4.
[86] A. Arnoult, D. Ferrand, V. Huard, J. Cibert, C. Grattepain, K. Saminadayar, C. Bourgognon, A. Wasiela, S. Tatarenko, J. Cryst. Growth 201/202 (1999) 715–718.
[87] Y.Y. Shan, K.G. Lynn, Cs. Szeles, P. Asoka-Kumar, T. Thio, J.W. Bennett, C.B. Beling, S. Fung, P. Becla, Phys. Rev. Lett. 79 (1997) 4473–4476.
[88] N.V. Agrinskaya, Sov. Phys. Semicond. 19 (1985) 199–200.
[89] Y. Marfaing, Prog. Cryst. Growth Charact. 4 (1981) 317–343.

AUTHOR INDEX

A

C

G

H

I

L

M

N

O

Q

R

S

X

Y

SUBJECT INDEX

E

F

G

H

I

Kolibri 360 GmbH
Ferdinand-Jühlke-Straße 7
99095 Erfurt, Deutschland
produktsicherheit@kolibri360.de